Clean Technologies Toward the Development of a Sustainable Environment and Future

Clean Technologies Toward the Development of a Sustainable Environment and Future

Physicochemical, Biochemical, and Biotechnological approaches

Edited by
Pradeep Verma and Maulin P. Shah

IWA PUBLISHING

Published by **IWA Publishing**
Unit 104–105, Export Building
1 Clove Crescent
London E14 2BA, UK
Telephone: +44 (0)20 7654 5500
Fax: +44 (0)20 7654 5555
Email: publications@iwap.co.uk
Web: www.iwapublishing.com

First published 2023

British Library Cataloguing in Publication Data
A CIP catalogue record for this book is available from the British Library

ISBN: 9781789063776 (paperback)
ISBN: 9781789063783 (eBook)
ISBN: 9781789063790 (ePub)

This eBook was made Open Access in December 2023.

Doi: 10.2166/9781789063783

Contents

The Editors

Professor Pradeep Verma

Professor Verma completed his Ph.D. from Sardar Patel University Gujarat, India in 2002. In the same year, he was selected as a UNESCO fellow and joined the Czech Academy of Sciences Prague, Czech Republic. He later moved to Charles University, Prague to work as postdoctoral fellow. In 2004 he worked as a visiting scientist at UFZ Centre for Environmental Research, Halle, Germany. He was awarded a DFG fellowship that provided him with another opportunity to work as a postdoctoral fellow at Gottingen University, Germany. Later in 2007 he returned to India and joined Reliance Life Sciences, Mumbai, and worked extensively on biobutanol production, which attributed a few patents to his name. Later he was awarded JSPS Post-Doctoral Fellowship Programme and joined the laboratory of Biomass Conversion, Research Institute of Sustainable Humanosphere (RISH), Kyoto University, Japan. Professor Verma has also been the recipient of various prestigious awards such as the Ron Cockcroft award by Swedish society, and UNESCO Fellow ASCR Prague. Recently for his contribution to the area of fungal microbiology, industrial biotechnology, and environmental bioremediations, he has been awarded the prestigious Fellow Award from Mycological Society of India (2020), P.C. Jain Memorial Award (MSI), and Biotech Research Society of India (2021). In 2020 he was also awarded fellow of Biotechnology Research Society of India (BRSI) and Fellow of Academy of Sciences of AMI India 2021 (FAMSc). Furthermore, he has also been awarded the JSPS Bridge Fellow award in 2022 and a short-term visit to Kyoto University, Kyoto, Japan to strengthen ties between the two laboratories.

Professor Verma in 2009 began his independent academic career as a reader and founder head at the Department of Microbiology, Assam University. In 2011 he moved to the Department of Biotechnology at Guru Ghasidas Vishwavidyalaya (a Central University), Bilaspur, and served as an associate professor till 2013. He is currently working as a professor at the Department of Microbiology, CURAJ (Central University of Rajasthan) and was also the former head and dean, School of Life Sciences. He is a member of various National & International societies/academies and has also completed two collaborated projects worth 150 million INR in the area of microbial diversity and bioenergy.

Professor Verma is a group leader of the Bioprocess and Bioenergy laboratory at the Department of Microbiology, School of Life Sciences, CURAJ. His area of expertise involves microbial diversity, bioremediation, bioprocess development, lignocellulosic, and algal biomass-based biorefinery. He holds 12 international patents in the field of microwave-assisted biomass pretreatment and bio-butanol production. He has more than 65 research articles in peer-reviewed international journals and contributed to several book chapters (32 published; 11 in press) in different edited books. He has also edited 11 books for international publishers such as Springer and Elsevier. He is a guest editor to several journals such as *Biomass Conversion and Biorefinery* (Springer) *Agriculture* (mdpi), *Frontier in Nanotechnology* (Frontiers), and *International Journal of Environmental Research and Public Health* (mdpi). He is also an editorial board member for the *Journal Current Nanomedicine* (Bentham Sciences) and an acting reviewer for more than 60 journals of different publishers such as Springer, Elsevier, RSC, ACS, Nature, Frontiers, mdpi, etc.

Bioprocess and Bioenergy Laboratory, Department of Microbiology, Central University of Rajasthan, Bandarsindri, Kishangarh, Ajmer, Rajasthan 305817, India

Email: vermaprad@yahoo.com; pradeepverma@curaj.ac.in
ORCID detail: http://orcid.org/0000-0003-2266-9437
Website: http://www.vermapradeep-lab.com/

Dr. Maulin P. Shah

He is an active researcher and scientific writer in his field for over 20 years. He received a B.Sc. degree (1999) in Microbiology from Gujarat University, Godhra (Gujarat), India. He also earned his Ph.D. degree (2005) in Environmental Microbiology from sardar Patel University, Vallabh Vidyanagar (Gujarat) India. His research interests include Biological Wastewater Treatment, Environmental Microbiology, Biodegradation, Bioremediation, & Phytoremediation of Environmental Pollutants from Industrial Wastewaters. He has published more than 250 research papers in national and international journals of repute on various aspects of microbial biodegradation and bioremediation of environmental pollutants. He is the editor of 120 books of international repute (RSC, Wiley, DeGruyter, Elsevier, Springer, and CRC Press).

Industrial Wastewater Research Lab
Division of Applied & Environmental Microbiology
Enviro Technology Limited
Ankleshwar 393002, Gujarat
India

Email: shahmp@beil.co.in
maulinshah1979@gmail.com
ORCID detail: (0000-0003-1880-5221)
Google scholar: https://scholar.google.co.in/citations?user=qmp4tR0AAAAJ&hl=en

doi: 10.2166/9781789063783_xvii

Preface

Water is an essential resource and crucial to sustaining life. Ever-increasing population, rapid urbanization, and industrialization have resulted in improper discharge of wastewater leading to hazardous impacts on human health and the environment. As a result, globally nations are adopting measures to retain water and improve the qualitative parameters of wastewater to make it harmless and reusable. These technologies are a fruitful approach for solving water shortages, removing contaminants, and recycling them back to the environment for various uses, for example, drinking, washing, etc.

Thus, considering the huge subject of wastewater treatment, this book focuses on various aspects such as bioremediation, soil management, water cleaning, and the role of biotechnology in environmental engineering. The book also emphasizes on the urgent need for developing new approaches for environmental cleaning, such as sensing and monitoring technologies for the detection of environmental pollution. The book is likely to serve as valuable reference material to understand the current challenges, research gaps, and advancements in wastewater treatment.

Additionally, the book looks at the use of nanotechnology in environmental cleaning, including waste management, and soil and water remediation. It discusses technologies available for carbon sequestration and their positive impact on creating a greener environment. Additionally, the book is focused on microbial-assisted remediation methodologies, for example, biodegradation of toxic dyes, and chemicals and also cellular sequestration and conversion of CO_2 into biomass and chemical products providing carbon neutral to negative alternatives to conventional fossil fuels.

The biodegradation of toxic dyes includes 1, 4-dioxane, which is an emerging organic pollutant and is produced as a by-product in several petrochemical-based industrial processes. It provides an overview and future perspective of microorganism-assisted reclamation of 1, 4-dioxane polluted sites and discusses the chemistry of dioxane, toxicity and the fate of dioxane in the environment, reclamation strategies, and different factors influencing reclamation, dioxane degrading monooxygenases, and microbial degradation pathways. The book also deals with the adverse impact of azo dyes and explains various mechanisms of dye degradation by microorganisms. Similarly, the impact of heavy metal on the environment, different microbes involved in the detoxification process and biotechnological solutions of heavy metal contamination have been discussed in detail.

Another aspect has included bacteria-assisted reclamation of petroleum hydrocarbon polluted sites. It also discusses the fate of petroleum hydrocarbons in the environment, and microbial hydrocarbon degradation pathways. Further, it focused on ecological contamination and gives deep insight into the status and advances of different computational techniques and omics approaches to enhance the process of bioremediation. The use of formulations or environmental enhancers is to biochemically strengthen selected bioremediation pathways.

The book also aims pedagogically to follow a problem-solving, data-based approach, starting by introducing the problem of Contaminants of Emerging Concern (CECs), and proposing a solution. For example, zooplankton and other microbes act to remove or reduce pollutants through the biological degradation of pollutants into nontoxic substances. Several parameters of bioremediation *in-situ* or *ex-situ* methods are discussed. It provides current information regarding the potential of biomass-based integrated biorefineries for wastewater treatment and resource recovery. Sustainable solutions can be applied to all types of industries including small and large, regardless of material, energy and water consumption levels. Thus, green technology can prove to be an asset in improving sustainability, enhancing green growth, and maintaining a balance between socioeconomic and environmental parameters, while preserving efficiency, productivity, and prosperity. The book aims to provide a comprehensive insight into the potential applications and also throws light on the recently developed advanced green technologies.

Acknowledgments

First of all, we would like to convey our gratitude to the publisher, IWA Publishing, for considering our thoughts to develop this book, which is the first of its kind covering various clean technologies. The current book became a reality because of the support from all researchers and academicians who have contributed to the book: the editor is thankful for their contributions. We would also like to thank all scholars for providing us with the necessary technical support and editorial assistance during the developmental stages of this book.

We are always thankful to God for his blessings and strength and motivation to always do well and contribute to the scientific world.

Professor Pradeep Verma
Dr. Maulin P. Shah

doi: 10.2166/9781789063783_0001

Chapter 1

Microbes and wastewater treatment

L. Dharani, R. Umapriya, J. Rohan, G. Surendran and P. Deepak*

Department of Chemical Engineering, KPR Institute of Engineering and Technology, Coimbatore, India
**Corresponding author: dharaniloganathan95@gmail.com*

ABSTRACT

Wastewater is considered a source of water, power, and enriching nutrient for plants. Wastewater treatment is among the most significant biotechnological procedures for treating municipal and industrial sewage across the world. Conventional wastewater treatment methods, however, have constraints, primarily because they are cost-intensive to achieve the aim of wastewater remediation. Microorganisms, on the contrary, outperform humans when it comes to sewage water purification. Their ability to decompose a wide range of organic chemicals and cycle components such as nitrogen, phosphorus, and carbon is unparalleled in ecology. These characteristics have been successfully used in microbial wastewater treatment plants. This chapter discusses the necessity for wastewater treatment, and the function of diverse microbes in wastewater treatment, including new and developing technologies that use microbes for wastewater treatment and purification, such as microbial fuel cells, bioremediation, and activated sludge processes along with the challenges and prospects of using microorganisms in wastewater treatment.

Keywords: microbes, wastewater, microbial fuel cell, bioremediation, municipal waste

1.1 INTRODUCTION

The most pervasive material in the natural world is water. Liquid, solid, and invisible vapor are the three different states in which water may exist. The topmost portions of the Earth's crust and soil layer are where water forms the oceans, seas, lakes, rivers, and subterranean bodies. In the arctic and alpine areas, water occurs as ice and snow cover in a solid condition. Water is present in the air as water vapour, water droplets, and ice in addition to inhabiting the biosphere to some extent. The composition of the many minerals that make up the Earth's crust and core contains enormous amounts of water. Water is a necessary component of practically every living thing's everyday life. Most regions of the world have recently started to experience acute water scarcity, necessitating the use of wastewater reuse techniques. The amount of available freshwater is insufficient to meet the planet's consumption demands, and the majority of it is found in polar areas as snow and ice accumulate (Shiklomanov, 1993).

Concern is also raised by the poisoning of the aquatic environment with chemicals and pharmaceuticals. The growing population requires a steady supply of clean water for drinking, hygienic purposes, irrigation, and other applications. The daily water needs are further impacted by the pathogenic bacteria, micro irritants, and spores within water bodies that exist naturally. Currently, in addition to preserving the water cycle, wastewater reusing is required to satisfy water needs and safeguard environmental health by adhering to severe wastewater treatment regulations (Larsson *et al.*, 2007).

Every day, both our homes and businesses require water. The water we use is extracted from lakes, rivers, and the earth (groundwater), and the majority of it returns to these places after we consume it and pollute it. Wastewater is the term for this used water. Serious contamination results if it is not handled before being released into rivers. A mixture of ground water, surface water, and storm water, along with liquid or water-borne waste from homes, institutions, and commercial and industrial organizations can be referred to as wastewater. It typically has a high concentration of oxygen-demanding wastes, disease-causing pathogens, organic compounds, micronutrients that promote plant development, inorganic compounds, minerals, and sediments. Toxic substances could also be present (Metcalf & Eddy, 2004) (Figure 1.1).

1.2 NEED FOR WASTEWATER TREATMENT

Water purification continues to place a high emphasis on protecting our planet and the waterways that provide water to people worldwide (McCarty *et al.*, 2011). It is not just an issue for impoverished nations. And it is crucial for a number of reasons (Figure 1.2).

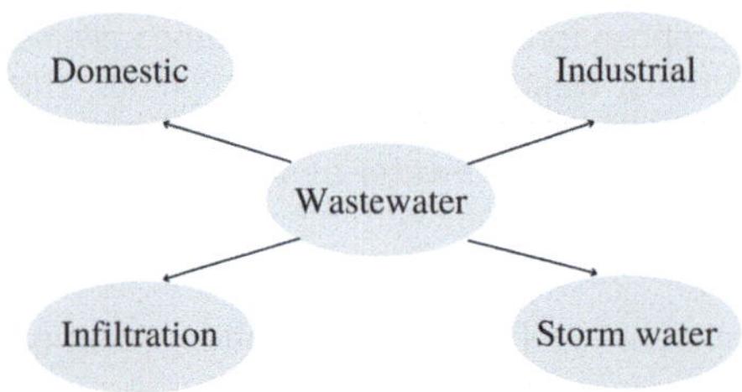

Figure 1.1 Classification of wastewater.

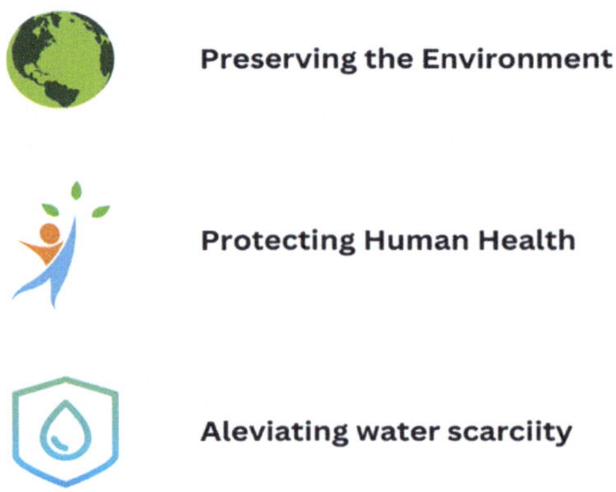

Figure 1.2 Need for wastewater treatment.

1.3 ROLE OF MICROBES IN WASTEWATER TREATMENT

Since microorganisms are arranged in species-rich structures in waste water treatment facilities, biodegradation of a larger spectrum of substrate is superior to that in pure culture (Cydzik-Kwiatkowska & Zielińska, 2016). The majority of the microorganisms used in biological wastewater treatment procedures are found in microbial aggregates, including biofilms, flocs, and granules. Extracellular polymeric substances (EPS), also known as complex high-molecular-weight mixtures of polymers, were found in activated sludge, biofilms, granulated sludge, and pure cultures after using a variety of electron microscopy techniques. By weak physicochemical interactions, EPS on the microbial surface is crucial in maintaining the microbial aggregates in a three-dimensional gel-like wet matrix. Hydrogen bonding, electrostatic interactions, hydrophobic contacts, and van der Waals interactions are examples of these weak forces. The metabolic waste products known as EPS build up on the bacterial cell surface (Liu & Fang, 2002). They are created by microbes in bioreactors after they consume organic substances found in wastewater. EPS builds up due to a variety of processes, including secretion, excretion, cell lysis, and sorption of waste water constituents. From an engineering perspective, EPS has certain crucial qualities including the ability to biosorb and the capacity to degrade. In addition to polysaccharides, proteins, nucleic acids, phospholipase, and other nonpolymeric molecules with smaller molecular weight, EPS is made up of a range of organic macromolecules. It has also been found that EPS contains organic compounds such humic chemicals and uronic acid (Pal & Paul, 2008).

1.4 COMMON MICROBES USED IN WASTEWATER TREATMENT

Microbes play a significant part in the process of recycling waste since they are the major actors and are primarily in charge of the degradability of inorganic organic compounds and the recycling of nutrients in the environment. Microbes are crucial throughout the fermentation process for recycling waste, treating wastewater, and producing alternative energy. The following group of bacteria is primarily present during the treatment of wastewater (Rani *et al.*, 2019).

1.4.1 Bacteria

They are largely in charge of extracting and converting these organic components in an effluent treatment, which is an essential part of the wastewater treatment process. Thus, these bacteria are crucial for the maintenance and proper operation of microbial treatment systems. Bacteria in wastewater treatment facilities are often present at concentrations of 10 ml. Typically, heterotrophic bacteria obtain energy from breaking down carbonaceous organic materials and utilize the same for cell formation and development. In the treatment of industrial wastewater, it was discovered that nearly 11 bacterial species were involved. They are *Escherichia coli*, *Klebsiella pneumoniae*, *Bacillus* sp., *Kosakonia oryzae*, and *Cronobacter sakazakii*.

1.4.2 Protozoa

They are larger than bacteria and are unicellular eukaryotic creatures. Protozoa are important in the process of treating wastewater because they feed on harmful bacteria. By digesting them, they eliminate floating bacteria as well as other suspended particles, giving wastewater an edge. Hence, the purity of wastewater effluent is improved. As with bacteria, certain protozoan species need very little oxygen to exist, while others can endure anaerobic environments. In aerobic activated sludge, the typical protozoan concentration ranges from 5×10 to 2×10. The many protozoa kinds found in wastewater treatment systems serve a variety of tasks. Flagellates (whose primary food source is the dissolved organic materials within effluent), ciliates (which eliminate floating bacteria and aid in pollutant clearing), crawling ciliates such as *Aspidisca* sp. and *Euplotes* sp. (which predominate activated sludge and indicate high treatment), *Chilodonella uncinata* are ciliated protozoan species that are frequently discovered in wastewater treatment plants (WWTPs).

1.4.3 Metazoa

Lagoon effluent in particular has a predominance of multicellular eukaryotics that are larger than most protozoa in size; however, their concentration in activated sludge is quite low (103/ml). Yet, they do highlight the problems of the therapeutic system. The rotifers and nematodes metazoan groups, which may be present in activated sludge, are known to serve distinct functions. They consume other microorganisms and clean the effluent. When hazardous amounts of effluents are present, they are the first to be impacted.

1.4.4 Filamentous bacteria

Long filaments of these bacteria are produced throughout their growth. The biomass of activated sludge often contains filamentous microorganisms. They are a regular part of the biomass of activated sludge and important for proper floc development. Their population is influenced by the nutritional circumstances (the wastewater system's dissolved oxygen (DO), pH, sludge age, temperature, the amount of nutrients that are readily accessible, and the amount of oil and grease). In activated sludge, filamentous bacteria of up to 25 distinct kinds have been identified. One of the major filamentous bacteria, *Nocardia* spp. causes foaming.

1.4.5 Algae

Algae are a class of photosynthesis-based organisms that are all around us and are important in the biological treatment of sewage. Algae play a variety of roles in the ecosystem because they may gather heavy metals, pesticides, poisonous organic and inorganic materials both within and outside of their cells. Algae improve the soil's nutrient content for plants and aid in their ecosystem mobilization. Excess nutrients may also be utilized in wastewater treatment systems to create biomass, which has a variety of uses, including the production of sustainable biofuel for the production of food (including proteins and carbohydrates), feed, and medicines (Rani *et al.*, 2019).

1.4.6 Fungi

Fungi are multicellular creatures which hydrolyze intricate organic compounds while contending with other microorganisms in a mix culture. They are distinguished by the ability to oxidize ammonia to nitrite and nitrate, which impede bacterial growth, and even decompose organic material at low pH atmosphere. Fungi capture and adsorb suspended materials using its fungal hyphae to acquire the energy and nutrients they require. It has also been shown that some fungi release enzymes that aid in the substrate breakdown process during wastewater treatment. *Sphaerotilus natans*, *Aspergillus*, *Penicillium*, *Fusarium*, *Absidia*, and so on are among the most prevalent fungal species linked to waste decomposition (Rani *et al.*, 2019).

1.5 MICROBIAL WASTEWATER TECHNIQUES

- Preliminary treatment involves removing bulky materials such as bottles, cans, and plastics that might obstruct downstream operations.
- Primary treatment: Elimination of grit, suspended particles, and oil.
- Secondary treatment: Usage of microorganisms in anaerobic or aerobic conditions to remove phenol, color, and oil while lowering biological oxygen demand (BOD).
- Tertiary treatment: Reverse osmosis, electrodialysis, and ion exchange are used for the effluent's ultimate purification and elimination steps.

Wastewater is treated biologically using techniques that have been around for a while to handle industrial effluents. These techniques, which use a membrane batch reactor, may be classified as aerobic and anaerobic reactor in batch mode. Moreover, there are anaerobic film reactors and filters (Rani *et al.*, 2019) (Table 1.1 and Figure 1.3).

Table 1.1 Benefits and drawbacks of wastewater purification techniques.

Treatment Methods		Advantages	Disadvantages
Aerobic system	CASP	• High reduction of BOD and pathogen • Operated at a wide array of organic load as well as hydraulic rates	• Energy consumption is high Sludge require further treatment and appropriate discharge
Aerobic system	OD	• Operation management is easy • Less sludge	• Effluent suspended solids concentrations are relatively high • Require larger land area
Aerobic system	MBR	• Treated water is clarified and can be applied directly to recycling	• Aeration limitation and membrane pollution
Aerobic system	TF	• Low concentration and maintenance cost	• Poor transparency of treated water
Aerobic system	RBC	• Low operating cost and less sludge generation	• Poor transparency of treated water
Anaerobic system		• Biological waste produced is less in quantity Low maintenance in terms of nutrient requirements Production of methane as an end product	• The growth rate of microorganisms is slow • Production of odour
Facultative system		• Require little energy • Easy to operate	• Periodic removal of settled sludge is required

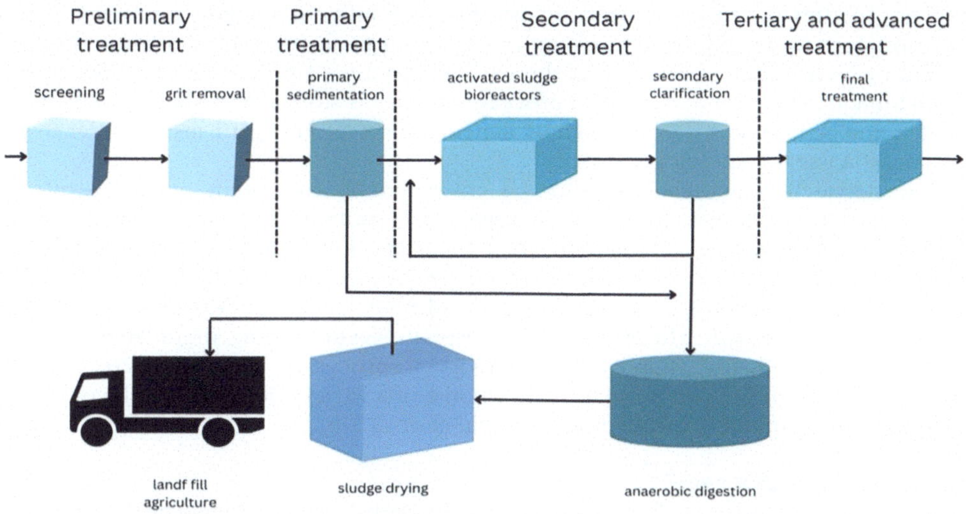

Figure 1.3 Typical wastewater purification system diagram.

1.5.1 Preliminary treatment

Large debris that might clog and hinder downstream processes, such as bottles and stones, are removed at the first stage of wastewater treatment. For screening and grit removal during initial treatment, tavern, ring, slashing, or bar screens that are slanted more on the side of inflowing water can be employed. These screens can catch particles as the sewage water passes past them. Following the removal of the mechanically or manually captured waste, fibrous materials may be additionally separated from the water. Grit removal, which entails getting rid of abrasive inorganic elements such as sand, gravel, and other heavy particulates, is essential to prevent obstruction and abrasive damage to the equipment or pipes that carry waste. There are additional varieties from grit channels, such as velocity channels, that reduce influent velocity and permit heavier abrasives to settle at ground level before removal (Chahal *et al.*, 2016).

1.5.2 Primary treatment

Processes used in primary treatment are intended in eliminating suspended particles together with lowering particle BOD levels. The majority of the time, it is referred to as the first step of treatment and eliminates 25–50% of BOD, 65% of oil and grease, and 50–70% of total suspended solids (TSSs). Sedimentation, which includes materials settling under the effect of gravity, and flocculation are two important physical mechanisms for separating solids from wastewater. The most typical clarifiers or sedimentation tanks are round or rectangular in form. Contrary to rectangular tanks, which have water entering through one end and draining at the other, circular tanks have water entering along the centre and distributing outward axially. The tank's weir is a crucial component. The weirs are therefore meticulously crafted physical barriers that control flow rate to be the same. The inclusion of coagulants or flocculants allows for the separation of dissolved organic materials in addition to basic primary treatment techniques. The metal salt flocculent clumps the suspended colloidal particles, which facilitates their separation via settling or filtration. Mostly dissolved materials will be present in the outflow water, also referred to as primary effluent. The primary effluent moves into secondary treatment after it has been cleared.

1.5.3 Secondary treatment

The secondary treatment processes used on the primary effluent include a variety of biological treatment techniques to get rid of minerals and soluble inorganic as well as organic solids. Carbon oxidation and nutrient removal are two of the various functional processes that take place during secondary treatment. The carbon oxidation process, which transforms organic materials into carbon dioxide, water, and cellular detritus, is crucially dependent on microorganisms. In order to grow and reproduce, bacteria use the energy generated.

Prior to wastewater being released into the environment, two main nutrients–nitrogen and phosphorous–must be removed. Inorganic nitrogen is eliminated by the biological processes of nitrification and denitrification. As denitrification turns the nitrate into nitrogen gas, nitrification turns ammonia into nitrate. Nitrous oxide, a strong greenhouse gas, can be produced as a result of incomplete nitrification or denitrification. To get rid of phosphorus, physiochemical and biological processes might be combined. The growth of phosphate-accumulating organisms that can store orthophosphate is specifically supported by biological processes such as increased biological phosphorus removal plants. Secondary treatments come in a variety of forms which could be applied in biological nutrient elimination procedures (Chahal *et al.*, 2016).

1.5.4 Activated sludge process

The biological removal of nutrients from wastewater is frequently accomplished by the activated sludge process (ASP). There are two main stages in an ASP. A heterogeneous and highly varied bacterial population breaks down pollutants at the initial stage while also converting the compounds into more streamlined and ecologically friendly by-products like nitrogen and carbon dioxide gas. The term 'activated sludge' refers to a heterogeneous microbial culture, as well as the biomass is frequently organized as microbial flocs, which are kept under suspension through mechanical mixing with aeration. An aeration tank along with aeration source make up the most essential ASP arrangement at this first stage. Incorporating anoxic and/or anaerobic regions that enhance overall nitrogen elimination through nitrification/denitrification and phosphorous assimilation has, however, resulted in numerous changes to this basic idea. The biomass is separated from the cleaned water in the second step of the ASP using secondary clarifier that employs gravity sedimentation. The cleared supernatant is transported for tertiary treatment, while the settled biomass is mainly recycled again to highest point of the ASP as return activated sludge (RAS). An ASP's main job is to biologically remove as well as stabilize nutrients, but it additionally possesses the ability to act as an effective therapeutic shield against diseases through predation along with connection, adsorption, and trapping to or inside the biological floc. Depending on how each WWTP operates and how the season affects pathogen density and treatment efficiency, there may be variations in accounts on the efficacy of ASP for pathogen elimination. It has been stated that between 1.5 and 2.5 log10 of *E. coli* have been eliminated, compared to between 1 and 3 log10 of *Cryptosporidium*. However, virus clearance differed between different WWTPs as well as appeared to vary for specific viral species. For instance, the reduction of rotavirus was between 7 and 8 log10, whereas the reduction of norovirus was between 1 and 6 log10.

1.5.5 Waste stabilization ponds

Large shallow reservoirs called waste stabilization ponds (WSPs) have been used to store wastewater for a longer duration to enable bacterial and algal communities to treat it biologically. WSPs provide a green treatment option with the advantages of low energy usage, low operating costs, and exceptionally efficient removal of organic debris and pathogens. In limited towns and villages, WSPs are commonly used as the only choice for sewage treatment, as an additional step of polishing after ASP, as one of several secondary treatments before discharge or reusing.

Pathogens are removed from WSPs by three main mechanisms: (1) unfavorable pond conditions (such as temperature, sunshine, and predator activity); (2) prolonged pond occupancy durations that result in natural mortality of microorganisms; and (3) particle binding and sedimentation. Pathogen eradication when the final discharge level of pathogens is both affected by pond depth, detention time, pond population, and pond geometry. By photooxidative DNA damage caused by sunlight and other physicochemical parameters such as temperature and pH, WSPs eliminate fecal coliforms, *E. coli*, and other dangerous microbes. *Cryptosporidium* suspended in a WSP has also been discovered to be inactivated by temperature and sunlight. Pathogens can be eliminated through predation by other microbes or zooplankton, particularly bacteria and protozoan parasites. WSPs demonstrated reductions about 2–4 log10, 3–6 log10, 1–2 log10, and up to 3 log10 for viruses, bacteria, protozoan cysts, and viruses, respectively. Protozoan removal for an Australian pond system was found to be highly seasonal, characterized by higher removal at the summer and autumn months (2.5–3 log10) while lower removal at the spring and fall months (0.5–1.2 log10). A number of facultative and maturation pools is the most common configuration, though there are many other pond design types. There is no pre-treatment employed, and the connection between the main facultative pond and the maturation ponds is fairly simple. An anaerobic pre-treatment stage is present before the facultative pond in more sophisticated plants. In the sections that follow, the various pond system types are discussed.

1.5.5.1 Anaerobic ponds

Anaerobic ponds eliminate organic mass while running without oxygen. They have brief retention durations and can reduce the organic load in wastewater by 40–70%. In these ponds, pathogens are primarily eliminated through sedimentation. Microorganisms often have low densities and hence slow settling rates. For sedimentation to occur, pathogens must be bounded onto heavier contaminants because the oocysts of *Cryptosporidium* settle at a rate of 30 mm/day. Large and rather dense helminth eggs easily settle by gravity and are eliminated in these ponds.

1.5.5.2 Facultative ponds

In facultative ponds, both aerobic and anaerobic activities occur. Pathogens are removed from these wetlands through a complex process that includes sedimentation, inactivation using sunlight, a pH level that is elevated, low carbon dioxide levels, and elevated oxygen concentrations. The photic, heterotrophic, and anaerobic zones are just a few of the many operational layers or areas that make up these wetlands. The most shallow zone is anaerobic and produces gases such as methane and carbon dioxide while removing organic material that has collected in sediment without using oxygen. The growth of algae is encouraged by carbon dioxide inside the heterotrophic region, providing air for heterotrophic aerobes to decompose organic materials. The uppermost layer, also known to be photic zone, where algae photosynthesis occurs at high rates, becomes extremely oxygenated. Heterotrophic bacteria use this oxygen in the aerobic breakdown of organic materials. Facultative ponds typically have retention periods of 5–30 days and are shallow, measuring between 1 and 2.5 m in depth.

1.5.5.3 Maturation ponds

Maturation ponds have a retention period of roughly 20 days and are 1–2 m deep. Although they also serve to extract nutrients, their primary use is to remove pathogens. It is more practical to avoid short circuits by using a succession of tiny maturation ponds as opposed to a single one. Because of the efficient pathogen elimination provided by solar radiation (UV penetration), a pH level that is elevated, high oxygen concentration, as well as low levels of nutrients, maturation ponds are typically shallower than other ponds. Maturation ponds can eliminate all protozoa and helminth eggs as well as 99% of coliforms. Both industrialized nations like Australia and developing nations such as India frequently employ them, but the mechanisms by which enteric viruses are eliminated are poorly understood and need further research.

1.5.5.4 High-rate algal ponds

A less common pond shape for treating primary wastewater is a rapidly growing algal pond (HRAP). To promote the development of green microalgae, which reduces the organic load and pathogen counts, these ponds are usually small and well-mixed. An additional advantage of HRAP is that it has the potential to inactivate pathogens; one study found that the infectivity of *Cryptosporidium* was reduced by 97% as a result. Using algae to remove toxins like heavy metals, alongside secondary treatment, HRAP also provides some tertiary therapy. Even though HRAP uses more energy compared to various systems, it still outperforms them in terms of expense and energy efficiency, especially when energy-efficient propellers are employed. Before HRAP is applied to main effluent, anaerobic ponds or clarifiers can be used as a pre-treatment process to remove particles from wastewater. If carbon is a limiting component in the effluent, CO_2 aeration can increase HRAP's effectiveness. Compared to WSPs, HRAP systems are a desirable alternative for urban or semiurban areas that are rapidly growing and require a decentralized treatment of waste solution that is affordable with minimal disturbance to land (Chahal *et al.*, 2016).

1.5.5.5 Tertiary treatment and disinfection

Tertiary treatment is the ultimate polishing step required to obtain recycled water of a suitable quality; it entails a number of biological and chemical processes. The treatment methods selected are influenced by the final use that is decided. When human interaction with the recycled water is infrequent, secondary treated effluent may be appropriate for use in applications like underground irrigation or woodlot irrigation. However, the need for tertiary cleaning rises as the danger of human exposure to the recycled water does. To reduce the quantity of pathogens to levels that comply with healthcare standards, the effluent must generally be correctly treated and/or disinfected. Risk analyses that take into account exposure pathways, exposure levels, infectious dosages, and disease outcomes produce these target numbers. Lime or alum can be used to precipitate out nutrients such as phosphorus, and, less frequently, microalgae (using HRAP) have also been shown to be successful at removing nitrogen and phosphorous. Pathogens can be physically eliminated using filtration methods such as flotation of dissolved air filtration, microfiltration for the removal of microorganisms and protozoans, as well as ultrafiltration to remove viruses. Larger pathogens like bacteria and protozoa can be effectively removed using membrane filtration techniques. The extra benefit of filtration is that it removes particles, which helps the subsequent disinfection procedures that are necessary to destroy any lingering bacteria.

The final and possibly most important step in tertiary treatment is the decontamination of the effluent prior to utilization (at least with respect to microbiological safety). UV rays and chlorine treatment are both widely used disinfection techniques. Chlorine is introduced to treated effluent for set periods of time to increase microbe exposure and inactivation. Then, any chlorine that is left over is neutralized before being discharged into the atmosphere or kept in aquifers. Traditionally, chlorine disinfection goals are determined using contact time, or CT, that is calculated as the combination of the chlorine dose (in mg/L) and the time (in minutes). As a result, the same CT can be acquired using either an increased dose/small period of time or a low dose/high period of time. Amount of free chlorine that is readily available, which is influenced by pH and temperature, has an impact on the CT. This is important because, compared to warm water, cold water necessitates significantly higher CTs for pathogen inactivation (e.g., a CT of 8 mg min/L for viruses at 5°C versus a CT of 3 mg min/L for viruses at 20°C). CTs of the main enteric pathogens have been determined for disinfection with chlorine of drinking water or wastewater, and these have been set in numerous guidelines. Chloramine is a significantly weaker oxidant than chlorine, and it requires high level of CTs to disinfect water to the same degree. Common gut bacterial pathogens with chlorine CTs of 1 mg min/L or less include *Salmonella*, *Campylobacter*, and *E. coli*. Viruses are inactivated by chlorine, despite having a little bit more resistance than intestinal bacteria. Certain protozoan parasites, most notably *Toxoplasma* are resistant to chorine.

Working with UV light is safer, involves fewer stages, and does not produce any disinfection by-products in comparison to working with chlorine and different ways of producing chlorine. Fair enough, building the required infrastructure for UV disinfection can be more costly than chlorination. When wastewater is treated with UV light, it is exposed to a UV-C illumination source, usually a UV lamp enclosed in a quartz sleeve inside of a steel conduit and attached to reinforced concrete. Although some viruses, most notably the adenovirus, can withstand UV radiation well, UV is very efficient toward bacteria alongside intestinal protozoans. Combining UV with chlorination can be particularly effective since it allows for the most economical use of each treatment method. For the elimination of viruses and *Cryptosporidium*, appropriate concentrations of chlorine are used. Other cleaning techniques have been created and researched as well. Peracetic acid has also been regarded as a powerful disinfectant due to its potent bactericidal, fungicidal, sporicidal, and virucidal properties. Ozone has been shown to be effective with viruses, protozoan cysts, and helminth eggs. Wastewater can be disinfected using conductive-diamond electrochemical oxidation (CDEO), and by preventing *E. coli* cells from congregating, CDEO and ultrasonic technology improved disinfection effectiveness. Yet it does not seem like this technique has been adopted for widespread commercial application (Chahal *et al.*, 2016).

1.6 MICROBIAL FUEL CELLS

A particular kind of bio-electrochemical fuel cell technology is the microbial fuel cell (MFC). It produces electricity by transferring electrons from reduced chemicals produced by microbial oxidation on the anode to oxidized compounds like oxygen upon this cathode into an auxiliary electrical circuit (Jatoi *et al.*, 2020).

1.6.1 MFC configuration

Two electrodes and a membrane separating these two compartments make up the majority of a microbial cell. As a result of microorganisms being oxidized at the anode location, electrons and protons are produced there and begin to transfer to the cathode. Unlike protons, which move through the membrane, electrons move across the circuit. When electrons and protons reach the cathode, they reduce the oxygen to create water. Most microbial cells are aided by mediators because they lack electrochemical activity (Jatoi *et al.*, 2020).

1.6.2 Mechanism of MFC

The electron transport process served as the foundation for the working of MFCs. Similar to how energy production is impacted, it also depends on a number of variables, such as the propensity of microorganisms to transfer electrons, the size of the electrodes' surfaces, the strength of the electrolyte, and the rate of the kinetic oxygen reaction. The three categories of these components are kinetic limits, ohmic restrictions, and transport restrictions. The MFC shows the worth of identifying limiting parameters and later changing to improve overall performance by treating wastewater through batch or continuous inputs. The transmission of electrons was impacted by the microbial cell wall and electron mediator. pH 7 and an impedance of more than 500 V are the ideal values (Jatoi *et al.*, 2020).

1.6.3 Wastewater from MFC

The substrate is crucial for any microorganism's growth and evolution since it provides nutrients. Similar to this, it is believed that the substrate is the main organic element of MFC responsible for energy generation. There are many different kinds of wastewater that have been employed, including home wastewater, swine wastewater, food processing wastewater, wastewater from the chocolate business, and wastewater from farm lands, municipal, and industrial sources. Microbial fuel are popular for how they operate in terms of converting substrate into energy and lowering wastewater-related environmental issues (Jatoi *et al.*, 2020).

1.7 MFCs WITH SYNTHETIC WASTEWATER AS SUBSTRATES

Carbon removal from wastewaters using MFCs has demonstrated high rates (>90%). Acetate, glucose, sucrose, and xylose are just a few of the many organic substrates employed in the MFCs as synthetic wastewaters for microbial oxidation in the anode chamber. For exo-electrogenic bacteria, acetate is the most straightforward and typical substrate employed in MFCs. Since acetate offers a straightforward metabolism, these bacteria can biodegrade it easily. Acetate is also a by-product of various metabolic processes for higher-order carbon sources. For instance, when wastewater sludge is digested anaerobically, carbonaceous matter is transformed into shorter-chain organic acids like acetic acid. In comparison to household wastewater and glucose substrates, MFCs with acetate substrates exhibited higher power densities. The power generated with acetate (506 mW/m^2, 800 mg/L) was up to 66% higher than that produced with butyrate (305 mW/m^2, 1000 mg/L) using a single-chamber MFC. A recent study examined the coulombic efficiency (CE) and power output of four different substrates. After butyrate (43.0%), propionate (36.0%), and glucose (15.0%), MFC fed with acetate displayed the highest CE (72.3%). The MFC based on acetate-induced consortia achieved more than two times the maximum electric power and half of the ideal external load resistance compared to the MFC based on consortia induced by a protein-rich wastewater when acetate was compared with it as a substrate in MFC. Another often researched substrate in MFDs is glucose. A glucose fed-batch MFC with 100 mM ferric cyanide as the cathode oxidant produced a maximum power density of 216 W/m^3. This study compared anaerobic sludge to glucose to assess its suitability as a fuel for MFC electricity production. Anaerobic sludge provided very little substrate to a baffle-chamber membrane-less MFC, and only a little amount of electricity (0.3 mW/m^2) could be produced. Yet, a maximum power of 161 mW/m^2 was produced when glucose was present in the same system. Acetate and glucose were used as MFC substrates in a different study, and the energy conversion efficiency (ECE) of each was examined. With acetate, the ECE was 42%; however, with glucose, it was only 3%, resulting in a low current and 14 power density. Glucose-fed due to electron loss by competing bacteria, MFC produced the lowest CE, but its relatively diversified bacterial structure allowed for the largest power density and significantly greater substrate use. The findings of numerous independent investigations demonstrate that complex wastewaters have power outputs that are five times lower or less than those of discrete substrates. The potential to handle high strength substrates is indicated by the substrate removal rates for fake wastewater, which can reach up to 8.9 kg COD(chemical oxygen demand)/m^3 reactor/day, compared to actual wastewater substrate removal rates, which vary from 0.5 to 2.99 kg COD/m^3 reactor/day (Gude, 2016).

1.8 MFCs WITH ACTUAL WASTEWATER AS SUBSTRATES

Municipal wastewaters are classified as low energy density carriers or feedstocks for MFCs due to their lower BOD concentrations, which are typically less than 300 mg/L. Due to the anaerobic conditions in the anode chamber, MFCs may also treat wastewaters with high energy density and BOD concentrations surpassing 2000 mg/L. These industrial waste streams include those from the food processing industry, breweries, dairy farms, animal feeding facilities, and other waste streams. With relatively low quantities of organic nitrogen, the effluent from the food processing industry is rich in organic acids and quickly biodegradable carbohydrates. Depending on the BOD and amount of water used in the processing, MFCs can generate electricity from the wastewater used to process food products in the range of 2–260 kWh/ton of product. The US milk dairy farms' low BOD wastewaters have the capacity to generate 46 MW of electricity, whereas the high BOD wastewater from the dairy industry has the ability to generate up to 1960 MW of electricity. Animal wastewaters from the livestock-related industry are frequently particularly high in organic material content (~100 000 mg/L COD for animal wastes) and may contain high levels of nitrogen-containing components, like proteins, and harder to degrade organic materials, like cellulose. In addition, 15 lipids may also be present in slaughterhouse wastewaters from the livestock-related business in addition to carbs, organic acids, and proteins. Although brewery wastewater concentrations vary, they are typically between 3000 and 5000 mg/L COD, making them around 10 times more concentrated than domestic wastewater. The fact that it has a high carbohydrate content (high energy density) and a low ammonium nitrogen concentration may make it a good substrate for MFCs. A maximum power density of 528 mW/m^2 was attained when 50 mM phosphate buffer was introduced to the wastewater during the treatment of beer brewery wastewater using air cathode MFC. In this instance, when both wastewaters were evaluated at equal intensities, the maximum power produced by brewery wastewater was less than that obtained by household wastewater (Gude, 2016).

1.9 BIOREMEDIATION

To decontaminate contaminated environments, a method known as bioremediation involves dissolving complicated contaminants into intermediates that are either safe for humans or other living organisms to consume. A crucial role of bioremediation is played by microorganisms, including fungi, bacteria, algae, and so on. Not all microorganisms have the ability to convert contaminants into simple, innocuous molecules that can be consumed (Ojhaa *et al.*, 2021).

1.9.1 Principle

Biodegradation is the foundation of the concept of biological restoration. Using the bioremediation method, environmental waste can be biologically decomposed to a safe form or at proportions beneath the corresponding limits of concentration. This is done under closely supervised conditions set by the relevant regulatory authorities. Utilizing organisms that are living, most commonly bacteria, bioremediation breaks down hazardous materials or chemicals to less harmful form that can be found in wastewater. It is an ecological treatment choice as well as feasible technology, but the environment can exhibit how successful it is. The essential microorganisms such as bacteria, mold, and flora have the biological ability of breaking down harmful pollutants in water and also clean it up. A moderately priced technology is available on-site. This technique requires the growth of microbial consortiums or promoted microflora. They can carry out required duties and are native to contaminated areas. The aforementioned colonies of bacteria can be produced in a variety of methods, such as by promoting growth, adding nutrients, including a terminal electron acceptor, or regulating the relative humidity and temperature. Throughout the course of bioremediation procedures, microorganisms consume these contaminants as food or energy sources. Under certain conditions, specific native microbes may already be at the location, while others may be collected and added in the purified medium via a bioreactor. In spite of the fact that bioremediation relies on the growth and activity of microorganisms, it is important to emphasize

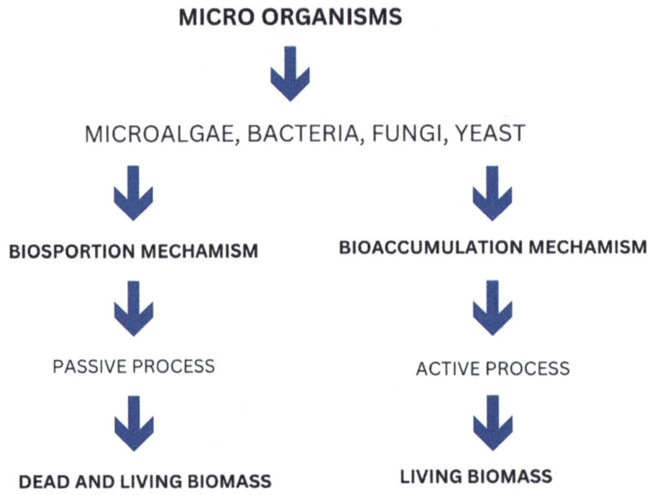

Figure 1.4 Methods of bioremediation for wastewater.

that the environmental factors that influence the growth and degradation rates of microbes have a significant impact on how effective bioremediation is. To locate the appropriate microorganisms in the right place for an effective degradation process under the required environmental conditions, the bioremediation method generally depends on doing so. It is doable to improve the biological mechanisms employed in bioremediation techniques and focus on removing harmful chemicals from water. This procedure results in waste mineralization and reduces the need for further treatment by converting waste into substances such as water, carbon dioxide, biomass, or other non-hazardous materials. The word 'bioremediation' refers to the management of different substances. Chemicals used in agriculture, petroleum-based components, and potential substances that are not biodegradable such as solvents that are chlorinated, chlorofluorocarbons, and various other industrial organic substances can all be broken down by microorganisms in addition to municipal waste and process water. The microorganisms may be transported to the contaminated location from elsewhere or they may be isolated and endemic to the polluted area. Living organisms convert contaminants through reactions that occur as part of their metabolic activities. The behavior of numerous species can also contribute to the biodegradation of a chemical (Ojhaa *et al.*, 2021) (Figure 1.4).

1.9.2 Methods of bioremediation of wastewater

1.9.2.1 Bacteria

There is a wide spectrum of bioremediation potential for bacteria from a financial and environmental standpoint. Heavy metal emissions from industrial uses become a big environmental problem such as Cd, Cr, Cu, and Hg. The hazardous heavy metal chromium is commonly used in the metalworking, tanning process, textile coloring, and electroplating manufacturing sectors. Several microorganisms, including *Desulfovibrio vulgaris*, *Arthrobacter*, *Pseudomonas* sp., *Serratia marcescens* and *Ochrobactrum* sp. have been seen to convert highly soluble and poisonous Cr (VI) into lower soluble and fewer toxic Cr. (III). *Arthrobacter psychrolactophilus* sp. 313 lowered the protein content of wastewater. They are utilized for the sewage from industrial facilities. In the course of the treatment, a variety of different bacteria, including *Pseudomonas* and *Bacillus subtilis*, were employed (Mishel *et al.*, 2023).

1.9.2.2 Applications of oxygenic photosynthetic bacteria (cyanobacteria in bioremediation)

Cyanobacteria are frequently used in wastewater purification for the removal of metal contaminants, petroleum-based chemicals, and colorants. Since cyanobacteria need nitrogen for their metabolic processes, they are usually quite effective at removing nitrate. To get rid of the nitrate-producing *Synechococcus* sp. strains and *Synechocystis minima* CCAP148014, distinct cyanobacterial species were used. Phosphorus and nitrogen were successfully extracted by the *Phormodium tenue* strain and the *Phormodium bohneri* strain. The cyanobacteria strains *Anabaena variabilis*, *Anabaena oryzae*, and *Tolypothrix ceytonica* were additionally treated using a mixture of domestic and commercial effluents. Organic detritus has been eliminated, respectively, by *A. variabilis* and *A. oryzae* were used to remove particles. Heavy metals such as copper, cobalt, manganese, zinc, and lead have been recovered from sewage wastewater by the plants *Anabaena subcylindrica* and *Nostoc muscorum*. Copper was obtained using *Nostoc* PCC 7936 and *Cyanospira* capsule. (II). Zn and Cu were both removed by *T. certonic*. Crude oil was effectively degraded by *Oscillatoria salina*, *Aphanocapsu* sp. *terenbans*, and *Plectonema*. *Nostoc ellipsosporum* strain B1453–7 and *Anabaena* sp. PCC 7120 both broke down lindane contaminants. It was discovered that *Oscillatoria formosa* NTDM 02 was particularly efficient at removing dye from the textile sector. Bioremediation using anoxygenic photosynthetic bacteria *Rhodobacter sphaeroides* Z08 was introduced to the PPB during the treatment of pharmaceutical wastewater. *R. sphaeroides* IL106 also eliminated phosphorus. The *Rubrivivax gelatinosus* strain SS51 and SY40 was able to break down the organic pollutants in the wastewater used to make latex. *Rhodobium marium* NW16 and *R. sphaeroids* KMS24 have been discovered to be effective heavy metal extractors. *R. palustris* WS17 was utilized to degrade pesticides. Used in the fading of colors is Purple non-sulfur Bacteria (PNSB), *R. palustris* ASI.2353 (Mishel *et al.*, 2023).

1.9.2.3 Algae

Algae are crucial to the process of naturally purifying water. They can be utilized for recovering precious metal ions such as gold and silver as well as for the sorption of toxic and radioactive metal ions. They contribute to the clean-up of nutrients by growing quickly and assimilating C, N, and P from wastewater. It is an alternate method of treating sewage effluent that is also cost-effective and environmentally friendly. Some researchers have acknowledged the practical application of microalgae for obtaining nutrients from various wastes to stop potential deterioration of the water quality of wastewater. Textile wastewater (TWW) contains organic dyes as well as the minerals (phosphate, nitrate, micronutrients, etc.) required for the growth of algae (a potential source of carbon). Clothing wastewater can be effectively bioremediated using microalgae that can utilize nutrients and dyes (TWW). Microalgae use the colors and nutrients in wastewater for their growth. Wastewater can be bioremediated using a culture of *Chlorella vulgaris* and *S. quadricauda*. Citric acid and diluted ethanol from the wastewater industry are produced using *C. vulgaris*. It accelerates the decline in effluent BOD and COD concentrations. *C. vulgaris* and *S. quadricauda* were used to remove nitrate. *S. quadricauda* successfully extracted the phosphate. During growing, *C. vulgaris* uses phosphorus to eliminate phosphate. Clothing wastewater is bioremediated using *C. vulgaris* strain UMACC 001 (TW). Both viable and non-viable algae have been used to remove color from dyes and wastewater. Two techniques, namely bioconversion and bioaccumulation, or biosorption operation, are used to bioremediate TWW utilizing microalgae. During the bioconversion process, these colors are ingested by microalgae as a source of carbon and converted to metabolites. Microalgae can nonetheless function as a biosorbent and attract the pigments to their surface. In simultaneous TWW bioremediation, both outcomes are conceivable. The aggregation of microalgae may be caused by adsorption, enzyme breakdown, or both. Microalgae have a high sorption potential due to traits like their enormous surface area along with powerful attraction to binding to azo dyes. The cells biosorb metal ions on the top of dead algae's cell membrane. It may be more effective, less costly, and environmentally safe to remove heavy metal ions from wastewater using an algae-based biosorption process. The removal of reactive dye from TWW has been proven to be possible using the nonviable biomass of *Spirogyra*. Basic dyes can be removed by living material from macroalgae such as *Caulerpa scalpelliformis*. Certain algae, like *Phormidium*, have the ability to bioconvert colors into simpler chemicals. Immobilized algae are used to draw color from the surface of textile pigments. Both *Spirulina* and *Chlorella* are effective at cleaning up sewage. Some of the microalgae that can convert azo dyes into simple amines include *Scenedesmus bijugatus*, *Chlorella vulgaris*, *Chlorella pyrenoidosa*, and *Oscillatoria tenuisin*. *Chlorella* sp. and *Scenedesmus* sp. microalgae G23 and *Cosmarium* sp. were commonly used to treat various types of wastewater because they are more efficient (Mishel *et al.*, 2023).

1.9.2.4 Fungi

Heavy metals are generally consumed by filamentous fungi. Utilizing the ability of fungal biomass, biosorbents may be used to remove radionuclides and heavy metals from contaminated waterways. White-rot fungi like *Pycnoporus sanguineus laccase* and *Phanaerochaete chrysosporium* can all devastate straw, sawdust, or corn cobs. Metal ions are taken up by *Penicillium*, *Aspergillus*, *Rhizopus*, *Mucor*, *Saccharomyces*, and *Fusarium*. Utilizing *Penicillium*, heavy metals can be bioabsorbed (Cr, Pb and As). Radionuclides can be bioabsorbed by the microbes *Penicillium*, *Rhizopus*, and *Saccharomyces* (U, Th, and Sr). For the bioremediation of distillery wastewater, *Trametes pubescens* MB 89, *Pycnoporus cinnabarinus*, and UD4 were used. White-rot fungus can degrade high-intensity phenolic wastes. Fungi known as white-rot fungi break down industrial waste. To decolorize and reduce the COD of molasses wastewater, it may be feasible to use the organisms *Funalia trogii* and *Pleurotus pulmonarius*. White-rot fungi, which include edible mushrooms like *Lentinula* and *Pleurotus*, as well as several yeast species can all be utilized to treat olive oil mill wastewater (OMWW). They reduce OMWW color, phenolics, and COD. White-rot fungi, such as *Coriolus versicolor* and *Funalia trogii*, *Geotrichum candidum*, *Lentinula* (*Lentinus*) *edodes*, and *Phanerochaete* sp., were employed for the remediation of OMWW. Most viable possibilities are fungi that play a key role in bioremediation. White-rot fungi are one of them and provide a number of benefits for environmental pollutant degradation. Endosulfan-degrading fungi such *Aspergillus terreus*, *Cladosporium oxysporum*, *Mucor thermohyalospora*, *Fusarium ventricosum*, *Phanerochaete chrysosporium*, and *Trichoderma harzianum* are utilized. Fungi of the Zygomycetes class and *Aspergillus*, *Mucor*, and *Penicillium* were employed to breakdown and detoxify textile effluent and crude oil. For the breakdown of polychlorinated biphenyls, *Penicillium chrysogenum* and *Fusarium solani* are utilized (PCB). Pinus is combined with *Rhizopogon roseolus*, *Suillus bovines*, and cadium extraction. A few plant-related fungi are used to decolorize textile industry waste (Mishel *et al.*, 2023).

1.9.2.5 Yeast

Yeast helps with absorption of heavy metals. In OMWW bioremediation, yeasts including *Trichosporon cutaneum* and *Saccharomyces* sp. are employed. Yeasts are successful at removing mono- and polyphenols from the body as well as lowering COD levels. Because harmful chromophores can be absorbed, collected, and degraded into simpler compounds, these are used to clean TWW. They can be used as biosorbents for dye biosorption and contain enzymes for pigment breakdown. Yeasts like *Candida krusei* and *Saccharomyces cerevisiae*, and so on, were used to break down colors and other compounds (Mishel *et al.*, 2023).

1.10 ACTIVATED SLUDGE PROCESS

The activated sludge system must also disperse the microbes using an efficient mixing system in order for them to come into direct contact with all of these substrates. The microbes must also mature into floes, three-dimensional aggregated microbial populations. This is perhaps most crucial. These flocs must possess excellent settling properties in order to successfully separate from the liquid supernatant in the clarifiers. It is necessary to give enough O_2, which is expensive, in order to mix the biomass in the reactor and supply what is needed by the aerobically respiring organisms that are in charge of breaking down these substrates. The respective advantages of the various widely used aeration techniques and the associated health risks related to aerosol production. The strength or concentration and composition of the wastes change significantly hourly, daily, and seasonally, and the ASP must be able to manage significant variations in flow rates that have a knock-on impact. The metabolic activity and rate of microbial development will be greatly impacted by similar variations in the temperatures of mixed liquor. The process's by-product, microbial detritus, also referred to as sludge, is still a resource that is vastly underutilized in most countries. Given all of these erratic factors, it can be considered astonishing that activated sludge facilities operate as efficiently as they do. Conventional plants' basic design has not changed much, but there is a general consensus that they will likely remain the most popular systems for aerobic waste treatment well into the 21st century due to their dependability, adaptability, and versatility. A rectangular basin with motorized surface agitators or submerged diffusers providing the mixing and aeration still makes up the bulk of aerobic reactors. The clarifier is still used to inoculate the inbound raw wastes with the majority of the separated biomass or sludge, which is separated from the liquid supernatant after the effluent mixed liquor is passed through it, the main factors influencing creativity and change in activated sludge plant design. When these systems were first developed, their primary objectives were to remove carbonaceous material from residential sources that contained naturally existing organic compounds and to produce a treated liquid effluent with low enough BOD and suspended particle levels to be securely discharged into another body of water. The so-called 30:20 standard for BODs and 55 as the allowable limit for treated effluent in conventional plants are adopted by many countries. The effluent had to meet standards for reducing NH_3 level later because it was believed that NH_3 was more hazardous to fish than NO^{3-} and as a consequence, plant design was altered to encourage nitrification. In response to the public's growing concern over eutrophication's long-term effects on the environment over the past 20 years, plants that can microbiologically extract both nitrogen (N) and phosphorus (P) are presently being created. It is expected that as governmental bodies impose stricter requirements for effluent quality in response to public pressure, these occasionally expensive devices will become more common. The addition of more reactor tanks alters the environment in a number of ways, especially in terms of the biomass's availability to oxygen, which encourages the development of particular physiological groups. For example, denitrification is enabled by the incorporation of anoxic zones with low DO levels but abundant NO^{3-} and the existence of anaerobic zones with low levels of both NO^{3-} and O_2 gives the bacteria required to break down phosphate (P). Activated sediment testing was done in batches in the laboratory. Two-liter laboratory dividers made of Plexiglas were used. To inoculate the wastewater, activated sludge from a plant that treats domestic sewage was used. Every day, the aeration was turned off to enable the sludge to settle. Then, the supernatant was taken out and fresh wastewater was poured into the column. Until a significant amount of modified sludge was produced, this procedure was repeated. Numerous studies were conducted to find out how the aeration time affected the activated sludge. One such example is a different column that already contained a predetermined quantity of sludge (3–4 g/L) was added to the pre-treated wastewater. One hour to twenty-four hours of confinement were examined. Adjustments were made to maintain the DO content at a minimum of 2 mgO_2/l. The characterization of the treated wastewater and sludge analysis were conducted after 60 min of settling (Seviour *et al.*, 1998).

1.11 CONCLUSION

Understanding the role performed by the microbial community structure of the organisms participating in the treatment processes is essential and crucial for having a better understanding and control measures of wastewater treatment operations. Enzymes produced by microorganisms that participate in the breakdown of environmental pollutants are typically quite selective in the substrates they catalyze (Jatoi *et al.*, 2020). Microorganisms can, however, develop new enzymes to manufacture energy and nutrients from other substrates or under new growth conditions following an acclimation period when they are exposed to new growth conditions or substrates (Gude, 2016). To use the potential of the microorganisms for bioremediation, it is crucial to control their activities

(Ojhaa *et al.*, 2021). To ensure affordability while maintaining public health protection, wastewater must be appropriately treated. This is a crucial factor to take into account because recycled water would not be accepted without it. Customers will utilize the least-expensive water available, though, if it is too pricey. Treatment to remove or inactivate pathogens is one of the biggest expenses involved in producing reused water. Although chemical contaminants are also significant, health authorities frequently concentrate on those that cause acute disease, particularly when wastewater is reused and prolonged human exposure is unlikely (Krantz and Kifferstein 1996). To provide the best possible treatment for reducing the danger from pathogens in wastewater, it is necessary to understand how pathogens behave during the wastewater treatment and disinfection processes as well as the variables that affect these processes. Association with particles is one of the main elements that affects how pathogens behave in water. Understanding the nature of the pathogen is crucial. This presents a chance to improve wastewater treatment while lowering treatment costs: reduced amounts of association can improve disinfection, whereas higher levels of association can improve removal via sedimentation processes. This strategy seems to combine the benefits of each of the technologies that were previously discussed while attempting to get around the problems of nutrient scarcity and a lack of petroleum-degrading bacteria. It turned out to be an interesting detour because it offers the possibility of an economically viable and scientifically advantageous method through the use of locally accessible support material and bacteria that have been isolated from prior contamination. Moreover, the local biota, nutrient availability, and other environmental parameters crucial for getting the greatest results have to be considered in an efficient restoration method. It was thought that bacteria might have an enhancing effect on the biodegradation of persistent organics in the bioremediation of persistent organics-contaminated soil and particularly groundwater environments by increasing the bioavailability of these pollutants to microorganisms. Last but not least, a combination of technologies that are strictly regulated and given enough time will prove to be quite significant. Biological treatment techniques have frequently been regarded as the most comprehensive, environmentally friendly, and economically advantageous treatment solutions.

REFERENCES

Chahal C., van den Akker B., Young F., Franco C., Blackbeard J. and Monis P. (2016). Pathogen and particle associations in wastewater: significance and implications for treatment and disinfection processes. *Advances in Applied Microbiology*, **97**, 63–119, https://doi.org/10.1016/bs.aambs.2016.08.001

Cydzik-Kwiatkowska A. and Zielińska M. (2016). Bacterial communities in full-scale wastewater treatment systems. *World Journal of Microbiology and Biotechnology*, **32**, 1–8, https://doi.org/10.1007/s11274-016-2012-9

Gude V. G. (2016). Wastewater treatment in microbial fuel cells—an overview. *Journal of Cleaner Production*, **122**, 1–55.

Jatoi S., Akhter F., Mazari S. A., Sabzoi N., Aziz S., Soomro S. A., Mubarak N. M., Baloch H., Memon A. Q. and Ahmed S. (2020). Advanced microbial fuel cell for waste water treatment—a review. *Environmental Science and Pollution Research*, **28**, 1–15.

Krantz D. and Kifferstein B. (1996). Water pollution and society. *Community Development Journal*, **69**, 66–74

Larsson D. G. J., de Pedro C. and Paxeus N. (2007). Effluent from drug manufacturers contains extremely high levels of pharmaceuticals. *Journal of Hazardous Materials*, **148**, 751–755, https://doi.org/10.1016/j.jhazmat.2007.07.008

Liu H. and Fang H. H. P. (2002). Extraction of extracellular polymeric substances (EPS) of sludges. *Journal of Biotechnology*, **95**, 249–256, https://doi.org/10.1016/S0168-1656(02)00025-1

McCarty P. L., Bae J. and Kim J. (2011). Domestic wastewater treatment as a net energy producer—can this be achieved?. *Environmental Science & Technology*, **45**, 7100–7106, https://doi.org/10.1021/es2014264

Metcalf and Eddy (2004). Wastewater Engineering: Treatment, Disposal, and Reuse. McGraw-Hill, New York.

Mishel P. F., Steffi P. F., Thirumalaiyammal B. and Anburaj R. (2023). Bioremediation: a green technology for environmental cleanup. *Metagenomics to Bioremediation*, **1**, 3–21, https://doi.org/10.1016/B978-0-323-96113-4.00004-4

Ojhaa N., Karnb R., Abbasc S. and Bhugrad S. (2021). Bioremediation of industrial wastewater: a review. *Earth and Environmental Science*, **796**, 1–30.

Pal A. and Paul A. (2008). Microbial extracellular polymeric substances: central elements in heavy metal bioremediation. *Indian Journal of Microbiology*, **48**, 49–64, https://doi.org/10.1007/s12088-008-0006-5

Rani N., Sangwan P., Joshi M., Sagar A. and Bala K. (2019). Microbes: a key player in industrial wastewater treatment. *Microbial Wastewater Treatment*, **5**, 83–102, https://doi.org/10.1016/B978-0-12-816809-7.00005-1

Seviour R. J., Lindrea K. E., Griffiths P. E. and Blackall L. L. (1998). The activated sludge process. *The Microbiology of Activated Sludge*, **2**, 44–75, https://doi.org/10.1007/978-94-011-3951-9_2

Shiklomanov I. A. (1993). Water in crisis: a guide to the world's fresh water resources. *World Freshwater Resources*, **2**, 11–15.

doi: 10.2166/9781789063783_0019

Chapter 2

Elucidation of omics approaches and computational techniques for wastewater treatment: A deep insight

Hina Bansal, Abhishek Saxena, Himanshu, Pankaj Kumar Singh and Archana Tiwari*

Amity Institute of Biotechnology, Amity University, Noida, Uttar Pradesh 201313, India
**Corresponding author: hinabansal@gmail.com*

ABSTRACT

Ecological contamination has been growing rapidly over the past few decades. Bioremediation, a microbial degradation process, is considered to be one of the most ecological and environment safe and financially friendly remediation methods for ecological parts polluted with chemical-based contaminants. Toward this pursuit, the traditional approaches are skimpy, and individuals must enthusiastically think about the current developments in high-throughput 'omics' technologies to remove hazardous and highly polluted contaminants from the environment. The sequencing of various microbial genomes and their functional studies along with metabolic pathway reconstruction and systems biology provide strategies to enhance the degradation and decomposition of pollutants. The computational techniques and omics approaches provide genome-based analysis of the microbial genomes including genomics, proteomics, and system biology along with various tools to determine the structural and functional aspects of microbes and their biodegradation pathways. The current chapter gives a deep insight into the status and advances on different computational techniques and omics approaches to enhance the process of bioremediation. It is a great step against the degradation of the environment and restoring natural resources.

Keywords: bioremediation, bioinformatics, omics, genomics, proteomics, systems biology, data mining

2.1 INTRODUCTION

Cleaner production (CP) is the process of successfully utilizing resources and energy, as well as removing hazardous raw materials and reducing the toxicity of all emissions and solid waste. It is the most proactive technique for reducing waste, negative environmental effects, and health concerns by using energy and natural resources in a more organized and efficient manner (Giannetti *et al.*, 2020). On the contrary, the obligation to conserve natural resources and maintain global ecosystems in order to support health and well-being today and in the future is known as environmental sustainability. In 2015, the United Nations highlighted the Agenda 2030 for sustainable development goals (SDG), which emphasized overcoming any deficiency linked to the policies, strategies, and applications that may otherwise put a roadblock in the path of achieving CP through SDG (UN, 2015). Therefore, there is a pressing need to establish more sophisticated research tools and techniques that may play an inter/multidisciplinary role to address the issue related to CP leading to achieving the future concept of sustainable development goals.

Currently, activities carried out by humans can create and generate a lot of toxic chemical waste because of the increase in industries and factories which release toxic waste into the environment. These activities have led to an increase in pollution in the land soil and water because of the toxic fumes released by this chemical waste that affects the air as well. Even after implementing various rules for managing, maintaining, and disposal of this waste, hundreds of such cases are being reported daily for mismanagement of toxic chemical waste (Umadevi *et al.*, 2015). This pollutant contaminating the environment has drawn attention across the world, taking into account their unwanted slow degrading compounds. Several xenobiotics and aromatic hydrocarbons along with choro- and nitro-aromatic compounds have raised concerns about the environment changing them into highly toxic mutagenic and cancer-causing agents among living organisms. A variety of microorganisms are to be considered as the best medium among other organisms for the procedure to bioremediate the environmental contaminants into the geological biochemistry due to their diverse, versatile character, and adaptability in adverse conditions (Fulekar & Sharma, 2008). These microorganisms have been continuously able to display a great range of contamination which happens to be the ability to degrade, restore natural and raw environmental conditions. It has been shown, however, that many toxins are abnormal, that is, microorganisms do not modify or modify some other metabolic compounds that can accumulate them into the environment. So that makes it important to study other forward pathways that are leading towards complete mineralizing of the pollutants (Mohanta *et al.*, 2015). Bioremediation is considered to have a great potential to recover and regenerate the environment from contamination and pollutants, but there has been a great lack of information regarding the factors which control the growth and metabolism among the considered microorganism in the pollution-ridden environment more often have been able to act as an obstacle to limit the implementation of the bioremediation process Garg (2020). The schematic representation of biodegradation process is presented in Figure 2.1.

2.2 BIOREMEDIATION

Bioremediation can be defined as the process in which microorganisms are made to rapidly remove the unwanted dangerous organic pollutants which affect the ecological levels in soil, sediment, chemical substances, and groundwater (Singh *et al.*, 2021). Biodegradation is a natural process to recycle waste to nutrients that can be used by other organisms. Degradation is caused by microbes that consume nutrients such as carbon, nitrogen, and phosphorus from pollutants that are subject to long-term adaptation turn toxins into a friendly environment. By utilizing the energy of biodegradation, people can reduce waste and purify environmental pollution. Digestion accelerates energy degradation and converts organic matter into energy-efficient products (Vishnoi & Dixit, 2019). The mechanism of bioremediation is presented in Figure 2.2.

Water pollution treatment continues to break down old energy for degradation and harm organic things so that the release of water into the environment does not cause pollution problems. Bioremediation uses microorganisms to clean up oil spills and other contaminants. In the case of bioremediation, therefore, it provides the technology to eliminate pollution by promoting the same destructive process that occurs in nature (safe, cheap, and appropriate treatment) (Chakraborty *et al.*, 2012; Rawat & Rangarajan, 2019). Bioremediation on infected sites usually works in two ways (a) support the growth of contaminants consumed by microorganisms in the contaminated area, (b) special microorganisms have been added that break down contaminants.

2.3 BIOREMEDIATION AND OMICS

Bioremediation can be considered as an umbrella concept that can take in and covers multiple stage binders belonging to different groups transport toxic waste from contaminated sites. There is a lot of

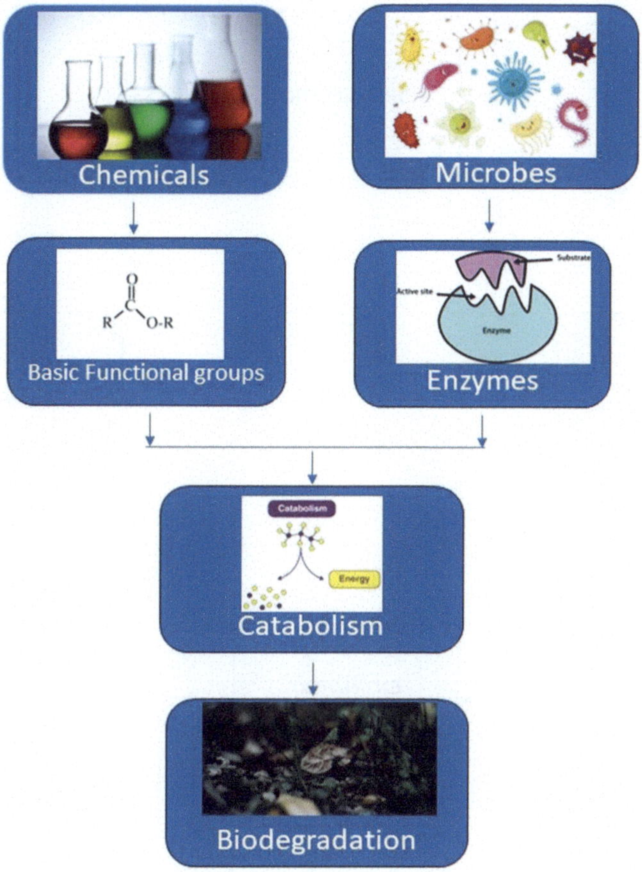

Figure 2.1 Schematic representation of biodegradation process.

information out there for tech support tools, and it is part of the process across all places. This includes distribution/management of chemical toxins, chemical composition, and environmental properties of these chemicals, microbial enzymes, metabolic agents, and degree of degrading action level (Rawat & Rangarajan, 2019).

Bioremediation, in contrast to these approaches, is a long-term and cost-effective way for removing hazardous contaminants from our environment using a diverse range of microorganisms. However, there are certain drawbacks to bioremediation, such as the fact that it takes a long time and has a narrow action range. As a result, significant work will be required to make the process quicker, more effective, and capable of acting on a wide spectrum of organic contaminants and heavy metals (Zhang *et al.*, 2010). *In-silico* bioremediation process is presented in Figure 2.3.

On-site examination of uncultivable microorganisms is now possible because of the development of culture-independent genomics tools. Other 'omics' platforms, like transcriptomics, proteomics, metabolomics, interactomics, fluxomics, and so on, can observe dynamic variations (in terms of mRNA, proteins, and metabolites) occurring in a cell over time. However, a single 'omics' study is insufficient; for example, transcriptomics can measure the total mRNA present in a cell at a particular moment but does not provide information on the amount of expressed proteins, their cellular location, biological activity, or other factors (Chen *et al.*, 2005). Bioremediation is a complicated biological process that

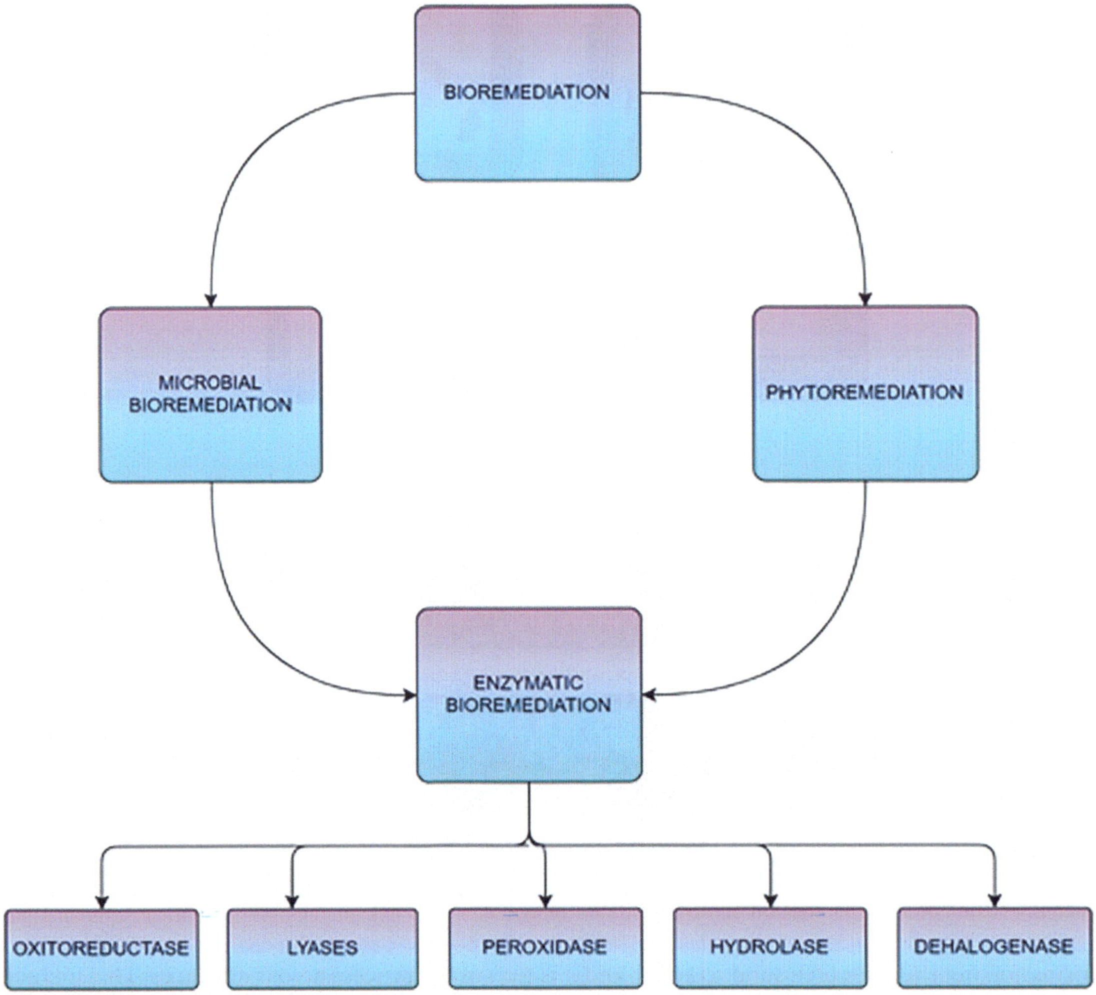

Figure 2.2 Overview of bioremediation highlighting enzymes.

involves several metabolic pathways and interactions between the entire microbial population present at a polluted location, rather than just individual bacteria. Multiple-omics research, rather than single 'omics' approaches, can give a better knowledge of the metabolic and regulatory mechanisms of microbial bioremediation, allowing such complex biological systems to be fully characterized.

2.4 BIOREMEDIATION AND GENOMICS

Non-molecular technology: Currently, most microbiology of bioremediation processes uses 'readability studies'. In this study, samples from contaminated environments and other contaminants are taken to the laboratory and the degree of deterioration or degradation is documented resulting in the immobilization of pollutants (Singh & Nagaraj, 2006). These studies assess the potential metabolic activity of microbe populations, but biomedical causative microorganisms or specific stimulants may or may not be considered when performing bioremediation.

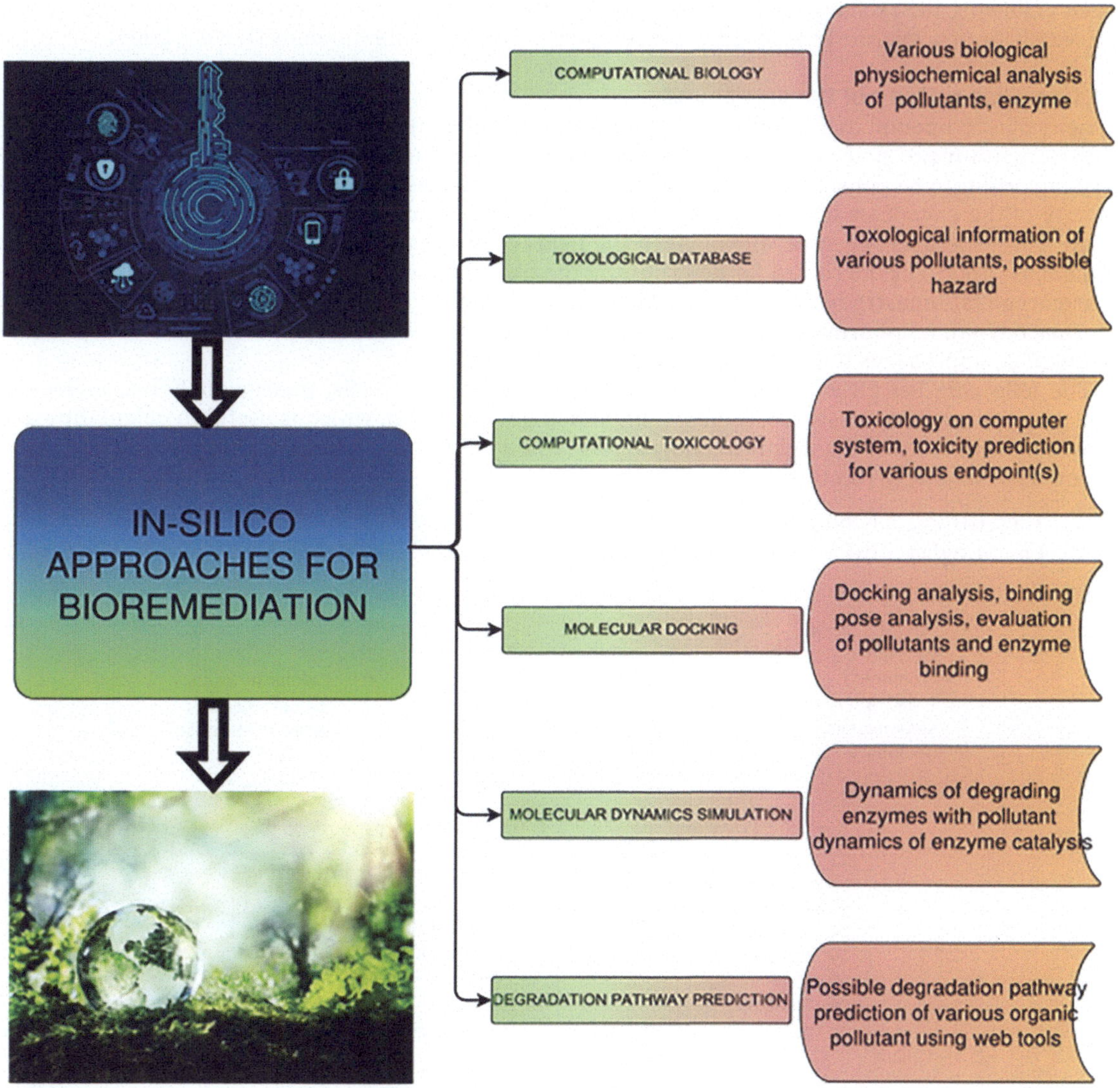

Figure 2.3 Brief outline of *in-silico* strategies for bioremediation process.

The developmental process of the body was thoroughly studied and bioremediation is usually treated because of the efforts of our ancestors. The isolation of pure culture and the voluntary work of such men remain important for the development of molecular analysis and interpretation. Recovery of criminal isolates of microorganisms. Bioremediation processes can be unique because, as described below, the monitoring of these isolates provides an opportunity to study not only the biodegradation process but also the physiological side that can control their biodegradation processes (Eyers *et al.*, 2004). Evolutionary methods are much more useful for dual-fuel adaptation of the biodegradation process than the step-by-step variation introduced by rational conception. Similarly, recent advances in genome switching between species allowing the exchange and reproduction of different traits of the same species have accelerated the search for new microorganisms that can compound-complex contaminants (Lovley, 2003).

2.4.1 *In-silico* toxicity of the compounds

Toxicity of a compound in the environment is important to be contained to a level where it is tolerable in the environment, however, knowledge about the levels of toxicity in the given compounds, are very important to be considered to further develop technologies and research to find out the level of toxicity of the chemicals before focusing on the bioremediation process (Parthasarathi & Dhawan, 2018). Thus, there has been development in the fields of omics and bioinformatics to make a system of database and tools developed for determining the bioremediation technology of any chemical. The pharmaceutical industry has developed and ingrained several *in-silico* procedures to understand the pharmacodynamic, pharmacokinetic, and toxicological profiles of compounds. Figure 2.4 presents the *in-silico* approaches to predict toxicology.

The following list gives a summary of different homology models, pharmacophores, molecular modeling approaches, databases, toolbox, methodology, and so on, which is useful in calculating and knowing the toxicity of the chemical compounds (Khan *et al.*, 2013).

Tools for toxicity prediction:

(1) HazardExpert (CompuDrug) (http://www.compudrug.com/)
(2) OncoLogic (USEPA) (http://www.epa.gov/oppt/st/pubs/oncologic.htm)

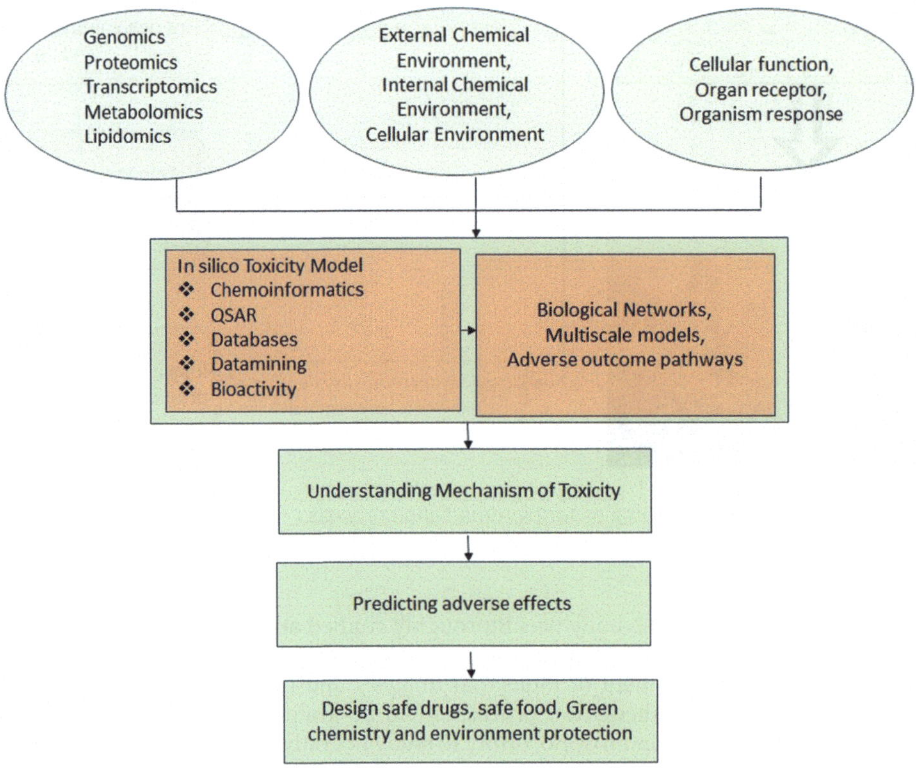

Figure 2.4 *In-silico* approaches to predict toxicology.

(3) MolCode Toolbox http://www.molcode.com
(4) ACD/Tox Suite (ToxBoxes) (http://www.acdlabs.com/products/pcadmet/tox/tox/)
(5) ADMET Predictor http://www.simulations-plus.com/Products.aspx?pID=13
(6) Derek (Lhasa Ltd) (http://www.thasalimited.org/)

Database for toxicity prediction

(1) Acutorbase https://acubase.amwaw.edu.pl
(2) SuperToxic http://bioinformatics.chante.de/supertoxic/index.php?site-home
(3) Terra-Base http://www.terrabaseinc.com/
(4) Carcinogenic Potency Database http://toxnet.nlm.nih.govicg-tinsis/htmlgenCPDB.htm

2.5 SYSTEM BIOLOGY APPROACH IN BIOREMEDIATION

In a contaminated environment, bioremediation entails applying a variety of chemical structures to a complex multispecies metabolic network. The complexity of such processes is becoming increasingly accessible to systems biology's conceptual framework and techniques. The accessibility of biodegradative microorganism genes, genomes, and metagenomes allow scientists to analyze and even forecast chemical fates using the comprehensive metabolic network that comes from linking all known biochemical transactions. The freely diffusible metabolic pool shapes the landscape of pan enzymes that microbial communities embody (Chakarborty *et al.*, 2012). Figure 2.5 is describing the three dimensions to define the activity of each bioremediation process (Roy & Kar, 2016). New computational resources are increasingly helping to develop higher biocatalysts due to the biodegradation and biotransformation of desired chemicals, which are beneficial to the developing field of synthetic biology. Scenarios play a role in which the body consists of several factors. Various physical conditions like flow voltage, electron acceptor, and other properties, diverge in catalysis. These abiotic surroundings over the configuration of the ancestral and bacterial population and the availability of a particular species. However, it is necessary to consider the different levels of combination that differentiate the occurrence of different types of catalytic activity in individual genomes in order to characterize the extensive application of these catalysts to attack humans (Lovley, 2003, Pinto *et al.*, 2021).

2.6 METAGENOMICS IN BIOREMEDIATION

It includes the investigation of the general genes of specific habitats delivering straighten trees to all basins of ecological genes without restriction linked to laboratory cultivation of microbial species.

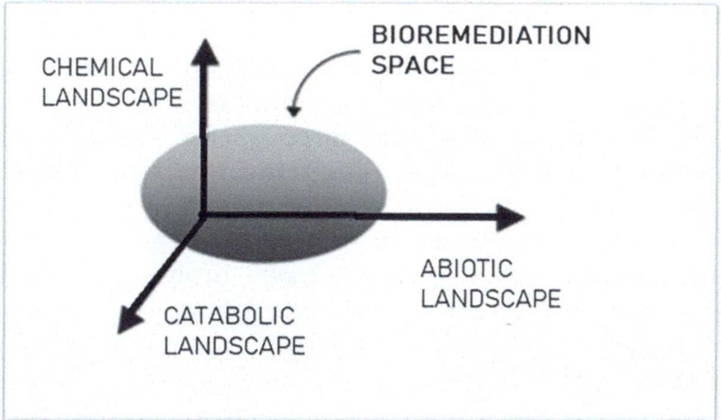

Figure 2.5 Bioremediation space showing chemical, catabolic, and abiotic landscapes.

Metagenomics offers an excellent opportunity to restore unknown processes or sequences in the environment. It is very hard to describe differences among DNA genetics from complex environmental problems, such as soil, fertilizer, soil, or soil, in terms of quality, litter, and full access to microbial species and other species (Thomas *et al.*, 2012). Research guidelines for metamorphic libraries can be adapted to study functions and/or different types. By applying interest, clones are examined or selected, and heterologous markers of the desired character are obtained (Cowan *et al.*, 2005). Library research is facilitated by autonomous collections system of the microbial colony, robotic pipelines, microtiter plates, database-assisted data management, and complex analysis focusing on multiple biomolecules based on the sensor array. Due to several constraints, such as entry size, heterologous expression of cloned cells, as well as post-transcriptional (relative to post-translational material) and the secretion of foreign proteins, 'hit' you are very limited.

However, the filtration and screening rate can be increased by the following ways:

- Using the host with different expressions.
- Develop strategies to use variety/social genomes to first choose the desired brand building and then the metagenomic libraries.
- Experience and humidity allow simultaneous testing practices of the cultures to filter out the different clones.
- Develop new areas of research for high-throughput screening.

Sequence-based analyses are mostly applied to identify members of known gene families or new versions of known protein functional classes. Extensive DNA input libraries offer more possibilities to recover a complete set of functional genes and combined them with potential metabolic functions among a specific microorganism (Daniel, 2005). Although shotgun sequencing may be useful for comparing different media, it is not known whether this sequence will provide information about certain genes and functional interactions. The 'omics' approaches involved in bioremediation is presented in Figure 2.6.

2.7 MICROARRAY ANALYSIS IN BIOREMEDIATION

The ability of a single-stranded DNA or RNA molecule to hybridize with a probe connected to a solid support is the basis for microarrays. The presence of single mismatch probes on the array ensures probe–target specificity, allowing sequence-specific signals to be distinguished from nonspecific signals (Gentry *et al.*, 2006). In environmental genetics, the following three chief classes of microchips have been established: Phylogenetic oligonucleotide arrays (POAs),

(1) Functional gene arrays (FGAs),
(2) Community genome arrays (CGAs).

The search form for POAs and FGAs types is dependent on the sequence program loaded from databases. These subclasses do not have access to functions related to unidentified functional activities and phylogenetic correlations. Given the large reservoir of molecular sequences, which is unknown to the environment, the main weakness lies in the 'traditional' microarray analysis (Gentry *et al.*, 2006). Conditions for using the infringement adjustment service, which allows you to find the clone that has genes that have been expressed (or repressed) because of the therapy. The next stage will be to integrate metagenomics with microarray using probes derived directly from the DNA environment without any or any culture step. As the competition progresses, use the information that the first one does not require much to hope for, but the metagenomics array (MGA) technology has enormous promise for high-throughput screening of natural settings, of the microbial communities being examined (Zhou, 2003). The purpose of this microarray element is to provide a direct link between biogeochemical processes of microbial communities and their functional activities in different environments.

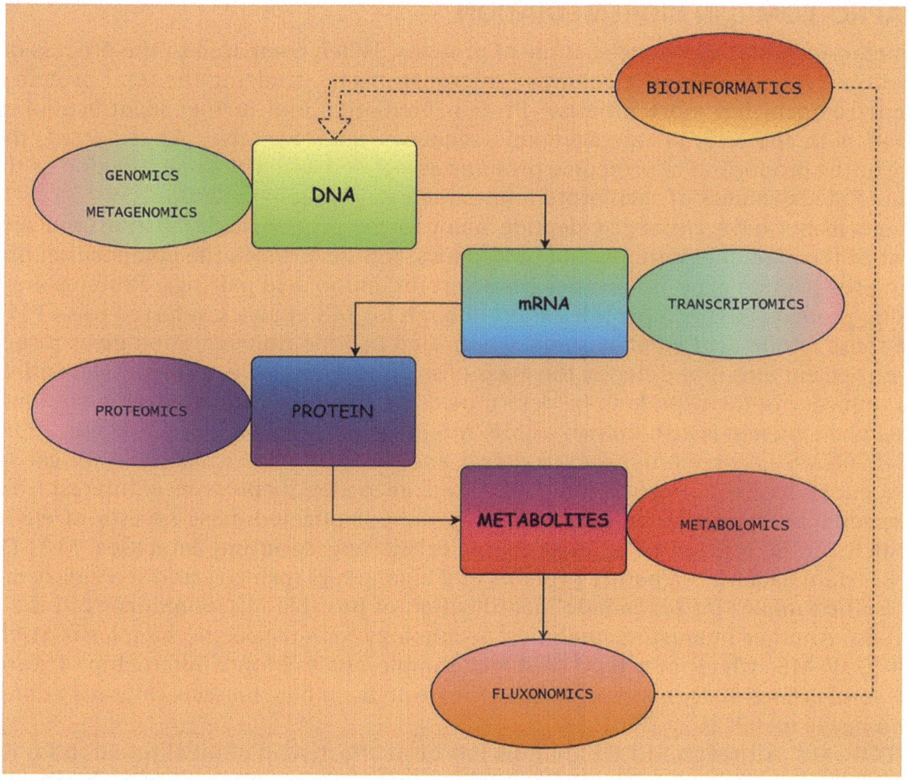

Figure 2.6 Schematic view of 'omics' approaches involved in bioremediation.

2.8 SINGLE CELL SEQUENCING APPROACH IN BIOREMEDIATION

DNA sequencing from single cells has provided new insights into the world of bacteria. Without creating a culture, it is becoming more common to identify bacterial species directly from clinical samples. Recent technological developments often lead to the almost complete completion of genomic data for this sometimes-inaccessible species. New bioinformatic technologies also facilitate the assembly of single-cell genomes. The usage of single-cell sequencing along with metagenomic analysis is proving to be an effective new strategy for the analysis of bacterial communities (Lasken, 2012).

2.9 NEXT-GENERATION SEQUENCING IN BIOREMEDIATION

The next-generation sequencing (NGS) platform can create enormous amounts of monoclonal and digital DNA data in parallel. An increasingly convenient solution at a lower sequence price tools will be provided to the environmental science community. For example, NGS has revolutionized environmental science and contributed to metagenomics and meta-transcriptomics (Bihari, 2013). In various molecular microbiology laboratories, state-of-the-art technologies are playing the same role in bioremediation and biodegradation as they enter the omics age. Gigabytes of monoclonal and digital DNA data can be generated simultaneously on modern NGS machines. As the cost of sequencing in the environmental science community drops, less-expensive tools and solutions are offered (Singh, 2006).

2.10 METAPROTEOMIC IN BIOREMEDIATION

The term 'metaproteomic' refers to the study of proteins. When compared to the process of genomics for the functional examination of a colony of microorganisms, studying the total protein content of a specific environment has more promise. Protein prediction and further identification are highly in facilitation with the relevant metagenomic sequence data available. As expected, there was a discrepancy in the proportion of proteome proteins associated with the same function as the genes in the function-related genomes of each protein function category (Singh, 2006).

MS advancements have enhanced peptide analysis for protein identification and aided in the development of the area of environmental proteomics. MS determines the composition of molecules by measuring the peptide mass that comes from certain amino acid pairings. Proteins are frequently digested using proteases to yield peptides short enough for MS analysis, referred to as PMF, in order to examine these amino acid combinations. Controlled peptide fragmentation gives a succession of overlapping fragment ions that differ by the mass of specific amino acid, allowing the entire or partial amino acid sequence of the peptide to be determined. Because one MS analyzer is often used to pick samples, this phenomenon is also known as MS/MS or tandem MS (Wilmes & Bond, 2006).

MALDI–TOF–MS is commonly used to detect and identify viruses, bacteria, fungal spores, and low-mass chemicals in environmental samples, as well as to identify proteins of interest from 2-D gels. Many site-specific bacteria may be detected using the complicated mass spectra of environmental samples, which can be utilized to develop characteristic fingerprinting databases. MALDI–TOFMS can identify certain bacterial signature proteins and biomarkers (primary and secondary metabolites) from site-specific samples for taxonomic identification of possible microorganisms in the context of bioremediation. Another interesting analytical technology for site-specific samples is MALDI–TOF–MS–SELDI–TOF–MS, which is a type of direct sample analysis on a microchip. The researchers looked at several differentially expressed signature proteins in blue mussels (*Mytilus edulis*) exposed to PAHs and heavy metals (Lay, 2001)

SELDI–TOF–MS. Although SELDI analysis has been effective in clinical research for discovering possible biomarkers, its repeatability and specificity have been questioned. Another new technique is ESI MS/MS used to detect different epoxide hydrolases, peroxisomal antioxidant enzymes, and sarcosine oxidase from a peroxisomal protein pattern associated with marine pollution exposure. Thus, the identification of proteins/enzymes from any site-specific bacterium using 2-DE and MS might open the door for cell-free bioremediation in the future (Mishra *et al.*, 2021).

2.11 META-TRANSCRIPTOMICS IN BIOREMEDIATION

Meta-transcriptomics refers to the analysis of the transcriptome collectively amongst all the microorganisms of a certain specific habitat. Researchers have led to the development of a transcriptomics approach based on the collection and analysis of microbial transcriptomes among marine and freshwater microbial organisms. Environmental transcriptomics procedures provide promising tools for the study while expressing natural genes for the organic microbial communities without biasing known sequences. This easily applicable high-throughput approach has not been tested for the process to analyze microbial communities in contaminated areas (Futschik *et al.*, 2018; Malik *et al.*, 2021).

2.12 METABOLOMICS IN BIOREMEDIATION

Metabolomics is the study of every primary and secondary metabolite with a low molecular weight (b1000 Da) found in and around cells developing under certain physiological circumstances. Metabolomics is the study of the end products, or metabolites, generated by a living organism under certain circumstances. Unlike previous studies that looked at a limited number of metabolites, metabolomics looks at all the metabolites in a biological system, therefore there is no bias in the metabolites analyzed. In a site-specific organism, however, metabolites are part of an in vivo metabolite flow that controls whole metabolic pathways. Furthermore, metabolism-based broad fluxes enable us to locate physiological regulatory situations in an organism. The major difficulty in metabolomics is figuring out how to get the most out of the hidden information in varied metabolite compositions. Hundreds of primary and secondary metabolites are released by a microbial cell during the course of its existence in response to environmental or cellular changes (Futschik *et al.*, 2018).

Metabolomics is a step beyond metabolite fingerprinting in that it tries to measure every single molecule in its functional role rather than merely building an inventory of the metabolites in a cell. Any metabolomics investigation must start with the measurement of all metabolites in a biological system. This may be accomplished via metabolite isolation and characterization methods that combine automation and miniaturization, such as sampling, extraction of particular chemical classes, storage temperature, sample preparation, and analysis. As a result, changes in the reporter organisms' metabolism during the encounter with any pollution are easily identified, and the end products, that is, metabolites, clearly showed the amount of contamination and feasible bioremediation procedures (Mashego *et al.*, 2007).

2.13 MOLECULAR DOCKING APPROACHES IN BIOREMEDIATION

Molecular docking can be defined as a technique that can be used to calculate the most appropriate binding position for a molecule to another molecule with a minimal amount of energy conformation while there is the formation of a molecular complex. Docking can be done between protein and a small molecule, protein–protein, protein–DNA, protein–RNA, protein–lipid, protein–sugar, polymer–ligand, and so on where a substrate is the bigger metabolite molecule and a smaller one is the ligand (Shockcor & Holmes, 2002). Molecular docking speeds up the selection process and improves the enzymatic properties of certain pollutants through rapid computational scrutiny. This technique offers the anomalous potential to predict the function of enzyme binding and degradation or adsorption of pollutants and provides clarity while depicting the relationship and communication between pollutants and enzyme molecules (Basharat *et al.*, 2020). This allows one to infer whether the active site of a particular enzyme contains a dye molecule, based on damage to the stearic acid residue and the nature of the active site pocket. The association between proteins and contaminants is verified by evaluating thousands of potential conformational docking poses. Scanning the entire surface of the protein attempts to blindly bind proteins to unknown active sites. The link also provides information about the regulation of brokers in the active pocket of the site and speculates on the productive or unfavorable link. The resulting coloration of the active site residues and the binding properties of the test compound is likely to be used for site-specific mutagenicity testing (Bhatt *et al.*, 2020).

2.14 CONCLUSION AND FUTURE PERSPECTIVE

CP is the process of successfully utilizing resources and energy, as well as removing hazardous raw materials and reducing the toxicity of all emissions and solid waste. It is the most proactive technique for reducing waste, negative environmental effects, and health concerns by using energy and natural resources in a more organized and efficient manner. Today, there is a pressing need to establish more sophisticated research tools and techniques that may play an inter/multidisciplinary role to address the issue related to CP leading to achieving the future concept of sustainable development goals. Bioinformatics and omics data technology have contributed a lot to the bioremediation process. A very large chunk of data has been extracted and is used for the *in-silico* analysis advancement. Omics approaches have been able to provide a medium for the analysis of different components of cells such as functional genes, proteins, metabolic interactions, and regulatory pathways. Bioinformatics has facilitated and fastened the cellular analysis process to understand the microbial working and their impact on treating the environment and bioremediation process. Thus, these multiple omics tools and databases are providing valuable information for the purpose to recognize the mechanism involved in the bioremediation process to use microbial technology to work against pollutants. In conclusion we can state that the future holds more understanding about the molecular mechanism and using it to manipulate cellular working by incorporating bioinformatics to demonstrate and the ultimate goal is to translate these computational data and processes to the nature against the different pollutants and the heavy metals.

REFERENCES

Basharat Z., Bibi M. and Yasmin A. (2020). Implications of molecular docking assay for bioremediation. In: Data Analytics in Medicine: Concepts, Methodologies, Tools, and Applications, Mehdi Khosrow-Pour, Steve Clarke, Murray E. Jennex (eds.), IGI Global, Hershey, Pennsylvania, pp. 1556–1577.

Bhatt P., Rene E. R., Kumar A. J., Zhang W. and Chen S. (2020). Binding interaction of allethrin with esterase: bioremediation potential and mechanism. *Bioresource Technology*, **315**, 123845, https://doi.org/10.1016/j.biortech.2020.123845

Bihari Z. (2013). Current trends in bioremediation and biodegradation: next-generation sequencing. *Journal of Bioremediation and Biodegradation*, **4**, e138, https://doi.org/10.4172/2155-6199.1000e138

Chakraborty R., Wu C. H. and Hazen T. C. (2012). Systems biology approach to bioremediation. *Current Opinion in Biotechnology*, **23**(3), 483–490, https://doi.org/10.1016/j.copbio.2012.01.015

Chen W., Mulchandani A. and Deshusses M. A. (2005). Environmental biotechnology: challenges and opportunities for chemical engineers. *AIChE Journal*, **51**(3), 690–695, https://doi.org/10.1002/aic.10487

Cowan D., Meyer Q., Stafford W., Muyanga S., Cameron R. and Wittwer P. (2005). Metagenomic gene discovery, past, present and future. *Trends in Biotechnology*, **23**(6), 321–329, https://doi.org/10.1016/j.tibtech.2005.04.001

Daniel R. (2005). The metagenomics of soil. *Nature Reviews Microbiology*, **3**(6), 470–478. 1160 PMID: 15931165, https://doi.org/10.1038/nrmicro1160

Eyers L., George I., Schuler L., Stenuit B., Agathos S. N. and El Fantroussi S. (2004). Environmental genomics: exploring the unmined richness of microbes to degrade xenobiotics. *Applied Microbiology and Biotechnology*, **66**(2), 123–130, https://doi.org/10.1007/s00253-004-1703-6

Fulekar M. H. and Sharma J. (2008) Bioinformatics for bioremediation. *Innovative Romanian Food Biotechnology*, **2**, 28–36.

Futschik M. E., Morkel M., Schäfer R. and Sers C. (2018). The human transcriptome. *Molecular Pathology*, **135**, 164.

Garg S. (2020). Bioremediation of agricultural, municipal, and industrial wastes. In: Waste Management: Concepts, Methodologies, Tools, and Applications, Mehdi Khosrow-Pour, Steve Clarke, Murray E. Jennex (eds.), IGI Global, Hershey, Pennsylvania, pp. 948–970.

Gentry T. J., Wickham G. S., Schadt C. W., He Z. and Zhou J. (2006). Microarray applications in microbial ecology research. *Microbial Ecology*, **52**(2), 159–175, https://doi.org/10.1007/s00248-006-9072-6

Giannetti B. F., Agostinho F., Eras J. J. C., Yang Z. and Almeida C. M. V. B. (2020). Cleaner production for achieving the sustainable development goals. *Journal of Cleaner Production*, **271**, 122127, https://doi.org/10.1016/j.jclepro.2020.122127

Khan F., Sajid M. and Cameotra S. S. (2013). In silico approach for the bioremediation of toxic pollutants. *Journal of Petroleum & Environmental Biotechnology*, **4**(161), 2.

Lasken R. S. (2012). Genomic sequencing of uncultured microorganisms from single cells. *Nature Reviews Microbiology*, **10**(9), 631–640. PMID: 22890147, https://doi.org/10.1038/nrmicro2857

Lay J. O. Jr. (2001). MALDI-TOF mass spectrometry of bacteria. *Mass Spectrometry Reviews*, **20**(4), 172–194. PMID: 11835305, https://doi.org/10.1002/mas.10003

Lovley D. R. (2003). Cleaning up with genomics: applying molecular biology to bioremediation. *Nature Reviews Microbiology*, **1**(1), 35–44, https://doi.org/10.1038/nrmicro731

Malik G., Arora R., Chaturvedi R. and Paul M. S. (2021). Implementation of genetic engineering and novel omics approaches to enhance bioremediation: a focused review. *Bulletin of Environmental Contamination and Toxicology*, **180**, 1–8.

Mashego M. R., Rumbold K., De Mey M., Vandamme E., Soetaert W. and Heijnen J. J. (2007). Microbial metabolomics, past, present and future methodologies. *Biotechnology Letters*, **29**(1), 1–16. PMID: 17091378, https://doi.org/10.1007/s10529-006-9218-0

Mishra M., Singh S. K. and Kumar A. (2021). Role of omics approaches in microbial bioremediation. In: Microbe Mediated Remediation of Environmental Contaminants A. Kumar, V.K. Singh, P. Singh, V.K. Mishra (eds.), Woodhead Publishing, Duxford, United Kingdom, pp. 435–445.

Mohanta T. K., Mohanta Y. K. and Mohanta N. (2015). Role of biotechnology in bioremediation. In: Handbook of Research on Uncovering new Methods for Ecosystem Management Through, Bioremediation S. Singh and K. Srivastava (eds.), IGI Global, Hershey, PA, pp. 399–432.

Parthasarathi R. and Dhawan A. (2018). In silico approaches for predictive toxicology. In: In Vitro Toxicology A.K. Dhawan and S. Kwon (eds.), Academic Press, pp. 91–109.

Pinto É. S. M., Feltes B. C., Pedebos C. and Dorn M. (2021). Modifying the catalytic preference of alpha-amylase toward n-alkanes for bioremediation purposes using in silico strategies. *Journal of Computational Chemistry*, **42**(22), 1540–1551, https://doi.org/10.1002/jcc.26562

Rawat M. and Rangarajan S. (2019). Omics approaches for elucidating molecular mechanisms of microbial bioremediation. In: Smart Bioremediation Technologies, P. Bhatt (ed.), Academic Press, pp. 191–203.

Roy K. and Kar S. (2016). In silico models for ecotoxicity of pharmaceuticals. In: In Silico Methods for Predicting Drug Toxicity. Humana Press, New York, NY, pp. 237–304.

Shockcor J. P. and Holmes E. (2002). Metabolomics applications in toxicity screening and disease diagnosis. *Current Topics in Medicinal Chemistry*, **2**(1), 35–51, https://doi.org/10.2174/1568026023394498, PMID: 11899064

Singh O. V. (2006). Proteomics and metabolomics: the molecular make-up of toxic aromatic pollutant bioremediation. *Proteomics*, **6**(20), 5481–5492, https://doi.org/10.1002/pmic.200600200

Singh O. V. and Nagaraj N. S. (2006). Transcriptomics, proteomics and interactomics: unique approaches to track the insights of bioremediation. *Briefings in Functional Genomics*, **4**(4), 355–362, https://doi.org/10.1093/bfgp/eli006

Singh A. K., Bilal M., Iqbal H. M., Meyer A. S. and Raj A. (2021). Bioremediation of lignin derivatives and phenolics in wastewater with lignin modifying enzymes: status, opportunities and challenges. *Science of the Total Environment*, **777**, 145988, https://doi.org/10.1016/j.scitotenv.2021.145988

Thomas T., Gilbert J. and Meyer F. (2012). Metagenomics–a guide from sampling to data analysis. *Microbial Informatics and Experimentation*, **2**(1), 1–12, https://doi.org/10.1186/2042-5783-2-3

Umadevi S., Emmanuel A. S., Ayyasamy P. M. and Rajakumar S. (2015). Computational approaches in waste management: special emphasis in microbial degradation. *Research & Reviews: Journal of Ecology and Environmental Sciences*, 22–27.

UN. 2015. Transforming our world: The 2030 Agenda for Sustainable Development. Resolution A/RES/70/1. New York. https://www.un.org/en/development/desa/population/migration/generalassembly/docs/globalcompact/A_RES_7 0_1_E.pdf (accessed 4th Jan 2020)

Vishnoi N. and Dixit S. (2019). Bioremediation: new prospects for environmental cleaning by fungal enzymes. In: Recent Advancement in White Biotechnology Through Fungi, A.N. Yadav, S. Mishra and A. Gupta (eds.), Springer, Cham, pp. 17–52.

Wilmes P. and Bond P. L. (2006). Towards exposure of elusive metabolic mixed-culture processes, the application of metaproteomic analyses to activated sludge. *Water Science and Technology*, **54**(1), 217–226, https://doi.org/10.2166/wst.2006.390

Zhang W., Li F. and Nie L. (2010). Integrating multiple 'omics' analysis for microbial biology: application and methodologies. *Microbiology*, **156**(2), 287–301, https://doi.org/10.1099/mic.0.034793-0

Zhou J. (2003). Microarrays for bacterial detection and microbial community analysis. *Current Opinion in Microbiology*, **6**(3), 288–294. PMID: 12831906, https://doi.org/10.1016/S1369-5274(03)00052-3

doi: 10.2166/9781789063783_0033

Chapter 3

Bioremediation: role of zooplankton in urban waters

*Tanusree Sengupta[1] and Arup Kumar Mitra[2]**

[1]Department of Zoology, Vivekananda College for Women, Barisha, Kolkata, West Bengal, India
[2]Department of Microbiology, St. Xavier's College (Autonomous), Kolkata, West Bengal, India
**Corresponding author: drakmitra01@sxccal.edu*

ABSTRACT

The various everyday developments of our urban society have led to the improvement of industries and agricultural growth causing the mismanagement of wastewaters coming out from these industries. This wastewater exerts many faceted pollution issues leading to the biomagnification and thus affecting human population. Bioremediation is one of the most effective methods of removing hazardous impurities from wastewater. Due to their role as bio indicators in showing the status of any environment, zooplankton have been used to remove the effects of hazardous pollutants from wastewater and thus aid in bioremediation for a while. Zooplankton and other microbes act to remove or reduce pollutants through biological degradation of pollutants into nontoxic substances. Several parameters decide the correct method of bioremediation among *in situ* or *ex situ* methods. Processes like bio stimulation, bioaugmentation, bioventing, bio piles and bio attenuation are required to develop a complete bioremediation process.

Keywords: bioremediation, wastewater, zooplankton, biomagnification, pollutants

3.1 INTRODUCTION

The growing urban population has led to the large-scale production of urban waste–be it untreated or partially treated. Since the unplanned urban society requires more and more amount of water, hence the requisition of water has increased and will increase in future also. Pollutants from several sources such as nutrients, pathogens, plastics, chemicals and so on are found in the urban wastes. The sewage produced every day is not treated and only one-third is treated resulting in waste accumulation and leading to the contamination of the local water bodies where there is constant inflow of untreated or partially treated sewage wastes. Due to the huge amounts of untreated sewage wastes, the search for clean water becomes really challenging and the entire habitat service provided by the lentic and lotic ecosystems are hindered (Prakash *et al.*, 2022).

The urban wastewater is usually disposed in several ways but in all cases the cost of treatment is preferred to be reduced (Ramachandra & Solanki, 2007). Due to the population explosion, most of the waterbodies are getting depleted to meet the various needs (agricultural–irrigational, industrial, and domestic) of the population leading to the deterioration of the water quality. Domestic sewage, agricultural run-offs, and industrial effluents are the sources from where the urban surface water bodies get polluted and cause several enteric diseases in humans.

Among the major pollutants polluting the soil, environment and water are heavy metals, hydrocarbons, polychlorinated biphenyls, industrial effluents, pharmaceutical chemicals, xenobiotics, polyaromatic hydrocarbons (PAHs), and radionuclides are the industrial contaminants that are threats to human health as well as various ecosystems. To minimize and eradicate these pollutants and their harmful effects, several techniques were used but they all seem to exert long-term dangers to the environment. Therefore, bioremediation was the only method that can be long-term and eco-friendly to eradicate contaminants with the help of microorganisms like zooplankton. In addressing water contamination issues in urban areas, bioremediation emerges as a sustainable and environmentally friendly approach that utilizes microorganisms such as zooplankton to effectively eliminate contaminants over a long period of time (Singh *et al.*, 2022). This method stands out as the sole viable alternative that combines long-term efficacy and ecological considerations (Singh *et al.*, 2022).

The crucial role of zooplankton as an important community in maintaining the ecological balance for other biotic components of the urban waters are discussed in the following portions. The role of zooplankton as bioindicators, how they assist in bioremediation process in wastewaters along with the parameters assisting in the process and how the zooplankton cumulatively help in wastewater treatments with other urban water organisms are discussed.

3.2 URBAN WATERS AND ZOOPLANKTON AS A PART OF ITS DYNAMIC POPULATION

Various works take place in an urban area like transportation of goods by various vehicles and various other anthropogenic events that lead to change in soil, air, and water composition polluting the corresponding areas. There were different methods of land use in the urban areas, which resulted in the production of land wastes both solid and liquid produced as a result of various anthropogenic reasons resulting from industries, vehicles, construction along with domestic, all created pollution of the urban waters, which is more deteriorated in quality and not fit for usage. Untreated industrial wastes get released in the urban waters and increases the percentage in wastes like heavy metals, inorganic compounds, and untreated industrial effluents resulting in highly polluted urban waters.

Zooplankton being an integral part of the urban freshwater ecosystem and its population is affected by the changes in water quality. They have an integral role in the material cycle of any waterbody where they reside by being a functional entity of the urban water ecosystem. They act as an important link in the aquatic ecosystem and the members belong to Protozoa, Cladocera, Copepoda, Rotifera, and so on. Studies reveal that the number of zooplankton reduces in waterbodies when sewage water gets mixed (Menezes-Sousa *et al.*, 2018). The plankton's role in pollutants dynamics as a tool for ecotoxicological studies (see *Orbital: The Electronic Journal of Chemistry*, pp. 346–354).

Several types of urban waters are present which show complex internal systems. Unlike urban lakes, urban and semi-urban ponds are very important in acting as natural tertiary waste disposal areas where natural sewage treatment is done. The zooplankton are found to be present in both polluted industrial and municipal waste waters where they maintain their own freshwater ecosystem (Vidali, 2001).

Zooplankton have a growth dynamic like that of algal bloom of high-rate algal ponds (HRAPs) and can furnish all the information regarding the habitat they live in. The zooplankton have individuals as small as (<20 μm) unicellular protozoa and as large as multicellular rotifers and micro-crustaceans (>200 μm) along with copepods and cladocerans (5 mm) (Alprol *et al.*, 2021).

3.3 ROLE OF ZOOPLANKTON IN PROVIDING SIGNIFICANT AND VALUABLE ROLE IN URBAN WATERS

Zooplankton remain a very important part of the aquatic ecosystems of any type of water. On the one hand they are an important biotic component in the aquatic food chain where they feed on the producers and in turn, they themselves are eaten by fishes. In comparison to the phytoplankton, zooplankton are one of the required biological components of environmental impact assessment (EIA) are more diverse and their variability in any ecosystem is dependent on seasons, patchiness, and diurnal-vertical migration (Figure 3.1).

In case of urban waters their function gets more pronounced as the urban waters need more attention since they get polluted due to the various anthropogenic hazards that not only pollutes the water, but also shift the homeostatic balance to a quality that becomes a threat to the aquatic flora and fauna residing in the same waterbody. Given that the growth and distribution of zooplankton are heavily influenced by various abiotic factors such as temperature, salinity, pH, pollution, stratification, as well as biotic factors such as resource competition, any alteration or disruption in these factors caused by environmental hazards can be detected through observable fluctuations in the zooplankton cycle and population sizes. Therefore, a slight change or malfunctioning of any of this due to other environmental hazards may be known by the changes in the zooplankton cycle and their population numbers.

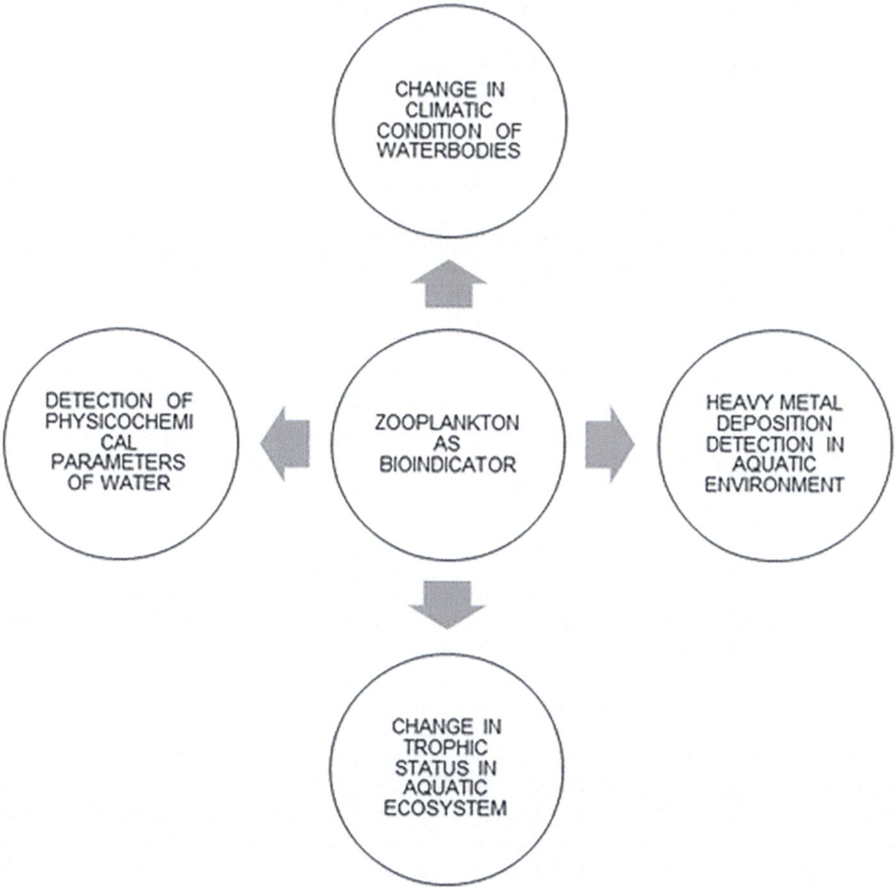

Figure 3.1 Role of zooplankton as ecosystem service provider in any aquatic biome.

The three basic stages of any traditional wastewater treatment are:

(1) Primary treatment in which the solids are removed.
(2) Secondary treatment in which aerobic degradation of soluble organic compounds are used to remove the effluents.
(3) Tertiary treatment in which the dissolved nutrients like nitrogen and phosphorus are removed to stop eutrophication of the incoming waters (Pous *et al.*, 2021).

Zooplankton usually act on the treatment stages to bioremediate.

3.4 ZOOPLANKTON AS BIOINDICATOR SPECIES

Any kind of major or minor changes in the environment and conditions which make the biogeography and health of the environment can be assessed and identified by bioindicators. Zooplankton are the best examples of bioindicators who can screen water quality and indicate the severity of water pollution (Hays *et al.*, 2005). Biological monitoring or biomonitoring is a method where the changes in any environment are measured by any kind of biological response. So, while enquiring about the water quality in any area aquatic biomonitoring can be of great help. Ecosystem components, especially the biotic ones are used to enquire about the abrupt environmental changes in the ecosystem from time to time. The biotic ecosystem components involved in biomonitoring process is known as bioindicators (Campos *et al.*, 2022) (Figure 3.2).

Any kind of major or minor changes in the environment and conditions which make the biogeography and health of the environment can be assessed and identified by bioindicators. Zooplankton are the best examples of bioindicators who provide information about their habitat and can screen the water quality to indicate the severity of water pollution (Ferdous & Muktadir, 2009). The zooplankton usually react to the slightest environmental changes which makes them good indicators of any pollution in the natural environment.

Zooplankton (rotifers, cladocerans, copepods (Bari *et al.*, 2021) being a part of the planktonic population of the aquatic ecosystem, are considered to have a very crucial role in ecotoxicology and serve as an essential tool in understanding the different minute pollution cycle side effects on any habitat (Mani & Kumar, 2014) as they have the following features:

- They are the trophic connection between producers and higher-level consumers in the food chain, so they also transfer pollutants along with energy in any polluted site.
- They can identify the slightest changes in the surroundings they live since in most cases they usually are the first to get affected.
- The recycling of organic matter and toxic hazardous wastes are simultaneously done by the zooplankton only.

The zooplankton found in freshwaters are mostly microscopic protozoans, rotifers, cladocerans and copepods. The zooplankton assemblage is used for water quality assessment. They usually are the second or third trophic level of aquatic trophic web where they themselves feed on algae and bacteria and are consumed by the fishes and many other invertebrates. Hence, any deterioration in the zooplankton level will directly affect the fish population and in turn other higher trophic levels.

There are certain advantages of using zooplankton as indicators for water quality assessment:

- The optimum size of zooplankton makes them easy to handle and identify.
- Processing and sampling are also done very easily.
- Their life cycle can be easily studied in a small time since they have a very short reproductive span.
- Since they are quite abundantly found, working with them as specimen becomes easy.
- They can identify changes in water quality and nature faster than any other bigger biota of the aquatic ecosystem (Campos *et al.*, 2022).

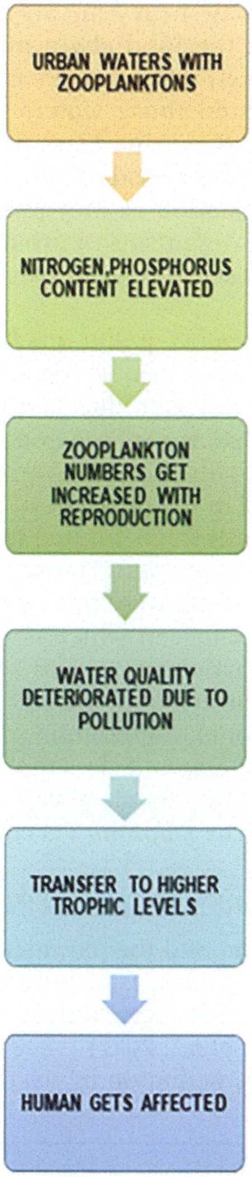

Figure 3.2 Zooplankton: indicators of water pollution: The urban waters have zooplankton as a biotic community, any adverse changes in them will be shown in the elevated nitrogen, phosphorus content balance which the zooplankton will increase their numbers through reproduction. The change will in course pollute the water and zooplankton being the intermediate link between aquatic trophic levels will affect the fish population which then in turn will harm human beings.

3.5 ZOOPLANKTON-ASSISTED BIOREMEDIATION IN WASTEWATERS

Bioremediation is an eco-friendly, cost-effective method which usually uses a group of harmless microbial communities to degrade pollutants in conditions where oxygen is used as an electron acceptor and water containing nutrients are infiltered.

Two types of bioremediations are witnessed–*in situ* and *ex situ*. The application of zooplankton to bioremediate urban wastewater shows a great potential to complement traditional wastewater treatment processes.

Due to anthropogenic activities, the toxic heavy metals accumulate in the soil and environment and finally into water where they enter the food chain and interfere with the normal biological mechanisms. Bioremediation is the best method to retrieve urban waters and to get rid of the harmful hazardous heavy metals that have polluted those waters. Toxic heavy metals such as lead (Pb), cadmium (Cd), mercury (Hg), chromium (Cr), zinc (Zn), uranium (Ur), arsenic (As), and selenium (Se) can disrupt metabolic processes and alter food chains. These metals enter ecosystems in small amounts initially and undergo biomagnification as they progress through successive trophic levels. This poses a threat to the health of living organisms in urban water bodies and can eventually affect humans through the food chain. Mostly reactive oxygen species (ROS) are formed due to heavy metal toxicity in organisms.

The ability of zooplankton to reduce/remove pollutants from waters depends on the suitability of aquatic conditions for their growth and metabolism which includes suitable temperature, pH, salinity, and so on. Zooplankton can degrade pollutants due to their metabolism. The bioremediation of industrial waste waters and municipal waste waters are done by a mixed community of microbes like zooplankton (protozoa) to form a high-quality effluent (Amin *et al.*, 2013).

The zooplankton activity gets inhibited in the presence of high concentration of organic matter. As a result of which the activity of the zooplankton is limited to tertiary treatment of wastes. Therefore, besides working with other natural treatments performing both primary and secondary treatments, wastewater clarification and nutrient polishing were done by zooplankton.

Daphnia is a Cladoceran plantonic crustacean which is very frequently used for ecotoxicological works. *Daphnia* usually remove small particles to aid in secondary systems, they also are known to reduce bacterial loss. Biological oxygen demand (BOD) is reduced by consuming the particulate organic wastes. Contaminants like ammonia, ammonium, nitrite, organic wastes, metals and so on also trigger daphnids. Studies show how *Daphnia* is suitable for tertiary wastewater treatment by reducing the suspended solid content (acting as clarifier), reducing the bacterial load (acting as a disinfectant). The BOD content is reduced by *Daphnia* via consumption of particulate organic matter.

3.6 PARAMETERS CONTROLLING BIOREMEDIATION IN WASTEWATERS BY ZOOPLANKTON

The periodic movement of the zooplankton and the parameters driving their erratic habits are prone to change under the varying ecological variables, especially in places where the water is connected to a very busy urban industrial shore and there is severe anthropogenic influence on the nutrient content affecting the aquatic ecosystem (Adhikari *et al.*, 2017).

The most essential parameters for bioremediation in waste waters are (Table 3.1):

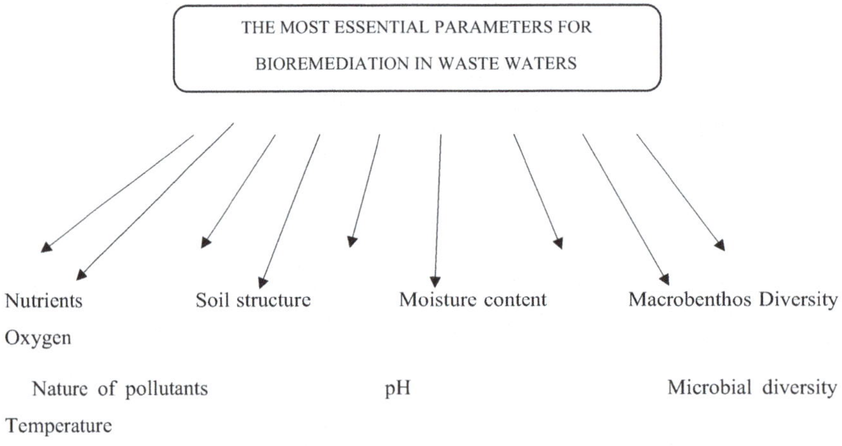

Table 3.1 Most essential parameters for bioremediation in wastewaters.

Nutrients	Contaminated areas do not possess enough nutrients for cellular metabolism and growth of microorganisms due to the presence of high amount of organic carbon which may be depleted during microbial metabolism. To enhance the bioremediation process, it is important to replenish any deficiencies in nutrients and supplement them accordingly. Nutrients such as nitrogen, phosphate, and potassium play a vital role in activating the cellular metabolism and promoting the growth of microorganisms involved in bioremediation.
Pollutants	Pollutants can be classified based on various characteristics. Here are some common types of pollutants: 1. **Physical State:** • Solid pollutants: Refer to particulate matter such as dust, soot, and debris. • Semi-solid pollutants: Include sludge, sediment, and other viscous materials. • Liquid pollutants: Refer to contaminated liquids like wastewater, oils, and chemical solutions. • Volatile pollutants: Refer to substances that easily evaporate into the air, such as volatile organic compounds (VOCs). 2. **Toxicity:** • Toxic pollutants: These substances have harmful effects on living organisms and ecosystems. • Non-toxic pollutants: Refer to substances that do not have significant toxic properties. 3. **Chemical Composition:** • Organic pollutants: Include substances derived from living organisms or containing carbon atoms. Examples include polycyclic aromatic hydrocarbons (PAHs) and pesticides. • Inorganic pollutants: Refer to substances that do not contain carbon atoms, such as heavy metals (e.g., lead, mercury) and chlorinated solvents. It is important to note that pollutants can exhibit multiple characteristics, and their impact on the environment can vary depending on factors such as concentration, exposure duration, and specific ecosystems involved.
Structure of soil	Different textures of sand, silt, and clay are present ranging from low-to-high contents. A well textured soil can facilitate aeration, nutrition and water to microorganism for bioremediation.
pH	The ideal pH range for microbial growth and the degradation of contaminants typically fall between 5.5 and 8.0. Within this range, microbial activity and enzymatic processes are most efficient, allowing for effective biodegradation of contaminants. However, it is important to note that the optimal pH range may vary depending on the specific microorganisms involved and the type of contaminants present. Some microbial species may thrive in more acidic or alkaline conditions, and certain contaminants may require specific pH ranges for effective degradation. Therefore, it is essential to consider the specific characteristics of the microbial community and contaminants when designing bioremediation strategies. Maintaining the pH within the optimum range can be achieved through various methods, including pH adjustment using suitable amendments, monitoring pH levels during bioremediation processes, and providing conditions favorable for microbial growth and activity.
Soil moisture content	The optimum soil moisture range is 25–28%. The soil water determines the dielectric constant of soil and other medium.

(Continued)

Table 3.1 Most essential parameters for bioremediation in wastewaters. (*Continued*)

Microbial diversity	The microbial species mentioned including *Pseudomonas, Aeromonas, Flavobacteria, Chlorobacteria, Corynebacteria, Acinetobacter, Mycobacteria, Streptomyces, Bacilli, Arthrobacter*, and *Cyanobacteria*, are known to play important roles in various environmental processes, including bioremediation. These microorganisms possess diverse metabolic capabilities and can contribute to the degradation and transformation of different contaminants in different environments. It is important to note that the effectiveness of these microbial species in contaminant degradation can vary depending on the specific contaminants, environmental conditions, and the presence of other microorganisms. Therefore, a comprehensive understanding of the specific contaminants and the microbial community present in each environment is crucial for developing effective bioremediation strategies.
Macrofauna diversity	An association of aquatic plants *E. crassipes, S. molesta, C. demersum* along with aquatic animals *A. woodiana* and *L. hoffmeisteri* degrades the ammonia content, nitrite, nitrate in domestic wastewaters. The association of aquatic plants such as *Eichhornia crassipes* (commonly known as water hyacinth), *Salvinia molesta* (commonly known as giant salvinia), *Ceratophyllum demersum* (commonly known as hornwort), along with aquatic animals like *Ampullariidae woodiana* (commonly known as golden apple snail) and *Limnaea hoffmeisteri* (commonly known as ramshorn snail) can contribute to the degradation of ammonia, nitrite, and nitrate in domestic wastewaters. Aquatic plants play a vital role in wastewater treatment by absorbing and utilizing nutrients like ammonia and nitrates as a source of growth. They help reduce nutrient levels in the water by assimilating these compounds into their tissues. Additionally, aquatic plants provide a suitable habitat for beneficial microorganisms, such as nitrifying bacteria, which aid in the conversion of ammonia to nitrite and nitrate. The presence of aquatic animals like *Ampullariidae woodiana* and *Limnaea hoffmeisteri* further contributes to the wastewater treatment process. These animals feed on organic matter, including decaying plant material and algae, promoting the breakdown of organic compounds, and reducing nutrient levels in the water. Overall, the combination of aquatic plants and animals in wastewater treatment systems can enhance the removal of ammonia, nitrite, and nitrate, leading to improved water quality. However, the specific effectiveness of this association may vary depending on factors such as environmental conditions, wastewater composition, and the presence of other organisms in the ecosystem.
Temperature	The biochemical reaction rates are affected by temperature, optimum temperature being 15–45°C.
Oxygen	The presence or absence of oxygen is a crucial factor in determining the type of bioremediation process: aerobic or anaerobic. In aerobic bioremediation, oxygen is available, and microorganisms utilize it as a terminal electron acceptor during the degradation of contaminants. Oxygen plays a vital role in the initial breakdown of hydrocarbons and other organic pollutants, as it serves as an essential component in the metabolic processes of aerobic microorganisms. These microorganisms, often referred to as aerobic bacteria, possess enzymes that require oxygen to efficiently oxidize and break down contaminants into less harmful by-products, such as carbon dioxide and water. On the contrary, in anaerobic bioremediation, the absence or limited availability of oxygen creates an environment suitable for anaerobic microorganisms. These microorganisms can still degrade certain contaminants in the absence of oxygen by utilizing alternative electron acceptors such as nitrate, sulfate, or carbon dioxide. Anaerobic bioremediation processes are particularly effective for certain types of contaminants, such as chlorinated solvents and heavy metals, which can undergo reductive transformations. It is important to consider the oxygen availability and the specific requirements of the contaminant and microbial community when designing and implementing bioremediation strategies. The initial breakdown of hydrocarbons and subsequent degradation processes can be significantly influenced by the presence or absence of oxygen in the contaminated site.

3.7 CUMULATIVE ROLE OF ZOOPLANKTON WITH OTHER ORGANISMS OF URBAN WATERS

The main aim of bioremediation is to discover a novel microorganism that can remediate pollutants from any polluted site. Different methods such as bioaccumulation, biodegradation pathways and different methods of biosorption are used to remove pollutants. The zooplankton resides in several other microorganisms some of which help the former to bioremediate. Daphnids, among other zooplankton, is very important since they improve the efficiency of wastewater treatment in ponds.

In several cases, zooplankton work with other microorganisms to be more potent at wastewater treatment. One of the examples is the natural-based alternative for treatment of waste waters by the cumulative effort of *Daphnia* (zooplankton) and bacterial/algal biofilm in a zooplankton-containing reactor. In this case, filtration capacity of the former acts with the nutrient removal capacity of the latter.

3.8 CONCLUSION

Therefore, it is quite evident that zooplankton are very crucial biotic members with respect to the proper health functioning of any aquatic ecosystem. They have a pivotal role in trophic chain as they transfer energy from planktonic algae to the upper levels such as larger invertebrates .The zooplankton are very quick in responding to the changes in the water they inhabit and their species composition indicate the water quality. Zooplankton themselves or with some other aquatic members are ecologically good at removing BOD, bacterial load, other organic matter, ammoniacal nitrogen, and so on from urban wastewaters. Being high in aesthetic value and low-cost maintenance requirement, zooplankton are the best bioindicators for isolated aquatic ecosystems. In the nature-based system where the integrated use of rotifer, cladocerans, with other zooplankton for treatment of wastewaters, sludge reduction is found to be very eco-friendly for the restoration of the environmental balance.

REFERENCES

Adhikari S., Goswami A. R. and Mukhopadhyay S. K. (2017). Diversity of zooplankton in municipal wastewater-contaminated urban pond ecosystems of the lower Gangetic plains. *Turkish Journal of Zoology*, **41**(3), 464–475, https://doi.org/10.3906/zoo-1601-12

Alprol A. E., Heneash A. M., Soliman A. M., Ashour M., Alsanie W. F., Gaber A. and Mansour A. T. (2021). Assessment of water quality, eutrophication, and zooplankton community in Lake Burullus, Egypt. *Diversity*, **13**(6), 268, https://doi.org/10.3390/d13060268

Amin A., Naik A. R., Azhar M. and Nayak H. (2013). Bioremediation of different waste waters–a review. *Continental Journal of Fisheries and Aquatic Science*, **7**(2), 7.

Bari J. B. A., Islam M. S., Nisa S. A., Tisha N. A., Mashkova I. and Khan N. S. (2021). Response of freshwater zooplankton as biological indicators to the aquatic chemical properties. *Current Environment*, **1**(1), 9–14, https://doi.org/10.54479/ce.v1i01.6659

Campos C. V. F. D. S., Oliveira C. Y. B., dos Santos E. P., de Abreu J. L., Severi W., da Silva S. M. B. C., Brito L. O. and Gálvez A. O. (2022). Chlorella-Daphnia consortium as a promising tool for bioremediation of Nile tilapia farming wastewater. *Chemistry and Ecology*, **38**(9), 873–895, https://doi.org/10.1080/02757540.2022.2120612

Ferdous Z. and Muktadir A. K. M. (2009). A review: potentiality of zooplankton as bioindicator.

Hays G. C., Richardson A. J. and Robinson C. (2005). Climate change and marine plankton. *Trends in Ecology & Evolution*, **20**(6), 337–344, https://doi.org/10.1016/j.tree.2005.03.004

Mani D. and Kumar C. (2014). Biotechnological advances in bioremediation of heavy metals contaminated ecosystems: an overview with special reference to phytoremediation. *International Journal of Environmental Science and Technology*, **11**, 843–872, https://doi.org/10.1007/s13762-013-0299-8

Menezes-Sousa D., Kasper D., Cavalcanti E. A. H., Torres J. P. M., Bonecker S. L. C. and Malm O. (2018). The plankton role in pollutants dynamics as a tool for ecotoxicological studies. *Orbital: The Electronic Journal of Chemistry*, 346–354.

Pous Rodríguez N., Barcelona A., Sbardella L., Hidalgo M., Colomer J., Serra Putellas T. and Salvadó Martín V. (2021). Zooplankton-based reactors for tertiary wastewater treatment: a pilot-scale case study. *Journal of Environmental Management*, **278**, 111538, https://doi.org/10.1016/j.jenvman.2020.111538

Prakash D., Kumar R., Rajan K., Patel A., Yadav D. K., Dhanker R. and Khudsar F. A. (2022). Integrated application of macrophytes and zooplankton for wastewater treatment. *Frontiers in Environmental Science*, **10**, 941841, https://doi.org/10.3389/fenvs.2022.941841

Ramachandra T. V. and Solanki M. (2007). Ecological assessment of lentic water bodies of Bangalore. *The Ministry of Science and Technology*, **25**, 96.

Singh N., Poonia T., Siwal S. S., Srivastav A. L., Sharma H. K. and Mittal S. K. (2022). Challenges of water contamination in urban areas. In: Current Directions in Water Scarcity Research, A. L. Srivastav, S. Madhav, A. K. Bhardwaj, E. Valsami-Jones (eds.), Elsevier, Vol. 6, pp. 173–202.

Vidali M. 2001. Bioremediation. an overview. *Pure and Applied Chemistry*, **73**(7), 1163–1172, https://doi.org/10.1351/pac200173071163

doi: 10.2166/9781789063783_0043

Chapter 4

Carbon sequestration: principle and recent advances

*Samantha Blessing and John M. Pisciotta**

Biology Department, West Chester University, West Chester, PA, USA
**Corresponding author: jpisciotta@wcupa.edu*

ABSTRACT

Cellular sequestration and conversion of CO_2 into biomass and chemical products can provide carbon neutral to negative alternatives to conventional fossil fuels. This chapter examines microorganisms and processes used for carbon sequestration and production of biofuels and other products. The benefits and challenges related to biological carbon sequestration from point sources, namely combustion-based power plants and wastewater treatment facilities, are addressed. Algal cultivation for synthesis of liquid biofuels offers an attractive, solar-power method for temporarily sequestering carbon plus nutrient pollutants from wastewater. Autotroph-derived biochar can provide for longer term carbon sequestration into soil. Microbial waste treatment can be paired to CO_2 reduction by methanogens for production of the gaseous biofuel methane. Emerging hybrid technologies enable biogas enhancement with further reduction of CO_2 and conversion into biomass. Other microbes assimilate carbon and release clean-burning hydrogen.

Keywords: carbon sequestration, algae, biodiesel, hydrogen

4.1 ATMOSPHERIC CARBON AND ITS SEQUESTRATION

Anthropogenic greenhouse gas (GHG) emissions have risen over the past century and today present an emerging climactic challenge for many regions of the planet (Mikhaylov *et al.*, 2020). Carbon dioxide is the most important carbon-based GHG released through human activity. Industrial combustion of fossil fuels including coal, natural gas, and oil for applications such as electrical power generation and internal combustion engine (ICE)-based transportation are society's leading sources of CO_2 emissions. With Earth's population surpassing 8 billion people in 2022, the demand for fossil fuels and petroleum-based products in energy and other sectors continues to accelerate. Concentrations of CO_2 in the atmosphere currently exceed 400 ppm and large developing nations, such as China, have emerged as the leading emitter of GHGs (Mikhaylov *et al.*, 2020).

GHG influenced climate change has adversely influenced many of Earth's natural processes. Sea ice coverage in the Arctic has diminished with rising temperatures as seawater levels have correspondingly risen, threatening many island and coastal communities (Mengel *et al.*, 2018). Atmospheric increases in CO_2 further elevate ocean acidity as CO_2 diffuses into sea water as carbonic acid (H_2CO_3), deleteriously impacting critical marine ecosystems like coral and oyster reefs (Doney *et al.*, 2020). As surface temperatures have increased so too have the frequency and intensity of storm events. It is broadly accepted that to avert the worst effect of climate change by mid-21st century, CO_2 emissions must be aggressively reduced. Carbon capture and sequestration from large point source emitters represents one attractive opportunity for reducing CO_2 emissions.

4.2 CONVENTIONAL CO_2 CAPTURE APPROACHES

Reduction of anthropogenic CO_2 emissions is a goal that can be achieved by various means, both natural and technological. Renewable energy systems, such as photovoltaics, wind turbines and hydroelectric methods, is a rapid growth area. Cost reductions in photovoltaics associated with economies of scale over the past two decades have enabled solar power plants to economically outcompete many polluting fossil plants, even in the absence of carbon taxes. Extended lifecycle operation of nuclear power plants ensures a low carbon source of baseload power in countries lacking ample fossil feedstocks. Some European nations have recently extended the operations of nuclear plants, despite concerns with radioactive waste storage, as natural gas supplies have been interrupted. Improved energy efficiency during electricity generation, transmission and use is critical for the reduction of carbon dioxide emissions.

Steel and concrete making are highly controlled, CO_2 intensive, heavy industries in which CO_2 reduction may be achieved using recent chemical advances. The so-called green chemical processes using novel catalysts to enhance mineral carbonization efficiency may offer advantages over more polluting traditional approaches (Hanifa *et al.*, 2023). For some carbon emission intensive industrial processes, like wastewater treatment, CO_2 production may be more difficult to avert chemically. At established wastewater treatment facilities, chemical capture of CO_2 can be challenging since the CO_2 is released over a broader area, rather than from just one flue stack or chimney. Carbon capture and sequestration describes processes used to avert increases in atmospheric CO_2 by recovering carbon from waste streams, typically process off gas. Existing CO_2 capture for sequestration technologies at power plants targets three main approaches: pre-combustion, post-combustion, and oxy-fuel.

Pre-combustion is typically used in industrial power plants where fossil fuel such as coal is converted into syngas composed of hydrogen gas (H_2) and carbon monoxide (CO), in the presence of water vapor, using a high-pressure gasifier (Ramesh *et al.*, 2021). The CO is run through a reaction that yields more CO_2, which is removed before combustion and captured (Ramesh *et al.*, 2021). The H_2 is then used to generate electricity as a fuel in a gas turbine (Molazadeh *et al.*, 2019). The pre-combustion process is typically used in integrated gasification combined cycle (IGCC) power plants. This method successfully sequesters CO_2; however, it comes with a high cost and added complexity which is a major limitation in this CO_2 removal approach. In post-combustion carbon capture, chemical absorption is employed to seize the CO_2 from flue gas after combustion has occurred (Prasad *et al.*, 2021). Fuel such as coal is burned with air to produce steam in a boiler that will power a turbine for electricity generation. In this approach, a chemical solvent, typically an amine, is used to react with the CO_2 in the flue gas at high temperature to create purified CO_2 that can be captured and stored (Erga *et al.*, 1995). Post-combustion can be used in powerplants other than IGCC and can be beneficial over a more diverse range of facilities (Erga *et al.*, 1995). Oxy-fuel combustion is a third widely used approach for carbon capture in which fuel is combined with pure oxygen to produce high yields of CO_2 (Raho *et al.*, 2022). Oxy-fuel combustion results in higher temperatures and is more favorable than other methods based on the higher concentration stream of CO_2 produced (Raho *et al.*, 2022). Chemical looping is a newer and potentially promising carbon capture method for in which carbon-based fuel is divided into separate oxidation and reduction reactions (Raganati *et al.*, 2021). In this approach, oxygen from air is taken by way of a carrier to the fuel (Raganati *et al.*, 2021).

4.3 CHEMICAL AND EMERGING CAPTURE METHODS

Chemical or physical CO_2 removal from industrial process off gases can occur through several methods. Adsorption and absorption are two commonly used approaches to capturing CO_2 from flue gas. In adsorption, CO_2 is collected onto an adsorber surface whereas in absorption, sometimes called chemical scrubbing, CO_2 is absorbed into an absorber medium. For effective adsorption, an adsorbent must adsorb the adsorbate with a regenerative ability based on a temperature or pressure condition (Raganati *et al.*, 2021). An adsorbent such as an amine reacts with the flue gas CO_2, capturing the CO_2 through intermolecular attraction between the molecules on the adsorbent and the CO_2 gas itself (Pires *et al.*, 2011). Physical adsorption, such as pressure and temperature swing adsorption (PTSA), can be employed by heating the adsorbent and regenerating CO_2 through depressurization (Ishibashi *et al.*, 1996). Chemical absorption of CO_2 into a liquid can be achieved through the temporary bonding of a liquid solvent such as monoethanolamine (MEA) to CO_2 through upward flow of gas through the packing of an absorber, providing a greater surface area for bonding (Raksajati *et al.*, 2018). The CO_2 enriched rich solvent leaves the bottom of the column, while the flue gas containing a lower concentration of CO_2 leaves from the top of the column (Raksajati *et al.*, 2013).

Emerging methods of CO_2 separation include membrane separation and cryogenic separation. Membrane gas separation (GS) is a means of removing CO_2 through employment of a membrane with specific permeability upon gas pressure application. Membrane GS technologies have certain drawbacks when treating low CO_2 concentration gas such that high quantities of flue gas need to be processed for efficient method optimization (Brunetti *et al.*, 2010). Potentially low CO_2 concentration coupled with low pressure of the flue gas require membranes with high selectivity to stay within International Agency Specifications, which tends to increase cost (Bounaceur *et al.*, 2006). Another consideration when using membrane GS is that the membranes must be resistant to harsh chemicals present in the flue gas so as not to disrupt the integrity of the membrane itself (Brunetti *et al.*, 2010). Periodic changing of membranes at the end of their life cycle could disrupt normal facility operations and lower the capacity factor of a power plant. In cryogenic separation, concentrated CO_2 is cooled and liquified through several steps to induce a phase change in the CO_2 from gas-to-liquid state (Meisen & Shuai, 1997). The low-temperature refrigeration equipment insulated piping and extra energy needed to operate cryogenic separation incurs additional cost. Disadvantages with many chemical and physical processes for carbon capture is that they are typically expensive and generally not environmentally sustainable (Molazadeh *et al.*, 2019). Captured carbon may be chemically or biochemically converted into different products or it can be stored underground in geologic formations (Moreira & Pires, 2016). A general overview of the possible fates of CO_2 produced through combustion at a large point source, such as a power plant, is shown in Figure 4.1.

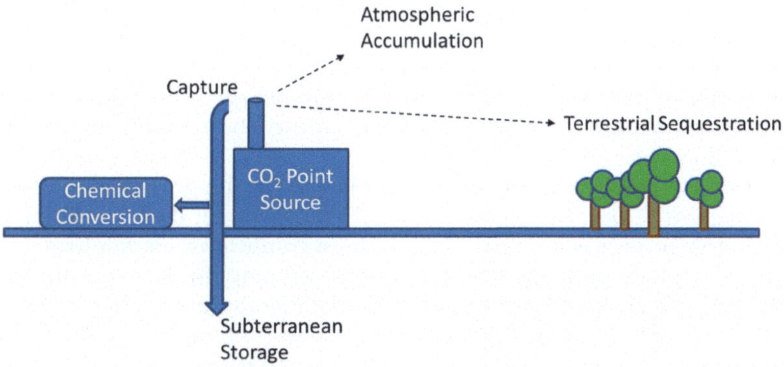

Figure 4.1 Overview of possible fates and disposition of CO_2 from a point source.

Emerging chemical CO_2 capture and sequestration methods show promise, but cost, complexity, and robustness are critical technoeconomic considerations with such industrial processes. One way to enhance the value proposition of a carbon sequestration process is to convert captured CO_2 into a more valuable product, such as biofuels or longer-lived bioplastics (Thevasundaram *et al.*, 2022). For this, some chemical or physical methods of CO_2 capture may be paired with carbon-fixing organisms, such as algae or cyanobacteria, for enhanced CO_2 sequestration and valorization (i.e., value upcycling) and/or water bioremediation. Biological carbon sequestration methods can also be used independently and may be particularly well suited for some industrial GHG emission sites, such as breweries and wastewater treatment plants where a constant stream of CO_2 is available under mesophilic conditions with few inhibitory or toxic chemicals. Wastewater treatment plants release copious CO_2 in addition to methane (CH_4), and nitrous oxide (N_2O), which are both potent GHGs (Asadi & McPhedran, 2021). Wastewater treatment plants contain diverse nutrients and other growth factors that can be used to promote the growth of CO_2 sequestering autotrophic cells. Even 100% unfiltered flue gas from coal-fired power plants can facilitate the cultivation of *Desmodesmus* spp. algae (Aslam *et al.*, 2017).

4.4 BIOLOGICAL CARBON CAPTURE AND SEQUESTRATION

Algae and higher plants naturally sequester inorganic carbon through photosynthesis. Algae is a general term that can be subdivided into three main groups: macroalgae, microalgae, and cyanobacteria. Macroalgae and microalgae are eukaryotic organisms whereas cyanobacteria are prokaryotes, many of which can survive extreme conditions. In marine environments macroalgae are informally referred to as seaweed for their rapid growth and apparent ubiquity. Macroalgae are more complex multicellular organisms ranging from tens of meters tall giant kelp of the Pacific to barely visible representatives of a few millimeters. Macroalgae include brown seaweed (Phaeophyceae), green seaweed (Chlorophyceae), and red seaweed (Rhodophyceae). Microalgae groups include diatoms, green algae (Chlorophyceae), and golden algae (Chrysophyceae) (Demirbas, 2010). Useful cyanobacteria, formerly referred to as blue-green algae, include members of the genera *Arthrospira* (*Spirulina*), *Nostoc*, and *Anabaena* (Li *et al.*, 2023). Some cyanobacteria are diazotrophic and can fix atmospheric nitrogen giving them an advantage in certain environments. Green and brown algae have most often been investigated for biofuel production. Novel animal feed applications for red algae have recently been described for inhibition of methanogens present in the gastrointestinal tracts of cows. Since ruminants are a source of the potent GHG methane, red algae can serve a role in addressing climate change by reducing emissions even though they are not commonly used to produce fuels directly.

Over 100 Pg of carbon is assimilated every year on Earth with around half by marine ecosystems (Beardall *et al.*, 2009). In the marine environment, around 75% is assimilated by marine phytoplankton and about 25% by algae and seagrasses (Beardall *et al.*, 2009). Algae are diverse, photosynthetic autotrophs that are critical to aquatic and marine ecosystems. They account for around 50% of the carbon assimilated into the biosphere each year (Field *et al.*, 1998). Algae use different methods of metabolism to assist in assimilation of CO_2. These including photoautotrophic metabolism as well as mixotrophic metabolism. In photoautotrophic species, light energy is harvested via photopigments to enable photosynthesis, transforming inorganic carbon into carbohydrates as CO_2 is assimilated via the Calvin–Benson (CB cycle).

In photoheterotrophic metabolism, organic carbon is used in the presence or absence of light. This method of metabolic activity can occur in certain algal species under dark conditions or in conditions of low-light where photosynthesis is not favorable. In these conditions, metabolism will typically occur through the pentose phosphate pathway (PPP) and organic compounds are assimilated from sources such as acetate, glucose, glycerol, and lactate, via different metabolic enzymes (Giordano *et al.*, 2005; Prasad *et al.*, 2021). Although this typically occurs in low-to-no light conditions, certain species can perform photoheterotrophy under normal light conditions (Ingram *et al.*, 1973; Prasad *et al.*, 2021). Interestingly, growth under low light conditions facilitates production of saturated fatty acids, which may be preferable as energy dense biofuel feedstocks (Constantopoulos, 1970). Certain photosynthetic bacteria are also capable of photoheterotrophic metabolism.

Mixotrophs are organisms, including certain algae that can metabolize using both photoautotrophic metabolism and heterotrophic metabolism. During mixotrophic growth, CO_2 and organic carbon are used together, in combination with both respiration and photosynthesis (Prasad *et al.*, 2021). Because this metabolic strategy permits the usage of both inorganic and organic carbon sources, high production of biomass is promoted (Venkata Mohan & Devi, 2014; Prasad *et al.*, 2021; Wang *et al.*, 2014). Aerobic respiration is employed to produce energy needed for organic carbon anabolism while photosynthesis is used for inorganic carbon assimilation into *de novo* organic material (Hu *et al.*, 2012; Prasad *et al.*, 2021). With multiple metabolic strategies employed within the organism, mixotrophs can produce higher yields per unit of energy compared to the other metabolic strategies independently (Prasad *et al.*, 2021; Yang *et al.*, 2000). Although mixotrophs show promise for carbon sequestration, they may entail a heightened cost of cultivation due to their higher carbon demands. However, mixotrophic cultivation of *Spirulina* using low-cost organic wastes, such as acetate, can increase growth rate and CO_2 assimilation by stimulating ribulose-1,5-bisphosphate carboxylase-oxygenase (RuBisCo) to enhance the overall utility of cyanobacteria for carbon sequestration (Li *et al.*, 2023). Geothermal emission off gases containing as high as 90% CO_2 have been used to cultivate cyanobacteria for carbon sequestration (Svavarsson *et al.*, 2017).

Oxygenic photosynthesis is the dominant mechanism algae use to convert light energy to chemical energy with water serving as the electron donor (Blankenship, 2010, 2014). Photopigments including chlorophyll and carotenoids enable algae to harness solar radiation (Blankenship, 2014). Chlorophylls absorb blue and red light, while carotenoids absorb blue-green wavelengths (Bari, 2004; Geider & Osborne, 1992). Certain cyanobacteria use phycobiliproteins such as phycoerythrin and phycocyanin to expand the fraction of the visible spectrum captured (Formighieri & Melis, 2015). Harnessed solar radiation triggers light-dependent reactions in the Z-scheme of oxygenic photosynthesis (Govindjee *et al.*, 2017).

Light energy absorbed by the photopigment antenna complex is transferred to photosystem II (PS-II), inducing water photolysis, which establishes a proton motive force (PMF) within the thylakoid lumen (Figure 4.2). For each water molecule oxidized, two light-energized electrons are carried through the photosynthetic electron transport chain (P-ETC) from PSII to cytochrome b6f via the lipophilic carrier plastoquinol. Additional protons are transported from the stroma into the lumen through Q-cycle proton pumping. Chemiosmotic dissipation of the PMF thus generated through the integral ATP synthase generates ATP through the overall process known as photophosphorylation. Electrons in the P-ETC are next transferred to photosystem I (PS-I) from cytochrome b6f via plastocyanin. At PS-I they are reenergized to a higher energy state by additional light enabling the reductive formation of NADPH (Bari, 2004). The precise amount of ATP to NADPH generated in many photoautotrophs depends on if the cell is utilizing non-cyclic (i.e., linear) versus cyclic photophosphorylation; the latter serving as a photoprotective mechanism during high light stress (Huang *et al.*, 2015).

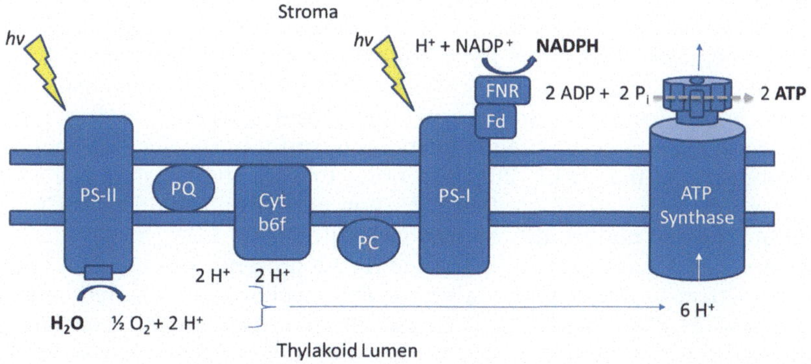

Figure 4.2 P-ETC showing conversion of light into chemical energy used to drive CO_2 fixation.

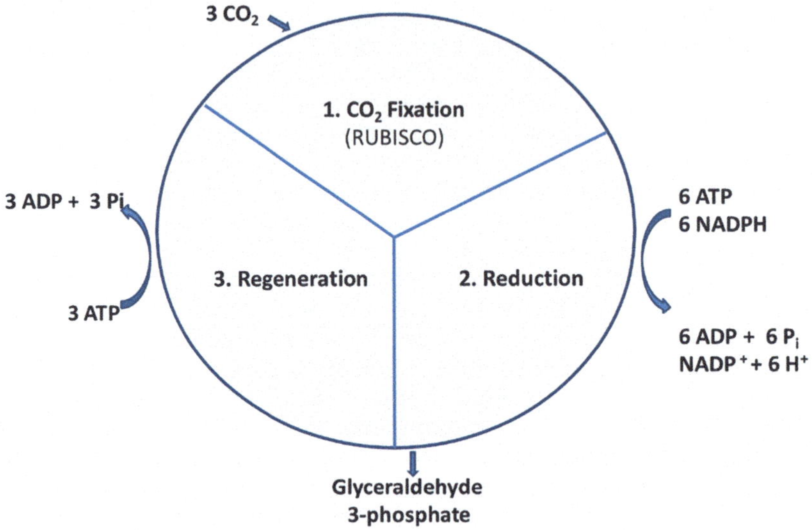

Figure 4.3 CB cycle-mediated biological CO_2 fixation.

CO_2 fixation during the light-independent (i.e., dark) reactions of the Calvin Benson cycle (CB cycle) is an anabolic process powered by the ATP and NADPH generated by the light reactions. The three primary stages of the light-independent reactions of CB cycle CO_2 assimilation involve CO_2 fixation, reduction and substrate regeneration (Nguyen & Hoang, 2016). In the CB cycle, carboxylation of ribulose-1,5-bisphosphate (RuBP) by RuBisCo produces two 3-phosphoglycerate (PGA) molecules (Furbank & Taylor, 1995). Energy from ATP and NADPH is utilized for subsequent conversion of PGA into glyceraldehyde-3-phosphate (G3P) as depicted in Figure 4.3 (Furbank & Taylor, 1995). Most of the G3P molecules are used for RuBP regeneration, but some G3P are used to synthesize other organic molecules, such as glucose that may be stored as starch (Furbank & Taylor, 1995). In contrast to many cyanobacteria, which tend to form starch as their energy storage material, oleaginous microalgae convert this fixed carbon into neutral lipids, namely triacylglycerides (TAGs), that can be converted to biodiesel (Li *et al.*, 2019).

4.4.1 Carbon capture mechanisms

CO_2 is essential for anabolism during photosynthesis, as it is the primary carbon substrate for microalgae to produce organic compounds using sunlight energy (Lam *et al.*, 2012). CO_2 for microalgae photosynthetic cultivation can be gained from the atmosphere at concentrations around 0.03–0.06% (v/v) or in concentrated form from point source flue gas of power plants in concentrations of approximately 6–15% (v/v) (Cheah *et al.*, 2015; Rahaman *et al.*, 2011). When considering microalgae as a means for carbon capture from point sources, pH, temperature, and algal stress must be considered. If the CO_2 concentration becomes too high, it becomes harmful to the algae by increasing the acidity of the local environment (Kumar *et al.*, 2011). In order to limit the negative effects associated with high CO_2 concentrations, microalgae contain a carbon capture mechanism (CCM) that allows for continued proliferation when photosynthesis may be limited.

Carbonic anhydrase (CA) activity can be limited in the presence of high CO_2 concentration. The effect on CA can be detrimental to the microalgae, as this mechanism is catalytically important for its part in alterations between CO_2 and HCO_3^- (Zhang, 2015). With high CO_2 concentration, the pH of the medium decreases due to increases in bicarbonate (HCO_3^-) and hydrogen ion (H^+) concentration.

The concomitant pH drop negatively affects the action of CA, resulting in a decrease in the CO_2 fixing abilities of the microalgae (Zhao & Su, 2014). Through different regulatory processes however, some algal species have adapted to the lower pH. These adaptations include the ability to increase the amount of ATP present while concurrently reducing organic carbon synthesis, increasing the ability of these cells to tolerate exceptionally high concentrations of CO_2 (Zhao & Su, 2014). Mechanisms involved in increasing tolerance to higher CO_2 concentrations include preventing acidification of the stromal compartment in the chloroplast and cytoplasm in order to keep RuBisCo functioning properly (Zhang, 2015).

Efforts to improve the efficiency of algal photosynthesis and thereby algal biofuel production have focused on genetic improvement of photopigments, components of the P-ETC and/or RuBisCo enzyme, which is inefficient due to its capacity for photorespiration in which O_2 rather than CO_2 is incorporated. CO_2 fixation in oxygenic phototrophs is reliant on RuBisCo and CA. Other important enzymes involved in carbon fixation process include phosphoribulokinase, phosphoenolpyruvate carboxylase, phosphoenolpyruvate carboxy-transfer enzyme, pyruvate carboxylase and malic enzyme (Bharti *et al.*, 2014). Similar to oleaginous microalgae, certain bacteria, including various *Rhodococcus* species, also synthesize fatty acids and triglycerides from acetyl-CoA using NADPH as a source for high energy reducing equivalents (Kumar *et al.*, 2018; Schirmer *et al.*, 2010).

Alkane fatty acid chain formation occurs by: (1) hexadecenoic acid elongation, (2) fatty acid reduction to aldehyde via fatty acid reductase, and (3) decarboxylation of aldehyde via aldehyde decarboxylates forming alkanes (Kumar *et al.*, 2018; Messner & Sleytr, 1992). *De novo* fatty acid biosynthesis in algae is similar to the process in plants (Rismani-Yazdi *et al.*, 2011; Tan & Lee, 2016). Fatty acid synthase (FAS) is the primary enzyme involved in the biosynthesis of fatty acids in many different algae, including model organism *Chlamydomonas reinhardtii* (Khozin-Goldberg & Cohen, 2011). The first step is the carboxylation of acetyl-CoA by CoA carboxylase (ACCase), forming malonyl-CoA (Reverdatto *et al.*, 1999). This is the rate-limiting step for fatty acid biosynthesis and genetic optimization for enhancement of ACCase activity can improve oil yield from algae (Post-Beittenmiller *et al.*, 1991, 1992). Next, malonyl-CoA binds acyl carrier protein (ACP) to form a metabolic scaffold malonyl-ACP via malonyl-CoA transacylase (Blatti *et al.*, 2013; Goncalves *et al.*, 2016). The chain is lengthened two-carbons at a time through integration of the acetyl group from acyl-ACP, then fully reduced by ketoreductase, dehydratase, and enoyl reductase (Blatti *et al.*, 2013). Once the chain is fully developed, the mature fatty acid, is transferred by acyl transferase onto glycerol-3-phosphate (Blatti *et al.*, 2013). A buildup of fatty acid acyl-ACP regulates the rate of fatty acid synthesis through ACCase feedback inhibition (Davis & Cronan, 2001; Tan & Lee, 2016). Thioesterases can reduce the inhibition caused by acyl-ACP build-up by converting acyl-ACP into acyl-CoA and releasing it into the chloroplast to become a component of TAGs (Chen & Smith, 2012). Acylation of glycerol-3-phosphate leads to production of diacylglycerol (DAG) in the chloroplast (Goncalves *et al.*, 2016). DAG is a precursor for membrane lipids, in *Chlamydomonas* (Goncalves *et al.*, 2016; Li Beisson *et al.*, 2015).

In microalgae, TAG neutral lipid biosynthesis occurs in the endoplasmic reticulum (ER)-derived sections of the chloroplast (Fan *et al.*, 2011). Thioesterase is responsible for the carbon flux between termination pathways, acting as a metabolic director for biosynthesis of lipids (Bonaventure *et al.*, 2003). The importance of thioesterase and its gatekeeping role for the destination of fatty acids has therefore made it a target for algal fatty acid bioengineering for improvement of biodiesel feedstock (Radakovits *et al.*, 2010). Genetic engineering has shown improvements in oil yields by modifying ACCase expression in plant cells, with similar findings in algae (Roessler, 1988). Genetic modification of thioesterases can prove valuable for fatty acid biosynthesis in microalgae since most synthesize C16–C22 chains with considerable saturation variation (Harwood & Guschina, 2009). These enzymes are responsible for chain termination, which determines the fatty acid end-product length. Since biodiesel production generally requires shorter-chain fatty acids, altering the chain length could prove valuable not only for biodiesel, but potentially for use in production of gasoline and jet fuel

(Radakovits *et al.*, 2010). Genetic engineering of thioesterases to select specifically for C8–C14 fatty acid products is one means of improving biofuel production from microalgae through modification of fatty acid biosynthesis (Blatti *et al.*, 2012).

4.4.2 Biological CO_2 sequestration from point sources

A consideration for algal technologies includes the use of microalgae for CO_2 capture from flue gas generated from power plants. In order to generate electricity, power plants will burn fossil fuels, which create the thick, heavy smoke that can be seen exiting stacks of different industrial plants. This flue gas contains high concentration CO_2 emissions that are being released directly into the atmosphere. Release of these chemical streams is adding to the GHG concentrations, and the greenhouse effect result from GHG increases. Coal is a primary fuel used in power plants. From combustion, this gas mainly contains CO_2, nitrogen (N_2), oxygen (O_2), and water vapor (Zhang, 2015). Concentrations of CO_2 up to 15% (v/v) are found in power plant flue gas, making it an excellent candidate carbon source for microalgae biomass production and coordinated CO_2 sequestration (Cheah *et al.*, 2015) The concentration being considerably higher than that of atmospheric CO_2 concentrations makes flue gas an excellent carbon source for the microalgae, creating a potential means of avoiding excess CO_2 emissions release. However, many microalgae will only grow at an optimal CO_2 concentration less than that of flue gas. The high concentrations of CO_2 in flue gas can inhibit growth of certain microalgae species. There are some species that are able to grow under the high CO_2 concentrations associated with flue gas; however, those species have been shown to have a decrease in CO_2 fixation and biomass generation (Zhang, 2015). This was also shown by Maeda *et al.* (1995) as optimal *Chlorella* sp. growth occurred around 10% CO_2 with tolerance possible at up to 100%. Due to these limitations, Zhang (2015) found that adaptation to the higher CO_2 concentrations is necessary for microalgae use in flue gas treatment and removal of CO_2, suggesting strains from lakes and ponds surrounding the power plants as a starting point for high CO_2-tolerating, combustion product exposed microalgae species. In addition to CO_2 concentration considerations, the additional products found in flue gas including NO_x and SO_x species along with certain heavy metals need to be analyzed further for their effects on microalgae. Recent research indicates optimal algal growth and productivity at 5% CO_2 and 30–100 ppm NO_x (Biscaia *et al.*, 2022).

Algal technologies have long been considered for usage in wastewater treatment. Algae are generally cultured during wastewater treatment in waste stabilization ponds (WSPs) or high-rate algae ponds (HRAP), typically in warmer climates (Molazadeh *et al.*, 2019). In a WSP, raw wastewater is fed into the pond and treated using algal and bacterial metabolic processes. The algae provide oxygen for the aerobic activity of the bacteria so that the bacteria can continue to metabolize the organic compounds in the wastewater. This system is much more cost effective than alternatives to treating wastewater, providing a continuous source of oxygen from the algae and the breakdown of organic matter from the bacteria. This process removes the biological oxygen demand (BOD) from such aerobic WSPs, and by heterotrophic bacteria in anaerobic WSPs (Molazadeh *et al.*, 2019). The HRAP in contrast use oxidation ponds and a photobioreactor with the algae providing oxygen for the bacteria, similar to a WSP and the bacterial ability to convert waste nutrients into usable forms for the algae (Molazadeh *et al.*, 2019).

Algae have been used in wastewater treatment, recovery of minerals and micronutrients and are being analyzed for usage in an integrated systems for and concurrent carbon sequestration. Considering the often-high nutrient concentrations in wastewater when combined with a CO_2 stream, algal-based systems are an attractive option for CO_2 capture and wastewater treatment as a synergistic system. With some intervention, these resource streams together can create an optimal growth medium for various algal species. To make such a system successful, innovations to optimize the C:N ratio in wastewater along with the need for a sourced carbon stream are necessary, as atmospheric CO_2 would not be able to adequately supply the necessary amount of CO_2 to the algae in such a system (Molazadeh *et al.*, 2019). One source being examined is the flue gas from power plants providing a continuous, concentrated CO_2 stream for ample algal carbon fixation. Woertz *et al.* (2009) found that adding a steady CO_2 stream to a wastewater supplied HRAP can successfully remove nutrients from the wastewater, while simultaneously building algal biomass that can subsequently be used as a feedstock for biofuels. Since many wastewater, treatment plants burn off their excess methane, the CO_2 thus generated from flaring of methane could serve as an on-site source of concentrated CO_2.

Flue gas from biogas flaring is an intriguing potential CO_2 source where the prevention of CO_2 from entering the atmosphere could be coupled with the treatment of wastewater to create a cohesive environmentally friendly clean-up mechanism. This can lower GHG emissions while simultaneously capturing nutrients like nitrogen and phosphorus from wastewater sources. Chaudhary *et al.* (2017) found a successful algal-based remediation system, with increased photosynthetic activity present in CO_2-fed municipal wastewater with algal cultures *C. vulgaris* and *S. obliquus*. Pond cultivation systems and flat panel bioreactors are the most cost-effective cultivation methods for growing algae on flue gas (Schipper *et al.*, 2021). The delivery method of CO_2 is an important consideration when optimizing maximum photosynthetic efficiency for cultivation of algae. Addition of CO_2 to the wastewater can cause pH alteration potentially affecting the growth rate of the microalgae. One means of containing the pH within the optimal level for illuminated, actively photosynthesizing, microalgae is periodic sparging with CO_2 from sources such as flue gas (Wang *et al.*, 2008). Using microalgae for wastewater treatment, CO_2 sequestering coordinated hybrid technologies must demonstrate cost-effective, environmentally friendly potential.

An advantage of algae-based technologies is that they are typically scalable. An algae–wastewater treatment model has shown promise for CO_2 assimilation and nitrogen uptake (Eze *et al.*, 2018). Up to 200 tons of CO_2 per ML of wastewater treated could potentially be captured (Green *et al.* 1995; Molazadeh *et al.* 2019) Adding flue gas can deliver needed CO_2 while addressing pH fluctuations that result from photoautotrophic growth of microalgae. Eze *et al.* (2018) found that phototrophs successfully assimilated nitrogen and CO_2 at pH of 8.1; however, as the pH increased CO_2 and nitrogen assimilation decreased. This pH increase was associated with ammonia volatilization and increased CO_3^- presence, decreasing autotrophic biomass production and nutrient usage as these forms of carbon and nitrogen are not readily utilized by microalgae (Eze *et al.*, 2018). The algal biomass may be used as a biofuel feedstock, neutralizing CO_2 emissions. Alternatively, the biomass could be buried or integrated as agricultural biochar for long-term carbon sequestration in soil (Mona *et al.*, 2021). The pairing of bioenergy production with longer term carbon storage for decades to millennia is sometimes called bioenergy with carbon capture and storage (BECCS) (Moreira & Pires, 2016). While around 99% of the roughly 2 billion tons of carbon sequestered annually is due to conventional land-use practices, like planting and encouragement of forests, there is a pressing need for scalable, cost-effective new methods (Naddaf, 2023).

4.5 ALGAL BIOFUELS

Microalgae have economic potential for environmentally favorable clean-up strategies of various waste streams. Microalgae are excellent candidates for CO_2 sequestration, and many can work synergistically with different bacteria. Using microalgae to perform multiple tasks, from CO_2 uptake to nitrogen recovery, in an integrated system could provide beneficial, cost-effective, and an environmentally friendly alternative to conventional methods of CO_2 capture and wastewater treatment. Once carbon fixation occurs in these organisms, the biomass has additional potential in that of biofuels. This not only allows for the reduction of CO_2 emissions into the atmosphere and potentially more efficient mechanisms for wastewater treatment, but also the ability to take that fixed carbon and use it for environmentally favorable biofuel production. Biodiesel is produced from algal neutral lipids when the fatty acids in TAGs are chemically converted into fatty acid methyl esters. Algal biomass can alternatively be converted into biocrude oil via hydrothermal liquefaction.

4.5.1 Biodiesel

Biodiesel can be produced from various neutral lipid feedstocks, including many fast-growing microalgae (Saad *et al.*, 2019). Conventional production of algal biofuel (i.e., biodiesel) requires destruction of the production cell during the extraction step. Newer processes have been developed that do not necessarily require destruction of the phototroph. Strategies using naturally occurring and genetically modified cyanobacteria are promising due to their ability to rapidly convert CO_2 into biomass and to produce alcohol (Zhang *et al.*, 2017). CO_2-concentrating cyanobacteria fix CO_2 into biomass and minimize rates of wasteful photorespiration attributable to the oxygenase activity of RuBisCo (Kamennaya *et al.*, 2012). The carboxysome of cyanobacteria concentrates CO_2 for RuBisCo, which has low carboxylation efficiency (Kamennaya *et al.*, 2012). The mechanism, which involves CA works by bicarbonate (HCO_3^-) symports and CO_2 transporters (Kamennaya *et al.*, 2012). Most of the CO_2 is converted into HCO_3^- via enzyme CA activities (Kamennaya *et al.*, 2012). Cyanobacteria are promising candidates for biofuels due to their high lipid content, efficient carbohydrate metabolism, mainly in the thylakoids, high photosynthetic capabilities, and growth rates, along with their low growth requirements (Quintana *et al.*, 2011). However, cyanobacterial lipids are mainly charged membrane phospholipids which are less suited for biodiesel production than neutral lipids.

The neutral lipid composition of many microalgae allows for the production of biodegradable biofuels, namely biodiesel (Saad *et al.*, 2019). The optimal growth conditions of algae are dependent on temperature, pH, CO_2 availability, light intensity, and nutrient availability (Li *et al.*, 2019). Most algal species have an optimum temperature between 20 and 30°C (Singh & Singh, 2015). Specifically, light, CO_2 and water are primary factors in biomass accumulation in algae and cyanobacteria (Khan *et al.*, 2017). Photoautotrophic systems are promising due to their usage of CO_2 as a sole carbon source and their minimal negative environmental impact (Mata *et al.*, 2010). The production strain screening process generally occurs in *in vitro* experiments, many turning to microfluidics or on-chip technology due to the higher costs of conventional experimental techniques (Saad *et al.*, 2019). Such techniques allow droplets to be captured and tested using many different configurations based on different culturing conditions along with DNA and oil extraction (Saad *et al.*, 2019). Selecting an appropriate cultivation system is equally important because it can have a direct impact on biofuel generation. Wild-type algae traditionally were used in algal biofuel production, but recent advances in genetic editing have led to the creation of novel algal strains with optimized carbon sequestration capabilities (Figure 4.4).

Processing of microalgae is simplified compared to macroalgae, but economical harvesting and dewatering of the cells can be difficult. Many microalgae such as *Chlorella pyrenoidosa*, accumulate significant intracellular neutral lipids that can be extracted and chemically converted into biodiesel. Nitrogen limitation provides one elegant means of inducing over three-fold lipid accumulation in *C. pyrenoidosa*, provided adequate light and carbon dioxide are available (Han *et al.*, 2013). Alternative approaches include genetic modification of components associated with both the light and light-independent catalysts of photosynthesis, an inherently inefficient process. Over 8 billion gallons of biodiesel is produced annually worldwide (Guo, 2020). Unfortunately, much of this is

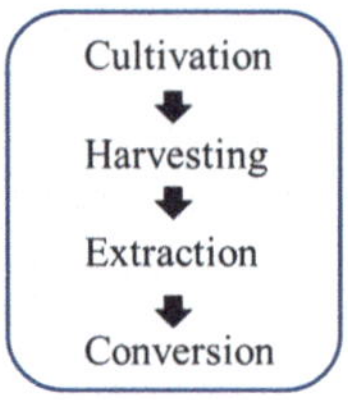

Figure 4.4 Flowchart overview of algal biodiesel production.

produced from lipids extracted from crop phototrophs used for foods, such as soybeans. Algal lipids can alternatively be used to produce algal biodiesel. The lipid content in many microalgae strains can be up to 50% of the organism's dry weight (Li *et al.*, 2019; Rawat *et al.*, 2013). The type of fatty-acid precursor a specific algal species produces affects various properties of biodiesel production including melting point, oxidative stability, cetane values and lubricity (Saad *et al.*, 2019). Biodiesel production is generally reliant on palmitoleic acid, palmitic acid, steric acid, linoleic acid, oleic acid, and fatty acids (Zheng *et al.*, 2013). Microalgae typically store solar energy as neutral lipids (i.e., TAGS) whereas many cyanobacteria accumulate starch which is more suited for ethanol production (John *et al.*, 2011; Möllers *et al.*, 2014). Accordingly, microalgae are considered useful feedstocks for biodiesel production. *Nanochloropsis* is a promising production strain capable of lipid accumulation of 47% of dry weight (Alishah Aratboni *et al.*, 2019; Pal *et al.*, 2011). *Chlorella* sp. BTA 9031 exhibited 25% of lipid accumulation per dry weight under 3% (v/v) CO_2 concentration (Mondal *et al.*, 2017), and *Chlorella vulgaris* SDEC-3M also accumulates significant neutral lipids (Qi *et al.*, 2019).

4.5.1.1 Cell cultivation and harvesting

Algal cultivation can be carried out in open system such as ponds or raceways or closed bioreactors (Chew *et al.*, 2018). Open systems are generally a closed-loop (raceway) or circular pond tank, a closed pond, or large shallow pond (Chew *et al.*, 2018). These ponds generally contain natural fresh or seawater, depending on the species being used (Anto *et al.*, 2020). Open systems are useful because they allow atmospheric CO_2 to be utilized and are generally lower cost than a closed system (Dharmaprabhakaran *et al.*, 2020). These open systems come with challenges such as contamination by grazers, bacteria, protozoa, or other microalgae species which can create toxic products that cannot be utilized (Narala *et al.*, 2016; Tan *et al.*, 2020; Ullah *et al.*, 2015). Open systems can be significantly affected by rainwater runoff which can cause alterations in salinity, pH, and increases in turbidity of water due to erosion (Tan *et al.*, 2020) along with other factors such as evaporation and temperature (Mata *et al.*, 2010; Narala *et al.*, 2016). A closed system can help to alleviate some of the challenges of an open pond such as contamination and evaporative losses (Carvalho *et al.*, 2006).

Closed systems employ photobioreactors which can be highly controlled allowing for ideal stirring and efficiency due to high light availability, delivering high-productivity yields (Lee & Lee, 2016; Liao *et al.*, 2018). Size and shape of the bioreactor can be flat-plate, tubular, vertical columnar, along with other configurations, providing more compact organization and more efficient land usage (Tan *et al.*, 2020). These systems can use acrylic or glass materials and can be constructed as bags, tanks, or towers (Anto *et al.*, 2020). Closed system bioreactors tend to have a higher productivity than open systems because of the controlled conditions used, but the overall cost of using an open system is more favorable (Leite *et al.*, 2013). For example, photobioreactors could be used for starting seed cultures that are subsequently scaled up into large ponds.

Harvesting of algae can be a challenge and care must be taken to efficiently remove the cells from the growth medium. Harvesting and dewatering of microalgae are energy-intensive processes. Methods such as flocculation, sedimentation, centrifugation, filtration, precipitation, sonication, and flotation have all been used with varying degrees of success. In many cases, dry biomass may be required, and dewatering must occur to remove the water content of the algal cells (Chen *et al.*, 2015). This process can occur through bulk harvesting and thickening (Alam *et al.*, 2017). In bulk harvesting, the biomass is separated from suspension, and the solid matter can be separated using gravity sedimentation, flocculation, or flotation (Lam *et al.*, 2019). Drying using sunlight may reduce input energy cost. Flocculation is a step that can be taken before using gravity sedimentation or flotation. Flocculants can be used to treat large quantities of microalgae, inducing flocculation (Alam *et al.*, 2017). In flocculation, aggregation of microalgae or other cells collide and adhere to one another into larger particles through charges on the cell surfaces (Khan *et al.*, 2017). This method is favorable due to the ability to harvest large quantities of cells from different taxonomic classifications (Wan *et al.*, 2015). Gravity sedimentation is used in species with high sedimentation rates or larger species such as *Spirulina* for harvesting in wastewater treatment applications (Brennan & Owende, 2010). This technique can be employed to separate algal cells of different types due to differences in settling velocity (Peperzak, 2003). Due to low microbial settling rates, it is not favorable for routine harvesting, but is a popular choice for wastewater treatment applications (Christenson & Sims, 2011; Greenwell *et al.*, 2009). In flotation, biomass is removed using gravity separation (Alam *et al.*, 2017). In this technique, gas bubbles are passed through a liquid suspension and attach to solid particles within the suspension (Nguyen & Hoang, 2016). This is an advantageous technique compared to others due to low microalgal density and buoyancy characteristics of certain species such as *Anabaena, Microcystis, Spirulina,* and *Nostoc* (Hanotu *et al.*, 2012; Laamanen *et al.*, 2016).

Thickening of the reduced volume cell slurry may be performed using more energy-intensive methods such as centrifugation, filtration, or ultrasonic aggregation (Brennan & Owende, 2010). Centrifugation is fairly cost-efficient and harvests many microalgal species rapidly (Khan *et al.*, 2017). It has proven 95% efficient at increasing the concentration of the cell slurry (Li *et al.*, 2017). Filtration can be used to harvest larger microalgae such as *Spirulina* and *Coelastrum* (Brennan & Owende, 2010), but it is not appropriate for large batch harvesting (Nguyen & Hoang, 2016). This allows for the collection of low-density algal cells using different membranes and pore sizes with the addition of a suction pump (Rastogi *et al.*, 2018). Flow filtration has shown a 70–89% efficiency for microalgae harvesting and cell structure preservation (Danquah *et al.*, 2009; Rastogi *et al.*, 2018). Ultrasonic aggregation employs acoustic force and heightened sedimentation for concentration of microalgal cells (Nguyen & Hoang, 2016).

4.5.1.2 Lipid extraction and conversion

Extraction of lipids from the slurry and dewatering must be carried out after harvesting to avoid decomposition (Kadir *et al.*, 2018). During dewatering, water is further removed from the biomass through a drying process either by use of sunlight, freeze-drying, fluidized drying, or spray drying, with spray drying being the most widely used technique (Khan *et al.*, 2017; Nguyen & Hoang, 2016). This allows the water content from the algae to be reduced, leaving dry biomass. Cell disruption is commonly employed during lipid extraction and is accomplished through mechanical, physical, chemical, and enzymatic phases (Lee *et al.*, 2012). Cell disruption is important when performing lipid extraction for biodiesel production (Johnson, 2008; Kadir *et al.*, 2018). Using a wet route to lipid extraction has been examined and has potential to improve production of valuable biofuel (Lakshmikandan *et al.*, 2020; Xu *et al.*, 2011). Using organic solvents and supercritical fluids are popular for high production yield in algal lipid extraction (Liu *et al.*, 2013; Rastogi *et al.*, 2018; Santana *et al.*, 2012; Soh & Zimmerman, 2011). Using solvents for extract on dry biomass is the most common method, but it can have a high-cost and high energetic requirement (Rastogi *et al.*, 2018). Because of this, wet biomass usage is an economically promising alternative (Chatsungnoen & Chisti, 2016; Grima *et al.*, 2012; Liu *et al.*, 2013). Lakshmikandan *et al.* (2020) found promising results using wet biomass of *C. vulgaris*, showing increased biomass and total lipid yield (22.5%) through extraction by mild pressure and heat shock (Lakshmikandan *et al.*, 2020).

Lipid extraction can be accomplished in four ways: using chemical solvents, supercritical CO_2, physiochemically, or biochemically (Nguyen & Hoang, 2016). Chemical solvents can be employed to break the chemical linkages within the cell, with minimal requirements for energy or heat (Nguyen & Hoang, 2016). Pressurized CO_2 can accomplish lipid extraction without utilizing toxic chemicals, but it is energetically demanding and expensive (Perrut, 2000; Tan & Lee, 2011). Various physical methods can be employed to cause cell disruption, but microwave is one promising technique (Lee *et al.*, 2012). The blending method is a simple method for fatty-acid extraction and separation of biodiesel and consists of the stages: filtration, drying, oil extraction, and biodiesel production (Khan *et al.*, 2017). Other methods that have been employed for biodiesel production include microemulsion and pyrolysis. Microemulsion involves using vegetable oils for biodiesel production, however, micro emulsification tends to have heavy carbon residues, inadequate combustion, random injector needle sticking, and high viscosity of lubricant oils (Tabatabaei *et al.*, 2019). Various impurities may need to be removed. Pyrolysis can also be employed using thermochemical mechanisms through application of heat to convert various species into chemical species in anoxic conditions (Tabatabaei *et al.*, 2019). This technique, however, has a high-cost due to complex equipment and the reaction conditions, and form a fuel more closely related to gasoline than to diesel (Tabatabaei *et al.*, 2019). Transesterification is the chemical conversion technique usually used in biodiesel production.

Transesterification occurs when short-chain alcohols (i.e., methanol, ethanol, propanol, or butanol) react with microalgae free fatty acids (triglycerides) using a catalyst (Tabatabaei *et al.*, 2019). Popular catalysts include acids such as HCl, H_2SO_4, H_3PO4, and sulfonic acid, alkali species such as KOH, NaOH, CH_3KO, CH_3ONa, or lipase enzymes (Demirbas, 2009; Fukuda *et al.*, 2001). Alkali species are favorable for use with methanol due to lower cost and higher reactivity than ethanol (Rastogi *et al.*, 2018; Robles-Medina *et al.*, 2009). Alkali-catalysts, such as KOH or NaOH, are important because they increase the rate of the reaction 4000 times over acid-based catalysts (Fukuda *et al.*, 2001). Alkali-catalysts can, however, cause some unfavorable issues including saponification (Tabatabaei *et al.*, 2019). In transesterification, TAGs combine with alcohol (methanol or ethanol), to form diglyceride followed by monoglyceride then glycerol, as three fatty acid methyl esters (FAME) are produced when methanol is used or three fatty acid ethyl esters (FAEE) when ethanol is used (Demirbas, 2010; Yusoff *et al.*, 2014). These FAME or FAEE biodiesel molecules can be combusted as fuel in regular diesel engines with certain benefits, like superior lubrication (Balat & Balat, 2010; Pragya *et al.*, 2013; Robles-Medina *et al.*, 2009). Large quantities of the glycerol waste product have accumulated as the biodiesel industry has matured. An example of the transesterification used for biodiesel production reaction is shown in Figure 4.5.

Ethanol and methanol are alcohols of choice for the transesterification process (Bouaid *et al.*, 2007). Methanol is the more widely used alcohol due to lower cost than ethanol, easy phase separation, mild conditions required for reactivity, along with easier extraction of methyl esters over ethyl esters (Yusoff *et al.*, 2014). Ethanol does have some advantages over methanol however, including easy attainment

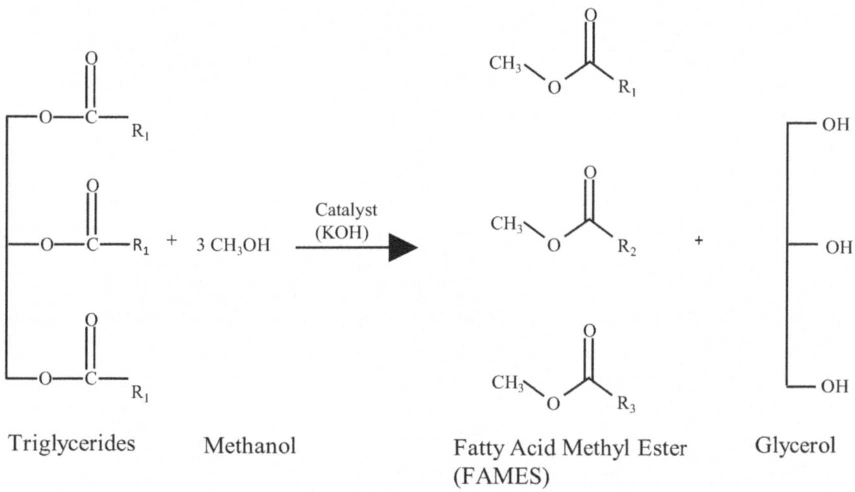

Figure 4.5 Depiction of the transesterification reaction used to produce fatty acid methyl esters (i.e., biodiesel) from algal TAG plus methanol in the presence of KOH catalyst. NaOH can alternatively be used to catalyze the transesterification reaction.

from agricultural sources such as sugar cane or corn, and produces cleaner products of ethyl esters opposed to methyl esters (Mendow *et al.*, 2011; Yusoff *et al.*, 2014). Ethanol has a lower safety risk and is easier to handle than methanol, making it a popular alcohol for transesterification (Yusoff *et al.*, 2014). Direct transesterification can be employed by combining oil extraction, esterification, and transesterification simultaneously, using alcohol for extraction (Liu *et al.*, 2017). Algal biodiesel has some drawbacks, such as low-temperature gelling during winter, capacity to degrade rubber seals and oxidation over time (Zuleta *et al.*, 2012). Ideally, synthesis and direct secretion of FAME could preserve the phototrophic producer cells.

4.5.2 Biocrude and triterpenes

Biocrude oil, alternatively called pyrolysis bio-oil, is produced through thermochemical conversion of organic material, such as algal cells. For this, biomass is thermally broken down and organic chemicals are recruited to restructure the products into biofuels (Naik *et al.*, 2010; Raheem *et al.*, 2018). This process can produce biochar, liquid fuel, or gaseous fuel (Adeniyi *et al.*, 2018). Pyrolysis is the first step in thermochemical conversion of solid fuel. In algal pyrolysis, algae decomposition occurs in anoxic conditions through thermochemical degradation causing the formation of gases and volatile products that condense to form liquid bio-oils at high temperatures (Aravind *et al.*, 2020). The solid product that remains is biochar, which can be used in gasification to form syn-gas which contains H_2 and CO (Saber *et al.*, 2016; Zainan *et al.*, 2015). This process may involve various mechanisms such as dehydration, decarboxylation, fragmentation, polymerization, and rearrangement reactions (Li *et al.*, 2019). Dewatering is required for pyrolysis because the feedstock must be dry in order for pyrolysis to occur, causing an increase in total energy consumption of the system (Saber *et al.*, 2016). Pyrolysis can occur by slow heating (10°C/min), fast heating (100°C/s), or by flash heating, generally done at 300–750°C/s (Li *et al.*, 2019). Slow pyrolysis generates larger quantities of solid char than fast or flash pyrolysis (Saber *et al.*, 2016). Fast or flash pyrolysis have up to an 80 wt.% dry feed yield of bio-oils, making them a promising pathway for bio-oil generation (Huber *et al.*, 2006). For bio-oil production, microwave-enhanced pyrolysis (MEP) is a popular method (Saad *et al.*, 2019; Zhang *et al.*, 2017). Microalgae pyrolysis-generated bio-oils have a composition mainly of aliphatic and aromatic hydrocarbons, along with oxygen-containing and nitrogen-containing compounds (Li *et al.*, 2019). The oxygen-containing and nitrogenous compounds are formed from carbohydrate and protein decomposition in the algal biomass (Li *et al.*, 2017). These oxygen-containing compounds can affect bio-oil by causing acidity, corrosion, and high viscosity and volatility, while nitrogen-containing compounds can contaminate catalysts and cause nitrogen-oxide formation during combustion (Czernik & Bridgwater, 2004; Oasmaa & Czernik, 1999; Wang *et al.*, 2013; Zainan *et al.*, 2015). The high oxygen content of pyrolysis oils causes them to have a lower heating value than oils generated through liquefaction (Duan & Savage, 2011; Peterson *et al.*, 2008). Oil products from pyrolysis are dark brown and viscous, containing over 300 different compounds including hydrocarbons, alcohols, acids, phenols, sugars, polyaromatics, nitrogenous compounds, and others, with straight-chain hydrocarbon and paraffin aromatics (up to C_{10}) preferred for transportation fuel (Li *et al.*, 2019). Catalysts can be recruited to lower the reaction temperature requirements and energy requirements in general (Hazrat *et al.*, 2015). Some catalysts that have been used in microalgae pyrolysis studies include zeolite, metal-loaded catalysts, and metal-organic frameworks (Yang *et al.*, 2019). The cost of pyrolysis is lower than that of liquefaction and many technologies are being commercially used (Yang *et al.*, 2019).

Hydrothermal liquefaction (HTL) can be used to process algal biomass waste into bio-oil. Essentially, biomass (algal slurries) is reacted in water at higher temperatures (200–400°C) and pressures (5–20 MPa) with or without the use of a catalyst to produce biocrude, gas, or char (Brand *et al.*, 2014; Chen *et al.*, 2015; Chiaramonti *et al.*, 2017; Ross *et al.*, 2010; Shuping *et al.*, 2010). Water is a popular solvent used in this technique, but the product is a very viscous bio-oil with a prominent oxygen content (Saber *et al.*, 2016). This has prompted the use of organic solvents such as ethanol, acetone, 2-propanol, methanol, among others, for a higher yield of bio-oil (Liu & Zhang,

2008; Ogi *et al.*, 1994; Yuan *et al.*, 2011). Catalysts, such as Na_2CO_3, KOH, CH_3COOH, and zeolite can be utilized to improve the bio-oil quality and yield (Duan & Savage, 2011; Jena *et al.*, 2012; Ross *et al.*, 2010). This process includes various mechanisms such as solvolysis, decarboxylation, depolymerization, hydrogenolysis, and hydrogenation (Demirbaş, 2008). HTL produces a higher yield of bio-oil than pyrolysis of up to 97% (Dote *et al.*, 1994; Saad *et al.*, 2019; Zou *et al.*, 2010). The main factors affecting HTL include temperature, loading, and residence time (Saber *et al.*, 2016). Bio-oil derived from this process is partially water soluble and of a higher energy content than pyrolysis due to lower oxygen content and water content, but also have a higher viscosity (Cheng *et al.*, 2017). Cost can be higher than that of pyrolysis due to the higher-pressure requirement (Huber *et al.*, 2006). It may be beneficial, however, in the use of wet biomass, eliminating the need for a drying step as in pyrolysis. Research by Jena and Das (2011) found HTL produced a higher bio-oil production with a lower energy requirement and solid char yield than pyrolysis, showing its potential advantages for bio-crude production.

Gasification can be used, in which case biomass is partially oxidated using a controlled quantity of air, oxygen, or steam at 700–1000°C, that produces a mixture of CO_2, H_2, CH_4, and CO (Demirbas, 2010; Naik *et al.*, 2010). The oxidized biomass can be used for heat production, electricity, or further processed (Li *et al.*, 2019). Direct combustion may be employed for the production of algal energy. In direct combustion, the algal biomass is first pre-treated through drying and mechanically grinding into smaller particles (Adeniyi *et al.*, 2018). Following pre-treatment, a furnace, boiler, or steam turbine is used for biomass oxygenation at ~1000°C (Saad *et al.*, 2019). Undesirable facets of bio-oil may include high oxygen, nitrogen, and water content, high viscosity and acidity, and thermal and chemical instability (Saber *et al.*, 2016). These characteristics are derived from the fundamental components of the algal biomass including carbohydrates, proteins, and fatty acids. Gasification of waste can facilitate CO_2 valorization with ethanol production through the process of microbial syngas fermentation (Stoll *et al.*, 2020). LanzaTech is one company that is currently commercially developing microbial syngas fermentation (Stoll *et al.*, 2020).

Triterpenes are chemical components of certain algae that can be converted into drop-in biofuels without necessarily destroying the cells. Triterpenes are relatively large organic molecules composed of six isoprene rings that are produced by various bacteria, fungi, animals and various phototrophic microorganisms, including some algae. They may serve as steroid precursors and are involved in the synthesis of biologically important molecules like cholesterol or ergosterol in fungi. Triterpenes have a wide variety of functions in microorganisms as well, some of which can factor towards the production of biofuels (Chacko *et al.*, 2019). For example, the hopanoid triterpenes present within the cell membrane of *Z. mobilis* help confer tolerance to high concentrations of ethanol by this ethanologenic microbe (Bringer *et al.*, 1985; Hermans *et al.*, 1991). Elevated ethanol tolerance might accordingly be conferred to other biofuel producing microorganisms, such as yeast, via the genomic modification with genes and pathways needed for triterpene biosynthesis. Triterpene molecules are energy dense hydrocarbons that can additionally serve as biofuels, or biofuel precursor. Triterpenes produced by microalgae or other photosynthetic microbes are of particular interest as potential biofuels.

Botryococcene is a leading algal triterpene biofuel candidate. Botryococcene is synthesized by *Botryococcus braunii,* with the *B. braunii* race B strain recognized as a particularly productive isolate since it can accumulate 89% botryococcene by dry weight (Brown *et al.*, 1969). This microalga synthesizes botrycoccene via an unusual biochemical mechanism distinct from the mechanism other eukaryotes utilize to synthesize triterpenes such as squalene (Niehaus *et al.*, 2011). One of the advantages of this green microalgae is that the botryococcene is secreted from the cells into the surrounding media (Suzuki *et al.*, 2013). This could potentially simplify collection or avert the need to sacrifice the producer strain which is normally necessary for TAG producers during the extraction step. Another in key advantage of botryococcene is that it can easily be processed in existing oil refineries and converted into jet fuel or a drop in 'green gasoline' biofuel replacement capable of powering conventional gasoline-powered trucks and automobiles (Chacko *et al.*, 2019).

One major disadvantage of *B. braunii* is that it has an exceptionally slow growth rate with a maximum doubling time of around a day and a half for one strain investigated under a variety of conditions (Yoshimura *et al.*, 2013). Related to its slow growth rate is the presence of a thick cell wall that may pose challenges if extraction is utilized. Different strategies have been taken to overcome this, one approach cultivates the microbe in a trickle-film photobioreactor for optimized metabolic production rates of botryococcene (Khatri *et al.*, 2013). Genetic modification of other phototrophs with the metabolic machinery akin to *B. braunii* is an alternative approach to circumvent the slow growth rate. Genetic modification of *Synechocystis* allowed this cyanobacterium to produce terpene hydrocarbon (Formighieri & Melis, 2014).

Microalgae may hold promise for various stages of environmental cleanup and CO_2 capture and sequestration (Bahr *et al.*, 2013). Although some limitations may be present, a cohesive system might be used on a large scale to clean up the areas impacted by nutrient pollution while addressing, on a limited scale, the effects of GHG emissions and climate change. The treated liquid effluent of wastewater treatment plants is often associated with downstream environmental issues, including phytonutrient pollution and eutrophication. Nutrient-fed blooms of macroalgae can disrupt coastal waterways, cause dead zones and hinder industries such as fishing and tourism. Clean up and removal of the carbon-rich algae waste material could be economically incentivized if it were to serve as a biofuel feedstock. One challenge associated with macroalgae like *Sargassum* seaweed is that it is relatively low in neutral lipids; however, HTL technology can convert *Sargassum* into bio-crude oil (He *et al.*, 2020). The large size of macroalgae simplifies harvesting, but low lipid content and specialized structures, such as recalcitrant holdfasts can present a processing challenge. Even without harvest, some macro algae can promote long-term carbon sequestration into the environment. Seagrasses and possibly macroalgae support complex microbial communities that influence long-term carbon sequestration in marine sediments (Mohapatra *et al.*, 2022).

4.6 BIOGAS

Wastewater and municipal solid waste contain carbon in varied organic forms that can be catabolized to produce methane microbiologically through a process called anaerobic digestion (AD) (Gijzen, 2002). A portion of the carbon is biologically sequestered as biomass with a high degree of biochemical efficiency while the rest is released as methane which can be used as a gaseous biofuel (Lemaire *et al.*, 2020). Diverse microbial communities naturally found with input waste establish a food web inside of the anaerobic digester, essentially an airtight tank, to break down the mixture of waste lipids, proteins, and carbohydrates into biogas plus a residual liquid called digestate (Chen *et al.*, 2020). Liquid digestate is a useful, phytonutrient-rich coproduct of biomethane production that can be used in place of nitrogen-rich chemical fertilizers in agriculture (Gijzen, 2002). Cultivation of phototrophs, such as oleaginous algae, on digestate and CO_2 from AD is a useful way to recover nutrients from wastewater while producing algal biomass as a biodiesel feedstock (Gijzen, 2020). Biogas produced from AD is not pure methane, but rather a mixture of microbially produced gases that contains a significant fraction of carbon dioxide. Depending on the organic waste material being processed and incubation conditions, the composition of biogas is typically about 60% methane and 40% CO_2 with trace amounts of ammonium and hydrogen sulfide (H_2S) (Liu *et al.*, 2020).

4.6.1 Anaerobic digestion

Bacteria, eukaryotic microbes, and methanogenic archaea all work together in syntrophic connection to metabolize the many polymers in waste into first CO_2, H_2, and volatile fatty acids (VFAs), eventually forming methane. Hydrolytic enzymes including proteases, lipases, cellulases, and amylases break down proteins, lipids, and carbohydrates, respectively. This initial polymer hydrolysis and fermentation phase represents the rate-limiting step of biogas formation. Metagenomic community analysis indicates *Proteobacteria*, *Firmicutes*, *Bacteroidetes*, and *Actinobacteria* as the most numerous bacterial groups, with *Methanosaeta* and *Methanosarcina* conspicuous among the archaea (Guo *et al.*, 2015). Polymer hydrolysis is considered the rate-limiting step during biogas formation. Much of the CO_2 and H_2 produced during fermentation is subsequently metabolized by acetogenic bacteria. These acetogens use the reductive acetyl coenzyme A pathway or Wood–Ljungdahl pathway to form acetate that is released into the digester (Müller, 2003). Bacteria do not directly produce methane as they breakdown organic wastes. Rather, they produce molecules that are consumed by two groups of methanogenic archaea. The methane produced during anaerobic digestion is created by archaea. Acetoclastic methanogens, like members of the genus *Methanosarcina*, consume the acetate secreted by acetogenic bacteria and convert the methyl group into methane (Ferry, 1997). Acetoclastic methanogens produce most of the methane released because relatively few other microorganisms can compete with them for acetate. The second group involved in CH_4 production in digesters are the hydrogenotrophic methanogens. They consume and oxidize hydrogen gas as their electron and energy source as they reduce carbon dioxide into methane in accordance with the following equation (4.1) (SanchoNavarro *et al.*, 2016).

$$CO_2 + 4H_2 \rightarrow CH_4 + 2H_2O \tag{4.1}$$

Acetogenic bacteria, namely the homoacetogens, within the digester compete for substrate CO_2 and H_2 with the hydrogenotrophic methanogens. This substrate competition tends to reduce the amount of methane produced via the hydrogenotrophic route. In contrast, acetate produced by the homoacetogens serves as the substrate for acetoclastic methanogens which produce most of the methane in wastewater treatment anaerobic digesters (Guo *et al.*, 2015). Hydrogenotrophic methanogens such as *Methanococcus maripaludis* are capable of rapid autotrophic growth as they assimilate CO_2 into biomass using energy released from the oxidation of H_2 (Lyu *et al.*, 2016).

In the future, hydrogenotrophic microorganisms could prove useful for the direct biological sequestration of CO_2 into long-lived consumer goods, like bioplastics, using H_2 produced from renewable energy sources, such as can be generated using wind turbines or solar photovoltaic panels (Ayol *et al.*, 2021; Pisciotta & Blessing, 2022). Recent genetic modification of *M. maripaludis* has resulted in autotrophic conversion of CO_2 into the bioplastic polyhydroxybutyrate (PHB) (Thevasundaram *et al.*, 2022). Conversion of CO_2 into long-lived consumer could help to offset the demand for petroleum-based plastics while slowing the carbon return rate back into CO_2. While methanogenic archaea appear to hold considerable promise for carbon capture, sequestration, and valorization, at present, methanogens are primarily used in mixed communities for the synthesis of biogas. The ordered sequence of waste conversion steps that take place within an anaerobic digester are shown in Figure 4.6.

Many types of anaerobic digesters have been developed over the past century. Mesophilic digesters operate at ambient temperatures and are the most common form but may suffer from a long hydraulic retention time (HRT). HRT is described as the amount of time the waste must pass through the digester for effective treatment and energy extraction with conversion to biogas. Thermal hydrolysis as a means of pretreatment of the input organic waste can substantially reduce HRT (Chen *et al.*, 2020). Thermophilic anaerobic digesters operate at higher temperatures and have shorter HRTs for untreated wastes but require an input of heat energy. Mesophilic digesters feature more ecologically diverse

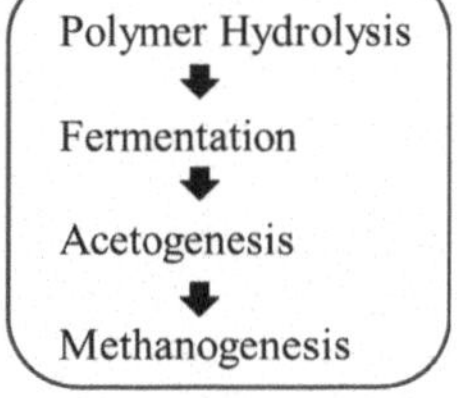

Figure 4.6 Stages of organic waste catabolism form biogas that occurs within an anaerobic digester.

microbial communities as compared to thermophilic digesters (Chen *et al.*, 2020). A key industrial advantage of thermophilic digesters for wastewater treatment over mesophilic digesters is enhanced pathogen inactivation by the former (Xu *et al.*, 2016). The resulting liquid digestate can be used as agricultural fertilizer. Solid municipal waste can also be anaerobically digested for biogas production at landfills for bioenergy production (Gijzen, 2002).

4.6.2 Biomethane enhancement

Several approaches, such as processing biogas with phototrophs using hybrid carbon sequestration systems, can elevate methane levels while reducing CO_2. Such hybrid approaches bolster conversion of diverse biomass wastes into useful methane concurrently removing CO_2 and H_2S from the product biogas (Bahr *et al.*, 2013; Nagarajan *et al.*, 2019). Since CO_2 is not an energy carrier, this increases, or rather enhances, the energy density of the biogas product. Application of electrical energy can increase methane production in a connected digester while concurrently altering the microbial community composition (Lee *et al.*, 2017). This relatively recently discovered bioelectrochemically driven phenomenon is referred to as electro-methanogenesis (Logan *et al.*, 2019). Bioelectrochemical technologies may be used to sequester CO_2 and improve the treatment of various waste and gas streams, including different types of refinery wastewaters (Ren *et al.*, 2013). Addition of the calcium-based mineral wollastonite was recently found to enhance CH_4 production while concurrently chemically sequestering CO_2 (Zhang *et al.*, 2019). Increases in CH_4 concentration to over 95% of biogas are possible through pairing wollastonite addition into bioelectrochemical digester systems (Zhang *et al.*, 2020).

4.7 BIOHYDROGEN

Carbon sequestered by autotrophic microbes is converted to cell biomass that can generate useful H_2 gas either directly by the autotroph or indirectly through biomass fermentation (Kumar *et al.*, 2013). Hydrogen presents an attractive fuel for multiple reasons. It is an energy-dense molecule that can be produced using a number of chemicals as well as biological methodologies. It is widely used in industry today for applications ranging from welding to the production of fertilizer from atmospheric nitrogen via the Haber–Bosch process (Rapson *et al.*, 2020). It can be combusted directly to power engines, including jet and rocket engines, or to provide heat (Baroutaji *et al.*, 2019). Another advantage of H_2 is that it can be converted into direct current (DC) electricity using fuel cells, such as are used in fuel cell vehicles. Since H_2 does not contain carbon, its oxidation either through combustion or using a fuel cell produces only water without climate warming carbon dioxide. For these reasons, governments are demonstrating heightened interest in renewably produced 'green' H_2 as a tool for the decarbonization of human society (Galvin, 2020). Microbiologically produced hydrogen can be produced through three main routes: dark fermentation, photofermentation, and biophotolysis.

4.7.1 Dark fermentation

Dark fermentation occurs under anaerobic conditions in sealed fermenters and is catalyzed by various anaerobic or facultative anaerobic heterotrophic bacteria. Members of the genera *Bacillus*, *Clostridia*, and *Enterobacter* have all been examined for their capacity to produce H_2 gas through dark fermentation (Sharma *et al.*, 2020). In most organisms, including *Clostridium* and *Bacillus* hydrogenase enzymes are employed to generate H_2 (Arunasri *et al.*, 2016). Hydrogenases catalyzed proton reduction as a means of oxidizing electron carriers resulting in H_2 formation. While loss of H_2 represents a net loss of energy from cells, over-reduction of the pool of electron carriers, such as NADH and ferrodoxin, could inhibit cell metabolism (Schut & Adams, 2009). Various types of hydrogenases have been identified including [FeFe]-type hydrogenases and [NiFe]-hydrogenase that can be genetically engineered into more robust bacterial species (Lamont & Sargent, 2016).

Hydrogenases are typically highly sensitive to atmospheric O_2 and are irreversibly inhibited by it. The Knall gas bacterium *Cupriavidus necator*, formerly known as *Ralstonia eutropha*, provides an interesting exception in that this microbe synthesizes an oxygen tolerant, membrane-bound [NiFe] hydrogenase which provides for sustained hydrogen production, even in the presence of oxygen (Goldet *et al.*, 2008). While not autotrophic, certain facultatively anaerobic Enterobacteriaceae family members, such as *Escherichia coli* and other mixed acid fermenters, are capable of producing hydrogen gas using a different enzyme, formate hydrogen lyase, which oxidizes formic acid into CO_2 plus H_2 (Yoshida *et al.*, 2005). During dark fermentation, carbohydrates are catabolized to CO_2 plus H_2. Dark fermentation requires no input of light energy and tends to decrease media pH through release of organic acids along with H_2. The process generally does not result in complete substrate catabolism and mineralization to inorganics, so the H_2 yields is suboptimal. Glycerol waste from biodiesel production can be further processed microbiologically to form hydrogen gas biofuel through fermentation (Rastogi *et al.*, 2018; Selembo *et al.*, 2009). Production of a mixture of fermentation end products such as volatile organic acids (VOAs) alongside H_2 by many dark fermenting microbes means a less optimal hydrogen yield is achievable by this method. Hybrid approaches are being investigated in which VOAs are used as substrates by other microbes for additional H_2 production via a light-stimulated photofermentation (Silva *et al.*, 2019).

4.7.2 Photofermentation

Certain anoxygenic phototrophic microorganisms create hydrogen gas through a light-driven process called photofermentation. Here, organic molecules, typically short-chain VOAs, including those produced by dark fermenters, serve as the electron donors with excitation energy provided by light (Silva *et al.*, 2019). The VOA-rich effluent from dark fermentation can be used to feed illuminated photofermenters for sustained and enhanced hydrogen production in a dual stage hybrid H_2 production system (Silva *et al.*, 2019). The most prominent group of microbes capable of carrying out photofermentation include the purple non-sulfur bacteria (PNS) with members of the genera *Rhodobact*er or *Rhodopseudomonas* most often evaluated for their ability to form H_2 biofuel by photofermentation (Sharma *et al.*, 2020). Their distinctive purple color is attributable to photopigments including carotenoids and bacteriochlorophyll which allows for the absorption of long wavelength, infrared radiation (Fowler *et al.*, 1997). Many PNS bacteria are diazotrophic and can produce H_2 by a second mechanism.

Nitrogenase is yet another enzyme capable of generating H_2 as a coproduct, along with ammonia in diazotrophic bacteria which facilitate biological cycling of nitrogen from the atmosphere. All of life requires bioavailable nitrogen for amino acid synthesis, mainly in the form of ammonia, which today comes from (1) atmospheric discharge, (2) biological binding, and (3) chemical synthesis via the Haber–Bosch process (Wolinska *et al.*, 2016). Conversion of gaseous N_2 into ammonia accessible to microorganisms and plants is the process of nitrogen fixation (N_2) and can only be used by certain microorganisms containing an enzymatic nitrogenase system (Wolinska *et al.*, 2016). The cellular energy required to drive nitrogenase activity can be provided by light-dependent as well as light-independent processes. When diazotrophs are grown in an inert atmosphere lacking dinitrogen, such as argon, H_2 production is enhanced. Much like most hydrogenases, nitrogenase is irreversibly inhibited by O_2 and sophisticated physiological mechanisms exist in microbes to separate O_2 from nitrogenase. Under anaerobic conditions, hydrogen gas is produced by nitrogenase alongside ammonia by way of the following biochemical reaction (Gu & Milton, 2020).

$$N_2 + 8e^- + 8H^+ + 16ATP \rightarrow 2NH_3 + H_2 + 16ADP + 16Pi. \tag{4.2}$$

4.7.3 Biophotolysis

Biophotolytic H_2 production is carried out by certain photoautotrophs capable of concurrently sequestering carbon as biomass. As such, biophotolysis holds the greatest promise as a means of cellularly sequestering substantial CO_2 while producing clean burning H_2. Provided the biomass was disposed of properly, for instance as agricultural biochar, this could enable a shift from net carbon neutral biofuels to net negative H_2-based biofuels (Mona *et al.*, 2021). Biophotolysis is similar to photofermentation in that both are light-driven, microbiological processes useful for generating H_2. However, in biophotolysis, an inorganic molecule, water, rather than an organic acid, serves as the primary source of electrons utilized for biological H_2 formation (Kosourov *et al.*, 2020). Biophotolysis and photofermentation are not mutually exclusive and some phototrophs, such as *Chlamydomonas reinhardtii*, can carry out both processes (Scoma *et al.*, 2014). In algae, under illuminated conditions, photosystem II-mediated water photolysis provides electrons to the P-ETC and under illuminated anaerobic conditions the green microalgal *C. reinhardtii* produces H_2 gas using primarily water-derived electrons (Scoma *et al.*, 2014). Cyanobacteria such as diazotrophic *Anabaena* spp. are capable of producing H_2, though this may generate from biophotolysis and as an end product of nitrogenase activity (Kosourov *et al.*, 2014). Hydrogen production appears to be one-way photoautotrophs shed excess electrons from the P-ETC under high light conditions or when CO_2, the usual electron acceptor, is limited (Cinco *et al.*, 1993). Downregulation of RuBisCo gene expression in the green algae *C. reinhardtii*, significantly increased H_2 production by 10-fold (Pinto *et al.*, 2013). This indicates when the normal electron acceptor is lacking or biochemically inaccessible, an alternative electron path must be provided via proton reduction to sustain metabolism. Hydrogenases catalyze the proton reduction reaction and formation of H_2 gas according to the following equation:

$$2e^- + 2H^+ \rightleftharpoons H_2 \quad (4.3)$$

Hydrogenases, in most cases, are reversible yet are usually exceptionally susceptible to irreversible oxygen-mediated inactivation (Fakhimi *et al.*, 2020). This presents something of a quandary precluding sustained biophotolytic H_2 production in wild-type algae when using H_2O as electron source. This is because oxygenic phototrophs generate O_2 as a metabolite from light-driven photolysis of water. A variety of approaches have been taken to minimize O_2 formation for sustained biophotolytic H_2 formation by algae. One method seeks to reversibly downregulate PS-II activity via sulfur limitation since this element is needed for this photosystem (Zhang *et al.*, 2002). While this approach can temporarily boost hydrogenase activity and thereby H_2 formation, it is not a long-term solution as exhaustible internal starch reserves are catabolized to supply electrons rather than water. Furthermore, sulfur limitation would tend to restrict cell growth, consistent with Liebig's law of the minimum. Since over 90% of the electrons used for proton reduction during H_2 formation may derive from water, sustained PS-II activity is an essential requirement for industrialization of this means of green hydrogen generation (Kosourov *et al.*, 2020). Co-culturing highly aerobic bacteria such as *Pseudomonas* alongside *Scenedesmus* or *Chlorella* green algae is one simple yet elegant method for quickly consuming O_2 through respiration to sustain H_2 production (Ban *et al.*, 2018). *Scenedesmus* is commonly found in wastewaters suggesting water treatment and biophotolytic H_2 could be industrialized in clear, covered tanks at treatment plants with the algae later harvested for lipid conversion to biodiesel or assimilated carbon and cellular nitrogen used as biochar and field fertilizer in agriculture for longer term carbon sequestration (Mona *et al.*, 2021). Recent research indicates biochar amended into some soil types can sequester the buried carbon for hundreds of years while conferring certain agricultural benefits to the land (Yin *et al.*, 2022).

4.8 CONCLUSION

Industrial emissions are recognized as key contributors to the enduring rise in atmospheric GHG levels and associated ecological issues, including global climate change. Carbon dioxide is the most important anthropogenic GHG because of the vast and growing amounts released from sources ranging from centralized electric power generation stations to distributed transportation in developed and developing nations (Kumar *et al.*, 2011). International political and scientific consensus has reaffirmed a commitment that CO_2 levels must be reined in to avert a worsening of the climate crisis in the 21st century (Mikhaylov *et al.*, 2020). Carbon capture and sequestration offers one attractive avenue by which the rise in CO_2 levels might be alleviated.

Point source emitters, such as fossil fuel-fired power plants and wastewater treatment plants, are foremost targets for CO_2 mitigation efforts through carbon capture and sequestration. Chemical and physical carbon capture methods ranging from chemical absorption to cryogenic separation have been or are currently being developed (Meisen & Shuai, 1997). Unfortunately, many such capture methods are prohibitively expensive and or unsustainable. Long-term carbon sequestration by pumping captured carbon into subterranean geologic formations is also costly and may not be a viable option in some locations or with all sediment types. Hybrid technologies that combine chemical, physical, and biological methods may be used to lower costs. Biological carbon capture using organisms, such as microalgae, for temporary or long-term sequestration occur naturally, are scalable, and can be used independent of chemical or physical CO_2 capture methodologies.

Biological carbon fixation offers an intriguing alternative since fast-growing autotrophs, such as cyanobacteria and microalgae, already possess effective solar-powered CO_2 fixation mechanisms, grow rapidly, and form many useful products and product precursors (Lam *et al.*, 2012). Carbon-neutral biofuels, like biodiesel made from the transesterification of extracted algal TAGs, can serve to temporarily sequester carbon up until the point of combustion. The use of such carbon neutral to near-neutral biofuels could help society to reduce the amount of CO_2 released from the combustion of petroleum and other fossil fuels as we transition to more renewable sources of energy.

Methanogenic archaea are mainly useful during waste and wastewater treatment for biogas methane production. In this regard, methanogens can help lower human dependence on fossil fuel natural gas. This is a critical consideration throughout Europe and other areas afflicted by natural gas supply disruptions. However, since methanogens can assimilate and reduce CO_2 into bioplastics, even in complete darkness, these archaea are emerging as potential living biological catalysts for cell-mediated CO_2 sequestration (Thevasundaram *et al.*, 2022).

Carbon negative and carbon-free biofuels can enable long-term carbon sequestration for an enhanced impact on CO_2 mitigation efforts. Certain microorganisms, including various algae, assimilate CO_2 and release clean-burning hydrogen gas formed through disparate biological routes (Kumar *et al.*, 2013). Some algae-derivative products, like agricultural biochar, can be employed in a carbon-negative manner to sequester biologically fixed carbon into amended soil over long periods. Future efforts should seek to optimize the carbon capture efficiency of biological processes while decreasing the cost of implementation of carbon sequestration technologies.

REFERENCES

Adeniyi O. M., Azimov U. and Burluka A. (2018). Algae biofuel: current status and future applications. *Renewable and Sustainable Energy Reviews*, **90**, 316–335. Available at: https://doi.org/10.1016/j.rser.2018.03.067

Alam M. A., Wang Z. and Yuan Z. (2017). Generation and harvesting of microalgae biomass for biofuel production. *Prospects and Challenges in Algal Biotechnology*, **1**, 89–111. Available at: https://doi.org/10.1007/978-981-10-1950-0_3

Alishah Aratboni H., Rafiei N., Garcia-Granados R., Alemzadeh A. and Morones-Ramírez J. R. (2019). Biomass and lipid induction strategies in microalgae for biofuel production and other applications. *Microbial Cell Factories*, **18**(1), 1–17. Available at: https://doi.org/10.1186/s12934-019-1228-4

Anto S., Mukherjee S. S., Muthappa R., Mathimani T., Deviram G., Kumar S. S., Verma T. N. and Pugazhendhi A. (2020). Algae as green energy reserve: technological outlook on biofuel production. *Chemosphere*, **242**, 125079. Available at: https://doi.org/10.1016/j.chemosphere.2019.125079

Aravind S., Kumar P. S., Kumar N. S. and Siddarth N. (2020). Conversion of green algal biomass into bioenergy by pyrolysis. A Review. *Environmental Chemistry Letters*, **18**(3), 829–849. https://doi.org/10.1007/s10311-020-00990-2

Arunasri K., Annie Modestra J., Yeruva D. K., Vamshi Krishna K. and Venkata Mohan S. (2016). Polarized potential and electrode materials implication on electro-fermentative di-hydrogen production: microbial assemblages and hydrogenase gene copy variation. *Bioresource Technology*, **200**, 691–698. Available at: https://doi.org/10.1016/j.biortech.2015.10.091

Asadi M. and McPhedran K. N. (2021). Greenhouse gas emission estimation from municipal wastewater using a hybrid approach of generative adversarial network and data-driven modelling. *Science of the Total Environment*, **800**, 149508. Available at: https://doi.org/10.1016/j.scitotenv.2021.149508

Aslam A., Thomas-Hall S. R., Mughal T. A. and Schenk P. M. (2017). Selection and adaptation of microalgae to growth in 100% unfiltered coal-fired flue gas. *Bioresource Technology*, **233**, 271–283. Available at: https://doi.org/10.1016/j.biortech.2017.02.111

Ayol A., Peixoto L., Keskin T. and Abubackar H. N. (2021). Reactor designs and configurations for biological and bioelectrochemical C1 gas conversion: a review. *International Journal of Environmental Research and Public Health*, **18**(21), 11683. Available at: https://doi.org/10.3390/ijerph182111683

Bahr M., Díaz I., Dominguez A., González Sánchez A. and Muñoz, R. (2013). Microalgal-biotechnology as a platform for an integral biogas upgrading and nutrient removal from anaerobic effluents. *Environmental Science & Technology*, **48**(1), 573–581. Available at: https://doi.org/10.1021/es403596m

Balat M. and Balat H. (2010). Progress in biodiesel processing. *Applied Energy*, **87**(6), 1815–1835. Available at: https://doi.org/10.1016/j.apenergy.2010.01.012

Ban S., Lin W., Wu F. and Luo J. (2018). Algal-bacterial cooperation improves algal photolysis-mediated hydrogen production. *Bioresource Technology*, **251**, 350–357. Available at: https://doi.org/10.1016/j.biortech.2017.12.072

Bari R. (2004). A novel approach for the suppression of photorespiration in C3 plants by gene transfer. Dissertation.

Baroutaji A., Wilberforce T., Ramadan M. and Olabi A. G. (2019). Comprehensive investigation on hydrogen and fuel cell technology in the aviation and aerospace sectors. *Renewable and Sustainable Energy Reviews*, **106**, 31–40. Available at: https://doi.org/10.1016/j.rser.2019.02.022

Beardall J., Ihnken S. and Quigg A. (2009). Gross and net primary production: closing the gap between concepts and measurements. *Aquatic Microbial Ecology*, **56**, 113–122. Available at: https://doi.org/10.3354/ame01305

Bharti R. K., Srivastava S. and Thakur I. S. (2014). Proteomic analysis of carbon concentrating chemolithotrophic bacteria *Serratia* sp. for sequestration of carbon dioxide. *PLoS ONE*, **9**(3), 1–10. Available at: https://doi.org/10.1371/journal.pone.0091300

Biscaia W. L., Miyawaki B., de Mello T. C., de Vasconcelos E. C., de Arruda N. M. and Maranho L. T. (2022). Biofixation of air emissions and biomass valorization—evaluation of microalgal biotechnology. *Applied Biochemistry and Biotechnology*, **194**(9), 4033–4048. Available at: https://doi.org/10.1007/s12010-022-03972-y

Blankenship R. E. (2010). Early evolution of photosynthesis. *Plant Physiology*, **154**(2), 434–438. Available at: https://doi.org/10.1104/pp.110.161687

Blankenship R. E. (2014). Molecular Mechanisms of Photosynthesis. John Wiley & Sons, Hoboken, New Jersey.

Blatti J. L., Beld J., Behnke C. A., Mendez M., Mayfield S. P. and Burkart M. D. (2012). Manipulating fatty acid biosynthesis in microalgae for biofuel through protein-protein interactions. *PLoS ONE*, **7**(9), 1–12. Available at: https://doi.org/10.1371/journal.pone.0042949

Blatti J. L., Michaud J. and Burkart M. D. (2013). Engineering fatty acid biosynthesis in microalgae for sustainable biodiesel. *Current Opinion in Chemical Biology*, **17**(3), 496–505. Available at: https://doi.org/10.1016/j.cbpa.2013.04.007

Bonaventure G., Salas J. J., Pollard M. R. and Ohlrogge J. B. (2003). Disruption of the *fatb* gene in Arabidopsis demonstrates an essential role of saturated fatty acids in plant growth. *The Plant Cell*, **15**(4), 1020–1033. Available at: https://doi.org/10.1105/tpc.008946

Bouaid A., Martinez M. and Aracil J. (2007). A comparative study of the production of ethyl esters from vegetable oils as a biodiesel fuel optimization by factorial design. *Chemical Engineering Journal*, **134**(1–3), 93–99. Available at: https://doi.org/10.1016/j.cej.2007.03.077

Bounaceur R., Lape N., Roizard D., Vallieres C. and Favre E. (2006). Membrane processes for post-combustion carbon dioxide capture: a parametric study. *Energy*, **31**(14), 2556–2570. Available at: https://doi.org/10.1016/j.energy.2005.10.038

Brand S., Hardi F., Kim J. and Suh D. J. (2014). Effect of heating rate on biomass liquefaction: differences between subcritical water and supercritical ethanol. *Energy*, **68**, 420–427. Available at: https://doi.org/10.1016/j.energy.2014.02.086

Brennan L. and Owende P. (2010). Biofuels from microalgae—a review of technologies for production, processing, and extractions of biofuels and co-products. *Renewable and Sustainable Energy Reviews*, **14**(2), 557–577. Available at: https://doi.org/10.1016/j.rser.2009.10.009

Bringer S., Hurtner T., Poralla K. and Sahm H. (1985). Influence of ethanol on the hopanoid content and the fatty acid pattern in batch and continuous cultures of *Zymomonas mobilis*. *Archives of Microbiology*, **140**(4), 312–316. Available at: https://doi.org/10.1007/bf00446969

Brown A. C., Knights B. A. and Conway E. (1969). Hydrocarbon content and its relationship to physiological state in the green alga *Botryococcus braunii*. *Phytochemistry*, **8**(3), 543–547. Available at: https://doi.org/10.1016/s0031-9422(00)85397-2

Brunetti A., Scura F., Barbieri G. and Drioli, E. (2010). Membrane technologies for CO_2 separation. *Journal of Membrane Science*, **359**(1–2), 115–125. Available at: https://doi.org/10.1016/j.memsci.2009.11.040

Carvalho A. P., Meireles L. A. and Malcata F. X. (2006). Microalgal reactors: A review of enclosed system designs and performances. *Biotechnology Progress*, **22**(6), 1490–1506. Available at: https://doi.org/10.1002/bp060065r

Chacko AR., Amster D. E., Johnson T. E., Newman S. R., Gladchuk A. V., Sohn C. J., Prunkard D. E., Yakelis N. A. and Freeman J. O. (2019). High-throughput screen for sorting cells capable of producing the biofuel feedstock botryococcene. *Organic & Biomolecular Chemistry*, **17**(12), 3195–3201. Available at: https://doi.org/10.1039/c8ob02589d

Chatsungnoen T. and Chisti Y. (2016). Optimization of oil extraction from *Nannochloropsis salina* biomass paste. *Algal Research*, **15**, 100–109. Available at: https://doi.org/10.1016/j.algal.2016.02.015

Chaudhary R., Dikshit A. K. and Tong Y. W. (2017). Carbon-dioxide biofixation and phycoremediation of municipal wastewater using *Chlorella vulgaris* and *Scenedesmus obliquus*. *Environmental Science and Pollution Research*, **25**(21), 20399–20406. Available at: https://doi.org/10.1007/s11356-017-9575-3

Cheah W. Y., Show P. L., Chang J.-S., Ling T. C. and Juan J. C. (2015). Biosequestration of atmospheric CO_2 and flue gas-containing CO_2 by microalgae. *Bioresource Technology*, **184**, 190–201. Available at: https://doi.org/10.1016/j.biortech.2014.11.026

Chen J. E. and Smith A. G. (2012). A look at diacylglycerol acyltransferases (DGATs) in algae. *Journal of Biotechnology*, **162**(1), 28–39. Available at: https://doi.org/10.1016/j.jbiotec.2012.05.009

Chen W.-H., Lin B.-J., Huang M.-Y. and Chang J.-S. (2015). Thermochemical conversion of microalgal biomass into biofuels: a review. *Bioresource Technology*, **184**, 314–327. Available at: https://doi.org/10.1016/j.biortech.2014.11.050

Chen R., Li Z., Feng J., Zhao L. and Yu J. (2020). Effects of digestate recirculation ratios on biogas production and methane yield of continuous dry anaerobic digestion. *Bioresource Technology*, **316**, 123963. https://doi.org/10.1016/j.biortech.2020.123963

Cheng F., Cui Z., Chen L., Jarvis J., Paz N., Schaub T., Nirmalakhandan N. and Brewer C. E. (2017). Hydrothermal liquefaction of high- and low-lipid algae: bio-crude oil chemistry. *Applied Energy*, **206**, 278–292. Available at: https://doi.org/10.1016/j.apenergy.2017.08.105

Chew K. W., Chia S. R., Show P. L., Yap Y. J., Ling T. C. and Chang J.-S. (2018). Effects of water culture medium, cultivation systems and growth modes for microalgae cultivation: a review. *Journal of the Taiwan Institute of Chemical Engineers*, **91**, 332–344. Available at: https://doi.org/10.1016/j.jtice.2018.05.039

Chiaramonti D., Prussi M., Buffi M., Rizzo A. M. and Pari L. (2017). Review and experimental study on pyrolysis and hydrothermal liquefaction of microalgae for biofuel production. *Applied Energy*, **185**, 963–972. Available at: https://doi.org/10.1016/j.apenergy.2015.12.001

Christenson L. and Sims R. (2011). Production and harvesting of microalgae for wastewater treatment, biofuels, and bioproducts. *Biotechnology Advances*, **29**(6), 686–702. Available at: https://doi.org/10.1016/j.biotechadv.2011.05.015

Cinco R. M., MacInnis J. M. and Greenbaum E. (1993). The role of carbon dioxide in light-activated hydrogen production by *Chlamydomonas reinhardtii*. *Photosynthesis Research*, **38**(1), 27–33. Available at: https://doi.org/10.1007/bf00015058

Constantopoulos G. (1970). Lipid metabolism of manganese-deficient algae. *Plant Physiology*, **45**(1), 76–80. Available at: https://doi.org/10.1104/pp.45.1.76

Czernik S. and Bridgwater A. V. (2004). Overview of applications of biomass fast pyrolysis oil. *Energy & Fuels*, **18**(2), 590–598. Available at: https://doi.org/10.1021/ef034067u

Danquah M. K., Ang L., Uduman N., Moheimani N. and Forde G. M. (2009). Dewatering of Microalgal culture for biodiesel production: Exploring polymer flocculation and tangential flow filtration. *Journal of Chemical Technology and Biotechnology*, **84**(7), 1078–1083. https://doi.org/10.1002/jctb.2137

Davis M. S. and Cronan J. E. (2001). Inhibition of *Escherichia coli* acetyl coenzyme A carboxylase by acyl-acyl carrier protein. *Journal of Bacteriology*, **183**(4), 1499–1503. Available at: https://doi.org/10.1128/jb.183.4.1499-1503.2001

Demirbaş A. (2008). Production of biodiesel from algae oils. *Energy Sources, Part A: Recovery, Utilization, and Environmental Effects*, **31**(2), 163–168. Available at: https://doi.org/10.1080/15567030701521775

Demirbas A. (2009). Progress and recent trends in biodiesel fuels. *Energy Conversion and Management*, **50**(1), 14–34. Available at: https://doi.org/10.1016/j.enconman.2008.09.001

Demirbas A. (2010). Use of algae as biofuel sources. *Energy Conversion and Management*, **51**(12), 2738–2749. Available at: https://doi.org/10.1016/j.enconman.2010.06.010

Dharmaprabhakaran T., Karthikeyan S., Periyasamy M. and Mahendran G. (2020). Algal biodiesel-promising source to power CI engines. *Materials Today: Proceedings*, **33**, 2870–2873. Available at: https://doi.org/10.1016/j.matpr.2020.02.775

Doney S. C., Busch D. S., Cooley S. R. and Kroeker K. J. (2020). The impacts of ocean acidification on marine ecosystems and reliant human communities. *Annual Review of Environment and Resources*, **45**(1), 83–112. Available at: https://doi.org/10.1146/annurev-environ-012320-083019

Dote Y., Sawayama S., Inoue S., Minowa T. and Yokoyama S. (1994). Recovery of liquid fuel from hydrocarbon-rich microalgae by thermochemical liquefaction. *Fuel*, **73**(12), 1855–1857. Available at: https://doi.org/10.1016/0016-2361(94)90211-9

Duan P. and Savage P. E. (2011). Upgrading of crude algal bio-oil in supercritical water. *Bioresource Technology*, **102**(2), 1899–1906. Available at: https://doi.org/10.1016/j.biortech.2010.08.013

Erga O., Juliussen O. and Lidal H. (1995). Carbon dioxide recovery by means of aqueous amines. *Energy Conversion and Management*, **36**(6–9), 387–392. Available at: https://doi.org/10.1016/0196-8904(95)00027-b

Eze V. C., Velasquez-Orta S. B., Hernández-García A., Monje-Ramírez I. and Orta-Ledesma M. T. (2018). Kinetic modelling of microalgae cultivation for wastewater treatment and carbon dioxide sequestration. *Algal Research*, **32**, 131–141. Available at: https://doi.org/10.1016/j.algal.2018.03.015

Fakhimi N., Gonzalez-Ballester D., Fernández E., Galván A. and Dubini A. (2020). Algae-bacteria consortia as a strategy to enhance H_2 production. *Cells*, **9**(6), 1353. Available at: https://doi.org/10.3390/cells9061353

Fan J., Andre C. and Xu C. (2011). A chloroplast pathway for the de novo biosynthesis of triacylglycerol in *Chlamydomonas reinhardtii*. *FEBS Letters*, **585**(12), 1985–1991. Available at: https://doi.org/10.1016/j.febslet.2011.05.018

Ferry J. G. (1997). Enzymology of the fermentation of acetate to methane by *Methanosarcina thermophila*. *BioFactors*, **6**(1), 25–35. Available at: https://doi.org/10.1002/biof.5520060104

Field C. B., Behrenfeld M. J., Randerson J. T. and Falkowski P. (1998). Primary production of the biosphere: integrating terrestrial and oceanic components. *Science*, **281**(5374), 237–240. Available at: https://doi.org/10.1126/science.281.5374.237

Formighieri C. and Melis A. (2014). Regulation of β-phellandrene synthase gene expression, recombinant protein accumulation, and monoterpene hydrocarbons production in *Synechocystis* transformants. *Planta*, **240**(2), 309–324. Available at: https://doi.org/10.1007/s00425-014-2080-8

Formighieri C. and Melis A. (2015). A phycocyanin·phellandrene synthase fusion enhances recombinant protein expression and β-phellandrene (monoterpene) hydrocarbons production in *Synechocystis* (cyanobacteria). *Metabolic Engineering*, **32**, 116–124. Available at: https://doi.org/10.1016/j.ymben.2015.09.010

Fowler G. J. S., Hess S., Pullerits T., Sundström V. and Hunter C. N. (1997). The role of βArg_{-10} in the B800 bacteriochlorophyll and carotenoid pigment environment within the light-harvesting LH2 complex of *Rhodobacter sphaeroides*. *Biochemistry*, **36**(37), 11282–11291. Available at: https://doi.org/10.1021/bi9626315

Fukuda, H., Kondo, A. and Noda, H. (2001). Biodiesel fuel production by transesterification of oils. *Journal of Bioscience and Bioengineering*, **92**(5), 405–416. https://doi.org/10.1016/s1389-1723(01)80288-7

Furbank R. T. and Taylor W. C. (1995). Regulation of photosynthesis in C3 and C4 plants: a molecular approach. *The Plant Cell*, **7**(7), 797–807. Available at: https://doi.org/10.1105/tpc.7.7.797

Galvin R. (2020). Yes, there is enough money to decarbonize the economies of high-income countries justly and sustainably. *Energy Research & Social Science*, **70**, 101739. Available at: https://doi.org/10.1016/j.erss.2020.101739

Geider R. J. and Osborne B. A. (1992). Algal Photosynthesis. Chapman & Hall, New York, USA.

Gijzen H. J. (2002). Anaerobic digestion for sustainable development: a natural approach. *Water Science and Technology*, **45**(10), 321–328. Available at: https://doi.org/10.2166/wst.2002.0364

Giordano M., Beardall J. and Raven J. A. (2005). CO_2 concentrating mechanisms in algae: mechanisms, environmental modulation, and evolution. *Annual Review of Plant Biology*, **56**(1), 99–131. Available at: https://doi.org/10.1146/annurev.arplant.56.032604.144052

Goldet G., Wait A. F., Cracknell J. A., Vincent K. A., Ludwig M., Lenz O., Friedrich B. and Armstrong F. A. (2008). Hydrogen production under aerobic conditions by membrane-bound hydrogenases from *Ralstonia* species. *Journal of the American Chemical Society*, **130**(33), 11106–11113. Available at: https://doi.org/10.1021/ja8027668

Goncalves E. C., Wilkie A. C., Kirst M. and Rathinasabapathi B. (2016). Metabolic regulation of triacylglycerol accumulation in the green algae: identification of potential targets for engineering to improve oil yield. *Plant Biotechnology Journal*, **14**(8), 1649–1660. Available at: https://doi.org/10.1111/pbi.12523

Govindjee Shevela D. and Björn L. O. (2017). Evolution of the Z-scheme of photosynthesis: a perspective. *Photosynthesis Research*, **133**(1–3), 5–15. Available at: https://doi.org/10.1007/s11120-016-0333-z

Green F. B., Lundquist T. J. and Oswald W. J. (1995). Energetics of advanced integrated wastewater pond systems. *Water Science and Technology*, **31**(12), 9–20. Available at: https://doi.org/10.2166/wst.1995.0448

Greenwell H. C., Laurens L. M., Shields R. J., Lovitt R. W. and Flynn K. J. (2009). Placing microalgae on the biofuels priority list: a review of the technological challenges. *Journal of the Royal Society Interface*, **7**(46), 703–726. Available at: https://doi.org/10.1098/rsif.2009.0322

Grima E. M., González M. J. and Giménez A. G. (2012). Solvent extraction for microalgae lipids. *Algae for Biofuels and Energy*, **1**, 187–205. Available at: https://doi.org/10.1007/978-94-007-5479-9_11

Gu W. and Milton R. D. (2020). Natural and engineered electron transfer of nitrogenase. *Chemistry*, **2**(2), 322–346. Available at: https://doi.org/10.3390/chemistry2020021

Guo M. (2020). The global scenario of biofuel production and development. *Green Energy and Technology*, **1**, 29–56. Available at: https://doi.org/10.1007/978-81-322-3965-9_3

Guo J., Peng Y., Ni B.-J., Han X., Fan L. and Yuan Z. (2015). Dissecting microbial community structure and methane-producing pathways of a full-scale anaerobic reactor digesting activated sludge from wastewater treatment by metagenomic sequencing. *Microbial Cell Factories*, **14**(1), 1–11. Available at: https://doi.org/10.1186/s12934-015-0218-4

Han F., Huang J., Li Y., Wang W., Wan M., Shen G. and Wang J. (2013). Enhanced lipid productivity of *Chlorella pyrenoidosa* through the culture strategy of semi-continuous cultivation with nitrogen limitation and pH control by CO_2. *Bioresource Technology*, **136**, 418–424. Available at: https://doi.org/10.1016/j.biortech.2013.03.017

Hanifa M., Agarwal R., Sharma U., Thapliyal P. C. and Singh L. P. (2023). A review on CO_2 capture and sequestration in the construction industry: emerging approaches and commercialised technologies. *Journal of CO2 Utilization*, **67**, 102292. Available at: https://doi.org/10.1016/j.jcou.2022.102292

Hanotu J., Bandulasena H. C. H. and Zimmerman W. B. (2012). Microflotation performance for algal separation. *Biotechnology and Bioengineering*, **109**(7), 1663–1673. Available at: https://doi.org/10.1002/bit.24449

Harwood J. L. and Guschina I. A. (2009). The versatility of algae and their lipid metabolism. *Biochimie*, **91**(6), 679–684. Available at: https://doi.org/10.1016/j.biochi.2008.11.004

Hazrat M. A., Rasul M. G. and Khan M. M. K. (2015). A study on thermo-catalytic degradation for production of clean transport fuel and reducing plastic wastes. *Procedia Engineering*, **105**, 865–876. Available at: https://doi.org/10.1016/j.proeng.2015.05.108

He S., Zhao M., Wang J., Cheng Z., Yan B. and Chen G. (2020). Hydrothermal liquefaction of low-lipid algae *Nannochloropsis* sp. and *Sargassum* sp.: effect of feedstock composition and temperature. *Science of the Total Environment*, **712**, 135677. Available at: https://doi.org/10.1016/j.scitotenv.2019.135677

Hermans M. A., Neuss B. and Sahm H. (1991). Content and composition of hopanoids in *Zymomonas mobilis* under various growth conditions. *Journal of Bacteriology*, **173**(17), 5592–5595. Available at: https://doi.org/10.1128/jb.173.17.5592-5595.1991

Hu B., Min M., Zhou W., Li Y., Mohr M., Cheng Y., Lei H., Liu Y., Lin X., Chen P. and Ruan R. (2012). Influence of exogenous CO_2 on biomass and lipid accumulation of microalgae *Auxenochlorella protothecoides* cultivated in concentrated municipal wastewater. *Applied Biochemistry and Biotechnology*, **166**(7), 1661–1673. Available at: https://doi.org/10.1007/s12010-012-9566-2

Huang W., Yang Y.-J., Hu H. and Zhang S.-B. (2015). Different roles of cyclic electron flow around photosystem I under sub-saturating and saturating light intensities in tobacco leaves. *Frontiers in Plant Science*, **6**, 1–9. Available at: https://doi.org/10.3389/fpls.2015.00923

Huber G. W., Iborra S. and Corma A. (2006). Synthesis of transportation fuels from biomass: chemistry, catalysts, and engineering. *Chemical Reviews*, **106**(9), 4044–4098. Available at: https://doi.org/10.1021/cr068360d

Ingram L. O., Van Baalen C. and Calder J. A. (1973). Role of reduced exogenous organic compounds in the physiology of the blue-green bacteria (algae): photoheterotrophic growth of an 'autotrophic' blue-green bacterium. *Journal of Bacteriology*, **114**(2), 701–705. Available at: https://doi.org/10.1128/jb.114.2.701-705.1973

Ishibashi M. (1996). Technology for removing carbon dioxide from power plant flue gas by the physical adsorption method. *Fuel and Energy Abstracts*, **37**(3), 229. Available at: https://doi.org/10.1016/0140-6701(96)89243-5

Jena U. and Das K. C. (2011). Comparative evaluation of thermochemical liquefaction and pyrolysis for bio-oil production from microalgae. *Energy & Fuels*, **25**(11), 5472–5482. Available at: https://doi.org/10.1021/ef201373m

Jena U., Das K. C. and Kastner J. R. (2012). Comparison of the effects of Na_2CO_3, $Ca_3(PO_4)_2$, and NIO catalysts on the thermochemical liquefaction of microalga *Spirulina platensis*. *Applied Energy*, **98**, 368–375. Available at: https://doi.org/10.1016/j.apenergy.2012.03.056

John R. P., Anisha G. S., Nampoothiri K. M. and Pandey A. (2011). Micro and macroalgal biomass: a renewable source for bioethanol. *Bioresource Technology*, **102**(1), 186–193. Available at: https://doi.org/10.1016/j.biortech.2010.06.139

Johnson L. A. (2008). Oil recovery from soybeans. *Soybeans*, 331–375. Available at: https://doi.org/10.1016/b978-1-893997-64-6.50014-7

Kadir W. N., Lam M. K., Uemura Y., Lim J. W. and Lee K. T. (2018). Harvesting and pre-treatment of microalgae cultivated in wastewater for biodiesel production: a review. *Energy Conversion and Management*, **171**, 1416–1429. Available at: https://doi.org/10.1016/j.enconman.2018.06.074

Kamennaya N., Ajo-Franklin C., Northen T. and Jansson C. (2012). Cyanobacteria as biocatalysts for carbonate mineralization. *Minerals*, **2**(4), 338–364. Available at: https://doi.org/10.3390/min2040338

Khan S., Siddique R., Sajjad W., Nabi G., Hayat K. M., Duan P. and Yao L. (2017). Biodiesel production from algae to overcome the energy crisis. *HAYATI Journal of Biosciences*, **24**(4), 163–167. Available at: https://doi.org/10.1016/j.hjb.2017.10.003

Khatri W., Hendrix R., Niehaus T., Chappell J. and Curtis W. R. (2013). Hydrocarbon production in high density *Botryococcus braunii* race B continuous culture. *Biotechnology and Bioengineering*, **111**(3), 493–503. Available at: https://doi.org/10.1002/bit.25126

Khozin-Goldberg I. and Cohen Z. (2011). Unraveling algal lipid metabolism: recent advances in gene identification. *Biochimie*, **93**(1), 91–100. Available at: https://doi.org/10.1016/j.biochi.2010.07.020

Kosourov S., Leino H., Murukesan G., Lynch F., Sivonen K., Tsygankov A. A., Aro E.-M. and Allahverdiyeva Y. (2014). Hydrogen photoproduction by immobilized N_2 -fixing cyanobacteria: understanding the role of the uptake hydrogenase in the long-term process. *Applied and Environmental Microbiology*, **80**(18), 5807–5817. Available at: https://doi.org/10.1128/aem.01776-14

Kosourov S., Nagy V., Shevela D., Jokel M., Messinger J. and Allahverdiyeva Y. (2020). Water oxidation by photosystem II is the primary source of electrons for sustained H_2 photoproduction in nutrient-replete green algae. *Proceedings of the National Academy of Sciences*, **117**(47), 29629–29636. Available at: https://doi.org/10.1073/pnas.2009210117

Kumar K., Dasgupta C. N., Nayak B., Lindblad P. and Das D. (2011). Development of suitable photobioreactors for CO_2 sequestration addressing global warming using green algae and cyanobacteria. *Bioresource Technology*, **102**(8), 4945–4953. Available at: https://doi.org/10.1016/j.biortech.2011.01.054

Kumar K., Roy S. and Das D. (2013). Continuous mode of carbon dioxide sequestration by *C. Sorokiniana* and subsequent use of its biomass for hydrogen production by *E. cloacae* IIT-BT 08. *Bioresource Technology*, **145**, 116–122. Available at: https://doi.org/10.1016/j.biortech.2013.01.137

Kumar M., Sundaram S., Gnansounou E., Larroche C. and Thakur I. S. (2018). Carbon dioxide capture, storage and production of biofuel and biomaterials by bacteria: a review. *Bioresource Technology*, **247**, 1059–1068. Available at: https://doi.org/10.1016/j.biortech.2017.09.050

Laamanen C. A., Ross G. M. and Scott J. A. (2016). Flotation harvesting of microalgae. *Renewable and Sustainable Energy Reviews*, **58**, 75–86. Available at: https://doi.org/10.1016/j.rser.2015.12.293

Lakshmikandan M., Murugesan A. G., Wang S., Abomohra A. E.-F., Jovita P. A. and Kiruthiga S. (2020). Sustainable biomass production under CO_2 conditions and effective wet microalgae lipid extraction for biodiesel production. *Journal of Cleaner Production*, **247**, 119398. Available at: https://doi.org/10.1016/j.jclepro.2019.119398

Lam M. K., Lee K. T. and Mohamed A. R. (2012). Current status and challenges on microalgae-based carbon capture. *International Journal of Greenhouse Gas Control*, **10**, 456–469. Available at: https://doi.org/10.1016/j.ijggc.2012.07.010

Lamont C. M. and Sargent F. (2016). Design and characterisation of synthetic operons for biohydrogen technology. *Archives of Microbiology*, **199**(3), 495–503. Available at: https://doi.org/10.1007/s00203-016-1322-5

Lee O. K. and Lee E. Y. (2016). Sustainable production of bioethanol from renewable brown algae biomass. *Biomass and Bioenergy*, **92**, 70–75. Available at: https://doi.org/10.1016/j.biombioe.2016.03.038

Lee A. K., Lewis D. M. and Ashman P. J. (2012). Disruption of microalgal cells for the extraction of lipids for biofuels: processes and specific energy requirements. *Biomass and Bioenergy*, **46**, 89–101. Available at: https://doi.org/10.1016/j.biombioe.2012.06.034

Lee J. Y., Park J. H. and Park H. D. (2017). Effects of an applied voltage on direct interspecies electron transfer via conductive materials for methane production. *Waste Management*, **68**, 165–172. https://doi.org/10.1016/j.wasman.2017.07.025

Leite G. B., Abdelaziz A. E. M. and Hallenbeck P. C. (2013). Algal biofuels: challenges and opportunities. *Bioresource Technology*, **145**, 134–141. Available at: https://doi.org/10.1016/j.biortech.2013.02.007

Lemaire O. N., Jespersen M. and Wagner T. (2020). CO_2-fixation strategies in energy extremophiles: what can we learn from acetogens? *Frontiers in Microbiology*, **11**, 1–8. Available at: https://doi.org/10.3389/fmicb.2020.00486

Li F., Srivatsa S. C., Batchelor, W. and Bhattacharya S. (2017). A study on growth and pyrolysis characteristics of microalgae using thermogravimetric analysis-infrared spectroscopy and synchrotron Fourier transform infrared spectroscopy. *Bioresource Technology*, **229**, 1–10. Available at: https://doi.org/10.1016/j.biortech.2017.01.005

Li F., Srivatsa S. C. and Bhattacharya S. (2019). A review on catalytic pyrolysis of microalgae to high-quality bio-oil with low oxygeneous and nitrogenous compounds. *Renewable and Sustainable Energy Reviews*, **108**, 481–497. Available at: https://doi.org/10.1016/j.rser.2019.03.026

Li P., Hu Z., Yin Q. and Song C. (2023). Improving the growth of *Spirulina* in CO_2 absorption and microalgae conversion (CAMC) system through mixotrophic cultivation: reveal of metabolomics. *Science of the Total Environment*, **858**, 159920. Available at: https://doi.org/10.1016/j.scitotenv.2022.159920

Liao Y., Bokhary A., Maleki E. and Liao B. (2018). A review of membrane fouling and its control in algal-related membrane processes. *Bioresource Technology*, **264**, 343–358. https://doi.org/10.1016/j.biortech.2018.06.102

Li Beisson Y., Beisson F. and Riekhof W. (2015). Metabolism of acyl-lipids in *Chlamydomonas reinhardtii*. *The Plant Journal*, **82**(3), 504–522. Available at: https://doi.org/10.1111/tpj.12787

Liu Z. and Zhang F.-S. (2008). Effects of various solvents on the liquefaction of biomass to produce fuels and chemical feedstocks. *Energy Conversion and Management*, **49**(12), 3498–3504. Available at: https://doi.org/10.1016/j.enconman.2008.08.009

Liu Y., Bisson T. M., Yang, H. and Xu Z. (2010). Recent developments in novel sorbents for flue gas clean up. *Fuel Processing Technology*, **91**(10), 1175–1197. https://doi.org/10.1016/j.fuproc.2010.04.015

Liu C.-Z., Zheng S., Xu L., Wang F. and Guo C. (2013). Algal oil extraction from wet biomass of *Botryococcus braunii* by 1,2-dimethoxyethane. *Applied Energy*, **102**, 971–974. Available at: https://doi.org/10.1016/j.apenergy.2012.08.016

Liu Y., Tu Q., Knothe G. and Lu M. (2017). Direct transesterification of spent coffee grounds for biodiesel production. *Fuel*, **199**, 157–161. https://doi.org/10.1016/j.fuel.2017.02.094

Liu T., Lim Z. K., Chen H., Wang Z., Hu S., Yuan Z. and Guo J. (2020). Biogas-driven complete nitrogen removal from wastewater generated in side-stream partial nitritation. *Science of the Total Environment*, **745**, 141153. Available at: https://doi.org/10.1016/j.scitotenv.2020.141153

Logan B. E., Rossi R., Ragab A. and Saikaly P. E. (2019). Electroactive microorganisms in bioelectrochemical systems. *Nature Reviews Microbiology*, **17**(5), 307–319. https://doi.org/10.1038/s41579-019-0173-x

Lyu Z., Jain R., Smith P., Fetchko T., Yan Y. and Whitman W. B. (2016). Engineering the autotroph *Methanococcus maripaludis* for geraniol production. *ACS Synthetic Biology*, **5**(7), 577–581. Available at: https://doi.org/10.1021/acssynbio.5b00267

Maeda K., Owada M., Kimura N., Omata K. and Karube I. (1995). CO_2 Fixation from the flue gas on coal-fired thermal power plant by microalgae. *Energy Conversion and Management*, **36**(6–9), 717–720. Available at: https://doi.org/10.1016/0196-8904(95)00105-m

Mata T. M., Martins A. A. and Caetano N. S. (2010). Microalgae for biodiesel production and other applications: a review. *Renewable and Sustainable Energy Reviews*, **14**(1), 217–232. Available at: https://doi.org/10.1016/j.rser.2009.07.020

Meisen A. and Shuai X. (1997). Research and development issues in CO_2 capture. *Energy Conversion and Management*, **38**, 37–42. Available at: https://doi.org/10.1016/s0196-8904(96)00242-7

Mendow G., Veizaga N. S., Sánchez B. S. and Querini C. A. (2011). Biodiesel production by two-stage transesterification with ethanol. *Bioresource Technology*, **102**(22), 10407–10413. Available at: https://doi.org/10.1016/j.biortech.2011.08.052

Mengel M., Nauels A., Rogelj J. and Schleussner C. F. (2018). Committed sea-level rise under the Paris agreement and the legacy of delayed mitigation action. *Nature Communications*, **9**(1), 1–10. Available at: https://doi.org/10.1038/s41467-018-02985-8

Messner P. and Sleytr U. B. (1992). Crystalline bacterial cell-surface layers. *Advances in Microbial Physiology Volume* **33**, 213–275. Available at: https://doi.org/10.1016/s0065-2911(08)60218-0

Mikhaylov A., Moiseev N., Aleshin K. and Burkhardt T. (2020). Global climate change and greenhouse effect. *Entrepreneurship and Sustainability Issues*, **7**(4), 2897–2913. Available at: https://doi.org/10.9770/jesi.2020.7.4(21)

Mohapatra M., Manu S., Dash S. P. and Rastogi G. (2022). Seagrasses and local environment control the bacterial community structure and carbon substrate utilization in brackish sediments. *Journal of Environmental Management*, **314**, 115013. Available at: https://doi.org/10.1016/j.jenvman.2022.115013

Molazadeh M., Ahmadzadeh H., Pourianfar H. R., Lyon S. and Rampelotto, P. H. (2019). The use of microalgae for coupling wastewater treatment with CO_2 biofixation. *Frontiers in Bioengineering and Biotechnology*, **7**, 1–12. Available at: https://doi.org/10.3389/fbioe.2019.00042

Möllers K. B., Cannella D., Jørgensen H. and Frigaard N.-U. (2014). Cyanobacterial biomass as carbohydrate and nutrient feedstock for bioethanol production by yeast fermentation. *Biotechnology for Biofuels* **7**(1), 1–11. https://doi.org/10.1186/1754-6834-7-64

Mona S., Malyan S. K., Sain, N., Deepak B., Pugazhendhi A. and Kumar S. S. (2021). Towards sustainable agriculture with carbon sequestration, and greenhouse gas mitigation using algal biochar. *Chemosphere*, **275**, 129856. Available at: https://doi.org/10.1016/j.chemosphere.2021.129856

Mondal M., Goswami S., Ghosh A., Oinam G., Tiwari O. N., Das P., Gayen K., Mandal M. K. and Halder G. N. (2017). Production of biodiesel from microalgae through biological carbon capture: a review. *3 Biotech*, **7**(2), 1–21. Available at: https://doi.org/10.1007/s13205-017-0727-4

Moreira D. and Pires J. C. M. (2016). Atmospheric CO_2 capture by algae: negative carbon dioxide emission path. *Bioresource Technology*, **215**, 371–379. Available at: https://doi.org/10.1016/j.biortech.2016.03.060

Müller V. (2003). Energy conservation in acetogenic bacteria. Applied and Environmental. *Microbiology*, **69**(11), 6345–6353. https://doi.org/10.1128/aem.69.11.6345-6353.2003

Naddaf M. (2023). Carbon capture nets 2 billion tonnes of CO_2 each year – but it's not enough. *Nature*, **1**, 1. Available at: https://doi.org/10.1038/d41586-023-00180-4

Nagarajan D., Lee D.-J. and Chang J.-S. (2019). Integration of anaerobic digestion and microalgal cultivation for digestate bioremediation and biogas upgrading. *Bioresource Technology*, **290**, 121804. Available at: https://doi.org/10.1016/j.biortech.2019.121804

Naik S. N., Goud V. V., Rout P. K. and Dalai A. K. (2010). Production of first and second generation biofuels: a comprehensive review. *Renewable and Sustainable Energy Reviews*, **14**(2), 578–597. Available at: https://doi.org/10.1016/j.rser.2009.10.003

Narala R. R., Garg S., Sharma K. K., Thomas-Hall S. R., Deme M., Li Y. and Schenk P. M. (2016). Comparison of microalgae cultivation in photobioreactor, open raceway pond, and a two-stage hybrid system. *Frontiers in Energy Research*, **4**, 1–10. Available at: https://doi.org/10.3389/fenrg.2016.00029

Nguyen M. and Hoang A. (2016). A review on microalgae and cyanobacteria in biofuel production. *Economics and Finance*, 1–36. https://enpc.hal.science/hal-01383026/document Pages

Nguyen V. H., Klai N., Nguyen T. D. and Tyagi R. D. (2016). Impact of extraction methods on bio-flocculants recovered from backwashed sludge of bio-filtration unit. *Journal of Environmental Management*, **180**, 344–350. Available at: https://doi.org/10.1016/j.jenvman.2016.05.051

Niehaus T. D., Okada S., Devarenne T. P., Watt D. S., Sviripa V. and Chappell, J. (2011). Identification of unique mechanisms for triterpene biosynthesis in *Botryococcus braunii*. *Proceedings of the National Academy of Sciences*, **108**(30), 12260–12265. Available at: https://doi.org/10.1073/pnas.1106222108

Oasmaa A. and Czernik S. (1999). Fuel oil quality of biomass pyrolysis oils state of the art for the end users. *Energy & Fuels*, **13**(4), 914–921. Available at: https://doi.org/10.1021/ef980272b

Ogi T., Minowa T., Dote Y. and Yokoyama, S.-Y. (1994). Characterization of oil produced by the direct liquefaction of Japanese oak in an aqueous 2-propanol solvent system. *Biomass and Bioenergy*, **7**(1–6), 193–199. Available at: https://doi.org/10.1016/0961-9534(94)00059-3

Pal D., Khozin-Goldberg I., Cohen Z. and Boussiba S. (2011). The effect of light, salinity, and nitrogen availability on lipid production by *Nannochloropsis* sp... *Applied Microbiology and Biotechnology*, **90**(4), 1429–1441. Available at: https://doi.org/10.1007/s00253-011-3170-1

Peperzak L. (2003). Phytoplankton sinking rates in the Rhine region of freshwater influence. *Journal of Plankton Research*, **25**(4), 365–383. Available at: https://doi.org/10.1093/plankt/25.4.365

Perrut M. (2000). Supercritical fluid applications: industrial developments and economic issues. *Industrial & Engineering Chemistry Research*, **39**(12), 4531–4535. Available at: https://doi.org/10.1021/ie000211c

Peterson A. A., Vogel F., Lachance R. P., Fröling M., Antal Jr., M. J. and Tester J. W. (2008). Thermochemical biofuel production in hydrothermal media: a review of sub- and supercritical water technologies. *Energy & Environmental Science*, **1**(1), 32. Available at: https://doi.org/10.1039/b810100k

Pinto T. S., Malcata F. X., Arrabaça J. D., Silva J. M., Spreitzer R. J. and Esquível M. G. (2013). Rubisco mutants of *Chlamydomonas reinhardtii* enhance photosynthetic hydrogen production. *Applied Microbiology and Biotechnology*, **97**(12), 5635–5643. Available at: https://doi.org/10.1007/s00253-013-4920-z

Pires J. C. M., Martins F. G., Alvim-Ferraz M. C. M. and Simões, M. (2011). Recent developments on carbon capture and storage: an overview. *Chemical Engineering Research and Design*, **89**(9), 1446–1460. Available at: https://doi.org/10.1016/j.cherd.2011.01.028

Pisciotta J. and Blessing S. (2022). Microbial bioelectricity generation and product electrosynthesis. In: Industrial Microbiology and Biotechnology, P. Verma (ed.), Springer, Singapore, pp. 505–554.

Post-Beittenmiller D., Jaworski J. G. and Ohlrogge J. B. (1991). In vivo pools of free and acylated acyl carrier proteins in spinach evidence for sites of regulation of fatty acid biosynthesis. *Journal of Biological Chemistry*, **266**(3), 1858–1865. Available at: https://doi.org/10.1016/s0021-9258(18)52372-3

Post-Beittenmiller D., Roughan G. and Ohlrogge J. B. (1992). Regulation of plant fatty acid biosynthesis. *Plant Physiology*, **100**(2), 923–930. Available at: https://doi.org/10.1104/pp.100.2.923

Pragya N., Pandey K. K. and Sahoo P. K. (2013). A review on harvesting, oil extraction and biofuels production technologies from microalgae. *Renewable and Sustainable Energy Reviews*, **24**, 159–171. Available at: https://doi.org/10.1016/j.rser.2013.03.034

Prasad R., Gupta S. K., Shabnam N., Oliveira C. Y., Nema A. K., Ansari F. A. and Bux F. (2021). Role of microalgae in global CO_2 sequestration: physiological mechanism, recent development, challenges, and future prospective. *Sustainability*, **13**(23), 13061. Available at: https://doi.org/10.3390/su132313061

Qi F., Pei H., Mu R., Ma G., Wu D. and Han Q. (2019). Characterization and optimization of endogenous lipid accumulation in *Chlorella vulgaris* SDEC-3m ability to rapidly accumulate lipid for reversing nightly lipid loss. *Biotechnology for Biofuels*, **12**(151), 1–11. Available at: https://doi.org/10.1186/s13068-019-1493-9

Quintana N., Van der Kooy F., Van de Rhee M. D., Voshol G. P. and Verpoorte, R. (2011). Renewable energy from cyanobacteria: energy production optimization by metabolic pathway engineering. *Applied Microbiology and Biotechnology*, **91**(3), 471–490. Available at: https://doi.org/10.1007/s00253-011-3394-0

Radakovits R., Jinkerson R. E., Darzins A. and Posewitz M. C. (2010). Genetic engineering of algae for enhanced biofuel production. *Eukaryotic Cell*, **9**(4), 486–501. Available at: https://doi.org/10.1128/ec.00364-09

Raganati F., Miccio F. and Ammendola P. (2021). Adsorption of carbon dioxide for post-combustion capture: a review. *Energy & Fuels*, **35**(16), 12845–12868. Available at: https://doi.org/10.1021/acs.energyfuels.1c01618

Rahaman M. S., Cheng L.-H., Xu X.-H., Zhang L. and Chen H.-L. (2011). A review of carbon dioxide capture and utilization by membrane integrated microalgal cultivation processes. *Renewable and Sustainable Energy Reviews*, **15**(8), 4002–4012. Available at: https://doi.org/10.1016/j.rser.2011.07.031

Raheem A., Prinsen P., Vuppaladadiyam A. K., Zhao M. and Luque R. (2018). A review on sustainable microalgae based biofuel and bioenergy production: recent developments. *Journal of Cleaner Production*, **181**, 42–59. Available at: https://doi.org/10.1016/j.jclepro.2018.01.125

Raho B., Colangelo G., Milanese M. and de Risi A. (2022). A critical analysis of the oxy-combustion process: from mathematical models to combustion product analysis. *Energies*, **15**(18), 6514. Available at: https://doi.org/10.3390/en15186514

Raksajati A., Ho M. T. and Wiley D. E. (2013). Reducing the cost of CO_2 capture from flue gases using aqueous chemical absorption. *Industrial & Engineering Chemistry Research*, **52**(47), 16887–16901. Available at: https://doi.org/10.1021/ie402185h

Raksajati A., Ho M. T. and Wiley D. E. (2018). Comparison of solvent development options for capture of CO_2 from flue gases. *Industrial and Engineering Chemistry Research*, **57**(19), 6746–6758. https://doi.org/10.1021/acs.iecr.8b00283

Ramesh M., Muthukrishnan M. and Khan A. (2021). Metal-organic frameworks and permeable natural polymers for reasonable carbon dioxide fixation. *Metal-Organic Frameworks for Chemical Reactions*, **1**, 417–440. Available at: https://doi.org/10.1016/b978-0-12-822099-3.00017-4

Rapson T. D., Gregg C. M., Allen R. S., Ju H., Doherty C. M., Mulet X., Giddey S. and Wood C. C. (2020). Insights into nitrogenase bioelectrocatalysis for green ammonia production. *ChemSusChem*, **13**(18), 4856–4865. Available at: https://doi.org/10.1002/cssc.202001433

Rastogi R. P., Pandey A., Larroche C. and Madamwar, D. (2018). Algal green energy – R&D and technological perspectives for biodiesel production. *Renewable and Sustainable Energy Reviews*, **82**, 2946–2969. Available at: https://doi.org/10.1016/j.rser.2017.10.038

Rawat I., Ranjith Kumar R., Mutanda T. and Bux, F. (2013). Biodiesel from microalgae: a critical evaluation from laboratory to large scale production. *Applied Energy*, **103**, 444–467. Available at: https://doi.org/10.1016/j.apenergy.2012.10.004

Ren L. *et al.* (2013). Treatability studies on different refinery wastewater samples using highthroughput microbial electrolysis cells (MECS), *Bioresource Technology*, **136**, 322–328. https://doi.org/10.1016/j.biortech.2013.03.060

Reverdatto S., Beilinson V. and Nielsen N. C. (1999). A multisubunit acetyl coenzyme A carboxylase from soybean. *Plant Physiology*, **119**(3), 961–978. Available at: https://doi.org/10.1104/pp.119.3.961

Rismani-Yazdi H., Haznedaroglu B. Z., Bibby K. and Peccia, J. (2011). Transcriptome sequencing and annotation of the microalgae *Dunaliella tertiolecta*: pathway description and gene discovery for production of next-generation biofuels. *BMC Genomics*, **12**(148), 1–17. Available at: https://doi.org/10.1186/1471-2164-12-148

Robles-Medina A., González-Moreno P. A., Esteban-Cerdán L. and Molina-Grima E. (2009). Biocatalysis: towards ever greener biodiesel production. *Biotechnology Advances*, **27**(4), 398–408. Available at: https://doi.org/10.1016/j.biotechadv.2008.10.008

Roessler P. G. (1988). Changes in the activities of various lipid and carbohydrate biosynthetic enzymes in the diatom *Cyclotella cryptica* in response to silicon deficiency. *Archives of Biochemistry and Biophysics*, **267**(2), 521–528. Available at: https://doi.org/10.1016/0003-9861(88)90059-8

Ross A. B., Biller P., Kubacki M. L., Li H., Lea-Langton A. and Jones J. M. (2010) Hydrothermal processing of microalgae using alkali and organic acids. *Fuel*, **89**(9), 2234–2243. Available at: https://doi.org/10.1016/j.fuel.2010.01.025

Saad M. G., Dosoky N. S., Zoromba M. S. and Shafik H. M. (2019). Algal biofuels: current status and key challenges. *Energies*, **12**(10), 1920. Available at: https://doi.org/10.3390/en12101920

Saber M., Nakhshiniev B. and Yoshikawa K. (2016). A review of production and upgrading of algal bio-oil. *Renewable and Sustainable Energy Reviews*, **58**, 918–930. Available at: https://doi.org/10.1016/j.rser.2015.12.342

Sancho Navarro S., Cimpoia R., Bruant G. and Guiot S. R. (2016). Biomethanation of syngas using anaerobic sludge: shift in the catabolic routes with the CO partial pressure increase. *Frontiers in Microbiology*, **7**, 1–13. Available at: https://doi.org/10.3389/fmicb.2016.01188

Santana A., Jesus S., Larrayoz M. A. and Filho R. M. (2012). Supercritical carbon dioxide extraction of algal lipids for the biodiesel production. *Procedia Engineering*, **42**, 1755–1761. Available at: https://doi.org/10.1016/j.proeng.2012.07.569

Schipper K., Al-Jabri H. M., Wijffels R. H. and Barbosa M. J. (2021). Techno-economics of algae production in the Arabian Peninsula. *Bioresource Technology*, **331**, 125043. Available at: https://doi.org/10.1016/j.biortech.2021.125043

Schirmer A., Rude M. A., Li X., Popova E. and del Cardayre S. B. (2010). Microbial biosynthesis of alkanes. *Science*, **329**(5991), 559–562. Available at: https://doi.org/10.1126/science.1187936

Schut G. J. and Adams M. W. (2009). The iron-hydrogenase of *Thermotoga maritima* utilizes ferredoxin and NADH synergistically: a new perspective on anaerobic hydrogen production. *Journal of Bacteriology*, **191**(13), 4451–4457. Available at: https://doi.org/10.1128/jb.01582-08

Scoma A., Durante L., Bertin L. and Fava F. (2014). Acclimation to hypoxia in *Chlamydomonas reinhardtii*: can biophotolysis be the major trigger for long-term H_2 production?. *New Phytologist*, **204**(4), 890–900. Available at: https://doi.org/10.1111/nph.12964

Selembo P. A., Perez J. M., Lloyd W. A. and Logan B. E. (2009). High hydrogen production from glycerol or glucose by electrohydrogenesis using microbial electrolysis cells. *International Journal of Hydrogen Energy*, **34**(13), 5373–5381. Available at: https://doi.org/10.1016/j.ijhydene.2009.05.002

Sharma S., Basu S., Shetti N. P. and Aminabhavi T. M. (2020). Waste-to-energy nexus for circular economy and environmental protection: recent trends in hydrogen energy. *Science of the Total Environment*, **713**, 136633. Available at: https://doi.org/10.1016/j.scitotenv.2020.136633

Shuping Z., Yulong W., Mingde Y., Kaleem I., Chun L. and Tong, J. (2010). Production and characterization of bio-oil from hydrothermal liquefaction of microalgae *Dunaliella tertiolecta* cake. *Energy*, **35**(12), 5406–5411. Available at: https://doi.org/10.1016/j.energy.2010.07.013

Silva F. T., Bessa L. P., Vieira L. M., Moreira F. S., de Souza Ferreira J., Batista F. R. and Cardoso V. L. (2019). Dark fermentation effluent as substrate for hydrogen production from *Rhodobacter capsulatus* highlighting the performance of different fermentation systems. *3 Biotech*, **9**(4), 1–9. Available at: https://doi.org/10.1007/s13205-019-1676-x

Singh S. P. and Singh P. (2015). Effect of temperature and light on the growth of algae species: a review. *Renewable and Sustainable Energy Reviews*, **50**, 431–444. Available at: https://doi.org/10.1016/j.rser.2015.05.024

Soh L. and Zimmerman J. (2011). Biodiesel production: the potential of algal lipids extracted with supercritical carbon dioxide. *Green Chemistry*, **13**(6), 1422. Available at: https://doi.org/10.1039/c1gc15068e

Stoll I. K., Boukis N. and Sauer J. (2020). Syngas fermentation to alcohols: reactor technology and application perspective. *Chemie Ingenieur Technik*, **92**(1–2), 125–136. Available at: https://doi.org/10.1002/cite.201900118

Suzuki R., Ito N., Uno Y., Nishii I., Kagiwada S., Okada S. and Noguchi T. (2013). Transformation of lipid bodies related to hydrocarbon accumulation in a green alga, *Botryococcus braunii* (race B). *PLoS ONE*, **8**(12). Available at: https://doi.org/10.1371/journal.pone.0081626

Svavarsson H. G., Valberg J. E., Arnardottir H. and Brynjolfsdottir A. (2017). Carbon dioxide from geothermal gas converted to biomass by cultivating coccoid cyanobacteria. *Environmental Technology*, **39**(16), 2097–2104. Available at: https://doi.org/10.1080/09593330.2017.1349840

Tabatabaei M., Aghbashlo M., Dehhaghi M., Panahi H. K., Mollahosseini A., Hosseini M. and Soufiyan, M. M. (2019). Reactor technologies for biodiesel production and processing: a review. *Progress in Energy and Combustion Science*, **74**, 239–303. Available at: https://doi.org/10.1016/j.pecs.2019.06.001

Tan J. S., Lee S. Y., Chew K. W., Lam M. K., Lim J. W., Ho S.-H. and Show P. L. (2020). A review on microalgae cultivation and harvesting, and their biomass extraction processing using ionic liquids. *Bioengineered*, **11**(1), 116–129. Available at: https://doi.org/10.1080/21655979.2020.1711626

Tan K. T. and Lee K. T. (2011). A review on supercritical fluids (SCF) technology in sustainable biodiesel production: potential and challenges. *Renewable and Sustainable Energy Reviews*, **15**(5), 2452–2456. Available at: https://doi.org/10.1016/j.rser.2011.02.012

Tan K. W. and Lee Y. K. (2016). The dilemma for lipid productivity in green microalgae: importance of substrate provision in improving oil yield without sacrificing growth. *Biotechnology for Biofuels*, **9**(1), 1–14. Available at: https://doi.org/10.1186/s13068-016-0671-2

Thevasundaram K., Gallagher J. J., Cherng F. and Chang M. C. (2022). Engineering nonphotosynthetic carbon fixation for production of bioplastics by methanogenic archaea. *Proceedings of the National Academy of Sciences*, **119**(23), 1–8. Available at: https://doi.org/10.1073/pnas.2118638119

Ullah K., Ahmad M., Sofia Sharma, V. K., Lu P., Harvey A., Zafar M. and Sultana S. (2015). Assessing the potential of algal biomass opportunities for bioenergy industry: a review. *Fuel*, **143**, 414–423. Available at: https://doi.org/10.1016/j.fuel.2014.10.064

Uria N., Ferrera I. and Mas J. (2017). Electrochemical performance and Microbial Community Profiles in microbial fuel cells in relation to electron transfer mechanisms. *BMC Microbiology*, **17**(1). https://doi.org/10.1186/s12866-017-1115-2

Venkata Mohan S. and Devi M. P. (2014). Salinity stress induced lipid synthesis to harness biodiesel during dual mode cultivation of mixotrophic microalgae. *Bioresource Technology*, **165**, 288–294. Available at: https://doi.org/10.1016/j.biortech.2014.02.103

Wan C., Alam Md. A., Zhao X. Q., Zhang X. Y., Guo S. L., Ho S. H., Chang J. S. and Bai F. W. (2015). Current progress and future prospect of microalgal biomass harvest using various flocculation technologies. *Bioresource Technology*, **184**, 251–257. Available at: https://doi.org/10.1016/j.biortech.2014.11.081

Wang B., Li Y., Wu N. and Lan, C. Q. (2008). CO_2 bio-mitigation using microalgae. *Applied Microbiology and Biotechnology*, **79**(5), 707–718. Available at: https://doi.org/10.1007/s00253-008-1518-y

Wang H., Male J. and Wang Y. (2013). Recent advances in hydrotreating of pyrolysis bio-oil and its oxygen-containing model compounds. *ACS Catalysis*, **3**(5), 1047–1070. Available at: https://doi.org/10.1021/cs400069z

Wang J., Yang H. and Wang F. (2014). Mixotrophic cultivation of microalgae for biodiesel production: status and prospects. *Applied Biochemistry and Biotechnology*, **172**(7), 3307–3329. Available at: https://doi.org/10.1007/s12010-014-0729-1

Woertz I., Feffer A., Lundquist T. and Nelson Y. (2009). Algae grown on dairy and municipal wastewater for simultaneous nutrient removal and lipid production for biofuel feedstock. *Journal of Environmental Engineering*, **135**(11), 1115–1122. Available at: https://doi.org/10.1061/(asce)ee.1943-7870.0000129

Wolinska A., Kuźnia, A., Zielenkiewicz U., Banach A., Izak D., Stępniewska Z. and Błaszczyk M. (2016). Metagenomic analysis of some potential nitrogen-fixing bacteria in arable soils at different formation processes. *Microbial Ecology*, **73**(1), 162–176. https://doi.org/10.1007/s00248-016-0837-2

Xu L., (Wim) Brilman D. W. F., Withag J. A. M., Brem G. and Kersten S. (2011). Assessment of a dry and a wet route for the production of biofuels from microalgae: energy balance analysis. *Bioresource Technology*, **102**(8), 5113–5122. Available at: https://doi.org/10.1016/j.biortech.2011.01.066

Xu C., Salsali H., Weese S. and Warriner K. (2016). Inactivation of *Clostridium difficile* in sewage sludge by anaerobic thermophilic digestion. *Canadian Journal of Microbiology*, **62**(1), 16–23. Available at: https://doi.org/10.1139/cjm-2015-0511

Yang C., Hua Q. and Shimizu K. (2000). Energetics and carbon metabolism during growth of microalgal cells under photoautotrophic, mixotrophic and cyclic light-autotrophic/dark-heterotrophic conditions. *Biochemical Engineering Journal*, **6**(2), 87–102. Available at: https://doi.org/10.1016/s1369-703x(00)00080-2

Yang C., Li R., Zhang B., Qiu Q., Wang B., Yang H., Ding Y. and Wang C. (2019). Pyrolysis of microalgae: a critical review. *Fuel Processing Technology*, **186**, 53–72. Available at: https://doi.org/10.1016/j.fuproc.2018.12.012

Yin J., Zhao L., Xu X., Li D., Qiu H. and Cao X. (2022). Evaluation of long-term carbon sequestration of biochar in soil with biogeochemical field model. *Science of the Total Environment*, **822**, 153576. Available at: https://doi.org/10.1016/j.scitotenv.2022.153576

Yoshida A., Nishimura T., Kawaguchi H., Inui M. and Yukawa, H. (2005). Enhanced hydrogen production from formic acid by formate hydrogen lyase-overexpressing *Escherichia coli* strains. *Applied and Environmental Microbiology*, **71**(11), 6762–6768. Available at: https://doi.org/10.1128/aem.71.11.6762-6768.2005

Yoshimura T., Okada S. and Honda M. (2013). Culture of the hydrocarbon producing microalga *Botryococcus braunii* strain *Showa*: optimal CO_2, salinity, temperature, and irradiance conditions. *Bioresource Technology*, **133**, 232–239. Available at: https://doi.org/10.1016/j.biortech.2013.01.095

Yuan X., Wang J., Zeng G., Huang H., Pei X., Li H., Liu Z. and Cong M. (2011). Comparative studies of thermochemical liquefaction characteristics of microalgae using different organic solvents. *Energy*, **36**(11), 6406–6412. Available at: https://doi.org/10.1016/j.energy.2011.09.031

Yusoff M. F., Xu X. and Guo Z. (2014). Comparison of fatty acid methyl and ethyl esters as biodiesel base stock: a review on processing and production requirements. *Journal of the American Oil Chemists' Society*, **91**(4), 525–531. Available at: https://doi.org/10.1007/s11746-014-2443-0

Zainan N. H., Srivatsa S. C. and Bhattacharya S. (2015). Catalytic pyrolysis of microalgae *Tetraselmis suecica* and characterization study using in situ synchrotron-based infrared microscopy. *Fuel*, **161**, 345–354. Available at: https://doi.org/10.1016/j.fuel.2015.08.030

Zhang X. (2015). Microalgae Removal of CO_2 From Flue Gas – Usea, IEA Clean Coal Center. IEA, London, UK. Available at: https://usea.org/sites/default/files/042015_Microalgae%20removal%20of%20CO2%20from%20flue%20gas_ccc250.pdf (Accessed: January 30, 2023)

Zhang L., Happe T. and Melis A. (2002). Biochemical and morphological characterization of sulfur-deprived and H_2-producing *Chlamydomonas reinhardtii* (green alga). *Planta*, **214**(4), 552–561. Available at: https://doi.org/10.1007/s004250100660

Zhang A., Carroll A. L. and Atsumi S. (2017). Carbon recycling by cyanobacteria: improving CO_2 fixation through chemical production. *FEMS Microbiology Letters*, **364**(16), 1–7. Available at: https://doi.org/10.1093/femsle/fnx165

Zhang Y., Zhang L., Liu H., Gong L., Jiang Q., Liu H. and Fu B. (2019). Carbon dioxide sequestration and methane production promotion by wollastonite in sludge anaerobic digestion. *Bioresource Technology*, **272**, 194–201. Available at: https://doi.org/10.1016/j.biortech.2018.10.004

Zhang Y., Jiang Q., Gong L., Liu H., Cui M. and Zhang J. (2020). In-situ mineral CO_2 sequestration in a methane producing microbial electrolysis cell treating sludge hydrolysate. *Journal of Hazardous Materials*, **394**, 122519. Available at: https://doi.org/10.1016/j.jhazmat.2020.122519

Zhao B. and Su Y. (2014). Process effect of microalgal-carbon dioxide fixation and biomass production: a review. *Renewable and Sustainable Energy Reviews*, **31**, 121–132. Available at: https://doi.org/10.1016/j.rser.2013.11.054

Zheng Y., Li T., Yu X., Bates P. D., Dong T. and Chen S. (2013). High-density fed-batch culture of a thermotolerant microalga *Chlorella sorokiniana* for biofuel production. *Applied Energy*, **108**, 281–287. Available at: https://doi.org/10.1016/j.apenergy.2013.02.059

Zou S., Wu Y., Yang M., Li C. and Tong J. (2010). Bio-oil production from sub- and supercritical water liquefaction of microalgae *Dunaliella tertiolecta* and related properties. *Energy & Environmental Science*, **3**(8), 1073–1078. Available at: https://doi.org/10.1039/c002550j

Zuleta E. C., Baena L., Rios L. A. and Calderón J. A. (2012). The oxidative stability of biodiesel and its impact on the deterioration of metallic and polymeric materials: a review. *Journal of the Brazilian Chemical Society*, **23**(12), 2159–2175. Available at: https://doi.org/10.1590/s0103-50532012001200004

doi: 10.2166/9781789063783_0075

Chapter 5

Exploiting hydrocarbon-degrading bacteria for reclamation of petroleum hydrocarbon polluted sites

*Anil Kumar Singh**

Department of Agriculture, Sant Baba Bhag Singh University, Khiala, Punjab 144030, India
**Corresponding author: anilimtech@gmail.com*

ABSTRACT

Petroleum hydrocarbon pollution has emerged as one of the major problem worldwide. Indiscriminate activities of petrochemical industry and accidental spills have contaminated both land and water bodies. Cleaning of hydrocarbon contaminated site and refurbishing it to uncontaminated condition is a challenging task. Bioremediation is often employed for restoration and reclamation of petroleum hydrocarbon-contaminated site as it is an environment-friendly technology. In bioremediation, metabolic capabilities of hydrocarbon utilizing microorganisms particularly bacteria are exploited to remove pollutant. In recent past, bacteria-assisted reclamation process has advanced to a new level and achieved several goals with the aid of extremophiles. However, there are several environmental factors that hinder practical application of this technology. This chapter provides an overview of bacteria-assisted reclamation of petroleum hydrocarbon-polluted sites. It also discusses chemistry of petroleum hydrocarbons, sources of contaminations, toxicity, fate of petroleum hydrocarbons in the environment, microbial hydrocarbon degradation pathways, bioremediation technologies generally employed for reclamation and different factors influencing reclamation.

Keywords: biodegradation, biosurfactant, hydrocarbonoclastic bacteria, extremophiles, enzymes

5.1 INTRODUCTION

Petroleum hydrocarbons are important global resources for human society. It fulfills a major portion of global energy demands and serve as a raw material for several industries. The global policies and strategies are often framed keeping in view of the petroleum hydrocarbon resources. Large demand for petroleum hydrocarbons has helped petrochemical industry to burgeon globally. Relentless expansion and activities of petrochemical industries added hydrocarbons to soil and water bodies as a toxic environmental pollutant. Petroleum hydrocarbon spills are very common in the marine environment. The Exxon Valdez oil spill in Alaska (1989) and the Deepwater Horizon oil spills in Gulf of Mexico (2020) are two famous industrial accidents that became the source of hydrocarbon pollution. A number of small incidences of petroleum hydrocarbon pollution are often not reported and ignored, particularly in the regions having minimal or unenforceable environmental laws.

Petroleum hydrocarbon is a mixture of simple and complex organic compounds. Majority of organic compounds present in the petroleum hydrocarbons are enlisted in persistency organic pollutant (POP) list and considered as priority environmental pollutant. Contamination of soil and water by petroleum hydrocarbons cause extensive undesirable impact on the environment. Pollution due to hydrocarbons renders adverse effects on humans as well as on livestock, wild animals, microorganisms and plants. The recalcitrant nature of hydrocarbons make it a difficult pollutant for natural degradation. Gradual accumulation of petroleum hydrocarbons in the environment alters population dynamics and disturbs interaction of organisms at the tropic level. This further aggravates to disturb the natural community structure of polluted ecosystem. Reclamation of contaminated site is a difficult assignment due to recalcitrant nature of petroleum. This has drawn attention of scientific community to study the chemical properties and environmental fate of petroleum hydrocarbons so as to strategize cost-effective and environment-friendly methods for restoration and reclamation of contaminated sites. Reclamation of petroleum hydrocarbon-contaminated sites and enhanced oil recovery are two main issues facing petrochemical industries.

The term reclamation of petroleum hydrocarbon-contaminated sites refers to the methods used for transforming toxic hydrocarbon-polluted site to non-hazardous or less hazardous site approximately to pre-polluted conditions. Properly reclaimed site must support growth of living organisms. The methods selected for remediation purpose play a crucial role in complete removal, cleaning, reclamation, and restoration of petroleum hydrocarbon-contaminated sites. A successful and effective reclamation must refurbish the natural flora and fauna of contaminated site.

Decision with selection of remediation method is influenced by environmental condition of contaminated sites, cost, and time constrains. Conventionally, remediation of petroleum hydrocarbon-contaminated sites can be achieved by physical, chemical, and biological processes. Very often physical and chemical processes fail to achieve desirable level of toxic neutralization due to generation of toxic intermediates. Transportation of toxic intermediates to surrounding areas further aggravates the contamination problem. Use of biological systems like microorganisms, plant and their products for reclamation purpose is a sustainable and cost-effective method. Reclamation of petroleum hydrocarbon-contaminated sites by biodegradation and bioremediation techniques uses metabolic capabilities of living organisms for rendering hazardous pollutant to non-hazardous form.

This chapter focuses on microorganism-assisted reclamation of petroleum hydrocarbon-polluted sites. It discusses the chemistry of petroleum hydrocarbons, toxicity and the fate of petroleum hydrocarbons in the environment, type of bioremediation technologies generally employed for reclamation, different factors influencing reclamation and microbial metabolic pathways of petroleum hydrocarbon degradation.

5.2 CHEMICAL NATURE OF PETROLEUM HYDROCARBONS

Fossil fuel, petroleum (in Latin means 'crude oil') is a dark viscous liquid produced by thermal decomposition of buried organic material over millions of years. Petroleum hydrocarbons are complex substances consisting of different proportions of hydrogen and carbon. Sometimes, they also contain other impurities such as sulfur, nitrogen, and oxygen (Varjani, 2017). Total petroleum hydrocarbon (TPH) is a term used for any mixture of hydrocarbons that are found in crude oil. There are several hundreds of these compounds, but not all occur in any one sample. Composition of crude oil may differ with location and age, as well as with depth of oil well (Varjani, 2017).

Crude oil can be categorized as heavy, medium, and light oil based on the relative proportion of the present molecular weight constituents (Varjani, 2017). Generally, crude oil is classified in four broad categories according to their polarity and polarizability: (1) saturates (aliphatics), (2) aromatics (ringed hydrocarbons), (3) resins and (4) asphaltenes (Chandra *et al.*, 2013). Saturates are the simplest hydrocarbons. They constitute the maximum fraction of crude oil constituents and consist of non-polar linear, branched, and cyclic hydrocarbons without double bonds. Ethane, methane, propane, and cyclohexane are common saturates present in petroleum hydrocarbons.

Aromatic hydrocarbons are slightly more polarizable than saturated and consist of aromatic rings, mostly substituted with different alkyl groups. The simplest known aromatic hydrocarbon is benzene (C_6H_6). The aromatic name refers to '*Aroma*', fragrance of compounds. Other common examples of aromatic hydrocarbon include toluene, ethylbenzene, and xylene.

Alternatively, resins and asphaltenes contain polar non-hydrocarbon compounds and have highly complex structure consisting of carbon, nitrogen, oxygen, and sulfur atoms (Chandra *et al.*, 2013). Resins consist of fused aromatic rings with branched paraffin. The resins are soluble in lighter alkanes such as pentane, but are insoluble in liquid propane. The resins get adsorbed on solids such as alumina, clay, or silica, and subsequently removed by using more polar solvents. The composition of resins varies depending upon the nature of precipitating liquid. There are three main types of resins: (a) C5 aliphatic, (b) C9 aromatic, and (c) dicyclopentadiene (DCPD) cycloaliphatic resins. The C5 aliphatic hydrocarbon resins are made from C-5 piperylene. The important aliphatic resins include 2-methyl-2-butene, *cis/trans* 1,3-pentadienes, cyclopentadiene, cyclopentene, and DCPD. Aromatic hydrocarbon resins such as indene, styrene, alpha-methylstyrene, methyindenes and so on are important C9 hydrocarbons.

Asphaltenes are large dark-brown complex molecules occurring as a colloidal dispersion in saturates and aromatics. They are soluble in light aromatic hydrocarbons such as benzene and toluene (Chandra *et al.*, 2013). Asphaltenes may also contain trace amounts of vanadium and nickel along with carbon, oxygen, hydrogen, nitrogen, and sulfur. Carbon:hydrogen ratio in asphaltenes is approximately 1:1.2. Distillation products of bitumens are more likely to have asphalt-like properties, so it gained the name asphaltene.

5.3 SOURCES OF PETROLEUM HYDROCARBON POLLUTION

Crude oil is present in underground pockets called oil reservoirs. Drilling is employed to extract crude oil from reservoirs. Drilled-out crude oil is transported to oil refineries for separating different fractions of petroleum hydrocarbons. Petroleum exploration, extraction from reservoirs, transportation to refineries and hydrocarbon processing are largely associated with petroleum hydrocarbon pollution (Ossai *et al.*, 2020). Accidental spillage of petroleum hydrocarbon during transportation, loading, and discharging of oil from tanks are a common cause of environmental pollution. According to the EPA Toxic Release Inventory report (2005), crude oil industry is one of the major sources of petroleum hydrocarbon pollution. More than 1.7–8.8 million metric tonnes of petroleum hydrocarbon is released annually into the marine environment globally due to human-made accidents and spills from oil tankers. Recently, Russia reported the worst accidental fuel leakage from Norilsk-Taimyr Energy Thermal Power Plant on May 29, 2020. Damage occurred due to poor maintenance of a fuel storage tank that resulted in escape of nearly 20 000 tonnes of diesel oil into the Ambarnaya River. Greenpeace Russia compared the potential environmental effects of the Norilsk spill to that of the 1989 Exxon Valdez oil spill (Khurshudyan, 2020).

The Niger Delta is considered as one of the most heavily petroleum hydrocarbon-contaminated regions in the world (Sam *et al.*, 2017). The Nigerian delta has over five decades of oil exploitation history with poor management practices. There are more than 2000 land-based petroleum hydrocarbon-contaminated sites in the Nigerian delta. According to Ambituuni *et al.* (2014) sabotage, pipeline vandalism, well blowout, and engineering failures are the main contributors of petroleum hydrocarbon pollution in the Nigerian Delta.

5.4 TOXICITY OF PETROLEUM HYDROCARBONS

Petroleum hydrocarbons are composed of a wide range of compounds with varying degree of toxicity for living organisms and the environment (Varjani, 2017). Hazardous effects of petroleum hydrocarbons are primarily dependent on the chemical nature, composition, and properties of the constituting compounds. Mode, intensity, and duration of exposure also have profound influence on the environment. Petroleum hydrocarbons have toxic effect on all known living organisms.

Plants exposed to hydrocarbons exhibit stunted growth and death under extreme exposure. Hydrocarbons either reduce or obstruct water and mineral uptake by plants (Nie *et al.*, 2011). Plants experiencing stress due to hydrocarbons have chlorophyll-deficient leaves and deformed roots. Such plants are more susceptible to pest and diseases. Generally, leaves and flowers exhibit chlorotic and necrotic spots. Petroleum hydrocarbons are known to kill or inhibit growth of many microbial species. Hydrocarbon toxicity changes composition of microbial communities. This change alters functionality and subsequently the ecosystem. The absence of microbial population or diversity is known to adversely influence biogeochemical networks in nature (Truskewycz *et al.*, 2019).

Human and animal exposure to petroleum hydrocarbon causes severe health issues depending upon the nature and duration of contact. Direct contact may cause eye irritation, skin irritation, blisters, headaches, dizziness, nausea, vomiting, and lung infections (Ossai *et al.*, 2020). Long-term exposure has more drastic effect on humans and animals. They are potentially known to be toxic to genetic, immune, and endocrine systems (Figure 5.1).

5.5 FATE OF PETROLEUM HYDROCARBON IN NATURE

Petroleum hydrocarbon undergoes slow weathering in nature (Figure 5.2). Weathering includes combination of physical, chemical, and biological processes (Varjani, 2017). Weathering converts

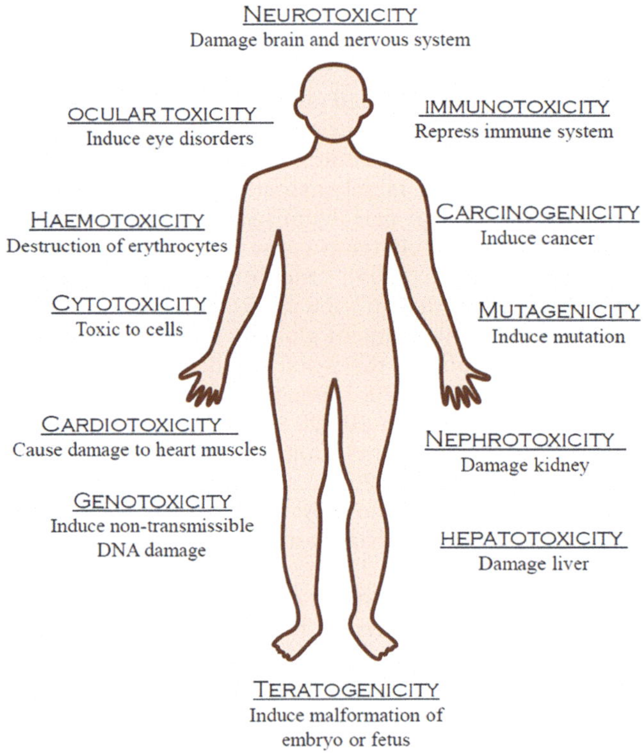

Figure 5.1 Toxicological effects of petroleum hydrocarbons on human health.

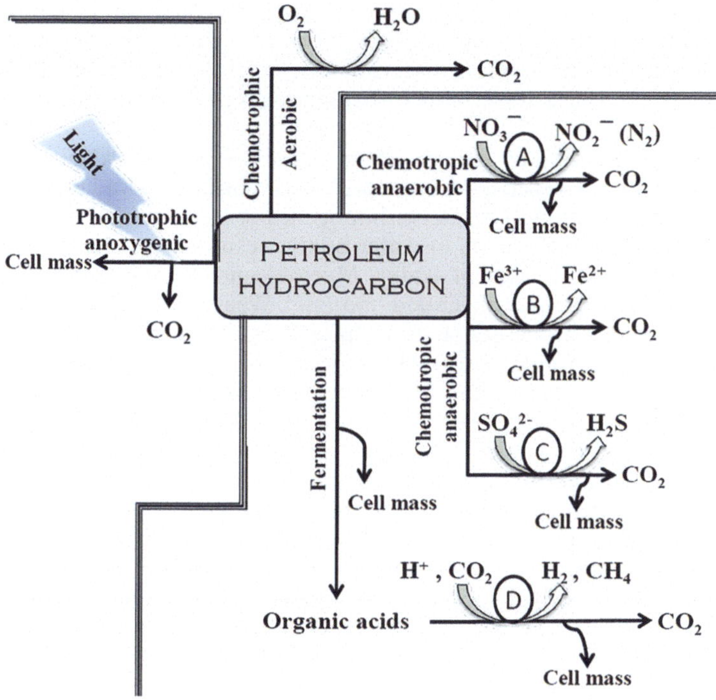

Figure 5.2 Degradation mechanisms for petroleum hydrocarbons in nature.

complex petroleum hydrocarbons to simpler forms or modifies it to less toxic forms. The ability of petroleum hydrocarbons to degrade under natural conditions is highly dependent on the surrounding environmental conditions (Widdel & Rabus, 2001). The nature, composition, physical, and chemical characteristics of petroleum hydrocarbons have significant influence on the weathering process (Varjani, 2017). Non-biological weathering of petroleum hydrocarbons occurs largely through hydrolysis and oxidation–reduction reactions. Hydrolysis of petroleum hydrocarbon requires H_3O^+, H^+, and OH^- ions. Metal ions present in loose association with soil catalyze hydrocarbon transformation and mineralization during oxidation–reduction reactions (Ossai *et al.*, 2020). Fraction of petroleum hydrocarbon with high dissolution exhibits more mobility in the environment. Dissolution is directly influenced by molecular size, structure, and polarity of hydrocarbons (Ossai *et al.*, 2020).

Aliphatic hydrocarbons are more volatile than aromatic hydrocarbons. Hence, aliphatic hydrocarbons are lost from the environment by a physical process called volatilization. Exposure to higher temperature converts lower molecular weight aliphatic hydrocarbons to vapors (Maletic *et al.*, 2013). These vapors enter atmosphere causing air pollution at the contaminated site. Among all weathering processes, volatilization is the rapid, immediate, and most effective process for removing lower molecular weight aliphatic hydrocarbons from contaminated sites (Ossai *et al.*, 2020). Light is also known to play an important role in petroleum hydrocarbon weathering process. Petroleum hydrocarbon absorbs light energy and undergoes photolytic splitting to form low molecular mass molecules (Widdel & Rabus, 2001).

Weathering by biological processes is the most prominent natural means of degrading petroleum hydrocarbons. Microorganisms carry out enzyme-specific degradation of petroleum hydrocarbons either in aerobic or anaerobic environmental conditions (Díaz, 2004). Very often, biological degradation is limited by bioavailability of hydrocarbons. Microorganisms produce surface-active substances like biosurfactant and bio-emulsifier to increase bioavailability of hydrocarbons. Microbial enzymatic activity converts hydrocarbon to less-hazardous or benign form that can be directly assimilated into biogeochemical cycles.

Petroleum hydrocarbon utilization by various microorganisms present in nature. Methanogenesis has to be coupled with fermentation reactions for complete hydrocarbon degradation (A, denitrification; B, ferric ion reduction; C, sulfate reduction; D, methanogenesis).

The presence of petroleum hydrocarbons in water bodies results in the formation of thin surface films or slicks that float on the surface of water. Marine aquatic environment often favors the formation of oil–water emulsion commonly called '*mousse*'. Weathering of petroleum hydrocarbons in aquatic environment includes processes such as spreading, evaporation, dissolution, dispersion, and emulsification (Rodriques & Totola, 2015). Heavy molecular fraction of petroleum hydrocarbon generally sinks to the bottom of water bodies. The movement of petroleum hydrocarbons from water bodies to coastal regions may expand the area under pollution (Rodriques & Totola, 2015).

Contamination of terrestrial environment by petroleum hydrocarbons has more serious impact on surrounding environment. Hydrocarbon gradually percolates through soil and reaches underground water resources. Petroleum hydrocarbons present in ground water moves later to increase the polluted area (Ossai *et al.*, 2020). Few fractions of petroleum hydrocarbon undergo hydrophobic interactions with soil organic matter and consequently get adsorbed on the soil particles (Ukalska-Jaruga *et al.*, 2019). Adsorption on soil particle decreases mobility of petroleum hydrocarbons. Volatile hydrocarbons become part of the soil gases (Ossai *et al.*, 2020). Hydrophobic nature of petroleum hydrocarbon implies it to form non-aqueous phase liquid (NAPL). NAPL tends to contaminate both soil and ground water (Logeshwaran *et al.*, 2018).

5.6 HYDROCARBON DEGRADING BACTERIA

Microorganisms are ubiquitous with diverse metabolic capability. However, only few microorganisms have ability to use petroleum hydrocarbon as carbon and energy source (Abbasian *et al.*, 2015). General ability of a microorganism to use petroleum hydrocarbon may be related to their encounter with these compounds or similar compounds during recycling of plant-derived material (Harwood & Parales, 1996).

In natural environment, microorganisms detoxify petroleum hydrocarbons by mineralization, transformation, and immobilization. However, among all microorganisms, bacteria are widely studied for hydrocarbon degradation. At any site, certain indigenous bacteria can successfully degrade or utilize certain components of petroleum hydrocarbon, while others fail to utilize available hydrocarbons (Varjani, 2017). This can be attributed to the fact that catalytic enzymes required for degradation of hydrocarbons is not available with all indigenous bacteria. Effective remediation of petroleum hydrocarbon-contaminated site requires joint action of multiple efficient bacteria. Several microorganisms like *Achromobacter, Acinetobacter, Alkanindiges, Alteromonas, Arthrobacter, Burkholderia, Bacillus, Dietzia, Enterobacter, Kocuria, Marinobacter, Mycobacterium, Pandoraea, Pseudomonas, Staphylococcus, Streptobacillus, Streptococcus,* and *Rhodococcus* are known as petroleum hydrocarbon degraders (Sarkar *et al.*, 2017; Xu *et al.*, 2018). Some microorganisms like *Alcanivorax, Cycloclasticus, Marinobacter, Oleispira,* and *Thallassolituus* are referred as obligate hydrocarbonoclastic bacteria (Naether *et al.*, 2013; Yakimov *et al.*, 2007). Hydrocarbonoclastic bacteria are a group of bacteria that are undetectable or exhibits low abundance in microbial community before hydrocarbon contamination (Radwan *et al.*, 2019). After experiencing hydrocarbon in there vicinity, hydrocarbonoclastic bacteria becomes dominant and play a crucial role in the degradation of petroleum hydrocarbons (Yakimov *et al.*, 2007). Hydrocarbonoclastic bacteria are believed to have noticeable influence on transformation and fate of petroleum hydrocarbons in the environment (Xu *et al.*, 2018).

Aerobic and anaerobic bacteria are known to play an important role in petroleum hydrocarbon removal from the contaminated site. Aerobic microorganisms are efficient in removing petroleum hydrocarbon from the environment. Oxygen is essential for aerobic respiration. In aerobic respiration, oxygen acts as final electron acceptor. Oxygen also activates substrate degradation by oxygenation reaction (Díaz, 2004). However, availability of oxygen for hydrocarbon degradation is not a ubiquitous phenomenon. Hydrocarbon present in aquatic sediments and submerged soils do not have enough oxygen concentration to support strict aerobic microorganisms. Strict anaerobes and facultative anaerobes bring about microbial degradation of hydrocarbon in the absence of oxygen. Anaerobes use alternative electron acceptors such as nitrate, sulfate, ferric ion (Fe-III), carbon dioxide, and so on (Díaz, 2004; Ossai *et al.*, 2020).

In terms of energy, aerobic degradation of hydrocarbon generates more energy than anaerobic mode of degradation. Anaerobic hydrocarbon degradation with nitrate and Fe-III as terminal electron acceptors is nearly as efficient as aerobic respiration. Contrary, sulfate reducers and methanogenic conditions are comparatively less efficient owing to less generation of energy (Field *et al.*, 1995). Fermentation of petroleum hydrocarbons results in the formation of intermediates. The presence of methanogens or sulfate-reducing bacteria aid in further degradation of intermediates generated during petroleum hydrocarbon fermentation (Gibson & Harwood, 2002). Hence, syntrophic existence of hydrocarbon-fermenting microorganisms and methanogens or sulfate-reducing bacteria is essential for complete degradation of petroleum hydrocarbon. Anaerobic degradation of petroleum hydrocarbon can also be carried out by photosynthetic bacteria. Energy assimilated from light is used for anaerobic degradation of petroleum hydrocarbon (Gibson & Harwood, 2002).

Bacteria exhibiting chemotaxis toward hydrocarbon plays a key role in remediation process by bringing cells into contact with degradation substrates (Parales *et al.*, 2008). Lanfranconi *et al.* (2003) reported a bacterium *Flavimonas oryzihabitans* chemotactic to gas oil and hexadecane. Another study by Smits *et al.* (2003) shows *Pseudomonas aeruginosa* PAO1 is chemotactic to hexadecane.

Petroleum hydrocarbon contains more than one toxic substance. Single bacterial species may not hold ability to degrade all toxic substances to benign form. Under these conditions, syntrophic bacterial consortia may provide suitable strategy for complete degradation (Díaz, 2004). In one of the study, Rizzo *et al.* (2018) reported that *Pseudomonas* strain A6, *Joostella* strain A8 and *Alcanivorax* strain A53 exhibited 38.6%, 26.8%, and 52.7% biodegradation efficiency of diesel oil, respectively. However, bacterial consortium of *Joostella* – *Alcanivorax* and *Joostella* – *Pseudomonas* exhibited biodegradation efficiency of 99.4% and 99.2%, respectively. Thus, highlighting the importance of bacterial consortium in biodegradation of petroleum hydrocarbons.

5.7 DEGRADATION PATHWAY OF PETROLEUM HYDROCARBON

Petroleum hydrocarbon is a complex mixture of alkanes, alkynes, cycloalkanes, and aromatic compounds. The aliphatic hydrocarbons are more easily degraded by the bacteria whereas the long chain and the branched or cyclic chain hydrocarbon are more difficult to degrade (Ossai *et al.*, 2020). Among all known hydrocarbons, aromatic compounds with benzene ring are more recalcitrant to microbial degradation. Benzene rings are thermodynamically stable, hence some of the aromatic compounds with benzene rings are persistent in the environment (Díaz, 2004). The bacterial degradation of aromatic hydrocarbons has more complex pathways as compared to alkanes, cycloalkanes, and alkynes (Varjani & Upasani, 2016). Bacteria assimilate and degrade petroleum hydrocarbon to generate ATP for fulfilling the daily energy requirements (Gibson & Harwood, 2002). The complete degradation of petroleum hydrocarbon in aerobic condition results in the formation of CO_2 and H_2O as by-products. The degradation tendency of the hydrocarbons is reported in the order of n-alkanes > branched alkanes > monoaromatics > cycloalkanes > polyaromatics (Tyagi *et al.*, 2011).

As evident from figure 9, microorganisms can utilize petroleum hydrocarbon by three possible ways namely, (a) phototrophic, anoxygenic; (b) chemotrophic, aerobic; and (c) chemotrophic, anaerobic.

Microbial degradation pathway of petroleum hydrocarbons can be grouped into two categories namely peripheral and central pathways (Figure 5.3). Peripheral pathways degrade or transform structurally diverse compounds into intermediates that can become part of central metabolic pathways (Harayama & Timmis, 1992). Generally, several peripheral pathways end up in a common product that further enters the central pathways. Peripheral pathways are often referred to as 'funnel' compounds into central pathways. Central pathways consist of a series of reactions leading to the formation of intermediates that can become part of Krebs cycle (Harayama & Timmis, 1992).

Aerobic reactions are mainly involved in bacterial degradation of alkanes. Aliphatic hydrocarbon degradation is catalyzed by monooxygenases and/or dioxygenase enzymes (Tyagi *et al.*, 2011). The degradation of aliphatic hydrocarbons begins by adding an oxygen atom to the terminal or sub-terminal carbon (Abbasian *et al.*, 2015). Oxygenation ultimately converts aliphatic hydrocarbons to primary or secondary alcohols. The enzyme alcohol dehydrogenase converts alcohol to aldehyde. Aldehydes are further converted to fatty acids. The fatty acid passes through β-oxidation pathway to produce acetyl CoA (Abbasian *et al.*, 2015). The process of converting alkane into acetyl CoA is often known as carboxylation reaction (Abbasian *et al.*, 2015). The acetyl CoA generated during carboxylation reaction enter the central metabolic pathway such as the Krebs cycle (Díaz, 2004). Degradation of n-alkanes and iso-alkanes follows similar pathways before becoming a part of the bacterial central metabolic system (Abbasian *et al.*, 2015; Varjani, 2017).

Degradation pathways of cycloalkanes are slightly different and complicated as compared to alkanes (Abbasian *et al.*, 2015). Generally, complete degradation of cycloalkanes is achieved through co-metabolism by bacterial consortium (Abbasian *et al.*, 2015). The degradation of cycloalkanes starts with oxygenation of alkyl side chain by cycloalkanes monooxygenase enzymes. Oxygenation of alkyl side chain of cycloalkanes results in the formation of cycloalkanols. The enzyme cycloalkanols dehydrogenase converts cycloalkanols to cycloalkanone. The cycloalkanone is acted upon by enzyme cycloalkanone monooxygenase to produce ε-caprolactone. Adipic acid is produced from ε-caprolactone by a reaction catalyzed by ε-caprolactone dehydrase. Adipic acid enters β-oxidation pathway to produce acetyl-CoA. Acetyl-CoA is utilized in the Krebs cycle (Abbasian *et al.*, 2015).

Degradation of alkynes by single bacterial species is very rare due to the presence of triple bonds between the carbon atoms. The degradation of alkynes starts by adding water in the presence of enzyme alkynes hydratase. Hydration of alkynes requires strong acid and Hg^{2+} ions. Hydration of alkynes produces ketone, which further gets converted to carboxylic acid (Abbasian *et al.*, 2015).

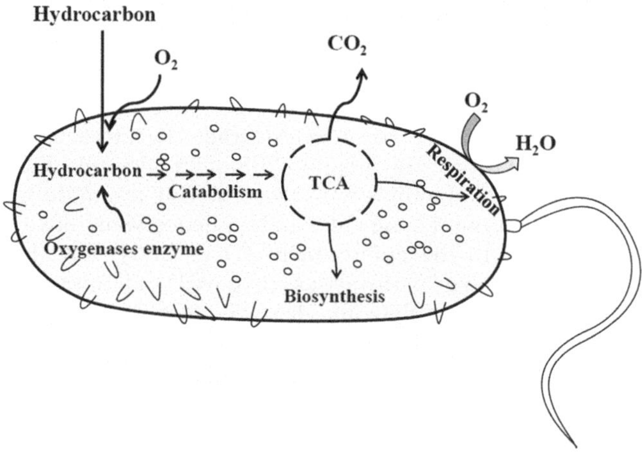

Figure 5.3 Schematic overview of potential pathways for petroleum hydrocarbons utilization by bacteria.

Formation of carboxylic acid requires strong permanganate oxidizing agent and carboxylase enzyme. The carboxylic acid is oxidized to produce acetyl-CoA which is further consumed in the Krebs cycle (Abbasian *et al.*, 2015).

Degradation of aromatic hydrocarbons is more complex due to the involvement of numerous enzymes. Most of the peripheral pathways of aromatic hydrocarbon degradation involve oxygenation reaction catalyzed by monooxygenases and/or hydroxylating dioxygenases (Abbasian *et al.*, 2015; Truskewycz *et al.*, 2019). Oxygenation results in generation of dihydroxy aromatic compounds. These aerobic degradation intermediates undergo ring cleavage to open the aromatic ring between hydroxyl groups or proximal to one of the two hydroxyl groups (Abbasian *et al.*, 2015). Ortho-cleavage or intradiol pathway (cleavage between two hydroxyl groups) is catalyzed by ring-cleavage enzymes intradiol dioxygenases (Peng *et al.*, 2008). Meta-cleavage or extradiol pathway (cleavage near one of the two hydroxyl groups) is catalyzed by extradiol dioxygenases. Ring cleavage reaction result in the formation of liner hydrocarbons like carboxylic acids and/or aldehydes (Abbasian *et al.*, 2015; Truskewycz *et al.*, 2019). These compounds are further degraded by pathway similar to aliphatic hydrocarbons degradation pathway. Peripheral pathways of anaerobic catabolism of aromatic compounds results in the formation of benzoyl-CoA. Under certain anaerobic conditions resorcinol or phloroglucinol may be formed during anaerobic catabolism of aromatic compounds (Heider & Fuchs, 1997).

5.8 GENETICS OF PETROLEUM HYDROCARBON BIODEGRADATION

Bacteria capable of degrading petroleum hydrocarbon have catabolic genes, transport genes, and regulatory genes, usually arranged together in a cluster (Meckenstock *et al.*, 2016). Catabolic genes encode for enzymes involved in catabolic pathway. The catabolic gene clusters are generally present on the mobile element like transposons and plasmids (Wilkes *et al.*, 2016). Mobile elements facilitate horizontal transfer of catabolic genes among microorganisms. Catabolic genes mobilization helps in rapid adaptation of microorganisms to new toxic environment. Product of transport genes help in uptake of petroleum hydrocarbon from the surrounding environment. Regulatory genes regulate expression of catabolic genes and transport genes under the influence of petroleum hydrocarbons present in the surrounding environment (Wilkes *et al.*, 2016; Varjani, 2017). Bacteria experiencing petroleum hydrocarbons in their vicinity receives external signals for activating degradation pathway. Depending upon physiological capability, efficiency of metabolic return and experienced toxicity, bacteria may initiate process of petroleum hydrocarbon degradation (Wilkes *et al.*, 2016).

Bacteria have evolved mechanism to exploit petroleum hydrocarbon as a source of energy and carbon. However, this mechanism is not always efficient enough to bring about bioremediation and subsequent reclamation of polluted sites. Bacteria have acquired degradation ability to gain in ecological advantage for survival under hostile conditions (Truskewycz *et al.*, 2019). Bioremediation of polluted site for reclamation purpose requires bacteria to have wide catabolic activity at optimal rate (Varjani, 2017).

Designing of recombinant bacteria for bioremediation purpose is often seen as a method to overcome the aforementioned hurdles. An efficient recombinant bacterium can be designed by manipulating the host cell and specific catabolic pathway. The rate of hydrocarbon removal as well as the range of substrate degraded by a catabolic pathway can be enhanced by manipulating the key enzymes and regulatory mechanism that control expression of catabolic genes (Timmis & Pieper, 1999).

Engineering metabolic pathway allows generation of novel hybrid pathways by combining catabolic components from different origins into the same host cell (Pieper & Reineke, 2000). Rational engineering of metabolic pathway may result in construction of bacteria with multiple pathways for degrading large number of toxic hydrocarbons and also to complete the incomplete pathways without generation of toxic intermediates (Díaz, 2004).

Bacteria surviving in petroleum hydrocarbon have to cope up with hostile environment. Extremophiles have inherent ability to thrive in hostile conditions like extreme pH, high temperature, high ionic strength, toxic hydrocarbons and heavy metals (Pieper & Reineke, 2000). Engineering extremophiles for creating hydrocarbon degrading bacteria can be an effective strategy for remediating a contaminated site (Díaz, 2004). For instance, consider a petroleum hydrocarbon-contaminated site with high salinity. Thus, engineering saline bacteria for hydrocarbon degradation can be a good option for remediating such contaminated site (Díaz, 2004). Several petroleum hydrocarbon sites contain a mixture of heavy metals and radioactive substances. In this condition, radiation-resistant bacteria can be engineered with hydrocarbon degradation genes to make them suitable for remediating such sites (Daly, 2000).

Biosafety is a major concern with release of recombinant microorganisms into the natural environment. Several microorganisms engineered for petroleum hydrocarbon degradation can thrive and survive only till pollutants are present. Once the pollutants have been consumed, there will be no carbon and energy source for survival of engineered microorganisms (Díaz, 2004). Transfer of genes from engineered microorganisms to native microbial population can be prevented by gene-containment circuits based on a toxin and its equivalent antidote (Torres *et al.*, 2004).

5.9 RECLAMATION OF PETROLEUM HYDROCARBON-CONTAMINATED SITES

Exploiting hydrocarbon-degrading bacteria for biodegradation of contaminates is simplest, environmentally safe, sustainable, and cost-efficient method for reclaiming site. Remediation of contaminated sites by petroleum hydrocarbon-degrading bacteria is slow process and often requires months to several years for successfully completing the process. Bacterial treatment for remediation can be *in situ* and *ex situ*.

Bioremediation is traditional and most sorted biological methods for cleaning and reclamation of contaminated site. Bioremediation of petroleum hydrocarbon-contaminated site has been known since 1940. However, it gained recognition after remediating the Exxon Valdez spill site in 1980s (Hoff, 1993). In recent past, bioremediation with selected hydrocarbon-degrading bacteria has attracted attention of researchers as well as policy makers. It exploits metabolic capability of petroleum hydrocarbon-degrading bacteria for neutralizing the toxic contaminates (Ossai *et al.*, 2020). The hydrocarbon-degrading microorganisms assimilate contaminates as energy source and exhibits growth and reproduction. Ample nutrient supply and overcoming all bacterial growth-limiting factors are essential for accomplishing remediation and reclamation of the site. Successful bioremediation oxidizes contaminates to CO_2 and H_2O while converting the site to a benign status (Yanti, 2018).

Bioattenuation exploits a variety of chemical, physical, and biological processes to remove, neutralize, transform, and reduce toxicity or concentration of petroleum hydrocarbons at the contaminated sites (Ossai *et al.*, 2020). Bioattenuation is accomplished through several processes like advection, dispersion, sorption, dissolution, stabilization, volatilization, abiotic, and biological transformation of petroleum hydrocarbons (Abatenh *et al.*, 2017). Indigenous microbial population degrades or transforms petroleum hydrocarbon to less toxic form depending upon their metabolic capabilities. Bioattenuation is useful at contaminated sites having low concentration of petroleum hydrocarbon, where adopting other remediation method is not feasible and time is not a limiting factor (Vásquez-Murrieta *et al.*, 2016).

Biostimulation is the addition of materials to support bacterial growth and enhance activity of hydrocarbon degrading enzyme for accomplishing the remediation of contaminated sites (Sarkar *et al.*, 2016). It involves addition or optimization of various parameters that limits remediation of hydrocarbon. Amending material may be biosurfactants, biopolymers, macro- and micro- nutrients. Cleaning petroleum hydrocarbons-contaminated site by biostimulation is the most successful and efficient '*in situ*' remediation method in comparison with other documented methods (Ossai *et al.*, 2020). Sarkar *et al.* (2016) enhanced degradation activity of indigenous microorganism by adding nitrate to the growth medium. Indigenous microorganism preferred utilization of both higher and middle-chain length hydrocarbons, and degraded 80% of TPH within 90 days (Sarkar *et al.*, 2016).

Bioaugmentation involves the addition of exogenous bacterial cultures, autochthonous bacterial consortium or genetically engineered bacteria for remediating petroleum hydrocarbon-contaminated site (Nwankwegu & Onwosi, 2017). Microorganisms selected for bioaugmentation must be adapted to the environmental site and have proven record to degrade hydrocarbon contaminants. Szulc *et al.* (2014) observed that bioaugmentation with consortium consisting of *Aeromonas hydrophila, Alcaligenes xylosoxidans, Gordonia* sp., *Pseudomonas fluorescens, Pseudomonas putida, Rhodococcus equi, Stenotrophomonas maltophilia,* and *Xanthomonas* sp. increased degradation efficiency by 89% in a 365-day treatment of diesel oil-contaminated soil.

Bioventing is a process of enhancing the rate of *in situ* remediation of petroleum hydrocarbon-contaminated site by injecting air into the soil (Thomé *et al.*, 2015). The addition of air stimulates and increases aerobic conditions for the growth of indigenous microorganisms and augments the catabolic degradation of the contaminants. Recently, Thomé *et al.* (2015), conducted bioventing for 60 days on diesel-contaminated soil and observed 85% remediation of the contaminated site. In another study by Agarry and Latinwo (2015), bioventing resulted in removal of 91.5% of the contaminants from brewery effluents amended diesel-contaminated soil within 28 days of remedial period.

Biosparging is *in situ* remediation process of increasing biological activity of indigenous microorganisms by injecting nutrients and air into the saturated zone (Azubuike *et al.*, 2016). Nutrients and air support growth of microorganisms and enhances degradation of petroleum hydrocarbon contaminates. Biosparging is effective in reducing the hydrocarbon adsorbed on the soil particles or trapped between capillary spaces of soil particles (Kao *et al.*, 2007). Biosparging at a petroleum oil spill site removed 70% of BTEX within 10 months of the remedial period (Kao *et al.*, 2007).

Bioslurry is an *ex-situ* method of remediation where excavated contaminated soil is treated in controlled bioreactors (Tuhuloula *et al.*, 2014). In the treatment process, nutrients are added to excavated soil present in the bioreactors. Nutrients support growth and enhance microbial degradation of petroleum hydrocarbons. The bioreactor used for treatment has means to monitor and regulate various essential parameters such as temperature, air flow and nutrients mixing rate. Remediation by bioslurry method is relatively faster with controlled emission. Major drawback with bioslurry method is longer treatment period, soil excavation, transportation facility, and pretreatment of soil (Ossai *et al.*, 2020). Tuhuloula *et al.* (2014) reported 79.35–99.73% of naphthalene removal from petroleum hydrocarbon-contaminated soil obtained from oil drilling site of Pertamina Petrochina in Indonesia with the help of microbial consortia (*Bacillus cereus* and *Pseudomonas putida*) in the slurry bioreactor.

Biotransformation is the process of modifying structure of hydrocarbon contaminants from one form to another with less toxicity or reduced persistence (Smitha *et al.*, 2017). Transformation can be carried out by bacteria, fungi, yeast, or genetically engineered microorganisms (Abatenh *et al.*, 2017). In natural environment, the process of transformation is very slow, nonspecific, and less productive. However, microbial biotransformation process is rapid and specific (Ossai *et al.*, 2020). Størdal *et al.* (2015) reported biotransformation of crude oil in the presence of marine copepods feces. In another study, Brakstad *et al.* (2004) reported depletion of n-alkanes and 4–5 ring PAH hydrocarbons from mineral oil due to biotransformation processes. Denaturing gradient gel electrophoresis bands indicated microorganisms responsible for transformation are related to the α- and γ-subdivisions of proteobacteria and to the chloroflexus–flavobacterium–bacteroides group (Brakstad *et al.*, 2004).

Landfarming is a remediation method that applies tilling, ploughing, and spreading of contaminated soil into a thin layer on the land surface (Brown *et al.*, 2017; Guarino *et al.*, 2017). Surface layering of contaminated soil increases and stimulates aerobic microbial activity. Efficiency of landfarming can be increased by the addition of oxygen, water, and nutrients (Brown *et al.*, 2017). Landfarming has been reported to be suitable for soil with low molecular weight contaminants and volatile organic compounds (Guarino *et al.*, 2017). Brown *et al.* (2017) reported that landfarming removed 53% of TPH from contaminated soil within 6 weeks. Combination of natural attenuation, landfarming, and bioaugmentation was applied by Guarino *et al.* (2017) to remove 86% of TPH from contaminated soil within 90 days. Landfarming practice alone achieved 70% reduction in TPH remedial period (Guarino *et al.*, 2017).

Windrow is an *ex-situ* method of remediation aiming to increase aeration and distribution of nutrients by periodic tilling and turning of hydrocarbon-contaminated soil with the addition of water

(Azubuike *et al.*, 2016). Aeration and hydration increases indigenous microbial activity and consequently degradation of contaminants. Indigenous microorganisms bring about cleaning by biotransformation, assimilation, and mineralization of petroleum hydrocarbons (Azubuike *et al.*, 2016). As compared to biopiling, windrow method exhibits higher rate of hydrocarbon removal. Windrow treatment is not suitable for remediating contaminated site that release volatile compounds. Toxic volatile compounds released during periodic turning and tilling can harm workers as well as surrounding atmosphere (Azubuike *et al.*, 2016). Al-Daher and Al-Awadhi (1998) performed remediation of petroleum-contaminated soil with windrow soil system. Windrow system was subjected to regular watering, tilling, and turning to increase aeration and microbial activities. After a period of eight months, nearly 60% reduction in the TPH was observed (Al-Daher & Al-Awadhi, 1998).

Compositing is the process of adding suitable microbial consortia and organic waste as soil supplement for enhancing the rate of bioremediation (Ossai *et al.*, 2020). Controlled addition of nutrients, periodic tilling and watering is also performed during compositing. Microbial activity during compositing increases soil temperature from 50°C to 65°C. High soil temperature aids in thermal degradation of petroleum hydrocarbon (Atagana, 2008). Petroleum hydrocarbon-contaminated soil was inoculated with sewage sludge and incubated in compost heaps for 19 months (Atagana, 2008). After incubation, petroleum hydrocarbons concentration got reduced to 99% and 17% in the sewage sludge compost and in the control lacking sewage sludge, respectively (Atagana, 2008).

Biopiling is an *ex-situ* method of remediation involving pilling of an excavated contaminated soil along with soil amendments to perform biostimulation. The basic biopiling system consists of treatment bed, aeration system, irrigation system, nutrient system, and leachate collection system (Kim *et al.*, 2018; Ossai *et al.*, 2020). Generally, piles of hydrocarbon-contaminated soil are covered with sheets to control runoff, evaporation, and volatilization. Covering of piles also increases solar heating. Important factors namely moisture, heat, nutrients, air, and pH are controlled for enhancing the microbial degradation. This method is suitable for treating a large volume of contaminated soil within limited space (Ossai *et al.*, 2020). Crude oil-contaminated soil of Kuwait was cleaned using sequential biowashing–biopile processes (Kim *et al.*, 2019). Biowashing was performed by an enrichment culture of the indigenous soil bacterial community.

Hemoglobin-catalyzed oxidation was used in biopile process to remove TPH from contaminated soil. The sequential biowashing and biopile process removed 86% of TPH (Kim *et al.*, 2019).

5.10 FACTORS INFLUENCING RECLAMATION OF PETROLEUM HYDROCARBON-CONTAMINATED SITE

Biological cleaning of petroleum hydrocarbon-contaminated site is influenced by several biotic and abiotic factors. Bacteria employed for cleaning of contaminated site is the most important biotic factor influencing the reclamation process. In the natural ecosystem, microbial community grows together in a synergistic relationship and produces bioactive metabolites (including oxidative and hydrolytic enzymes) to implicate the mineralization of petroleum hydrocarbons (Truskewycz *et al.*, 2019). The selection of bacteria or bacterial consortia is decided keeping in view the environmental conditions and available resources (Haghollahi *et al.*, 2016). Abiotic factors that influence remediation are summarized in Figure 5.4. Factors influencing bacterial growth are important abiotic factors that influence the reclamation process.

5.10.1 Bioavailability

Bacteria can easily utilize water-soluble compounds as they are easily bio-available. Petroleum hydrocarbon has low bio-availability due to poor water solubility. Few bacteria have ability to increase hydrocarbon solubility by producing surface-active agents (like biosurfactant or bioemulsifiers) in the surrounding environment. A combination of biosurfactant-producing bacteria and hydrocarbon-utilizing bacteria improves remediation efficiency. Mnif *et al.* (2017) reported that the addition

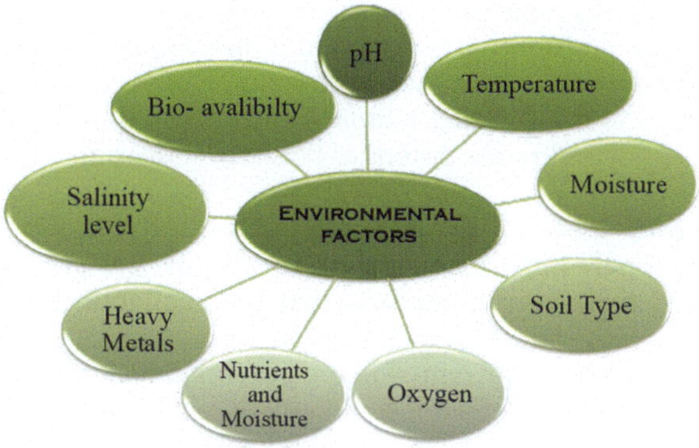

Figure 5.4 Environmental factors influencing reclamation of petroleum hydrocarbon contaminated site.

of biosurfactant-producing bacteria (*B. subtilis* SPB1) and hydrocarbon-degrading bacteria (*A. radioresistens* RI7) to diesel-contaminated soil improved the remediation of the contaminated soil by 32.67%.

5.10.2 pH

The pH of surrounding environment is an important factor influencing growth and enzyme activity. Each enzyme exhibits optimum activity at specific pH. Generally, neutral pH and slightly alkaline pH is known to support degradation of petroleum hydrocarbon. Neutral to slightly alkaline pH is known to support carboxylation reaction (Varjani, 2017).

5.10.3 Temperature

Temperature influences petroleum hydrocarbon degradation by affecting the rate of microbial metabolism and composition of the microbial community. Optimum temperature is essential for good bacterial growth and consequently degradation of petroleum hydrocarbons. Temperature has profound effect on microbial enzyme activity (Varjani, 2017). Extreme temperature (too high or too low) retards and in some scenario even inhibits growth of bacteria. Higher temperature increases the rate of hydrocarbon metabolism to a maximum, particularly in the range of 30–40°C. Generally, above 40°C, bacterial degradation of hydrocarbon is inhibited as it has adverse effect on growth. However, thermophilic hydrocarbon utilizing bacteria are known in literature that can accomplish cleaning activity even at higher temperatures (Nzila, 2018).

Temperature also influences degradation of hydrocarbons by altering physical nature and chemical composition of petroleum hydrocarbons. High temperature decreases viscosity, increases solubility, increases mobility, and enhances volatilization (Nzila, 2018). High temperature has a significant influence on the bioavailability of petroleum hydrocarbons, especially PAHs, by decreasing their viscosity, leading to an increase in their diffusion coefficients, therefore, improving their availability to microorganisms (Nzila, 2018).

5.10.4 Oxygen

Aerobic degradation of petroleum hydrocarbons is more effective in cleaning of contaminated site than anaerobic degradation (Mille *et al.*, 1988). The initial step in degradation of hydrocarbon under chemotrophic aerobic condition involves oxidation of substrate by oxygenases (Fritsche & Hofrichter, 2000). Molecular oxygen is also utilized to produce alcohol during peripheral degradation pathway (Varjani, 2017). Oxygen concentration is the rate-limiting variable in the biodegradation of petroleum hydrocarbon in the soil. Mille *et al.* (1988) reported that hydrocarbon degradation occurred in sediments where the dissolved oxygen concentration was 8 ppm and the redox potential was nearly +150 mV. The degradation rate decreased nearly 2–3 times when the dissolved oxygen concentration was 2–3 ppm and the redox potential was +30 mV. No degradation was observed under suboxic conditions having dissolved oxygen of 0.2–0.3 ppm and redox potential between −180 and −200 mV (Mille *et al.*, 1988).

5.10.5 Nutrient and moisture

Microorganisms require proper nutrition and moisture for growth. Petroleum hydrocarbons provide carbon, oxygen, and hydrogen to bacteria, but the availability of additional elements such as nitrogen, phosphorus, sodium, and so on is limited in contaminated sites (Ossai *et al.*, 2020). Biostimulation is the best example of nutrients-mediated enhancement of hydrocarbon degradation. Petroleum hydrocarbon-contaminated marine sediments having limited availability of nitrogen and phosphorus generated 1.10 ± 0.03 μmol CO_2/g wet sediment/day. The addition of inorganic nitrogen and phosphorus increased CO_2 production to 18.40 ± 1.04 μmol CO_2/g wet sediment/day (Singh *et al.*, 2014).

The moisture content of soil should be within an optimum range for accomplishing hydrocarbon degradation (Truskewycz *et al.*, 2019). The decrease in moisture concentration impacts soil microbial communities by hampering their interactions with each other and the environment, which may lead to a reduction in abundance, diversity, and structure. Excess moisture hinders oxygen transfer and retards hydrocarbon degradation (Truskewycz *et al.*, 2019).

5.10.6 Salinity

Most of the petroleum degradation is reported for non-saline environmental conditions. High salt concentration in the environment causes osmotic stress on bacteria and inhibits synthesis or activity of several enzymes involved in the degradation of hydrocarbon (Truskewycz *et al.*, 2019). High salinity is also known to reduce oxygen concentration in the environment, thus hindering bacterial growth. Degradation of petroleum hydrocarbons present in the marine environment requires halo-tolerant bacteria. Cao *et al.* (2020) reported that *Exiguobacterium* sp N4-1P exhibited maximum degradation efficiency at 15 g/L NaCl concentration.

5.10.7 Soil type

Soil type, pore-size distribution, and soil texture are some less investigated factors affecting remediation. Soil type affects remediation by influencing moisture, oxygen, and nutrient level of the soil. Fine-grained soils like clay have low permeability and retarded oxygen and nutrients mobility in the soil. Maintaining optimum soil moisture is also difficult in fine-grained soils due to small pore sizes and high surface areas (Haghollahi *et al.*, 2016). Hydrocarbon gets strongly adsorbed on the surface of clay soil particles, thus decreasing bioavailability of hydrocarbons to bacteria (Haghollahi *et al.*, 2016). Haghollahi *et al.* (2016) reported that remediation of hydrocarbon-contaminated sandy soil is more feasible than clay soil. Soils containing both clay and sand have good remediation feasibility than soil containing only clay.

5.10.8 Heavy metal contamination

Very often, hydrocarbon-contaminated site is co-contaminated with a toxic level of heavy metals. Osuji and Onojake (2006) reported the presence of heavy metals (Pb, Cd, V, Cu, and Ni) at Ebocha-8 oil spill site located in the Niger Delta region of Nigeria. The presence of heavy metal retards and sometimes even inhibits mineralization of hydrocarbons. Heavy metals are known to inhibit several metabolic pathways, such as the enzymatic and respiratory processes of many bacteria, and create additional stress to hydrocarbon-degrading microorganisms (Khudur *et al.*, 2019). The presence of lead at contaminated site decreases the rate of petroleum hydrocarbon degradation by reducing the number of hydrocarbon-degrading bacteria and activities of dehydrogenase enzyme. Khudur *et al.* (2019) observed that lead restricted microbial remediation of hydrocarbon-contaminated soils.

5.11 CONCLUSIONS AND FUTURE DIRECTION

Petroleum hydrocarbons are toxic pollutants that have contaminated both soil and water resources. Cleaning of petroleum hydrocarbon-contaminated site is of utmost importance for researchers due to its hazardous and persistent nature. Applying petroleum hydrocarbon-degrading bacteria for remediating contaminated site is widely considered as an environment-friendly approach of reclamation. A large number of hydrocarbon-degrading bacterial species have been reported for cleaning of contaminated sites. However, global success of bacterial-assisted reclamation is curtailed during practical applications. Several strategies have been proposed and formulated to make the process effective and efficient under all possible environmental conditions. Still there are lots of hurdles that need to be overcome for making bacterial-assisted restoration of petroleum hydrocarbon-contaminated site a global success.

Designing of novel and biocompatible surface-active agents for enhancing the bioavailability of hydrocarbon for bacteria is the primary requirement for on-field success. Modern omics approaches (genomics, transcriptomics, proteomics metabolomics, and fluxomics) can increase our understanding of bacterial metabolism in polluted environment and help in designing or formulating effective and efficient remediation strategies. Synthetic biology can be utilized to construct genetically engineered bacteria with broad hydrocarbon degradation ability. Exploring new and extreme location with high-throughput screening method can increase and enrich functional bacterial resources of petroleum hydrocarbon- assimilating bacteria.

REFERENCES

Abatenh E., Gizaw B., Tsegaye Z. and Wassie M. (2017). Application of microorganisms in bioremediation – review. *Journal of Environmental Microbiology*, **1**(1), 2–9.

Abbasian F., Lockington R., Mallavarapu M. and Naidu R. (2015). A comprehensive review of aliphatic hydrocarbon biodegradation by bacteria. *Applied Biochemistry and Biotechnology*, **176**, 1–30, https://doi.org/10.1007/s12010-015-1603-5

Agarry S. and Latinwo G. K. (2015). Biodegradation of diesel oil in soil and its enhancement by application of bioventing and amendment with brewery waste effluents as biostimulating-bioaugmentation agents. *Journal of Ecological Engineering*, **16**(2), 82–91, https://doi.org/10.12911/22998993/1861

Al-Daher R. and Al-Awadhi N. (1998). Bioremediation of damaged desert environment using the windrow soil profile system in Kuwait. *Environment International*, **24**(1/2), 175–180, https://doi.org/10.1016/S0160-4120(97)00133-5

Ambituuni A., Amezaga J. and Emeseh E. (2014). Analysis of safety and environmental regulations for downstream petroleum industry operations in Nigeria: problems and prospects. *Environmental Development*, **9**, 43–60, https://doi.org/10.1016/j.envdev.2013.12.002

Atagana H. I. (2008). Compost bioremediation of hydrocarbon-contaminated soil inoculated with organic manure. *African Journal of Biotechnology*, **7**(10), 1516–1525.

Azubuike C. C., Chikere C. B. and Okpokwasili G. C. (2016). Bioremediation techniques – classification based on site of application: principles, advantages, limitations and prospects. *World Journal of Microbiology and Biotechnology*, **32**(11), 180, https://doi.org/10.1007/s11274-016-2137-x

Brakstad O. G., Bonaunet K., Nordtug T. and Johansen Ø. (2004). Biotransformation and dissolution of petroleum hydrocarbons in natural flowing seawater at low temperature. *Biodegradation*, **15**, 337–346, https://doi.org/10.1023/B:BIOD.0000042189.69946.07

Brown D. M., Okoro S., Van Gils J., Van S. R., Bonte M., Hutchings T., Linden O., Egbuche U., Bruun K. B. and Smith J. W. N. (2017). Comparison of landfarming amendments to improve bioremediation of petroleum hydrocarbons in Niger Delta soils. *Science of the Total Environment*, **596–597**, 284–292, https://doi.org/10.1016/j.scitotenv.2017.04.072

Cao Y., Zhang B., Zhu Z., Song X., Cai Q., Chen B., Dong G. and Ye X. (2020). Microbial eco-physiological strategies for salinity-mediated crude oil biodegradation. *Science of the Total Environment*, **727**, 138723, https://doi.org/10.1016/j.scitotenv.2020.138723

Chandra S., Sharma R., Singh K. and Sharma A. (2013). Application of bioremediation technology in the environment contaminated with petroleum hydrocarbon. *Annals of Microbiology*, **63**, 417–431, https://doi.org/10.1007/s13213-012-0543-3

Daly M. J. (2000). Engineering radiation-resistant bacteria for environmental biotechnology. *Current Opinion in Biotechnology*, **11**, 280–285, https://doi.org/10.1016/S0958-1669(00)00096-3

Díaz E. (2004). Bacterial degradation of aromatic pollutants: a paradigm of metabolic versatility. *International Microbiology*, **7**(3), 173–180.

Field J. A., Stams A. J. M., Kato M. and Schraa G. (1995). Enhanced biodegradation of aromatic pollutants in cocultures of anaerobic and aerobic bacterial consortia. *Antonie Leeuwenhoek*, **67**, 47–77, https://doi.org/10.1007/BF00872195

Fritsche W. and Hofrichter M. (2000). Aerobic degradation by microorganisms. In: Environmental Processes – Soil Decontamination, J. Klein (ed.), Wiley-Vch, Weinheim, Germany, pp. 146–155.

Gibson J. and Harwood C. S. (2002). Metabolic diversity in aromatic compound utilization by anaerobic microbes. *Annual Review of Microbiology*, **56**, 345–369, https://doi.org/10.1146/annurev.micro.56.012302.160749

Guarino C., Spada V. and Sciarrillo R. (2017). Assessment of three approaches of bioremediation (natural attenuation, landfarming and bioaugmentation – assisted landfarming) for a petroleum hydrocarbons contaminated soil. *Chemosphere*, **170**, 10–16, https://doi.org/10.1016/j.chemosphere.2016.11.165

Haghollahi A., Fazaelipoor M. H. and Schaffie M. (2016). The effect of soil type on the bioremediation of petroleum contaminated soils. *Journal of Environmental Management*, **180**, 197–201, https://doi.org/10.1016/j.jenvman.2016.05.038

Harayama S. and Timmis K. N. (1992). Aerobic biodegradation of aromatic hydrocarbons by bacteria. In: Metal Ions in Biological Systems, H. Sigel and A. Sigel (eds.), Marcel Dekker, New York, Vol. **28**, pp. 99–156.

Harwood C. S. and Parales R. E. (1996). The β-ketoadipate pathway and the biology of self-identity. *Annual Review of Microbiology*, **50**, 553–590, https://doi.org/10.1146/annurev.micro.50.1.553

Heider J. and Fuchs G. (1997). Anaerobic metabolism of aromatic compounds. *European Journal of Biochemistry*, **243**(3), 577–596, https://doi.org/10.1111/j.1432-1033.1997.00577.x

Hoff R. Z. (1993). Bioremediation: an overview of its development and use for oil spill cleanup. *Marine Pollution Bulletin*, **26**, 476–481, https://doi.org/10.1016/0025-326X(93)90463-T

Kao C. M., Chen C. Y., Chen S. C., Chien H. Y. and Chen Y. L. (2007). Application of in situ biosparging to remediate a petroleum hydrocarbon spill spite: field and microbial evaluation. *Chemosphere*, **70**, 1492–1499, https://doi.org/10.1016/j.chemosphere.2007.08.029

Khudur L. S., Shahsavari E., Webster G. T., Nugegoda D. and Ball A. S. (2019). The impact of lead co-contamination on ecotoxicity and the bacterial community during the bioremediation of total petroleum hydrocarbon-contaminated soils. *Environmental Pollution*, **253**, 939–948, https://doi.org/10.1016/j.envpol.2019.07.107

Khurshudyan I. (2020). Arctic fuel spill prompts Russia's Putin to declare emergency and slam slow response. Washington Post. Retrieved 6 June 2020.

Kim J., Lee A. H. and Chang W. (2018). Enhanced bioremediation of nutrient-amended petroleum hydrocarbon contaminated soil over a cold climate winter: the rate and extent of hydrocarbon biodegradation and microbial response in a pilot-scale biopile subjected to natural seasonal freeze-thaw temperatures. *Science of the Total Environment*, **612**, 903–913, https://doi.org/10.1016/j.scitotenv.2017.08.227

Kim T., Hong J.-K., Jho E. H., Kang G., Yang D. J. and Lee S.-J. (2019). Sequential biowashing-biopile processes for remediation of crude oil contaminated soil in Kuwait. *Journal of Hazardous Materials*, **378**, 120710, https://doi.org/10.1016/j.jhazmat.2019.05.103

Lanfranconi M. P., Alvarez H. M. and Studdert C. A. (2003). A strain isolated from gas oil-contaminated soil displays chemotaxis towards gas oil and hexadecane. *Environmental Microbiology*, **5**, 1002–1008, https://doi.org/10.1046/j.1462-2920.2003.00507.x

Logeshwaran P., Megharaj M., Chadalavada S., Bowman M. and Naidu R. (2018). Petroleum hydrocarbons (PH) in groundwater aquifers: an overview of environmental fate, toxicity, microbial degradation and risk-based remediation approaches. *Environmental Technology & Innovation*, **10**, 175–193, https://doi.org/10.1016/j.eti.2018.02.001

Maletic S. P., Dalmacija B. D. and Roncevic S. D. (2013). Petroleum hydrocarbon biodegradability in soil – implications for bioremediation. In: Hydrocarbon, V. Kutcherov and A. Kolesnikov (eds.), INTECH Croatia, Vol. **3**, pp. 43–64.

Meckenstock R. U., Boll M., Mouttaki H., Koelschbach J. S., Tarouco P. C., Weyrauch P., Dong X. and Himmelberg A. M. (2016). Anaerobic degradation of benzene and polycyclic aromatic hydrocarbons. *Journal of Microbiology and Biotechnology*, **26**, 92–118.

Mille G., Mulyono M., El Jammal T. and Bertrand J. C. (1988). Effects of oxygen on hydrocarbon degradation studies in vitro in surficial sediments. *Estuarine, Coastal and Shelf Science*, **27**, 283–295, https://doi.org/10.1016/0272-7714(88)90057-1

Mnif I. S., Sahnoun R., Ellouz-Chaabouni S. and Ghribi D. (2017). Application of bacterial biosurfactants for enhanced removal and biodegradation of diesel oil in soil using a newly isolated consortium. *Process Safety and Environmental Protection*, **109**, 72–81, https://doi.org/10.1016/j.psep.2017.02.002

Naether D. J., Slawtschew S., Stasik S., Engel M., Olzog M., Wick L. Y., Timmis K. N. and Heipieper H. J. (2013). Adaptation of the hydrocarbonoclastic bacterium *Alcanivorax borkumensis* SK2 to alkanes and toxic organic compounds: a physiological and transcriptomic approach. *Applied and Environmental Microbiology*, **79**, 4282–4293, https://doi.org/10.1128/AEM.00694-13

Nie M., Wang Y., Yu J., Xiao M., Jiang L., Yang J., Fang C., Chen J. and Li B. (2011). Understanding plant-microbe interactions for phytoremediation of petroleum-polluted soil. *PLoS ONE*, **6**, 17961, https://doi.org/10.1371/journal.pone.0017961

Nwankwegu A. S. and Onwosi C. O. (2017). Bioremediation of gasoline contaminated agricultural soil by bioaugmentation. *Environmental Technology & Innovation*, **7**, 1–11, https://doi.org/10.1016/j.eti.2016.11.003

Nzila A. (2018). Current status of the degradation of aliphatic and aromatic petroleum hydrocarbons by thermophilic microbes and future perspectives. *International Journal of Environmental Research and Public Health*, **15**(12), 2782, https://doi.org/10.3390/ijerph15122782

Ossai I. C., Ahmed A., Hassan A. and Hamid F. S. (2020). Remediation of soil and water contaminated with petroleum hydrocarbon: a review. *Environmental Technology & Innovation*, **17**, 100526, https://doi.org/10.1016/j.eti.2019.100526

Osuji L. C. and Onojake C. M. (2006). Field reconnaissance and estimation of petroleum hydrocarbon and heavy metal contents of soils affected by the Ebocha-8 oil spillage in Niger Delta, Nigeria. *Journal of Environmental Management*, **79**, 133–139, https://doi.org/10.1016/j.jenvman.2005.06.004

Parales R. E., Ju K. S., Rollefson J. and Ditty J. L. (2008). Bioavailability, transport and chemotaxis of organic pollutants. In: Microbial Bioremediation, E. Diaz (ed.), Caister Academic Press, Norfolk, pp. 145–187.

Peng R. H., Xiong A. S., Xue Y., Fu X. Y., Gao F., Zhao W., Tian Y. S. and Yao Q. H. (2008). Microbial biodegradation of polyaromatic hydrocarbons. *FEMS Microbiology Reviews*, **32**, 927–955, https://doi.org/10.1111/j.1574-6976.2008.00127.x

Pieper D. H. and Reineke W. (2000). Engineering bacteria for bioremediation. *Current Opinion in Biotechnology*, **11**, 262–270, https://doi.org/10.1016/S0958-1669(00)00094-X

Radwan S. S., Khanafer M. M. and Al-Awadhi H. A. (2019). Ability of the so-called obligate hydrocarbonoclastic bacteria to utilize nonhydrocarbon substrates thus enhancing their activities despite their misleading name. *BMC Microbiology*, **19**, 41, https://doi.org/10.1186/s12866-019-1406-x

Rizzo C., Rappazzo A. C., Michaud L., De Domenico E., Rochera C., Camacho A. and Lo Giudice A. (2018). Efficiency in hydrocarbon degradation and biosurfactant production by *Joostella* sp. A8 when grown in pure culture and consortia. *Journal of Environmental Sciences*, **67**, 115–126, https://doi.org/10.1016/j.jes.2017.08.007

Rodriques E. M. and Totola M. R. (2015). Petroleum: from basic features to hydrocarbons bioremediation in oceans. *Open Access Library Journal*, **2**, e2136, https://doi:10.4236/oalib.1102136

Sam K., Coulon F. D. R. and Prpich G. (2017). Management of petroleum hydrocarbon contaminated sites in Nigeria: current challenges and future direction. *Land Use Policy*, **64**, 133–144, https://doi.org/10.1016/j.landusepol.2017.01.051

Sarkar J., Kazy S. K., Gupta A., Dutta A., Mohapatra B., Roy A., Bera P., Mitra A. and Sar P. (2016). Biostimulation of indigenous microbial community for bioremediation of petroleum refinery sludge. *Frontiers in Microbiology*, **7**, 1407, https://doi.org/10.3389/fmicb.2016.01407

Sarkar P., Roy A., Pal S., Mohapatra B., Kazy S. K., Maiti M. K. and Sar P. (2017). Enrichment and characterization of hydrocarbon-degrading bacteria from petroleum refinery waste as potent bioaugmentation agent for in situ bioremediation. *Bioresource Technology*, **242**, 15–27, https://doi.org/10.1016/j.biortech.2017.05.010

Singh A. K., Sherry A., Gray N. D., Jones D. M., Bowler B. F. and Head I. M. (2014). Kinetic parameters for nutrient enhanced crude oil biodegradation in intertidal marine sediments. *Frontiers in Microbiology*, **5**, 160, https://doi.org/10.3389/fmicb.2014.00160

Smitha M. S., Singh S. and Singh R. (2017). Microbial biotransformation: a process for chemical alterations. *Journal of Bacteriology & Mycology: Open Access*, **4**(2), 47–51.

Smits T. H., Witholt B. and van Beilen J. B. (2003). Functional characterization of genes involved in alkane oxidation by *Pseudomonas aeruginosa*. *Antonie Van Leeuwenhoek*, **84**, 193–200, https://doi.org/10.1023/A:1026000622765

Størdal I. F., Olsen A. J., Jenssen B. M., Netzer R., Altin D. and Brakstad O. G. (2015). Biotransformation of petroleum hydrocarbons and microbial communities in seawater with oil dispersions and copepod feces. *Marine Pollution Bulletin*, **101**, 686–693, https://doi.org/10.1016/j.marpolbul.2015.10.029

Szulc A., Ambrożewicz D., Sydow M., Ławniczak Ł., Piotrowska-Cyplik A., Marecik R. and Chrzanowski Ł. (2014). The influence of bioaugmentation and biosurfactant addition on bioremediation efficiency of diesel-oil contaminated soil: feasibility during field studies. *Journal of Environmental Management*, **132**, 121–128, https://doi.org/10.1016/j.jenvman.2013.11.006

Thomé A., Reddy K. K., Reginatto C. and Cecchin I. (2015). Review of nanotechnology for soil and groundwater remediation: Brazilian perspectives. *Water, Air, & Soil Pollution*, **226**, 1–20, https://doi.org/10.1007/s11270-014-2243-z

Timmis K. N. and Pieper D. H. (1999). Bacteria designed for bioremediation. *Trends in Biotechnology*, **17**, 200–204, https://doi.org/10.1016/S0167-7799(98)01295-5

Torres B., García J. L. and Díaz E. (2004). Plasmids as tools for containment. In: Plasmid Biology, G. Phillips and B. Funnell (eds.), ASM Press, Washington DC, pp. 589–601.

Truskewycz A., Gundry T. D., Khudur L. S., Kolobaric A., Taha M., Aburto-Medina A., Ball A. S. and Shahsavari E. (2019). Petroleum hydrocarbon contamination in terrestrial ecosystems – fate and microbial responses. *Molecules*, **24**(18), 3400, https://doi.org/10.3390/molecules24183400

Tuhuloula A., Altway A., Juliastuti S. R. and Suprapto S. (2014). Biodegradation of soils contaminated with naphthalene in petroleum hydrocarbons using bioslurry reactors. *IOP Conference Series: Earth and Environmental Science*, **175**, 012014, https://doi.org/10.1088/1755-1315/175/1/012014

Tyagi M., da Fonseca M. M. R. and de Carvalho C. (2011). Bioaugmentation and biostimulation strategies to improve the effectiveness of bioremediation processes. *Biodegradation*, **22**, 231–241, https://doi.org/10.1007/s10532-010-9394-4

Ukalska-Jaruga A., Smreczak B. and Klimkowicz-Pawlas A. (2019). Soil organic matter composition as a factor affecting the accumulation of polycyclic aromatic hydrocarbons. *Journal of Soils and Sediments*, **19**, 1890–1900, https://doi.org/10.1007/s11368-018-2214-x

Varjani S. J. (2017). Microbial degradation of petroleum hydrocarbons. *Bioresource Technology*, **223**, 277–286, https://doi.org/10.1016/j.biortech.2016.10.037

Varjani S. J. and Upasani V. N. (2016). Biodegradation of petroleum hydrocarbons by oleophilic strain of Pseudomonas aeruginosa NCIM 5514. *Bioresource Technology*, **222**, 195–201, https://doi.org/10.1016/j.biortech.2016.10.006

Vásquez-Murrieta M. S., Hernández-Hernández O. J., Cruz-Maya J. A., Cancino-Díaz J. C. and Jan-Roblero J. (2016). Approaches for removal of PAHs in soils: bioaugmentation, biostimulation and bioattenuation. In: Environmental Sciences. Soil Contamination – Current Consequences and Further Solutions, M. L. Larramendy and S. Soloneski (eds.), INTECH Croatia, 329–342.

Widdel F. and Rabus R. (2001). Anaerobic biodegradation of saturated and aromatic hydrocarbons. *Current Opinion in Biotechnology*, **12**, 259–276, https://doi.org/10.1016/S0958-1669(00)00209-3

Wilkes H., Buckel W., Golding B. T. and Rabus R. (2016). Metabolism of hydrocarbons in n-alkane utilizing anaerobic bacteria. *Journal of Microbiology and Biotechnology*, **26**, 138–151.

Xu X., Liu W., Tian S., Wang W., Qi Q., Jiang P., Gao X., Li F., Li H. and Yu H. (2018). Petroleum hydrocarbon-degrading bacteria for the remediation of oil pollution under aerobic conditions: a perspective analysis. *Frontiers in Microbiology*, **9**, 2885, https://doi.org/10.3389/fmicb.2018.02885

Yakimov M. M., Timmis K. N. and Golyshin P. N. (2007). Obligate oil-degrading marine bacteria. *Current Opinion in Biotechnology*, **18**, 257–266, https://doi.org/10.1016/j.copbio.2007.04.006

Yanti M. D. (2018). Bioremediation of petroleum contaminated soil: a review. *IOP Conference Series: Earth and Environmental Science*, **118**, 1–7, https://doi.org/10.1088/1755-1315/118/1/012063

doi: 10.2166/9781789063783_0093

Chapter 6

Recent advancement in microbial remediation of heavy metals from industrial effluents

K. Shalini[1], V. C. Padmanaban[2], T. Aswani[3] and S. Mariaamalraj[4,*]

[1]Department of Biotechnology, P.S.R. Engineering College, Sivakasi, India
[2]Department of Bioengineering, National Institute of Technology, Agartala, India
[3]Department of Biotechnology, School of Applied science, Reva University, Bangalore, India
[4]Associate Professor, Department of Biotechnology, Sethu Institute of Technology, Pulloor, Kariapatti, India
*Corresponding author: amalbiotech@gmail.com

ABSTRACT

Currently, heavy metal contamination has become a greatest challenge because, they are not degradable. They get accumulated in biotic and abiotic environment. Different strategies are being tested to remediate heavy metal pollution; however, an ecofriendly approach to remove them from the environment is much needed. Bioremediation is one such alternative to the conventional method. Among the utilization of plants and microbes for the heavy metal remediation, microbial remediation is found to be cost effective and low waste generating strategy. Many microorganisms acquired different mechanisms to adapt themselves and remove or detoxify the heavy metals. Besides, many biological agents like biosurfactants, can also be utilized for metal removal from an aqueous environment. This chapter focuses on the impact of heavy metal on the environment, different microbes involved in detoxification process, and biotechnological solutions of heavy metal contamination.

Keywords: heavy metals, bioremediation, microbial system, bioactive molecules, detoxification

6.1 INTRODUCTION

Due to population growth and economic development, pollution levels are rapidly rising. Due to the considerable increase in the quantity of industrial waste that is released into the ecosystem, especially soil and water, heavy metals have accumulated, particularly in urban areas. Because they cannot be converted into nontoxic forms and have a huge detrimental effect on the environment, heavy metals are a major health problem for everyone. Additionally, they are dispersed carelessly into the ground and rivers. There is very little risk to the public or security when heavy metals or their by-products are removed naturally. The body can store heavy metals, which are persistent. They might also bio-magnify and accumulate further up the food chain. To create a better environment for people, contaminated land and water bodies must be cleaned up to get rid of trace elements and heavy metals. The removal of heavy metal ions from liquid waste has lately been made possible by a number of novel methods. Reverse osmosis, ion-exchange, chemical precipitation, electrochemical treatment, and electrochemical processing are some of these procedures. These methods aim to decrease the bioavailability of metals by toxicizing heavy metals, although most of these methods are useless at doses below 100 mg. However, the majority of techniques are costly, using living, green plants or microorganisms, whereas bioremediation is a promising technique that cleans up soil, surface water, and groundwater of pollutants. Metals can also be detoxified by microbes through bio-sorption, bioleaching, bio-mineralization, and so on. The term 'bio-sorption/bioaccumulation' refers to the most common biotechnological technique for removing metal ions from wastewater. Pollutants may adhere to the cellular walls of microorganisms. The biosorption process looks to be more feasible for large-scale application than the bioaccumulation process since the addition of nutrients is necessary for bacteria to actively take up heavy metals, which increases the waste's biological or chemical oxygen need. In addition, heavy metal toxicity and other environmental variables make it particularly challenging to maintain a healthy population of bacteria. While poor nations appeared to focus more on sporadic, more expensive ex-situ treatments, developed economies appeared to embrace environmentally benign in-situ bioremediation techniques, such as monitored natural attenuation, more frequently.

6.2 TOXICITY OF HEAVY METALS

6.2.1 Arsenic

Drinking water may become contaminated using arsenical insecticides, naturally occurring mineral deposits, or the erroneous removal of arsenical synthetics. If arsenic is actively swallowed during suicide attempts or accidentally consumed by children, it can also result in acute poisoning episodes (Mazumder, 2008). Arsenic is a protoplastic toxin because it fundamentally damages cells in the sulfydryl bunch and prevents mitosis, cell enzymes, and respiration (Gordon & Quastel, 1948).

6.2.2 Lead

The main industrial processes, food and tobacco products, drinking water, and home sources are where lead exposure occurs. Plumbing fixtures, pewter pitchers, storage batteries, toys, and faucets are other sources of lead in addition to lead paint and gasoline (Thurmer *et al.*, 2002). Lead metal poisoning in living cells is caused by ionic and oxidative stress processes. Oxidative stress in living cells is caused by an imbalance between the production of free radicals to detoxify reactive intermediates and the production of antioxidants to stop damage or repair it. The key contributor to lead toxicity's ionic mechanism is the capacity of lead metal ions to displace other bivalent cations like Ca^{2+}, Mg^{2+}, and Fe^{2+} as well as monovalent cations like Na^{+}. This eventually causes disruptions in cell metabolism. The ionic mechanism of lead toxicity has a substantial impact on a number of biological processes, including cell adhesion, intra- and intercellular communication, protein folding, maturation, apoptosis, ionic transport, neurotransmitter release, and enzyme control. Lead can have an impact on protein kinase C, which regulates brain excitation and memory storage, even at pico molar quantities (Flora *et al.*, 2008).

6.2.3 Mercury

Mercury is highly toxic and bioaccumulative. The distribution of mercury in aquatic habitats is the subject of numerous studies since it has a negative impact on marine ecology. According to Chen *et al.* (2012), mining, waste incineration, municipal wastewater discharges, agricultural, and industrial wastewater discharges are the main man-made sources of mercury pollution. Methylmercury is a neurotoxin that accumulates neurotoxic compounds, including serotonin, aspartate, and glutamate in addition to damaging microtubules, mitochondria, lipid peroxidation, and microtubules (Patrick, 2002). The brain continues to be one of mercury's main targets, despite the fact that it can harm any organ and cause problems with nerves, organs, and muscles. It's possible to disturb both the membrane potential and the calcium balance within cells.

6.2.4 Cadmium

According to the ATSDR classification, cadmium is the sixth most dangerous heavy metal. It may come into contact with animals or people while working or in the environment because it is a by-product of the manufacture of zinc. People will continue to consume this metal throughout their lives, causing it to accumulate in their bodies. It was first used as a pigment and tin substitute in the paint industry during World War I. When cadmium attaches to metallothionein or another protein high in cysteine, its concentration rises three times. The accumulation of cysteine–metallothionein complex produces hepatotoxicity in the liver and nephrotoxicity in the renal tissue of the kidney (Lin *et al.*, 2005).

6.2.5 Chromium

Chromium is the eighth most common element in the cosmos. Chromium can be present in the environment in oxidation levels ranging from Cr^{2+} to Cr^{6+}, according to Rodríguez *et al.* (2007). Some examples of chromium's natural sources are coal and oil combustion, ferro chromate-derived petroleum, pigment oxidants, catalysts, chromium steel, fertilizers, oil well drilling, and tanneries used for metal plating. Because of its low membrane permeability, trivalent chromium Cr(III) is typically safe to breathe in. Hexavalent chromium, or Cr(VI), on the other hand, is more successful at rupturing cell membranes through openings for isoelectric and isostructural anions, such as $SO_4{}^{2-}$ and $HPO_4{}^{2-}$ channels, and these chromates are subsequently taken up through phagocytosis. Strong oxidant Cr(VI) is capable of being reduced to yield the temporary, distinct forms of pentavalent and tetravalent chromium (III). It is believed that intracellular reduction of Cr[VI] is a detoxifying process when reduction occurs somewhere other than the target site because glutathione stabilizes the pentavalent form. When the biological reductants thiols and ascorbate react with the metal Cr(VI), reactive oxygen species such the superoxide ion, hydrogen peroxide, and hydroxyl radical are produced. As a result, the cell experiences oxidative stress, which damages proteins and DNA (Hasin *et al.*, 2010).

6.2.6 Aluminum

According to Pazirandeh *et al.* (1995), aluminum is the third most common element in the planet's crust. Aluminum is a naturally occurring element that can be found on the earth, in water, and air. Mining and processing of aluminum have a greater negative impact on the environment. High concentrations of aluminum can result in osmoregulatory failure by depleting plasma and hemolymph ions, which is extremely detrimental to aquatic life, especially gill-breathing fish. Fish contain monomeric aluminum, which prevents an enzyme needed for ion absorption from working (Mishra *et al.*, 2017). Seaweed and crawfish are just two examples of the aquatic life that is affected by aluminum toxicity. Aluminum inhibits various enzymes that are active on DNA and RNA because it has a stronger affinity for these molecules than hexokinase, phosphodiesterase, alkaline phosphatase, and phosphoxidase. The metabolism of iron, calcium, phosphorus, fluorine, and other elements in living beings is impacted by aluminum. Aluminum has been found to be particularly harmful to hemopoietic, osseous, and neurological cells (Flora *et al.*, 2008).

According to the iron is the second most prevalent metal on Earth's surface. When the ingested iron does not attach to the protein, a large number of dangerous free radicals are created. This has a detrimental effect on the amount of iron in mammalian cells and bodily fluids. The digestive system and the circulating free iron corrodes bodily fluids. The body is saturated with iron once the rate-limiting step of absorption has been passed. These flexible irons can reach the cells in the heart, liver, and brain. Free iron raises the acidity of the metabolism by generating hydrogen ions, changing ferrous iron to ferric iron, and interfering with oxidative phosphorylation. The lipid peroxidation caused by free iron causes severe harm to microsomes, mitochondria, and other cellular organelles (Li, 2010). Iron-mediated tissue damage has been linked to cellular oxidizing and reducing processes as well as their toxicity towards intracellular organelles like mitochondria and lysosomes. A high iron intake produces several different free radicals, some of which are thought to have the ability to damage cells.

6.3 IMPACT OF HEAVY METALS ON SOIL

Heavy metals damage the soil biota by obstructing crucial microbial processes and lowering the number and activity of soil microorganisms. However, the ability of bacterial populations and fungi, notably arbuscular mycorrhizal (AM) fungi, to tolerate heavy metals over a long term may be crucial for the restoration of a damaged ecosystem (Mora *et al.*, 2005). The toxicity of heavy metals was influenced by their chemical compatibility with the enzymes of soil organisms. The microbial community in soil will suffer as a result. Important indicators of soil quality include the diversity and activity of soil microbes, which are essential for the detoxification of harmful compounds, the maintenance of soil structure, the management of plant pests, and the recycling of plant nutrients. The microbiological properties of the soil, such as respiration rate and enzyme activity, which seem to be highly effective markers of soil pollution, are generally adversely affected by an increase in metal content. In lead-contaminated soil (Pb), there was a slight change in the microbial composition of the soil.

6.4 IMPACT OF HEAVY METALS ON PLANTS

Because they interact with food chains and are absorbed by plant roots, heavy metals constitute a major threat to both animal and human health. Temperature, moisture, organic matter, pH, and the availability of nutrients are only a few of the numerous elements that affect how easily heavy metals are absorbed and deposited in plant tissue. Chlorosis, restricted plant development, low yields, reduced nutrient uptake, abnormalities in plant metabolism, and a reduced capacity of leguminous plants to fix molten nitrogen are all effects of heavy metal buildup. Low Pb concentrations may inhibit some crucial plant processes as photosynthesis, mitosis, and water absorption. Toxic warning signs include dark green leaves, fading older leaves, stunted foliage, and brown short roots.

6.5 IMPACT OF HEAVY METALS ON AQUATIC SYSTEMS

Aquatic species that are exposed to heavy metals, which are toxic in small doses, will experience oxidative stress. As a result, these pollutants have a big impact on ecotoxicology. Furthermore, since microbes cannot degrade metals, they stay in the marine ecosystem forever. When levels of heavy metal contamination rise, the diversity of aquatic life is restricted, and the ecological balance of the aquatic ecosystem may suffer. Particulate matter that finally settles down and becomes a part of the sediment is usually responsible for retaining heavy metals that are deposited into aquatic systems. Therefore, in aquatic ecosystems, surface sediment acts as the main reservoir or sink for metals and other contaminants.

Bioremediation is regarded as a significant, practical, eco-friendly, and cost-effective technology for the removal of heavy metals in contaminated sites in order to address the serious health consequences on humans caused by the accumulation of heavy metals in water and other sources. Both aerobic and anaerobic methods of bioremediation can be used to convert or degrade harmful components into harmless materials using microbial enzymes.

6.6 BIOREMEDIATION

Through a metabolic response, biological organisms are used in the bioremediation process to remove or neutralize an environmental contaminant. The term ‘biological’ in this sense refers to both the ‘remediation’ (the process of fixing the issue) and microscopic organisms like fungi, algae, and bacteria. Microbial enzymes can be utilized in both aerobic and anaerobic bioremediation processes to transform or degrade hazardous components into innocuous ones. The microorganisms utilized in bioremediation may be imported and introduced, or they may exist naturally in the contaminated area. Bioremediation is the most widely used technique for lowering, detoxifying, mineralizing, or changing more dangerous contaminations into less deadly toxins. Pesticides, agrochemicals, heavy metals, xenobiotics, hydrocarbons, atomic waste, colors, polymers, and slime are a few of the toxins in this group (Figure 6.1).

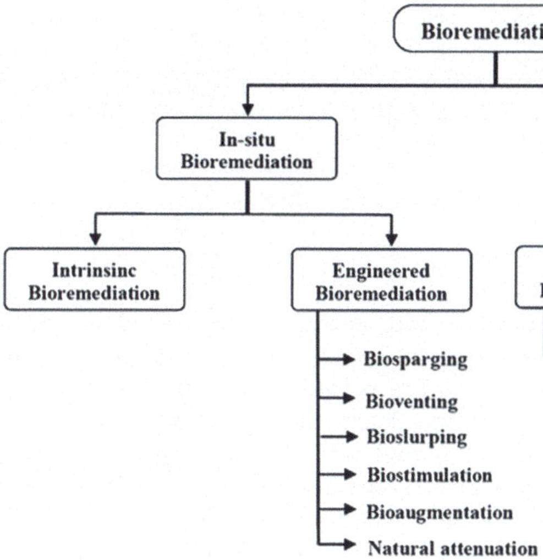

Figure 6.1 Classification of Bioremediation

Various methods are employed during the bioremediation process. The fundamental steps in the bioremediation process include bio-stimulation, attenuation, augmentation, and venting.

6.6.1 Bio stimulation

To encourage the activity of nearby microorganisms, this type of method entails adding specific nutrients to the area (soil or ground water). Its primary focus is on bacterial and fungal communities that are local or spontaneously occurring by initially supplying trace minerals, development aids, and fertilizers. Additional environmental requirements including pH, temperature, and oxygen are also addressed in order to accelerate their metabolic rate and pathway (Adams *et al.*, 2015; Kumar *et al.*, 2011). A modest amount of pollutant can also act as a stimulant by turning on the operons for the bioremediation enzymes. In order to support local microorganisms, this form of strategic corridor is typically maintained by supplying nutrients and oxygen.

6.6.2 Bio attenuation (natural attenuation)

The process of removing pollution concentrations from the environment is known as bio attenuation or natural attenuation. Advection, dispersion, dilution, diffusion, volatilization, sorption/desorption, chemical reactions (ion exchange, complexation, abiotic transformation), as well as biological processes like aerobic and anaerobic biodegradation, animal and plant absorption, and sorption/desorption, combine physical and biological processes to achieve this. Natural attenuation has a broader definition that encompasses ideas like intrinsic repair and biotransformation (Mulligana & Yong, 2004). According to Li *et al.* (2010), nature can remediate the environment following chemical pollution in four different ways.

(1) Certain chemicals are consumed by tiny insects or bacteria that live in soil and groundwater. Once the chemicals have been completely absorbed by the organism, they can be transformed into safe gases and water.
(2) Chemicals can adhere to dirt and retain it there by binding or sorbing to it. This method does not clean up the pollutants, but it may stop them from leaking out of the site and poisoning the groundwater.

(3) Pollution can contaminate clean water as it passes through land and groundwater. As a result, the pollution is diminished.
(4) Some chemicals, like oil and solvents, have the ability to evaporate, or change from liquids to gases inside the soil. If these gases are permitted to leak into the air close to the ground's surface, sunlight might kill them.

If the rate or degree of natural attenuation is insufficient, bioremediation will be hastened through bio stimulation or bio augmentation.

6.6.3 Bio augmentation

Bio augmentation is the manual addition of an organic culture to an environment, such as a bioreactor, in order to handle sewage or other contaminated pollutants. In most instances, this environment already has a microbial population, but bio augmentation is intended to enrich this population and increase its capacity to lower the level of contamination. Pre-grown microbial cultures can be added to make bioreactors and other treatment equipment available to immediately reach their peak levels of efficiency. The apparent justification for this microorganism addition is to boost the efficacy and efficiency of the processes taking place inside the bioreactor or in another industrial application. At a lower cost, the contaminants are eliminated more swiftly and efficiently. This makes it simpler for organizations to keep that crucial equilibrium, preserving profits while pursuing ecological sustainability. Increased levels of live microbes and microorganisms in the treatment area are how bio augmentation works. Each microorganism can only manage this transformation up to a certain rate, but these microorganisms are already working to convert contaminants into less harmful compounds.

6.6.4 Genetically engineered microorganisms in bioremediation

Genetically engineered (GE) bacteria, recombinant DNA, and RNA technology have all been used for efficient bioremediation. To improve bioremediation procedures, microbial genes have been changed to produce new metabolic pathways. GE microorganisms might be the preferred technique because of the distinctive features of their metabolic pathways. The use of GE bacteria, a cutting-edge technology, to remove heavy metals and toxic waste from contaminated places has attracted public attention. Additionally, it has aided in the detoxification of chemicals like heavy metals that are resistant to treatment. With the help of their metal regulatory genes, bacteria can convert hazardous forms of heavy metals into less toxic ones. The production of metallothioneins (MT) by GE microorganisms can speed up the accumulation of heavy metals. A possible technique for preparing GEO for bioremediation that might successfully remove environmental contaminants is recombinant DNA technology. Chakrabarty gave a description of the development of the first DNA technology for the bioremediation of contamination from petroleum. This method can be used to clean up regions that have become contaminated as a result of heavy metals, chlorinated hydrocarbons, pesticides, petroleum hydrocarbons, and explosives. DNA shuffling, a potent mutagenesis method can produce novel enzyme activity and biocatalysts that can degrade polyaromatic hydrocarbons and chlorinated ethane more quickly.

6.6.5 Bioventing

Bioventing is a technique for in-situ remediation that employs microbes to break down organic pollutants that have been adsorbed on soil in the unsaturated zone. Bioventing improves air or oxygen flow into the unsaturated zone and, if necessary, supplies nutrients by boosting the activity of local bacteria and replicating in-situ biodegradation of hydrocarbons in soil. During bioventing, direct air injection may be utilized to provide oxygen into dirt containing any lingering contaminants. Bioventing primarily aids in the breakdown of adsorbed fuel residuals, but it also aids in the degradation of volatile organic compounds (VOCs), as vapors move slowly through biologically active soil. Instead of a lack of nutrients, a lack of oxygen and other electron acceptors (i.e., substances that gain electrons during biodegradation) (i.e., electron donors) often limits the rate of natural deterioration. In conventional bioventing equipment, oxygen is delivered to underground wells by an electric blower. Bioventing, as opposed to soil vapor vacuum extraction, only uses modest airflow rates to supply the oxygen necessary to sustain microbial activity. Bioventing wells are used in passive bioventing systems to supply the subsurface with oxygen through spontaneous air exchange. When the pressure inside a vent well is lower than the atmospheric pressure, a one-way valve fitted on the well allows air to enter. The valve closes, trapping the air in the well and improving soil oxygenation when atmospheric pressure drops below the subsurface pressure (due to a shift in barometric pressure).

6.6.6 Biopile

In a bio-pile, excavated soils are enclosed for treatment after being mixed with soil additions and shaped into compost piles. The basic bio-pile system consists of an irrigation/nutrient system, an aeration system, a leachate collection system, and a treatment bed. To facilitate biodegradation, temperature, nutrients, oxygen, and pH are all regulated. To move air and nutrients through the earth, an irrigation/nutrient system is buried beneath the surface. The height of a soil mound can reach 20 feet. To regulate runoff, evaporation, and volatilization as well as to encourage sun heating, they may be covered with plastic. If the soil includes VOCs, the air leaving the soil may be treated to eliminate or destroy them before they are released into the atmosphere. Normally, the excavated material is processed for three to six months before being either put back where it came from or disposed of.

6.7 THE SEVERAL SPECIES OF ORGANISMS UTILIZED IN BIOREMEDIATION

The core of bioremediation is the co-metabolism of a single or a collection of microorganisms. Numerous investigations revealed that a wide variety of organisms were capable of biosorbing heavy metals. Prokaryotes are less sensitive to heavy metals than eukaryotes are, and whether an organism is a eukaryote or a prokaryote affects how it reacts to heavy metals. Active metal extrusion, intracellular chelation (in eukaryotes) by a variety of metal-binding peptides, and transformation into less dangerous chemical species are the three potential interaction pathways. Microorganisms must enzymatically attack the pollutants and change them into harmless chemicals for bioremediation to be successful. Aerobes, anaerobes, and fungi are a few of the microorganisms that participate in the enzymatic degradation process.

For remediation to be successful, a contaminated site may contain several different types of contaminants, necessitating the use of numerous microorganisms. Some bacteria can degrade the molecules of petroleum and use the carbon and energy they contain as a source. However, the organisms utilized must be carefully selected because they can only survive in the presence of a particular group of chemical pollutants, which varies based on the chemical makeup of the polluting agents. The ability of the particular microbe to combine the hydrocarbon with molecular oxygen and create the intermediates that eventually enter the cell's main metabolic route for generating energy determines the efficiency of the degradation process.

Numerous bacteria can consume oil, and many of them produce potent surface-active chemicals that can emulsify oil in water and facilitate its removal. Among the bacteria that can degrade petroleum compounds are *Pseudomonas, Aeromonas, Moraxella, Beijerinckia, Flavobacteria, Chlorobacteria, Nocardia, Corynebacteria, Modococci, Streptomyces, Bacilli, Arthrobacter, Cyanobacteria*, and various yeasts. *Pseudomonas putida* MHF 7109 may be isolated from microbial consortia present in cow dung and is capable of biodegrading a number of petroleum hydrocarbons, including benzene, toluene, and o-xylene (BTX) (Figure 6.2).

A quick and reversible passive adsorption mechanism is biosorption. The physical interactions between the metals and the functional groups on the cell surface, such as ion exchange, adsorption, complexation, precipitation, and crystallization, keep the metals in place. A few factors that can

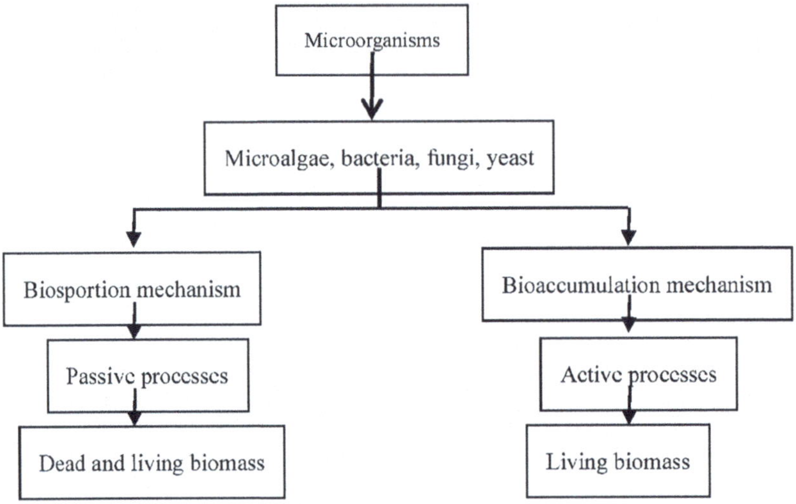

Figure 6.2 Microorganisms' function in the bioremediation process.

affect the biosorption of metals are pH, ionic strength, biomass concentration, temperature, particle size, and the presence of other ions in the solution. Biosorption is a process that can take place with both living and dead biomass since it is not dependent on cell respiration. Passive absorption, on the other hand, plays a much smaller and less clear-cut role in bioaccumulation, which includes both intra- and extracellular processes. Therefore, the only purpose of living things is bioaccumulation.

Bioremediation can be carried out using microbial bioreactors, which are also employed in a number of other processes, such as the management of solid waste, the treatment of waste water and industrial effluent, the treatment of soil and land, and the reduction of air pollution. A number of environmental parameters, including temperature, pH, moisture, the mix and concentration of pollutants, influences the growth and activity of microorganisms in bioreactors and macronutrients. Environment-related factors such as temperature, pH, oxygen availability/electron, and salinity have an impact on growth, the metabolic activities of bacteria, and to a lesser extent, the behaviour of the pollutant, such as solubility and volatility. When maximizing a process for biodegradation, it is crucial to take into account the effects of environmental factors and adapt the process in response to the relevant environmental conditions.

6.7.1 Temperature

Microorganisms can survive in a wide range of temperatures, including their lowest, optimal, and maximum temperatures. Additionally, every microorganism always performs the bio therapy it requires at the temperature that is optimal for its biochemistry. Both microbial development and enzyme-catalyzed processes in microorganisms are impacted by temperature extremes (too high or too low) (Sharma, 2011). When the temperature rises within a reasonable range, the rate of the bioremediation processes accelerates due to an increase in microbial metabolism. Many chemicals become more soluble at higher temperatures, increasing fluidity, and diffusion rates. For instance, warming increases both the solubility and subsequent bioavailability of contaminants like PAHs and heavy metals. Temperature is therefore an essential component for bioreactors to run as effectively as possible and deliver the greatest outcomes for bio treatment. Temperature regulation is usually considered when designing specialized bioreactors.

6.7.2 pH

pH has an impact on microbial growth and metabolic functions similarly to temperature. pH affects the ionic properties of microbial cells and subsequently microbial development. Microorganisms have a minimum, maximum, and optimal pH range in which they can develop. For example, most bacteria grow best at a pH of 6–7.5, though some (acidophiles) prefer an acidic or alkaline pH. (alkaliphiles). In general, fungi thrive at pH levels lower than bacteria do. The operating pH of the reactor must be adjusted to offer the ideal pH range for growth and enzyme activities. pH has an impact on the bioremediation of contaminants by altering their behaviour. For instance, pH has an impact on the redox and solubility of metals, and different forms and valences have different impacts on microbes. Alkaline pH favors metal ion precipitation, and metal solubility rises as middle pH falls (Gargouri *et al.*, 2011; Liu *et al.*, 2017). Metal attachment to the exterior of microbial cells frequently requires lower pH values. The metal ions are more soluble when acid-producing microorganisms are present. Buffers are used in the media preparation process as well as the bioreactor process to achieve the ideal pH conditions.

6.7.3 Nutrients

Microorganisms require nutrients for their growth and metabolism. The creation of energy and the process of biosynthesis both need several components. More than any other, carbon is the most crucial component of all living things. Depending on the type of microorganism, other components required for a healthy nutritional bioreactor environment include hydrogen, oxygen, nitrogen, sulfur, phosphorus, iron, calcium, and magnesium (Srivastava *et al.*, 2014). Reactor medium has every macro- and micronutrient need. When pollutants are co-metabolized, microorganisms get a major source of energy in addition to the pollutants they are currently utilizing as a source of energy.

6.7.4 Moisture

To enable microbial growth and catalysis, water is necessary. Since cellular chemical reactions take place in aquatic environments, water is required to maintain the right osmotic pressure for microbial development. The amount of water that can promote the growth of microorganisms is referred to as (aW). Osmotolerant bacteria can develop at a range of low aW, but most bacteria need water activity of 0.98 or higher to flourish (Naik & Duraphe, 2012).

6.7.5 Electron acceptors

Electron acceptors, such as oxygen in aerobic bacteria and NO_3^{1-}, SO_4^{2-}, and Fe (III) oxides in anaerobic microbes, also have an impact on the biodegradation processes.

6.7.6 Factors related to the reactor design

Bioreactors must create the right environment for microbial growth and metabolic activities to occur. Reactor design must take into account the reactor's size, configuration, and mode of operation. The reactor should provide optimal physical, biological, and physical–chemical conditions in order to achieve the best biological remediation processes. To ensure that the physical entity of the bioreactor is advantageously suited to the biological system that conducts the bioreactions, it is essential to construct the bioreactor under favorable physical conditions for the transit of gases, liquids, and solids throughout time (Mandenius, 2016). On the contrary, it is crucial to guarantee that the biophysical and biological processes that are taking place are doing so in settings that are representative of real-world conditions at their peak efficiency.

Both dry and slurry materials can be used to feed polluted samples for remediation into the reactor (Azubuike *et al.*, 2016). Hydrophobic pollutants are frequently resistant to microbial oxidation, especially if they are bonded to soil substrate. Therefore, their transition to liquid is what prevents them from degrading too much. Hexachlorocyclohexane (HCH) was discovered to degrade in slurry batch bioreactors with the least amount of mass transfer resistance (Quintero *et al.* 2007). Due to their extensive use in biotechnology, bioreactors have developed quickly, but there are still issues that need to be resolved regarding the safety and speed of bioprocesses.

The stability and efficiency of the process could be compromised by poor bioreactor design and building that results in insufficient mixing. In addition to keeping the conditions in the reactor consistent and ensuring adequate contact between the media reactants and the microbial culture, mixing prevents thermal stratification. It is impossible to overstate how crucial mixing is in bioreactors; inefficient mixing reduces the effectiveness of the bacteria process. It is essential to identify and optimize the hydraulic retention times (HRT) needed in the bioreactor to accomplish the required remediation goals. Shorter HRTs prevent microorganisms from successfully degrading the pollutant and may cause microbial washout from the system, whereas longer ones cause low substrate concentration that reduces the population of microbes.

6.7.7 Organism-related factors

Variables that affect organisms include population size, composition, and interactions between and within species. The most diverse living things on earth, microbes, have developed a wide range of metabolic pathways that enable them to endure in a wide range of ecological settings, including those that can expose them to xenobiotics. Numerous habitats have been used as sources of microorganisms for bioremediation, including aerobic, anaerobic, acidic, alkaline, and low-to-high temperature (Tekere *et al.*, 2005). Only a few types of microbes and fungi have demonstrated that they are effective pollutant degraders. In the natural world, complex interactions between microbial populations are frequently used to degrade pollutants. For pollutant remediation in bioreactors, either a single or a combination of microbial colonies is used. When bioagumentation is used, the newly imported organisms must get along with the local population. Because different microorganisms frequently have varying metabolic capacities, to find the best degraders, it is required to compare multiple strains of different microbial players.

The ability of newly identified white-rot fungal strains to digest PAHs was examined and compared. It was shown that these fungi did not accumulate the metabolite quinone, which builds up as a dead end metabolite in *P. chrysosporium*, and that they exhibited degradation capacities that were higher or equivalent to the model, highly regarded *P. chrysosporium*.

Environments, which are polluted, are a source of microbes that have developed a tolerance for or adaptation to the contaminant. However, it is also feasible to choose microbes that have been shown to possess particular innate physiological traits, such as the capacity to degrade well-known substrates that resemble significant xenobiotics structurally or the capacity to adapt to particular environmental conditions. This has been demonstrated in numerous experiments that used microorganisms to degrade pollutants.

6.7.8 Pollutant-related factors

Physical and chemical characteristics of the pollutant, such as its solubility, volatility, molecular complexity, concentration, and toxicity, might influence bioremediation in bioreactors. Most studies on pollutant biodegradation have focused on how various concentrations, combined pollutants, solubility, and molecular structure can affect microbial bioremediation. Alkanes degrade more slowly than PAHs than branched chain alkanes, low molecular weight aromatics, and cycloalkanes (Tekere *et al.*, 2005). The fact that some toxins are refractory, or resistant to degradation, and only degrade gradually even in the correct microbial population and environmental conditions, should also be kept in mind.

6.7.9 Mechanism of bioremediation

The process of bioremediation involves a wide range of mechanisms, including electrostatic attraction, redox reactions, adsorption, complexation, ion exchange, and precipitation. Microorganisms may start redox reactions that lead to metal movement or immobilization, affecting the bioremediation process.

Fe, As, Cr, and Hg are examples of heavy metals that go through stages of oxidation and reduction. When an element is converted from its fixed, soluble form in sediments to its mobile, soluble state, bioremediation is considerably simpler. Mobilization can also have detrimental impacts when harmful metal ions are redistributed and discharged from the solid phase of the sediments into the solution phase (Fomina and Gadd 2014). Heavy metals can now enter microbial metabolic pathways and are more bioavailable. The crucial process of heavy metal bio methylation in soil and water can change the toxicity, volatility, and mobility of heavy metals. By allowing the elimination of cells' volatile methylated species, it also significantly contributes to detoxification (Bolan *et al.*, 2014).

The microbial breakdown of organic molecules, which hastens the release of these ions, is another secondary form of metal mobilization. According to research by Wengel *et al.* (2006), *Schizophyllum commune* emits both dissolved organic waste and heavy metals. A vital technique for chelating metal ions is the bacteria's excretion of metabolites like carboxylic acids and amino acids.

To survive, bacteria have evolved defences against metal ions and methods of remediation (Mustapha & Halimoon, 2015). Metals including Cu, Zn, Pb, Cd, and Cr can be swiftly removed using bacterial detritus (Özer & Özer, 2003). The amount of heavy metal ions and the kind of bacteria have an effect on the efficacy of bio sorption because different bacteria have different cellular structures in terms of peptidoglycans such as poly-*N*-acetylglucosamine and *N*-acetylmuramic acid (Hassan *et al.*, 2010). The primary physical interaction between metal ions and bacterial material takes place through the bacterial cell wall. The ability to bind metals to or within the cell wall is conferred by the overall negative charge produced by anionic functional groups (such as amine, hydroxyl, carboxyl, sulfate, and phosphate) found in both Gram-positive and Gram-negative bacteria (in peptidoglycan, teichoic acids, and teichuronic acids) (Sherbet 1978).

Marine ecosystems have a wide variety of different types of algae. The three species of algae – Phaeophyta, Rhodophyta, and Chlorophyta – brown, green, and red algae – have the highest bio sorption capacity, according to studies on phytoremediation. Charge, chemical composition, and algal biomass type and form all have an impact on metal ion biosorption in different ways (Brinza *et al.*, 2007; Oyedepo, 2011). Whether they are in live or dead form, different algae have been used singly or in combination, in batches or columns, in in-situ remediation (Abbas *et al.*, 2014; Romera *et al.*, 2007). Numerous algal proteins have the potential to contain metal sites, such as amine, hydroxyl, carboxyl, sulfate, and phosphate, which work in complex ways to remove heavy metals. During ion exchange, heavy metal ions take the place of the calcium, magnesium, and sodium ions in the cell membrane. Mycoremediation is a reasonably priced method that does not generate any hazardous waste. The proper fungal species must be found and used for the target heavy metal or other pollutants for mycoremediation to be successful. Fungi can efficiently collect heavy metals in their fruit bodies in order to reduce or entirely eradicate their presence in the environment (Ogbo & Okhuoya, 2011). The bio sorption process is mediated by adsorption, ion-exchange, complexation, and interactions with the cell walls of fungal organisms, which contain chitin, proteins, glucans, lipids, pigments, and polysaccharides as well as functional groups like hydroxyl, carboxyl, amino, sulfate, or phosphate. The longevity of the fungi, the chemical composition of the elements, and the presence or absence of the fungi after sequestration all affect how easily accessible heavy metals and other contaminants will be in the media in the future. According to research by Damodaran *et al.* (2011), *Saccharomyces cerevisiae* can absorb up to 65–79% of the Pb and Cd found in polluted soil.

6.8 CONCLUSION

Globally, heavy metal poisoning has serious negative effects on the environment. Several specialized techniques – microorganism-based and hybrid – currently employed to lessen heavy metal pollution have been assessed in this study. Because it is dedicated to adhering to the biogeochemical cycle's principles, which have an effect on the environment, soil structure, and fertility, the ecological microbiome is significant. The future tendency is to promote the treatment of heavy metals by combining multiple strategies to make use of the best exploratory convention, which maximizes the benefits and minimizes the drawbacks. Similar to this, there have not been many heavy metal bioremediation field applications. Research for future innovation upgrade should be directed in this direction to analyse the viability of contemporary waste treatment on an overall scale and with the use of on-site treatment frameworks. Then try to pinpoint any field designing application issues.

REFERENCES

Abbas S. H., Ismail I. M., Mostafa T. M. and Sulaymon A. H. (2014). Biosorption of heavy metals: a review. *Journal of Chemical Science & Technology*, **3**(4), 74–102.

Adams G. O., Fufeyin P. T., Okoro S. E. and Ehinomen I. (2015). Bioremediation, biostimulation and bioaugmentation: a review. *International Journal of Environmental Bioremediation & Biodegradation*, **3**, 28–39, https://doi.org/10.12691/ijebb-3-1-5

Azubuike C. C., Chikere C. B. and Okpokwasili G. C. (2016). Bioremediation techniques – classification based on site of application: principles, advantages, limitations and prospects. *World Journal of Microbiology and Biotechnology*, **32**, 1–18, https://doi.org/10.1007/s11274-016-2137-x

Bolan N., Kunhikrishnan A., Thangarajan R., Kumpiene J., Park J., Makino T. and Scheckel K. (2014). Remediation of heavy metal (loid)s contaminated soils – to mobilize or to immobilize? *Journal of Hazardous Materials*, **266**, 141–166, https://doi.org/10.1016/j.jhazmat.2013.12.018

Brinza L., Dring M. J. and Gavrilescu M. (2007). Marine micro and macro algal species as biosorbents for heavy metals. *Environmental Engineering & Management Journal (EEMJ)*, **6**(3), 237–251.

Chen C. W., Chen C. F. and Dong C. D. (2012). Distribution and accumulation of mercury in sediments of Kaohsiung river mouth, Taiwan. *Apcbee Procedia*, **1**, 153–158, https://doi.org/10.1016/j.apcbee.2012.03.025

Damodaran D., Suresh G. and Mohan R. (2011). Bioremediation of soil by removing heavy metals using *Saccharomyces cerevisiae*. 2nd International Conference on Environmental Science and Technology, Singapore.

Flora S. J. S., Mittal M. and Mehta A. (2008). Heavy metal induced oxidative stress & it's possible reversal by chelation therapy. *Indian Journal of Medical Research*, **128**(4), 501–523.

Fomina M. and Gadd G. M. (2014). Biosorption: current perspectives on concept, definition and application. *Bioresource Technology*, **160**, 3–14, https://doi.org/10.1016/j.biortech.2013.12.102

Gargouri B., Karray F., Mhiri N., Aloui F. and Sayadi S. (2011). Application of a continuously stirred tank bioreactor (CSTR) for bioremediation of hydrocarbon-rich industrial wastewater effluents. *Journal of Hazardous Materials*, **189**(1–2), 427–434, https://doi.org/10.1016/j.jhazmat.2011.02.057

Gordon J. J. and Quastel J. H. (1948). Effects of organic arsenicals on enzyme systems. *Biochemical Journal*, **42**(3), 337–350, https://doi.org/10.1042/bj0420337

Hasin A. A., Gurman S. J., Murphy L. M., *et al.* (2010). Remediation of chromium (VI) by a methane-oxidizing bacterium. *Environmental Science and Technology*, **44**, 400–405, https://doi.org/10.1021/es901723c

Hassan S. H., Awad Y. M., Kabir M. H., Oh S. E. and Joo J. H. (2010). Biotechnology Cracking New Pastures. Bacterial Biosorption of Heavy Metals. *MD Publications Pvt. Ltd, New Delhi*, pp. 79–110.

Kumar A., Bisht B. S., Joshi V. D. and Dhewa T. (2011). Review on bioremediation of polluted environment: a management tool. *International Journal of Environmental Sciences*, **1**(6), 1079–1093.

Li C. H., Wong Y. S. and Tam N. F. (2010). Anaerobic biodegradation of polycyclic aromatic hydrocarbons with amendment of iron (III) in mangrove sediment slurry. *Bioresource Technology*, **101**, 8083–8092, https://doi.org/10.1016/j.biortech.2010.06.005

Lin C. C. and Lin H. L. (2005). Remediation of soil contaminated with the heavy metal (Cd^{2+}). *Journal of Hazardous Materials*, **122**, 7–15, https://doi.org/10.1016/j.jhazmat.2005.02.017

Liu S. H., Zeng G. M., Niu Q. Y., Liu Y., Zhou L., Jiang L. H. and Cheng M. (2017). Bioremediation mechanisms of combined pollution of PAHs and heavy metals by bacteria and fungi: a mini review. *Bioresource Technology*, **224**, 25–33, https://doi.org/10.1016/j.biortech.2016.11.095

Mandenius C. F. (2016). Bioreactors: Design, Operation and Novel Applications. John Wiley & Sons, Weinhem, Germany.

Mazumder D. G. (2008). Chronic arsenic toxicity & human health. *Indian Journal of Medical Research*, **128**(4), 436–447.

Mishra G. K. (2017). Microbes in heavy metal remediation: a review on current trends and patents. *Recent Patent Biotechnology*, **11**(3), 188–196.

Monalisa M. and Kumar P. H. (2013). Effect of ionic and chelate assisted hexavalent chromium on mung bean seedlings (*Vigna radiata* L. Wilczek. var k-851) during seedling growth. *Journal of Stress Physiology & Biochemistry*, **9**(2), 232–241.

Mora A. P., Calvo J. J. O., Cabrera F. and Madejon E. (2005). Changes in enzyme activities and microbial biomass after 'in situ' remediation of a heavy metal-contaminated soil. *Applied Soil Ecology*, **28**, 125–137, https://doi.org/10.1016/j.apsoil.2004.07.006

Mulligana C. N. and Yong R. N. (2004). Natural attenuation of contaminated soils. *Environment International*, **30**, 587–601, https://doi.org/10.1016/j.envint.2003.11.001

Mustapha M. U. and Halimoon N. (2015). Microorganisms and biosorption of heavy metals in the environment: a review paper. *Journal of Microbial and Biochemical Technology*, **7**(5), 253–256, https://doi.org/10.4172/1948-5948.1000219

Naik M. G. and Duraphe M. D. (2012). Review paper on-parameters affecting bioremediation. *Advance Research in Pharmaceuticals and Biologicals*, **2**(3), L77–L80.

Ogbo E. M. and Okhuoya J. A. (2011). Bio-absorption of some heavy metals by *Pleurotus tuber-regium* Fr. Singer (an edible mushroom) from crude oil polluted soils amended with fertilizers and cellulosic wastes. *International Journal of Soil Science*, **6**(1), 34–48, https://doi.org/10.3923/ijss.2011.34.48

Oyedepo T. A. (2011). Biosorption of lead (II) and copper (II) metal ions on *Calotropis procera* (Ait.). *Science Journal of Purel & Applied Chemistry*, **2011**, 1–7.

Özer A. and Özer D. (2003). Comparative study of the biosorption of Pb (II), Ni (II) and Cr (VI) ions onto *S. cerevisiae*: determination of biosorption heats. *Journal of Hazardous Materials*, ***100***(1–3), 219–229, https://doi.org/10.1016/S0304-3894(03)00109-2

Patrick L. (2002). Mercury toxicity and antioxidants: part I: role of glutathione and alpha-lipoic acid in the treatment of mercury toxicity. (mercury toxicity). *Alternative Medicine Review*, **7**(6), 456–472.

Pazirandeh M., Chrisey L. A., Mauro J. M., *et al.* (1995). Expression of the *Neurospora crassa* metallothionein gene in *Escherichia coli* and its effect on heavy-metal uptake. *Applied Microbiology and Biotechnology*, **43**, 1112–1117, https://doi.org/10.1007/BF00166934

Quintero J. C., Lu-Chau T. A., Moreira M. T., Feijoo G. and Lema J. M. (2007). Bioremediation of HCH present in soil by the white-rot fungus *Bjerkandera adusta* in a slurry batch bioreactor. *International Biodeterioration & Biodegradation*, ***60***(4), 319–326, https://doi.org/10.1016/j.ibiod.2007.05.005

Rodríguez M. C., Barsanti L., Passarelli V., Evangelista V., Conforti V. and Gualtieri P. (2007). Effects of chromium on photosynthetic and photoreceptive apparatus of the alga *Chlamydomonas reinhardtii*. *Environmental Research*, ***105***(2), 234–239, https://doi.org/10.1016/j.envres.2007.01.011

Romera E., González F., Ballester A., Blázquez M. L. and Munoz J. A. (2007). Comparative study of biosorption of heavy metals using different types of algae. *Bioresource Technology*, ***98***(17), 3344–3353, https://doi.org/10.1016/j.biortech.2006.09.026

Sharma S., Singh R. and Singh P. (2011). Microorganism as a tool of bioremediation technology for cleaning environment: a review. *Proceedings of the International Academy of Ecology and Environmental Sciences*, **4**(1), 1–6.

Sherbet G. V. (1978). *Biophysical characterisation of the cell surface*. Academic press, London.

Srivastava J., Naraian R., Kalra S. J. S. and Chandra H. (2014). Advances in microbial bioremediation and the factors influencing the process. *International Journal of Environmental Science and Technology*, ***11***, 1787–1800, https://doi.org/10.1007/s13762-013-0412-z

Tekere M., Read J. S. and Mattiasson B. (2005). Polycyclic aromatic hydrocarbon biodegradation in extracellular fluids and static batch cultures of selected sub-tropical white rot fungi. *Journal of Biotechnology*, **115**(4), 367–377, https://doi.org/10.1016/j.jbiotec.2004.09.012

Thurmer K., Williams E. and Reutt-Robey J. (2002). Autocatalytic oxidation of lead crystallite surfaces. *Science*, ***297***(5589), 2033–2035, https://doi.org/10.1126/science.297.5589.2033

Wengel M., Kothe E., Schmidt C. M., Heide K. and Gleixner G. (2006). Degradation of organic matter from black shales and charcoal by the wood-rotting fungus *Schizophyllum commune* and release of DOC and heavy metals in the aqueous phase. *Science of the Total Environment*, ***367***(1), 383–393, https://doi.org/10.1016/j.scitotenv.2005.12.012

doi: 10.2166/9781789063783_0107

Chapter 7

Clean production approaches in industries: a case study on pulp and paper production facility applications

Günay Yildiz Töre[1] and Melisa Can[2]*

[1]Department of Environmental Engineering, Çorlu Faculty of Engineering, Tekirdağ Namık Kemal University, Tekirdağ, Türkiye
[2]Environmental Engineering Department, Institute of Natural and Applied Sciences, Tekirdağ, Turkey
**Corresponding author: gyildiztore@nku.edu.tr*

ABSTRACT

The sustainability of life and economy is directly dependent on natural resources, and the effects of the use of these resources are felt on both ecology and the economy. As scarce and valuable resources such as raw materials, energy, and water constitute the basic inputs of the economy as well as life. Resource use not only causes increasing pressure on natural resource reserves and adversely affects the environment, but also affects national and international trade and market prices. The efficient and sustainable use of raw materials, energy, and water by both reducing the impact on the environment and producing more value by using less input also contributes to the sustainability of production and therefore the economy. Especially in the second half of the 20th century, the destruction of environmental values brought about by rapid technological and industrial developments and the rapid decrease of non-renewable resources continue to accelerate today.

The rapid industrialization that started with the Industrial Revolution in the 18th century and the unstoppable increase in the world population caused production and consumption to increase at the same rate. In parallel with the changes in industrialization and lifestyles, the wastes have shown a logarithmic increase over time and the local environmental problems due to these wastes have gained a global dimension. The uncontrolled discharge of these wastes continued for years until it was understood that there was a limit to the capacity of nature to accept the wastes generated after production and consumption. Thinning of the ozone layer, global warming, acid rain, toxic and hazardous waste discharges in quantities far above the absorption capacity to various natural receiving environments can be counted within this scope. One of the reasons for the increase in environmental problems has been the very high cost of investments based on the end-of-pipe approach to environmental protection. For this reason, over time, industrial organizations have focused on 'clean production techniques and practices,' which is an approach to prevent waste at its source. It has been observed that the industries that have started *Clean Production Practices* and allocate a budget for investments when necessary, amortize this cost in a short time as it saves water and energy from the cost of waste treatment, and therefore the application area is expanding day by day. It has been understood that *clean (sustainable) production* can be applied to all type of industries, including small and large, regardless of material, energy and water consumption levels. Observations show that this approach offers the potential to reduce resources by an average of 10–15% without costly investments.

Therefore, in this study, the clean production (CP) approaches in different industries will be discussed with details and CP practices of the pulp and paper production facility will be evaluated in this context.

Keywords: clean production and clean technologies, sustainability, waste, waste reduction

7.1 INTRODUCTION

Especially in the second half of the 20th century, the destruction of environmental values and the rapid decrease of non-renewable resources, brought about by technological and industrial developments, are accelerating today. Wastes that emerged in parallel with industrialization and changes in lifestyles showed a logarithmic increase over time and local environmental problems caused by these wastes have gained a global dimension. Thinning of the ozone layer, global warming, acid rains, various natural receiving environments – toxic and hazardous waste discharges in quantities far above their absorptive capacity can be counted in this context.

The first approach developed to prevent the destruction of environmental values and which has been used intensively until today was the elimination of pollutants after they have emerged. This approach, referred to as pollution control (or 'end-of-pipe'), can be defined as the elimination or removal of pollutants using various environmental technologies after they emerge. The removal of the pollutant with such an approach brings with it a high investment requirement. The financial burdens that conventional treatment and disposal facilities bring to the investor cause people or institutions to refrain from environmental investments in some areas and countries. One of the reasons why environmental problems are increasing is the very high cost of investments in environmental protection.

As a result of the continuous increase in the amount of waste generated in the process and the cost of treatment, the continuous decrease in the discharge standards of the receiving environment in parallel with the increasing environmental awareness in the public has led the institutions and sectors that produce products and services to seek cheaper solutions to this problem. In addition, increasing environmental awareness in the last 20–30 years has led consumers living in developed countries to increasingly prefer products and processes that are less harmful to the environment in production, use, and post-use processes.

The studies that started after this new orientation have revealed that even with simple precautions to be taken, these losses can be prevented and waste generation can be reduced as a result of the more effective use of raw materials that become waste before they can be turned into a useful product in the production process. This was followed by approaches such as increasing efficiency in the production and service sectors, replacing the raw materials used for production with those that are less harmful to the environment, and reducing the water and energy needs required in the production and usage processes. The result is waste reduction, recycling, reuse, more environmentally responsible design of products and services, and so on based on several research studies and 'pollution control' approaches have been replaced by 'clean production' approaches (Demirer, 2003).

For all these reasons, many countries have started to focus on clean production (CP) practices compared to the pollution control (end-of-pipe) approach. In a study conducted in 2004 by the Center for European Economic Research active in Germany, the current situation of end-of-pipe and CP approaches in the Organization for Economic Cooperation and Development (OECD) countries was compared (Figure 7.1). As can be seen in Figure 2.1, it is seen that the CP approach is at the forefront especially in developed countries, especially in Japan and France.

7.1.1 Clean (sustainable) production concept and approach

The concept of 'clean production' is defined by the United Nations Environment Program (UNEP) as 'reducing the risks on people and the environment through the continuous application of a holistic and preventive environmental strategy to products and processes' (UNEP (United Nations Environment Programme), 1996).

Contrary to the 'pollution control' approaches that try to solve environmental problems after they arise, CP approaches require that environmental issues be included in the planning processes as a parameter in the design phase of industrial, urban, agricultural, and all kinds of human activities. Contrary to conventional pollution control approaches, the CP approach aims to prevent/reduce pollution before it occurs. Pollution control approaches adopt production and design stages as constant

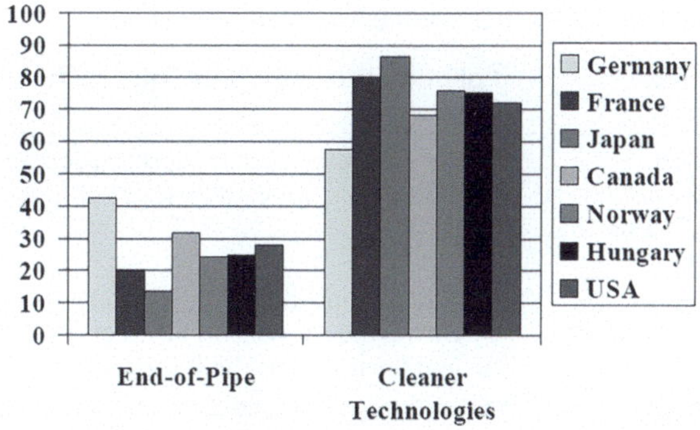

Figure 7.1 Current state of end-of-pipe and CP approaches in some OECD countries. *Source*: Frondel *et al.* (2004).

factors, and see pollution as an inevitable result of these stages. After the pollution has emerged, it tries to find a solution to this problem. Therefore, these approaches focus on better identification of pollution and treatment and disposal of waste, bringing significant additional costs to organizations. Contrary to conventional pollution control approaches, the CP approach aims to prevent/reduce pollution before it occurs (Table 7.1).

On the contrary, CP approaches see pollution and waste as a result of inadequacy, inefficiency, and ineffectiveness in the stages of design, resource use, and production processes and it aims at solving the problem by providing the necessary developments at these stages. Therefore, it not only reduces waste generation, but also provides economic benefits. CP is closely related to the concept of sustainability, as well as the development of environmental-friendly new products, processes, systems, and services that are compatible with natural processes (Glavic & Lukman, 2007; Kjaerheim, 2005).

In addition, UNEP states that the CP approach has an important role in fulfilling the responsibilities of countries related to international agreements (http://www.unep.org/pdf/dtie/CP_MEA_and_Cleaner_Production.pdf).

Since pollution control approaches aim only to comply with the laws and regulations in force, the attempts of the organizations to contribute to their environmental performance are limited to a change in the laws and regulations. Such an approach not only ignores many potential developments, but also causes unpreparedness in planning and implementing control and prevention practices in the event of radical changes in environmental pollution, and environmental improvements are only possible with high-cost initiatives. On the contrary, CP approaches are the environmental initiatives of the enterprises themselves. It aims to continuously monitor and increase its performance, and thus prevents these improvement requirements from being limited to static facts such as laws and regulations. Organizations that adopt and implement pollution prevention approaches will not have difficulty in complying with stricter laws and regulations that will come into force in the future, as they will increase their environmental performance to a much higher level than those required by these laws and regulations.

7.1.2 Development of CP concept

The United Nations Environment Program – Department of Technology, Industry and Economics (UNEP-DTIE) started the Clean Production Program in 1989 and took the first important step towards raising awareness on the issue, establishing an institutional structure and promoting sustainable development efforts by demonstrating its benefits. Since then, the concept of 'clean production', which has been adopted by many countries, institutions, and organizations, has gained a global character. For example, many references were made to the concept of CP, which is mentioned as an important

Table 7.1 Differences between approaches to prevent environmental pollution.

Conventional Pollution Control Approach	Pollution Prevention/CP Approach
Adopts the classical end-of-pipe treatment approach. It focuses not on the problem itself and its source, but on the removal of the resulting pollution.	Based on prevention at source approach: Deals with the problem itself and its source and tries to eliminate the root cause
Environmental improvement practices for pollution control require additional costs.	Pollutants and wastes are considered as potential sources that can be converted into useful products or by-products and are tried to be managed within the framework of circular economy.
Identification and application of pollution control technologies, waste managers, and so on is the responsibility of environmental experts.	All employees, including design and process engineers responsible for production, play an active role and take responsibility in all processes related to the improvements planned for pollution control, prevention or elimination.
Environmental improvements are achieved by carrying out technological applications that provide pollution control at the technical level.	Planned improvements to prevent pollution require the integration of non-technical and standardized managerial practices (ISO 9001, ISO 14001, ISO 45001, ISO 50001, etc.) with the use of technology.
Improvements in pollution control are primarily determined by taking into account the needs of customers, suppliers, or consumers.	The development or optimization of production processes is based on minimizing the effects on human health and the environment in producing products that will respond to the needs of customers, suppliers, or consumers by multi-dimensional analysis.

strategy to implement the concept of sustainable development in the Agenda 21 Program at the Rio Summit in 1992 (UNEP (United Nations Environment Programme), 2002).

When the examples of various countries are examined, the development of the concept of CP in a country generally started with raising awareness on the subject and continued with capacity building studies including exemplary practices in the production and service sectors. Efforts were made to spread CP practices through partnerships and information sharing networks created, followed by the establishment of financial mechanisms and the implementation of necessary policy reforms. It has been reported by UNEP that the typical process of the development of the concept of CP in a country can take place from the bottom up, but it can sometimes be realized from the bottom up or in different ways due to local, cultural, and similar reasons (UNEP (United Nations Environment Programme), 2002) (Figure 7.2). As a result of CP initiatives and efforts initiated by UNEP/United Nations Industrial Development Organization (UNIDO), the number of National Clean Production Centers (NCPC) established since 1994 has reached 58 and this number is increasing day by day. NCPCs have carried out important works for the development of the concept of CP in their countries/regions with their capacity building works, handbooks they published, trainings they implemented, demonstration projects and other activities. In Turkey, the UNIDO Eco-efficiency (Clean Production) Program has been carried out by TDFT under the responsibility of UNIDO since 2008 as a sub-program within the scope of 'Developing Turkey's Capacity to Adapt to Climate Change'. One of the goals of this program is the establishment of a national Ecoefficiency (Clean Production) Center (The UNEP Working Group for Cleaner Production in the Food Industry, 2004).

The development of the institutional structuring in CP, the trainings given, and the increase in technical infrastructure and competence have also caused a rapid increase in the number of projects carried out on the subject all over the world. Many information sharing networks have been created in order to convey the CP activities implemented today to all stakeholders. These include UNEP/UNIDO National Network of Clean Production Centers (http://www.unep.fr/scp/cp/network/ncpc.htm), Western Network for Sustainability and Pollution Prevention (http://www.westp2net.org/), United

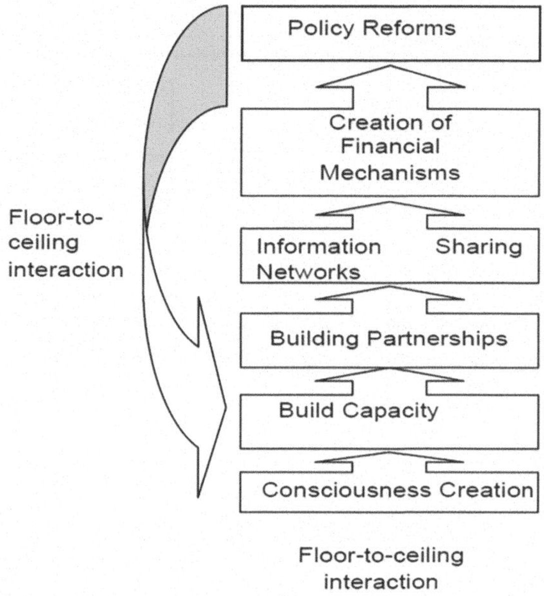

Figure 7.2 Typical process of development of CP concept in a country.
Source: UNEP (United Nations Environment Programme) (2002).

States Environmental Protection Agency (USEPA) Pollution Prevention Information Network (http://www.epa.gov/opptintr/ppic/) and others (http://www.cleanerproduction.com/Directory/networks.htm; http://www.serd.ait.ac.th/teenet/cleaner.htm). All these have made a serious knowledge on the subject available to the stakeholders.

At this point, different concepts, which are mentioned together with the concept of CP and overlap with the CP approach, have also been developed. One of the most interesting of these approaches is industrial symbiosis (industrial ecology). Industrial symbiosis represents that two or more economic processes that are physically close to each other and normally work independently of each other, establishing long-term partnerships to improve both environmental performance and competitiveness. In this respect, with the environmental management systems that can be designed for industrial zones similar to the Organized Industrial Zone (OIZ), by-products, residues, or wastes produced by one enterprise can be used as raw materials for another enterprise. In this way, it not only prevents environmental problems of industrial origin, but also provides economic returns.Consequently, after the pollution has occurred, the end-of-pipe approach, which consists of treating or eliminating with various equipment or technologies, has been replaced by methods that will prevent the generation of wastes from production processes with simple precautions to be taken. These methods were followed by approaches such as reducing the material content of the products, replacing the raw materials used for production processes with less harmful and less toxic ones to the environment, and reducing the water and energy needs required in the production and usage processes. This subject has become popular over time and it has become popular in terms of waste reduction, recycling, reuse, creation of products, and services with a more sensitive design on environmental damage, and so on. The number of researches and developed technologies on these issues has increased and has come to be known as 'Pollution Prevention' and, in parallel, 'Clean Production.'Clean production: it encompasses the continuous and regular implementation of a preventive and integrated environmental strategy applied to production processes, products, and services. CP also ensures that natural resources are

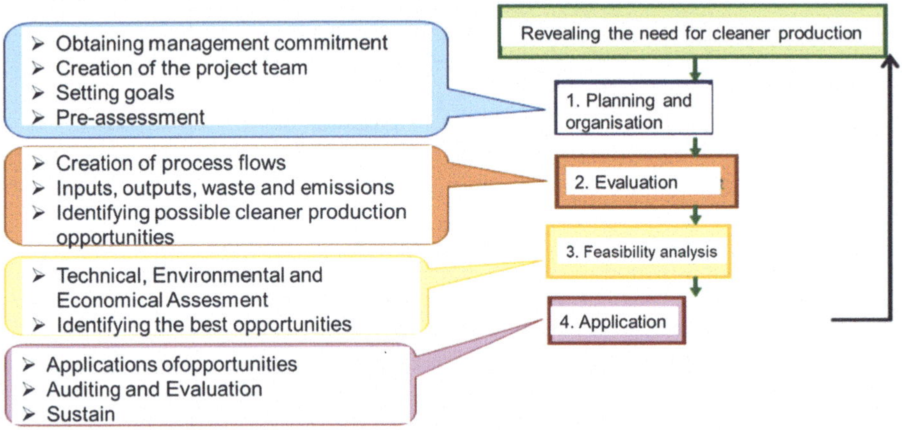

Figure 7.3 Typical CP methodology.

used more efficiently and that the risks of waste and pollution occurring from this process are likely to harm human health are reduced. In addition, it tries to eliminate environmental problems with the applications and technologies developed at the source of the production process, not with the end-of-pipe method at the end of the production process. Typical CP methodology is given in Figure 7.3.

7.1.3 CP to sustainable production

The rapid increase in studies conducted in different disciplines on the concept of 'sustainability' has also increased the use of different concepts and terminologies. CP is an interdisciplinary subject by its nature, so the concept and terminology developed by a particular discipline is used in different contexts by different disciplines. The emergence of a rapidly increasing number of concepts due to geographical and cultural differences causes various difficulties in understanding the studies carried out in the field of CP and in collaborating the concept and terminology. Although the concept of CP is still being used by many related organizations in the industry, sustainable production concept is also spreading rapidly in recent years.

The concept of CP has evolved into the concept of sustainable production in the last 5 years globally (Narayanaswamy & Stone, 2007). The concept of CP is still used by many related institutions (http://www.unido.org/index.php?id=o4460). Although it continues to be used, the concept of 'sustainable production' is rapidly spreading (http://sustainableproduction.org/). Sustainability can be defined that development of it has started recently (Clark, 2007; Luskin & Del Matto, 2007), an economically viable by preserving energy and natural resources with processes and systems and provide constructive and social benefits for all stakeholders in a safe and healthy environment for employees, consumers, and society in the production process.

The concept of CP has been used for the last 20 years. Different audiences have used the same concept in different contexts, in earlier studies, 'integrated pollution prevention and control,' 'life cycle analysis,' 'supply chain management,' and 'industrial ecology were used. The concepts of CP came to be used under the concept of 'umbrella.' However, these concepts are included in the subgroups of the concept of sustainable production, not CP, in the article of Glavic and Lukman (2007).

7.1.4 CP and eco-efficiency

Eco-efficiency is based on the principle of using less natural resources and energy and producing less waste for the same amount of production, with the use of highly efficient production technologies or managerial tools. The concept of eco-efficiency addresses not only environmental pollution and depletion of natural resources, but also many different areas such as 'protection of natural resources,' 'industrial efficiency,' and 'economic development.' In short, eco-efficiency can be defined as providing both environmental and economic benefits in parallel by increasing efficiency in production. In the eco-efficiency approach, 'end-of-pipe (pollution control) practices,' which refer to the disposal and treatment of waste after it is formed, is replaced by the selection of materials (raw materials and auxiliary materials), product design, procurement, production, transportation, and so on of environmental effects. It leaves it to an attitude that covers the processes and enables it to be projected in a wide framework. Eco-efficiency has goals and methods that conceptually overlap with the CP approach. Both concepts aim to minimize 'natural resource and energy consumption', 'use of toxic and dangerous chemicals,' and 'waste, wastewater and emission generation' in the production process (TDFT (Turkey Technology Development Foundation), 2011).

7.1.5 CP and industrial ecology (symbiosis)

Environmental management systems and CP practices not only increase the environmental performance of industrial organizations, but also positively affect their economic performance and corporate prestige. Although these practices are extremely important and effective, environmental performance can be improved to a certain extent since they remain within the company's boundaries. Providing additional gains in terms of environmental performances requires being able to go beyond company boundaries and often inter-firm cooperation. Industrial ecology is the presentation of a new conceptual framework for understanding the effects of industrial systems on the environment. This new framework sets out the ultimate goal of sustainable development, strategies to reduce, and then implement the environmental impacts of products and processes.

Industrial ecology is an approach used for the design of industrial products and processes by evaluating activities in terms of product competition and environmental interactions, in other words, the physical, chemical, and biological interactions between the industry and the environment. Because environmental problems are systematic, industrial applications require a systematic approach between human activities and environmental–ecological processes. A systematic approach involves identifying and solving problems, which makes it easier (Das, 2005; Davarcıoğlu, 2017).

At this point, different concepts, which are mentioned together with the concept of CP and overlap with the CP approach and have many aspects, have also been developed. One of the most interesting of these approaches is 'industrial symbiosis (industrial ecology)'. This approach has been implemented in many countries today. Industrial symbiosis, which first came to the fore in 1989, is based on the analogy between industry, natural life, and ecological systems and it is based on each other both economically and in terms of using each other's products and wastes (matter and energy) and it symbolizes the whole network of industrial processes. Industrial symbiosis represents that two or more economic processes, normally work independently of each other and preferably physically close to each other, coming together and establishing long-term partnerships that will increase both environmental performance and competitiveness, and working in solidarity. In this respect, with environmental management systems that can be designed for industrial zones similar to the organized industrial zone (OIZ), by-products, residues, or wastes produced by one enterprise can be used as raw materials for another enterprise. In this way, it not only prevents industrial environmental problems, but also provides economic returns. Especially implementation and investment projects in the field of waste evaluation (product from waste, energy from waste, etc.) continue at an increasing pace in many countries (Özbay, 2005).

7.2 CP BENEFITS/GAINS

CP approach is an approach that has the ability to provide permanent solutions to environmental pollution problems, which are on the agenda, by preventing pollution before it occurs. The number of organizations that are developing in this context is increasing day by day in the world. CP practices have many benefits for facilities (TDFT (Turkey Technology Development Foundation), 2011).

7.2.1 Economic gains

On the one hand, CP practices integrated into processes will reduce the cost of the industry as it increases efficiency and reduces the use of energy, natural resources, and raw materials. On the other, since it minimizes the wastes to be generated, disposal costs are also avoided. Less energy, natural resource, and raw material consumption is a beneficial practice for the industry as well as the environment.

7.2.2 Compliance with regulations

Industries that adopt the CP approach are more likely to not only comply with today's environmental regulations, but also meet the requirements of regulatory rules that are expected to become more stringent in the future.

7.2.3 Compliance with legal sanctions

Implementations to be realized within the scope of CP will prevent legal sanctions that may occur due to incompatibilities that may arise against the demands of the regulations and environmental accidents that may take place.

7.2.4 Motivation of employees

Employees are the biggest factor in the success of an organization. Employees of an organization that adopts and implements CP strategies will be motivated by the satisfaction of working with an organization that respects society and the environment, and they continue to work in a more motivated way by embracing CP values.

7.2.5 Environmental benefits

As the studies carried out within the scope of CP destroy the wastes discharged to the environment at their source before they occur, they prevent the formation of waste water, emissions, and other waste types originating from processes. Another environmental benefit is that these applications minimize or eliminate the use of energy, raw materials, natural resources, and harmful substances.

7.2.6 Increasing institution and product image

An organization's care about the damage it causes to the environment due to the processes taking place within its body and taking actions for compensation in this regard ensures that it receives the support of the society and employees. In this way, it increases the chance of increasing its market share and being superior to its competitors.

7.2.7 Reducing possible risks against occupational health and safety

As a result of the practices carried out, working conditions are improved and developed, preventing possible occupational accidents, while also preventing workers from being exposed to pollution and dangerous substances.

7.3 OBSTACLES IN CP PRACTICES

Although CP technologies, whose application area has expanded in recent years, have many benefits for businesses, many different difficulties may be encountered during the implementation phase (Demirer & Mirata, 1999).

7.3.1 Economic challenges

Although a significant portion of pollution prevention practices can be implemented with no cost and/or very low-cost measures. However, the inability of organizations to find financing for these less costly or higher cost measures to be implemented in the future is one of the most important obstacles for CP practices. To overcome this obstacle, the benefits of CP technology applications should be well understood by the organizations.

7.3.2 Barriers to implementation and management

(a) **Management indifference:** The viability, success, and sustainability of each new approach to be implemented in an organization is possible if the senior management embraces the project and tries to embrace it by providing various trainings and necessary equipment to the employees.

(b) **Financing:** It is possible to have maintenance, repair, renewal, and equipment changes and additions over time in order to implement and ensure the continuity of the applications. Financial investments are required for all these situations. At this point, a budget should be allocated for CP practices in order to avoid problems.

(c) **Product quality:** The technology to be applied may cause changes in the raw materials and/or processes during the production stages, and the fear that this may lead to changes in product quality, decrease in market share, and bad reputation is an obstacle to the application of CP technologies.

(d) **Employee resistance:** Changes to be experienced in processes may lead to changes in the working routines of employees, and naturally, employees who have difficulty getting used to this situation may have difficulties during implementation. Adequate awareness and equipment will help overcome this obstacle.

(e) **Continuity:** Maintenance, repair, renewal, or add-on needs that may occur in plans and projects from the implementation phase of the applications should be foreseen in the plan and project stages, the necessary capital and awareness in this regard are necessary for the sustainability and longevity of the application.

7.4 COMPONENTS, TOOLS, AND METHODS OF CLEAN (SUSTAINABLE) PRODUCTION

7.4.1 Clean (sustainable) production components

Applications that can be realized within the scope of clean (sustainable) production can be classified under three main headings (Regional Activity Centre for Cleaner Production (CP/RAC), 2000):

- Waste reduction and resource consumption at source reduction
- Reuse and/or recycling
- Product modifications

Focusing on the source of pollution and reducing resource consumption is the basic principle of clean (sustainable) production (Figure 7.4), and in this context, the components of CP are summarized in Figure 7.5.

7.4.1.1 Reduction of waste at source

Administrative precautions constitute one of the simplest methods of clean (sustainable) production and do not bring any investment costs and can be implemented immediately after the determination of the opportunities. Keeping the water valves closed, not running the equipment in vain, optimizing the dosage of chemicals, and so on through prevention of water, energy, and other resource losses are among the examples that can be given by means of administrative methods. These methods require a particular focus on the management and training of employees. Better process control covers by checking optimum process operating conditions in terms of sources consumption, production, and waste generation whether it is suitable for the level includes proper arrangement. It covers parameters such as temperature, time, pressure, pH, process speed, and keeping them as close to optimum levels as possible. Better process control requires more advanced monitoring and management than administrative precautions. It covers parameters such as temperature, time, pressure, pH, process

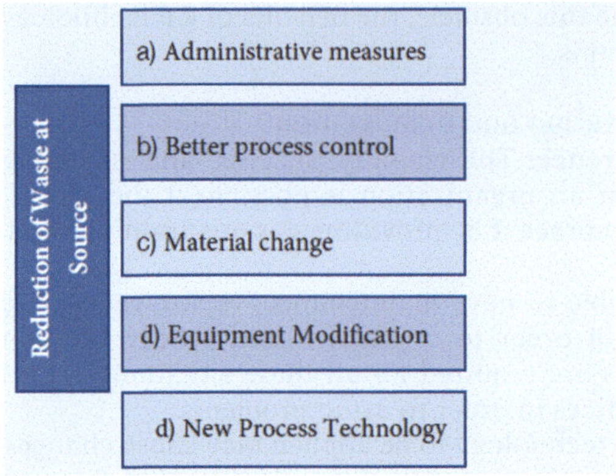

Figure 7.4 Practices for reducing wastes at source.
Source: Vietnam Cleaner Production Center (VNCPC) (2000).

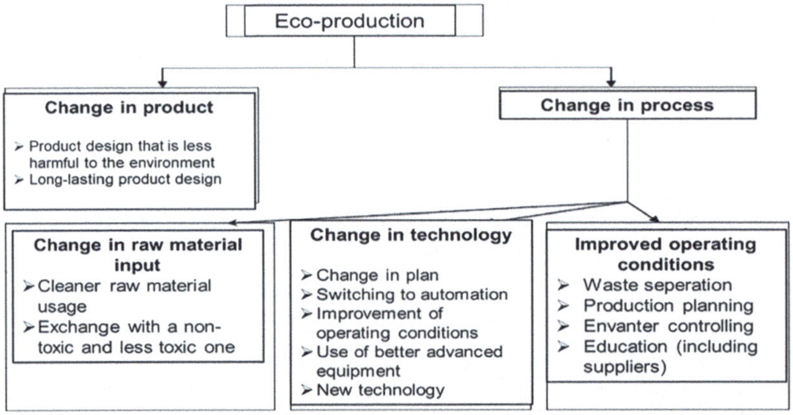

Figure 7.5 Components of eco-production.

speed and keeping them as close to optimum levels as possible. Better process control requires more advanced monitoring and management than administrative precautions. Material change, production efficiency from cost and quality, a higher quality will increase without compromising use of the material. Generally, there is a direct relationship between the quality of the materials and the quality and quantity of the products. In addition, material replacement includes replacing existing materials with those with better environmental properties. For example, the replacement of a paint containing hazardous chemicals with an environmentally friendly alternative, also ensures that the treatment requirements and costs are eliminated or reduced. Equipment modification is the improvement of existing equipment to create less waste and provide more efficient production processes. Adjusting engine speeds, reactors, tanks, optimization of their volumes, isolation of hot and cold surfaces, are some examples of equipment modification. Process change involves the use of modern and more efficient techniques and technologies. This application should be considered carefully as it requires a higher initial investment cost than other clean (sustainable) production applications. On the contrary, potential gains and increases in quality ensure that the investment returns in a short time, making it easier for the company to switch to more up-to-date and contemporary production processes. Such applications often provide to improve product and production quality (Vietnam Cleaner Production Center (VNCPC), 2000).

7.4.1.2 Reuse/recycle

Waste that cannot be avoided can be recycled within the company or as a by-product. On-site recycling or reuse covers collection of wastes and their reuse in the same or different parts of production. The reuse of washing water originating from one process in another cleaning process is an example that can be given to this application. Produced by-products covers waste sold to consumers or other companies as input after collection (if necessary after a certain processing). For example, waste yeast from the brewery can be used as animal feed, fish production and as a food additive (Vietnam Cleaner Production Center (VNCPC), 2000).

7.4.1.3 Product modification

Product specifications to reduce pollution from products replacement is among the main principles of clean (sustainable) production. Changing the product requires rethinking the product and its requirements. Improved product design allows for substantial reductions in material consumption and use of hazardous materials. Lightening the products, reducing the wall thickness, designs that make the product more recyclable, and so on can be given as an example of this approach. Packaging change is one of the important applications in this regard. The main point of view here is to minimize the amount of packaging material while guaranteeing the protection of the product. One of the most common examples in this field is the use of recycled cardboard instead of plastic foam to protect fragile materials (Vietnam Cleaner Production Center (VNCPC), 2000).

7.4.2 Clean (sustainable) production tools and methods

To implement the CP approach, different methods that can be applied to products and processes have been studied. As a result of these studies, many tools have been put forward to identify CP opportunities, and the tools that should be used change according to the application area. The choice of these tools depends on the problem in the business and the work to be done. One or more of the tools can be used depending on the nature of the study and the problem (Özbay, 2005).

The main tools and methods that make up the elements of CP practices are given in Figure 7.6.

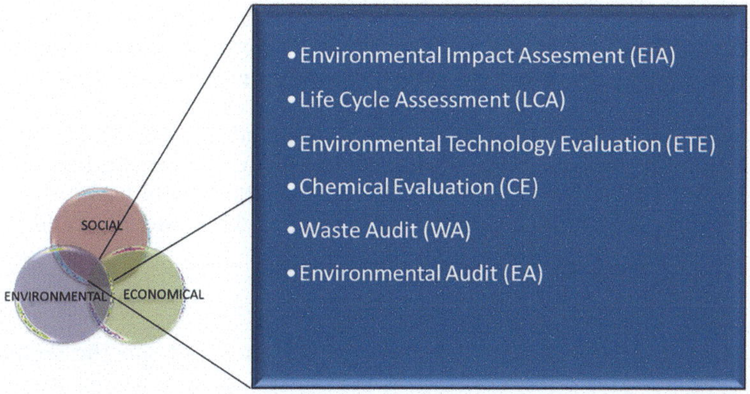

Figure 7.6 Main tools and methods of CP.

7.4.2.1 Environmental impact assessment

Environmental impact assessment (EIA) is a process that determines the significant impacts of a particular project or development on the environment (REC, 2011). EIA refers to the studies to be carried out in determining the positive and negative effects of the planned projects on the environment, in determining and evaluating the measures to be taken in order to prevent or minimize the negative effects in a way that does not harm the environment, the selected location and technology alternatives, and the monitoring and control of the implementation of the projects. In this context, EIA, which is used as a tool in CP practices, provides significant benefits within the framework of the following purposes:

- Identification of adverse environmental impacts expected to occur.
- Adding the necessary measures to the project in order to prevent these negative environmental effects.
- Determination of the project acceptability by the public in addition to the economic aspect also environmental.
- Determination of additional studies and monitoring mechanisms regarding the significant environmental impacts of the project.
- Ensuring public participation in decision-making processes.
- Helping all groups related to the realization of the project and its environmental impacts to understand their roles, responsibilities and relations with each other in this project.
- Ensuring that the process proposed in the project is BAT.

7.4.2.2 Life-cycle assessment

Life-cycle assessment (LCA) is a tool used to prevent or reduce these effects by analyzing the environmental impacts that occur in the process from the production of products and services to their use and disposal. The resources used throughout the entire life cycle of a product or service, as well as the environmental impact of emissions to air, water, and soil, and improvement opportunities can be systematically evaluated by means of LCA. The life-cycle stages cover the extraction and processing of raw materials, production, storage, and distribution of products, use, recycling, and disposal (cradle-to-cradle). LCA is mainly used in the development of environmental labeling criteria, raw materials of products, production processes, equipment, and so on. It is used to prevent/reduce environmental impacts through replacement and/or redesign. LCA studies are the activities in which the activity in question is monitored as 'cradle to grave,' 'cradle to door,' 'cradle to cradle,' and 'door to door' depending on the stage of a life cycle of all the stages, all inputs and intermediate and end products conducted for a defined time and place. LCA study is not a decision mechanism it has the quality of helping the decisions to be made. LCA is mainly used in the development of environmental labeling criteria, raw materials of products, production processes, equipment, and so on. It is used to prevent/reduce environmental impacts by changing and/or redesigning. Also useful for determining at what stage of a product's lifecycle a more detailed waste control exercise is required. The LCA approach is an important tool used in decision making in CP applications. However, the determination of the study limits, the quality and reliability of the collected information, and the approaches used in the analysis of this information, can give very different results depending on the differences in the viewpoints of the people and institutions conducting the study. Therefore, it would be beneficial to use it together with other tools (Özbay, 2005).

7.4.2.3 Environmental technology evaluation

Environmental technology evaluation (ETE) examines the effects of specific technology on human health and natural systems and resources. ETS is part of the technology assessment, which can be defined as the determination of the effects of a technology on human health and natural systems and resources. ETE includes (Özbay, 2005) environmental impact assessments of various facilities and projects, qualitative and quantitative determination of discharges resulting from the use of various technologies, determining the risks of various technologies on human health and environmental values using qualitative and quantitative methods.

7.4.2.4 Chemical evaluation

Chemical evaluation (CE) includes the analysis of the toxic effects and amounts of the chemicals used in the production stages by using various information sources and databases. Pursuant to CP, it is necessary to follow the approach of using chemicals that are the least harmful to human health and the environment in production (REC, 2011). For example, Materials Safety Data Sheets and the International Program on Chemical Safety are widely used information sources to determine the hazards of a chemical on human health and environmental quality. By using these resources, it is possible to choose the least harmful one for human health and environmental values from more than one chemical that can be used for the same job. CE is part of the risk assessment.

7.4.2.5 Waste audit

Requiring a balance of materials for each process and processes, waste audit (WA) includes identifying all inputs required for an industry, plant or process and all wastes generated from them. In waste inspection studies, all wastes generated during the production stages, the sources, quality and quantity of these wastes and the possibilities of reducing them are determined. Waste and emissions audit are terms used synonymously with waste and emissions prevention assessment and waste reduction audit. Waste inspection determines the following for CP: (REC, 2011);

- Input/output (mass balance) inventories of processes,
- Source, quality, and quantity of wastes generated,
- Efficiency and weak points of the current process,
- Waste minimization targets.

As a result of waste inspection, losses are reduced and/or prevented, thereby increasing process efficiency. A waste control application

- Defines the source, amount, and types of wastes produced.
- Gathers information on key processes used, raw materials, products, water use and waste generation.
- Identifies deficiencies and weak management points in the process.
- Determines the targets required for CP.
- Allows the development of inexpensive waste management plans.
- Helps to increase process efficiency by increasing the level of knowledge of the personnel working in the workplace about the processes used.

7.4.2.6 Environmental audit

The purpose of environmental audit (EA) is to determine the amount and character of waste streams originating from the production process or services and to make decisions about what should be done to reduce pollution. It is the most frequently used and most important application tool of CP. Since environmental audit is a very effective tool, there are also types developed for different purposes such as waste audit (WA), energy audit, and risk audit. Environmental auditing is an important tool for both business and government as its compliance with environmental policy and standards is evaluated. This tool of CP can enable industries to get rid of penal sanctions by detecting the measures to be taken regarding the environment in a timely manner (Özbay, 2005).

7.5 CP PRACTICES APPLIED IN DIFFERENT INDUSTRIES

7.5.1 Textile production facility: CP practices

7.5.1.1 In Korea (Asia Pacific Economic cooperation (APEC), 2006)

The main production divisions of the textile industry are fiber production (natural, artificial, synthetic, nonwoven), yarn production, fabric production (weaving, knitting), finishing (bleaching, dyeing, printing, finishing), ready-made garments. Pollution parameters in production include toxic chemicals discharged to air, water, and soil, as well as noise, energy consumption, consumption of natural resources, and waste. Textile finishing and dyeing companies are the ones that cause the most environmental pollution. In addition to the consumption of natural resources, noise, energy consumption and fiber dust parameters, which are also mentioned in textile finishing enterprises, yarn and weaving enterprises, toxic chemicals are used. Toxic chemicals pollute the soil with solid waste, environmental water with waste water, and air with flue gas.

In the study conducted by the Asia-Pacific Economic Cooperation (APEC) on CP case studies for selected industries in Asian and Pacific countries in 2006, two textile industries operating in Korea were examined. A group of experts assessed Wowcom Corp., a textile dyeing company in Korea, for process assessment, where possible CP technologies were identified. The general goal of the Clean Production Program was to achieve environment-friendly dyeing process with higher energy efficiency for the global competitiveness of these companies.

7.5.1.1.1 Evaluating dyeing processes

The problems identified before CP practices are listed below:

- The dyeing liquor is not measured quantitatively during the continuous process that generates waste.
- Since there is no inverter in large pumps, there is unnecessary energy consumption.
- Insulation of pipes is insufficient.
- Since the ventilation system is insufficient in the area where the tentering machine is located, it is operated continuously and unnecessary energy is consumed.
- Irregularities in the dyeing process due to the formation of unsuitable dye combinations cause a decrease in efficiency.
- Insufficient ventilation and lighting near the jigger machine causes a decrease in work efficiency and product quality.

Implemented application: CP options determined for these identified problems are listed as follows:

- For the proper management of chemicals, pumps capable of quantitative measurement have been placed in areas where chemicals are used, and an investment of $2900 has been made for this (*reduced wasted water by 6%, reduced energy consumption by 9%, reduced material usage by 5%, payback period: 4 months*).
- To reduce unnecessary energy consumption in the jet dyeing machine, an inverter application was made to control the main motor and an investment of approximately $26 000 was made in this regard (*reduced energy consumption, reduced air emissions, payback period: 34 months*).
- To prevent unnecessary energy consumption due to the lack of insulation in the pipes, pipe insulation was made and an investment of $2500 was made for this (*reduced energy consumption, reduced air emissions, payback period: 0, 6 months*).
- A new system was installed to reduce ventilation runtime when the tentering machine is not in use, with an investment of $4200 (*reduced energy consumption, reduced air pollution, payback period: 4 months*).
- In the dyeing process, dyeing recipes were optimized for each color, and an investment of $1700 was made for this (*reduced energy and chemical consumption, payback period: 1 month*).

- In the section where the jigger machine is located, arrangements were made on the ceiling to allow more sunlight to enter and to increase the lighting in the environment, and an investment of $2700 was made for this (*increased in labor productivity and improved safety of the work place, payback period: immediately*).

Benefits: As a result of all these improvements, with an investment of $40 000, an annual economic benefit of approximately $103 400 was provided to this industry, with a payback period of minimum 0.6 months (*insulation of pipes*) and maximum 34 months (*invertor usage*).

7.5.1.1.2 Evaluating waste minimization and cost reduction

The problems identified before CP practices are listed below:

- The cooling waters in the industry are not recycled.
- Energy loss occurs due to the high-speed operation of the fans when the tentering machine is in the standing position.
- Due to the lack of insulation in the steam distribution pipes within the facility, heat loss and thus energy loss occur.
- Even when the anti-felting machine is in the standby position, steam flow occurs and therefore unnecessary energy consumption occurs.
- Since dyes with low compatibility are mixed in dyeing prescriptions, the color quality decreases, which leads to a decrease in product quality.
- Uncontrolled liquor flow management in the dyeing process causes a decrease in product quality.
- Leakage occurs due to aging of pipes throughout the facility.

Implemented application: CP options determined for these identified problems are listed as follows:

- An investment of approximately $5800 was made for the reuse of the cooling water for the wool washing process before the dyeing process in the facility (*reduced wastewater generation, payback period: 5 months*).
- An investment of approximately $500 was made in order to prevent unnecessary energy losses, the tentering machine is equipped with a sensor that will allow the modification of operating conditions (*reduced gas emission, reduced energy consumption, payback period: immediate*).
- An insulation unit has been installed throughout the facility to prevent heat and energy losses, with an investment of approximately $2700 was made (*reduced energy consumption, payback period: 3 months*).
- An automatic valve system is installed to stop the steam supply when the anti-felting machine is on standby with an investment of approximately $1250 was made (*reduced energy consumption, payback period: 4 months*).
- For highly compatible dyestuff mixtures to be applied in dyeing recipes, the adsorption behavior of each dyestuff was determined by examining without any investment was made (*higher product quality, higher reproducibility of color, payback period: immediate*).
- A flow detector was installed to control the liquor flow in the dyeing process with an investment of approximately $10 000 was made (*reduced wastewater generation, payback period: 4 months*).
- Aged pipes throughout the facility were replaced with new ones with an investment of approximately $15 000 was made (*reduced wastewater generation, payback period: 5,5 months*).

Benefits: As a result of all these improvements, with an investment of $32 250, an annual economic benefit of approximately $135 200 was provided to this industry, with a payback period of minimum 3 months (*insulation unit*) and maximum 5,5 months (*pipe replacement*).

7.5.1.2 In Peru (Asia Pacific Economic cooperation (APEC), 2006)

In this study, a Peruvian textile company that deals with fiber spinning, knitting, fabric dyeing, and finishing was investigated. During CP studies, energy was focused on as the most important CP opportunities. These issues included thermal energy generation, energy recovery of hot gases released from the chimneys of boilers, inefficiencies in the combustion process in boilers, the presence of leaks in switches and steam pipelines, and high power consumption due to compressed air leakage in distribution lines.

Implemented application: CP options determined for these identified problems can be listed as follows:

- Administrative practices such as optimization of the combustion process, continuous control of fuel quality and keeping the boiler clean were carried out by making an investment of $1000 in order to eliminate the inefficiency in energy-generating boiler systems (*reduced fuel consumption by 28 800 gal/yr. Reduced gas emission of effect conservatory (GEI) and polluting gases of the atmosphere (GCA), payback period: 3 months*).
- With an investment of $11 000, the steam traps were reviewed to reduce heat losses in the steam distribution line and during condensation, eliminating the identified losses and optimizing the use of electricity to prevent compressed air losses during dry cleaning (*increased boiler efficiency by 16.5, 7.8, and 33.5%. Reduced CO_2 emissions by 850 tons/yr, payback period: 5 months*).

Benefits: As a result of all these improvements, with an investment of $12 000, an annual economic benefit of approximately $27 400 was provided to this industry, with a payback period of minimum 3 months (*optimization and maintenance*) and maximum 5 months (*isolated steam pipelines*).

7.5.1.3 In Turkey (TDFT (Turkey Technology Development Foundation), 2011; Alkaya et al., 2011)

In this study, a facility with a fabric production capacity of 15 00 000 meters/month, which produces fabrics (polyester, cotton, and lycra-based fabrics) for textile dyeing and finishing, women's outerwear group, was examined within the scope of Unido Eco-Efficiency Program.

Before the CP application, it was determined that a total of 300 000 m^3 of water was consumed annually depending on the production amount in the facility and 80–85% of the total water use of the facility was realized in the dyeing and finishing processes. In addition, it has been determined that the facility meets its energy needs with natural gas and electricity and has an annual consumption of approximately 1 300 000 m^3 natural gas and 4 250 000 kWh electricity. It has been determined that an old technology water-softening equipment used in the water-softening processes that form the heart of the facility is used, this equipment is operated manually and the amount of water consumed during the regeneration of the device is 15 m^3/rej and the amount of salt consumed is 450/kg/rej.

Implemented application: CP options determined in this facility making textile dyeing and finishing processes can be listed as follows:

- To better control the water consumption values for each process and to determine the optimum water consumption values in these processes, flow meters are installed at the process inlets and outlets.
- To prevent unnecessary water consumption, washing times were reduced and the overflow washing valves in the overflow washing used in finishing were closed.
- In the dyeing section of the facility, where a high amount of water is consumed, the inlet–outlet valves in the cooling water sections of the dyeing machines were renewed in order to reduce water losses and leakages.
- Dryer cooling water was collected and sent to the soft water pool for reuse.
- To determine the optimum flow rate of the water used in the fabric opening machine, a flow meter is mounted on the machine line.
- Reuse of fabric fluff burning machine cooling water was directed to the soft water pool.

- The old technology water softener is fully automatic and the new system, which also increases the quality of the water produced, has been replaced with a water-softening unit.
- A monitoring system has been set up to monitor the water savings on a monthly basis on the basis of the product produced.
- Heat exchangers have been installed at the entrance of the wastewater treatment plant of the enterprise to provide heat recovery from the wastewater.
- A system has been set up to provide heat recovery from the high-temperature flue gas released to the atmosphere from the RAM unit.

Benefits: As a result of all these improvements, with an investment of $34 659, an annual economic benefit of approximately $345 161 was provided to this industry, with a payback period of <2 months. While 111.7–129.4 L of water was consumed per kg of product produced before the project, water consumption was reduced to 50.9 L/kg of product with the project. Unnecessary consumption of approximately 162 000 m^3 of water per year is prevented and total water consumption of the facility was reduced by 54%. Moreover, as a result of the improvements in the water-softening system, 192 tons of salt (NaCl) was saved annually. Furthermore, with the heat recovery systems from waste water and flue gas, the energy used for hot water production was also saved and in this sense, 22% increase in efficiency was achieved and annual natural gas and electricity consumption was reduced by 4 780 000 kWh in total. Thus, the total CO_2 emission of the facility was reduced by 879.6 tons/year.

7.5.2 Rubber production facility CP practices in New Zealand (Asia Pacific Economic cooperation (APEC), 2006)

Rubber has been used for centuries, indispensable in our daily lives, from the shoes we wear to the wheels and windscreen wipers in our cars, from diving suits to hoses and soccer balls. The gasket that integrates the natural gas pipes with the stove, the bellows that protects the axle shafts in our cars, the gloves used in the operating room, the insulators that surround the conductive copper wire, the raincoats that protect us from the rain and the insulation materials used in house roofs are made of rubber. The reason why rubber is used in such a wide range is its properties such as flexibility, softness, durability, stickiness, and waterproofing. Compounds such as zinc oxide and lead oxide are used in the vulcanization process of rubber. These chemicals can poison living things by mixing with the soil, streams and water bodies in the form of liquid waste. The rate of waste material generated in the rubber processing process is approximately 25 times that of the rubber produced.

In the study conducted by the Asia-Pacific Economic Cooperation (APEC) on CP case studies for selected industries in Asian and Pacific countries in 2006, a rubber manufacturing company, implemented CP introduced by the Target Zero program, was examined. The objective of this program is to enhance efficiency in every aspect of the business from manpower right down to production. The problems identified before CP practices are listed below:

- The calender machine, which is an important part of production, is operated in an inefficient way in terms of waste generation, raw material use, and fuel use in the production of certain products.
- The plastic strips used in banding and wrapping the products are used once and thrown away.
- Leaks are detected in most of the faucets connected to the water mains throughout the facility.
- The extruder machine cooling waters drawn from the aquifer are discarded after using the water once.
- Blank cards attached to the cards used for tracking the production stages of rubber products throughout the facility cause unnecessary waste generation and resource consumption.
- Unnecessarily large amounts of paper are routinely used to print production line flow reports on site throughout the facility.
- Cardboard packaging is not recycled throughout the facility.

- Low efficiency monophosphor lamps are used throughout the facility in terms of energy efficiency.
- Steel strips, which are used as a component of the packaging in the palletized packaging of raw materials, are disposed of landfill and are not recycled.

Implemented application: CP options determined for the above identified problems are listed as follows:

- A special attachment was fitted to the calender machine to reduce waste generation, manual machining, and edge misalignment efficient operation of the machine with an investment of approximately \$20 000 was made (*reduced waste rubber and usage of raw materials and fuel, increased machine productivity, payback period: <1 month*).
- The plastic strips used in banding and wrapping the products were recycled without any investment being made (*reduced raw material consumption, payback period: immediate*).
- By applying a regular maintenance program throughout the facility, unnecessary water consumption was prevented and savings were achieved with an investment of approximately \$1000 was made (*reduced water consumption, Payback period: 4 months*).
- The amount of wastewater was reduced by feeding the cooling water used in the extruder mill back to the groundwater with an investment of approximately \$1500 (*reduced wastewater generation, payback period: 2 years*).
- Blank cards attached to the cards used for tracking the production stages of rubber products throughout the facility were eliminated without any investment was made (*reduced cardboard wastes, reduced resource consumption, payback period: immediately*).
- Unnecessary paper usage and resource consumption have been reduced by rationalizing the report printing format throughout the facility without making any investment (*reduced paper usage, reduced landfill disposal, payback period: immediately*).
- Cardboard packages are segregated within the facility to ensure recycling without making any investment (*reduced landfill disposal, reduced raw material usage, payback period: immediately*).
- Monophosphor lamps with low energy efficiency throughout the facility were gradually replaced with triphosphor lamps without making any investment (*reduced amount of mercury disposed, reduced energy consumption, payback period: immediately*).
- Steel strips, which are used as a component of the packaging in the palletized packaging of raw materials are recycled without making any investment (*reduced resource requirement, reduced landfill disposal, payback period: immediately*).

Benefits: As a result of all these improvements, with an investment of \$22 500, an annual economic benefit of approximately \$ 297 010 was provided to this industry, with a payback period of minimum <1 month (*fitted a special attachment to the calender machine*) and maximum 2 years (*recovered the extruder machine cooling water*).

7.5.3 Fertilizer manufacturer facility CP practices in New Zealand (Asia Pacific Economic cooperation (APEC), 2006)

Fertilizer industry is the branch of the chemical industry that produces phosphorus, nitrogen, potassium, and other plant nutrients required for plants in agriculture, either simply or in combination. Products in the fertilizer industry; nitrogen fertilizers, phosphate fertilizers, potash fertilizers, mixed fertilizers, and compound fertilizers are divided into five groups. Various processes are applied by using different raw materials according to the type of fertilizer to be produced in the fertilizer industry. Some products can be used as a raw material for another product. In the industry, some of the raw materials are imported, and some of them are produced in-house as intermediate products. Intermediate products used in the fertilizer industry are sulfuric acid, phosphoric acid, and nitric acid.

The Ravensdown Fertiliser Co-op site at Hornby in Christchurch is one of the three sites in the Ravensdown group that manufactures phosphate fertilizers for New Zealand's agricultural industry, has been a participant in the Target Zero Program. The Sustainable Management Fund of the Ministry for the Environment, Meridian Energy Ltd (formerly ECNZ), and the local authority and electricity company in each area sponsored the program. The objective of this program is to enhance efficiency in every aspect of the business from manpower right down to production. The problems identified before CP practices are listed below:

- The lamps that illuminate the roads inside the facility are left on during the day. Since electricity is generated on site, electricity is traditionally considered a free good and consumes unnecessary resources.
- The drums are not recycled due to the large amount of lithium grease residues under the drums used in production.
- The cooling water supplied by drilling in the acid plant is treated after use and then discharged to the ground. In addition, the drilling water used in acid-plant plate heat exchangers is stored in an open pond, which forms algae before use and requires chemical treatment.

Implemented application: CP options determined for the above identified problems are listed as follows:

- A $279 investment was made for the timing system, especially for the automatic on and off of road lighting systems throughout the facility (*reduced electricity consumption and increased bulb life, payback period: 9 months*).
- For an investment of $65, grease suction is made efficient by adding a 10 mm thick metal plate to the suction pump at the bottom of the drums (the extra weight allows more grease to be sucked in by the pump) (*increased amount of grease removed by 9.45 kg/drum, reduced grease waste from 22% to 4.5%, allowed recycling of drums, payback period: 3 weeks*).
- Existing bore water ponds were used for the acid plant wastewaters, make-up water for the wet scrubber of the facility was supplied from these ponds and also four new closed-hole water tanks were installed to prevent algal growth. All these were done with an investment of $185 344 (*compliance to regional council regulations, reduced bore water consumption and chemical use, payback period: several years*).

Benefits: As a result of all these improvements, with an investment of $185 344, an annual economic benefit of approximately $12 983 was provided to this industry, with a payback period of minimum 3 weeks (*by adding a 10 mm thick metal plate to the suction pump*) and maximum several years (*acid plant cooling water reuse*).

7.5.4 Leather processing facility CP practices in Croatia (Greco Initiative & Regional Activity Centre for Cleaner Production (CP/RAC), 2008)

Chemicals such as sulfites, acids, alkalis, and chromium are used in leather production. If these chemicals are not properly managed or removed, they cause adverse effects for the environment. Organic substances from raw leather materials such as leather or hide cutting, sawdust, scraps, hair, hair protein and keratin, dissolved in wastewater, lead to a decrease in oxygen in receiving environments such as lakes and rivers. If wastewater and solid wastes are not treated and/or managed well before discharge, they also harm flora and fauna.

Within the scope of the study, the following subjects were determined as the focal points of the CP activities carried out in the facility:

- Determination of actual volumes and pollution parameters of tannery waste.
- Choosing the most appropriate methods to reduce water consumption and pollution.
- Developing the most suitable waste and sludge treatment alternatives.

- Developing a financing program for the implementation of an environmentally friendly water management program by reducing operating and maintenance costs.
- It is aimed to establish standards, regulations, and pricing guides for discharge and sludge disposal.

Implemented application: The clean (sustainable) production possibilities determined for the facility within the framework of these focal points are listed below:

- Salt forging: reduction of chloride concentration at the tannery outlet (total reduction in 0.3–0.4%, chemical reduction with 5% for washing and soaking, reduction in pollution parameters are 0.6%, 4%, 3%, and 25% for SS, BOD_5, COD, and chloride, respectively).
- Recovery of residues: reduction of organic pollution and sludge volume by reusing residues through reduction of sulfite and lime pollution (*total reduction in 1.4%, chemical reduction with 9–10% for liming, reduction in pollution parameter for sulfide is 9–10%*). 16% of leather residues are converted into useful products (biogas, compost).
- **Hair removal:** reduction of organic pollution, sludge volume, raw material, and water consumption by reducing the amount of hair on the material before the waste is discharged into the sewer system (*chemical reduction with 14–15% for lime, 100% for NaHS (%72 Na_2S), 6–7% Na_2S (%62–67), reduction in pollution parameters are 6.7%, 41%, 25%, 25%, 18%, 35%, and 2% for TSS, SS, COD, BOD_5, sulfide, N-total and N-NH_4, respectively*). 47% of waste by weight has been reduced.
- **Chromium reduction:** reduction of chromium concentration in sludge and wastewater through the recovery of chromium from the tanning process (*chemical reduction cannot be provided but reduction in pollution parameters is determined as 98–99 for Cr*). The amount of chromium in the sludge has been reduced below the legal limits for regular storage.

Waste treatment and sludge treatment methods implemented are as follows:

- Sulfite removal
- Use of efficient flocculants and coagulants
- Sludge treatment, disposal, and reuse
- Laboratory analysis of discharge and sludge composition

Benefits: With the project, a model that can be applied by other tanneries in Croatia has been developed in order to reduce environmental pollution in leather production. A remarkable example has been set as a result of reducing environmental impacts and production costs through practices such as salt forging, hair reduction, and chrome recovery. As a result of the implementation of CP practices only in the hair removal process, BOD_5 in the wastewater was reduced to 25% with an investment on equipment and construction works of 49.384 € + annual operating costs of 14 523€. In addition, a total of 85 000€/year savings was achieved, with a chemical savings of €22 252/year and a savings of €53 526/year in sludge disposal costs. Total payback period was calculated as 1 year. On the contrary, chromium process with the chromium recovery system has an economic value gained with an annual investment cost at 218 352€, the operating cost is 70 253–80 077€ and a payback period of 5–7 years (when indirect savings costs for chromium recovery are included, for example, storage cost for sludge, reduction of discharge amount – payback period is calculated as 1–2 years). With this application, 111 512€ was gained by recovering 26 400 kg of chrome. However, here, basic sludge treatment methods in different sludge types are discussed in reducing the chromium content (below 1000 mg/L), and hazardous waste storage systems are not considered.

In addition, end-of-pipe measures were also included in the project and solid waste management strategy took place at the end of the project and uncontrolled waste, which was specified as 14 567 kg/year, was also disposed of.

7.5.5 Canned food production facility CP practices for water and energy saving in Egypt (Greco Initiative & Regional Activity Centre for Cleaner Production (CP/RAC), 2008)

In this study, two facilities were examined, one in İskenderiye and the other in Kaha, employing approximately 600–650 people. Two selected facilities, one private and one public, are one of the largest canned food producers in Egypt. The production line of both plants includes juice, jam, frozen vegetables, canned beans, and tomato paste. Canning facilities and a cooling unit are located on the premises of the facilities, and production is seasonal at both plants. Both facilities have high water and energy consumption and they discharge 780 680 and 520 000 m^3/year wastewater to the sewerage network, respectively.

As a result of the preliminary examinations carried out within the scope of CP practices in both facilities, it was determined as a priority to address the following issues:

- At the beginning of the energy-related problems in facilities, steam leaks in processes, steam lines, and traps (steam traps) take place.
- In addition to these, facilities experience heat losses due to poor steam line insulation, steam and water losses occur due to the can sterilization unit and the inability to manage condensate water well.
- As for water problems, a very high amount of water is wasted due to open cooling cycles and water leaks, some malfunctioning cooling towers, insufficient water recovery systems, water dripping from taps and hoses, water consumption for washing vegetables, washing equipment, and floors.

Implemented application: To prevent energy loss, the steam escape points on the steam lines are insulated. Steam and water losses occur due to not taking the necessary precautions in the can sterilization unit and the condensate water cycle, this situation is monitored by water meters placed in the company's 13 separate water-using processes, the establishment of a cooling tower for the bottled fruit juice line for the recycling and recovery of cooling water, the water loss. With the improvement of the storage system and the addition of hose caps that allow water to slow down, water usage has been reduced.

The following measures have been taken to conserve energy:

- Insulation of open steam pipes (investment cost for facility-1: 38 009€, payback period: 19 months, annual cost saving: 24 504€, facility-2: 38 009€, payback period: 10 months, annual cost saving: 22 307€).
- Replacing leaky steam traps (investment cost for facility-1: 4277€, payback period: 8 months, annual cost saving: 6182€, facility-2: 4430€, payback period: 5 months, annual cost saving: 9777€).
- Replacing leaky steam valves (investment cost for facility-1: 14 379€, payback period: 36 months, annual cost saving: 4790€, facility-2: 11 891€, payback period: 17 months, annual cost saving: 8400€).
- Establishment of pressure regulators in the sterilization system (investment cost for facility-1: 13 329€, payback period: 10 months, annual cost saving: 16 373€, facility-2: 13 822€, payback period: 4 months, annual cost saving: 47 231€).
- Installation of condensate water recovery system (investment cost for facility-1: 10 154€, payback period: 44 months, annual cost saving: 2772€, facility-2: 12 183€, payback period: 14 months, annual cost saving: 10 190€).
- Improved boiler efficiency (investment cost for facility-1: annual cost saving: 4734€, facility-2: annual cost saving: 10 603€).

The following measures have been taken for water saving:

- Use of hose nozzles (investment cost: 2587€, payback period: 7 months, annual cost saving: 2754€).
- Improvement of the water collection system (investment cost: 1499€, payback period: 5 months, annual cost saving: 7344€).
- Cooling tower investment (investment cost: 29 953€, payback period: 12 months, annual cost saving: 26 438€).

Benefits: As a result of the implementation of the above-mentioned options, 15 278 m^3/year steam savings were achieved in facility-1 and 18 125 m^3/year in facility-2. Fuel consumption was reduced by 40% in facility-1 and up to 34% in facility-2. As a result of all these improvements, with an investment of \$80 148 for facility-1 and \$61 288 for facility-2, an annual economic benefit of approximately \$59 355 for facility-1 and \$108 508 for facility-2 was provided to this industry, with a payback period of minimum 4 months and maximum 44 months.

7.5.6 Oil and soap facility CP practices in Egypt (Greco Initiative & Regional Activity Centre for Cleaner Production (CP/RAC), 2008)

The seed varieties processed in the facility, which processes an average of 68 000 tons of seeds per year, include sunflower, corn, soybean, and cotton seeds. The main by-products produced in the facility, which produces up to 24 000 tons of first-class edible oil per year, are approximately 40 000 tons of dry pulp (packaged in bags as animal feed) per year and up to 1800 tons of soapy substances annually.

The processing of oil at the plant is carried out in five stages:

(1) Storage after the seeds are separated from the broken seeds.
(2) Seed preparation, extraction of 50% of the crude oil content and obtaining the seed pulp containing 30% oil.
(3) Sending the seed cake to the solvent extraction section where a solvent–oil mixer is located and forming pulp (containing 2% oil), extracting the crude oil in a mixer with a three-stage evaporation system and removing the pulp from the solvent, and finally, the hexane used as a solvent is recovered in the system.
(4) Refining of oil neutralized with caustic soda (for soapy substances), washed, separated by centrifugation and separated from its odor, color, and aromatic flavoring substances.
(5) Bottling and packaging of the oil produced.

The following clean (sustainable) production opportunities have been determined as a priority within the scope of the project, whose facility is carried out through clean (sustainable) production inspection:

- Reducing steam losses due to damaged pipes and valves and lack of adequate insulation.
- Reuse of broken seeds and hulls from processes in the hull removal unit.
- Reducing fuel leaks and losses.
- Treatment and reuse of refining wastewater with high organic matter content.
- Reducing oil losses in chemical processes and storage/packaging areas in the refining unit.
- Reduction of oil losses in refining due to leakage.

Implemented application: Activities realized within the scope of CP practices:

(1) Simple productivity and good business practices:
    - Preventive maintenance program (maintenance and repair of steam lines and traps, repair of leaks, broken valves, damaged water and steam pipes, etc.) and reducing steam/hot water losses and process optimization have been put into practice (*investment cost: 4500€, payback period: 6 months, annual cost saving: 9000€, annual saving: 34 tons/year*).

- The oils in the packaging unit were collected and recycled. These oils were pumped into the collection tank and recovered at the refinery for processing and thus production increased (*investment cost: 750€, payback period: <1 month, annual cost saving: 10 500€, annual saving: 1392 tons/year*).

(2) Process change:
- Fine grains from the preparation unit are reused. A unique design has been implemented for the recycling of sunflower seed parts. At this stage, the seed pieces were changed in such a way that they were quickly directed to the sections with fresh seeds that provided high yields. Seed breaking capacity has been increased by reusing the seed pieces (*investment cost: 3000€, payback period: 1 month, annual cost saving: 36 000€, annual saving: 120 tons/year*).

(3) Changing chemicals:
- Caustic soda instead of liquid caustic soda solution was used during neutralization and thus caustic soda losses were reduced. As a result of the use of liquid caustic soda, a decrease of up to 47% in neutralization costs, an increase in the quality of soapy substances, corrosion reduction in the amount of caustic soda, reduction in caustic soda losses and improvement in working conditions have been observed (*investment cost: No investment, payback period: immediately, annual cost saving: 75 000€*).

(4) Water and energy saving:
- Development of steam network, rehabilitation of steam pipes, improvement of boiler combustion settings and improvement of boiler feedback water treatment, recovery of condensed steam, replacement of defective/broken valves, replacement/repair of steam traps and pipes, improvement of hot water and steam pipes measures such as isolation. As a result of these applications, a decrease in steam consumption, removal of a boiler from the production line (saving in the use of diesel oil), reduction in water consumption and maintenance costs have been achieved (*investment cost: 9000€, payback period: <1 month, annual cost saving: 165 888€, annual saving: 3600 tons/year reduction in steam consumption, 1728 tons/year for fuel consumption savings, and 28 800 tons/year reduction in water consumption and maintenance cost*).

(5) Recovery and reuse:
- Before the project, the shells were sold as animal feed. With the application, the shells were transferred to the preparation unit where they will be processed with the help of a screw conveyor, and then the processing started and the broken seeds and seed shells were recovered. As a result of the recovery of crushed seeds, the production of oil and pulp has been increased (*investment cost: 2700€, payback period: <1 month, annual cost saving: 138 975€, annual saving: increase in oil production, 78 tons/year for oil and 595 tons/year for pulp, respectively*).
- As a result of the application of the recovery of soapy substances, the pollution load of the wastewater has been reduced (*investment cost: 1500€, payback period: 4 months, annual cost saving: 4320€, annual saving: 29 tons/year*).

(6) Treatment of wastewater:
- Process water from refining has been treated and the remaining water has been used for irrigation near the factory. As a result of waste water separation, improvement was achieved in waste water disposal (*investment cost: no investment cost, payback period: immediately, annual cost saving: 5400€, annual saving: 13 464 m^3/year*).

Benefits: As a result of low-cost or no-cost applications, the plant has achieved significant benefits. Maintenance costs decreased by 10%, water consumption decreased to 46%, wastewater treatment requirement decreased by 66%, fuel consumption of the boiler decreased to 48% and annual oil, pulp, and soapy matter valued at 207 795€ recovery has been achieved. Compliance with legal discharge limits was also ensured with the implementations.

7.5.7 Beverage facility CP practices

7.5.7.1 In New Zealand (Asia Pacific Economic cooperation (APEC), 2006)

Investigated facility is a major manufacturer of fruit and vegetable juice concentrates and other processed apple and vegetable products in New Zealand. The facility manufactures clear and cloudy apple juice concentrates, as well as pear, kiwifruit, and carrot concentrates. The facility operates all year round with high levels of production in the winter months.

The main possibilities determined for the facility within the framework of below focal points are listed below:

- Fuel consumption between 125 000 and 187 500 L/h at the facility and a borehole water with 156 900 L/hour extraction permit was detected.
- It was determined that approximately 150 000 L/year fruit concentrate loss occurred in the facility and this lost stream was not returned to the main process.

Implemented application: The clean (sustainable) production possibilities determined for the facility within the framework of these focal points are listed below:

- To reduce the amount of water drawn and consumed from the drilling well in the facility and to ensure the limit values determined by the competent authorities in resource consumption, to recover the wastewater from the final cooling towers and evaporators of the fruit juices and the use of belt filters in pasteurization and heat exchanger plates in cooling towers (*investment cost: $67 287, payback period: 19 years, annual cost saving: $3536*).
- The recovery of lost sugar-rich product was maximized by using decanters capable of capturing all the retained material to prevent 23 000 L of sugar-rich product from entering the wastewater stream to trade waste and to achieve the predicted productivity increase for 150 000 L of additive product (*investment cost: $400 000, payback period: 1.8 years, annual cost saving: $226 036*).

Benefits: As a result of all these improvements, with an investment of $467 287, an annual economic benefit of approximately $226 036 was provided to this industry, with a payback period of minimum 1.8 years (*decanter usage*) and maximum 19 years (*re-used waste bore water*).

7.5.7.2 In Turkey (TDFT (Turkey Technology Development Foundation), 2011; Alkaya et al., 2011)

CP activities were carried out in a facility with 15 000 m² closed and 2000 m² open area, which is one of the leading producers of the non-alcoholic beverage (soft drink) sector in Turkey. The facility offers different products to the soft drink market such as fruit nectar, fruit drink, carbonated drink, and 100% fruit juice. In addition, there is a milk production line in its facilities.

The fruits coming under the Fruit Concentrate Production section of the facility are washed and after pre-treatment, are pasteurized and then are converted into fruit concentrate. Groundwater is used in the cooling process during the concentrate production phase. At this stage of production, where approximately 346 000 m³ of water is used annually, there is no recycling or reuse activity conducted.

In the soft drink production industry, fruit concentrate is combined with water and other additives to turn it into a soft drink. As in the concentrate production, groundwater is used for cooling purposes during the beverage production phase and approximately 173 000 m³ of water is consumed annually at this stage. In this production line, which causes one of the most intense water consumption of the facility, any recycling or reuse of water is not performed.

Table 7.2 Annual water consumption and savings in the facility.

Process	Before CP Applications (m^3/year)	After CP Applications (m^3/year)	Annual Water Consumption Savings (%)
Fruit concentrate production			
Fruit washing	11 500	11 500	–
Cooling	519 000	18 000	96
Cleaning	72 000	36 000	50
Juice production			
Cooling	173 000	28 000	84
Cleaning	36 000	36 000	–
Carbonated drink Production	55 000	55 000	–
Process operations (steam production, pasteurization, etc. ancillary processes)	14 000	14 000	–
Other (domestic use, in the product, etc.)	180 000	180 000	–
Total	851 500	378 500	56

Source: TDFT, Turkey Technology Development Foundation (2011).

The wastewater produced by the facility, which consumes a large amount of water, is sent to the central wastewater treatment plant of the organized industrial zone (OIZ) to be treated. Especially in the summer months, due to the intensification of activities in fruit processing, water consumption, and the amount of wastewater sent to the treatment plant increases. This situation forces the existing capacity of the OSB treatment system. On the contrary, intensive water consumption is an important cost factor for the facility. For these reasons, it was necessary to take precautions in the cooling processes, which are primarily responsible for water consumption in the facility, and water saving was taken into account as the focus of CP.

Implemented application: Two separate systems were implemented for the recovery and reuse of cooling waters used in fruit concentrate and soft drink production lines. Instead of existing open cooling systems (once-through cooling), two similar closed-loop cooling systems of different capacities have been installed, consisting of cooling tower, stainless-steel water pump, stainless-steel pipes/connections, inverter and control panel units. The water consumption determined before and after the CP practices in the facility is given in Table 7.2.

7.5.8 Dairy production facility CP technology practices (Kotan & Bakan, 2007)

In this facility, which produces dairy products, the milk purchased daily from the surrounding villages is first pasteurized. The pasteurization process is carried out by heating the milk in metal tanks to a certain temperature for a while with water vapor and suddenly decreasing it to a lower temperature. The facility decided to prevent heat losses by improving the insulation in the steam line to the pasteurization tank.

Implemented application: The company is a production facility that starts the process by pasteurizing freshly brought milk on a daily basis. This pasteurization process is carried out by heating the milk with water vapor in metal tanks and then freezing it with the help of a sudden reduction in temperature when the desired temperature is reached. The heat loss experienced during heating with this steam is prevented by insulation systems.

Benefits: With the insulation renewals carried out, fuel savings of 40% were achieved. In this way, the amount of pomace burned in a year has been reduced by 87 tons. This reduction in the amount of burned pomace also reduced air pollutant emissions such as NO_x, CO, CO_2, and organic substances formed in the burning process at the same rate. As a result of all these improvements, an annual economic benefit of approximately $4000 was provided to this industry with a payback period of immediately with an investment of $100.

7.5.9 Sugar production facility clean (sustainable) production practices (Greco Initiative & Regional Activity Centre for Cleaner Production (CP/RAC), 2008)

7.5.9.1 Facility-1 in Fes, Morocco

Sugar factories operate seasonally, intensive water consumption in production is estimated to be 80 000 m^3/day. In sugar factories, water is used for processes such as washing, conveying of beets, cooling, diffusion of sugar during extraction, as well as for washing floors, tanks, and machines. Washing and transport waters constitute 50% of organic pollution and 90% of suspended solids in wastewater. 200 000 tons of solid waste is emitted annually around the factory. Due to the intense energy production in the factory, 120 000 tons of CO_2 is released into the atmosphere annually. The factory was established at the end of the 1960s. The economic aspect of consumption or wastewater discharge was not considered. By the 1980s, the company became more sensitive to environmental problems and especially to the issue of water saving. As a matter of fact, the drought that occurred in Morocco during this period caused a decrease in the supply of water difficulties and increased production costs. Faced with this problem, the facility determined the focus of CP as water saving and reducing pollution after environmental inspection.

Implemented application: The following CP practices have been implemented within the facility:

- By adding a sugar beet cleaning process to the beet loading and unloading line, some of the soil in the untreated sugar beet was separated at the initial stage, thus reducing both the amount of waste and the amount of soil going to the filter and sludge pool.
- Necessary equipment changes were made in order to separate the organic substances that affect the quality from the hydraulic transport water.
- Recycling of beet transport water at the level of washing, removing stones and washing roots has been achieved, thus reducing the overflowing water during this process.
- The water that feeds the processes inside the facility is filtered, the quality of the water is increased by lime and aluminum sulfate treatment, and its recycling is ensured.
- The recovery of hot water overflowing from the processes inside the facility and its cooling by spraying (pulverized) was ensured.
- Cooling water was recycled in the gas scrubbing section, the liquid cycle section with CO_2 pumps, and the small cooling tower section.
- A plate heat exchanger system was installed to heat the printing water starting from the hot water, thus making it possible to use the energy of the hot water. With this application, steam consumption was reduced and heat recovery was ensured by passing all hot water through the heat exchanger.

Benefits: As a result of the technological improvements realized, the pollution load of the process wastewater and the amount of water usage in the processes have decreased. In this facility with a capacity of 3000 tons/day, 60% water savings were achieved with the investments made with an investment of $204 000 with a payback period of 26 months, while wastewater generation was reduced at the same time. Thus, it also contributed to reducing the production costs of the facility and reducing its environmental impacts.

7.5.9.2 Facility-2 in Slovenia

Sugar production and processing factories include processes that intensively consume water and energy, as well as solid wastes and emission of CO_2 wastes. In this study, water reduction possibilities as a CP focus on a sugar factory in Slovenia were evaluated (Kotan & Bakan, 2007; Zver, 2005).

Benefits: With the improvements made, the use of water has been reduced to approximately 54 250 m^3/h (69% of total water consumption) and in parallel, a decrease in waste discharge has been achieved, while the discharge standards have been met. As a result of all these improvements, an annual economic benefit of approximately 51 144€/day was provided to this industry with a payback period of 5 days with an investment of 2500€.

7.5.10 Metal coating and painting facility in CP practices

7.5.10.1 Facility-1 (MPM Publications, 2007)

The focus of CP in the investigated facility was the rinsing and filtrate areas of the cyanide zinc coating process.

Implemented application: The rinsing process was conducted by reducing the amount of leachate formation and other chemicals used during the rinsing and leaching of the cyanide zinc coating. This was done by reducing the amount of chromium 6 used in the current state of the rinsing baths and rinsing tanks.

Benefits: As a result of the improvements made in the rinsing process, 65% reduction in water savings per product and up to 60% reduction in the use of some chemicals has been achieved. The decrease in the use of chemicals provided a reduction in the pollution load in the process wastewater. These improvements, with an investment of 7316 TL an annual economic benefit of approximately 18 875 was provided to this industry, with a payback period 4,7 months. It was reported that an annual reduction of 3000 TL was achieved in water costs.

7.5.10.2 Facility-2 (Demirer, 2009)

Implemented application: In the facility, the surface preparation process before painting was carried out by the workers using thinner for certain parts. With the applications carried out within the scope of the project, a more environmentally friendly production system was started, which includes degreasing, rinsing, oxylan chemical and drying steps instead of cleaning and cadmium coating processes. Labor force and transportation costs were also saved with applications where significant savings were achieved in the use of chemicals such as thinner, cadmium oxide (CdO) and sodium cyanide (NaCN).

Benefits: As a result of the improvements made in facility-2 are given in Table 7.3.

Table 7.3 Improvements in metal processing industry (Facility-2) with CP.

Facility Sector	Metal Processing Industry: Metal Coating and Painting
Chemical consumption subject to the project	Thinner: 9000 kg/year cadmium oxide (CdO): 1230 kg/year sodium cyanide (NaCN): 5219 kg/year
Chemical savings	Thinner: 7650 kg/year (85%) cadmium oxide (CdO): 1230 kg/year (100%) sodium cyanide (NaCN): 5219 kg/year (100%)
Project budget	UNIDO contribution: $25 000, Facility contribution: $4500
Total investment cost	$29 500
Thinner savings	14 645 $/year
Elimination of service procurement	15 329 $/year
Labor savings	17 947 $/year
Payback period	7 months

7.6 CASE STUDY ON CP PRACTICES AT PULP AND PAPER PRODUCTION FACILITY IN TURKEY

The paper industry is the industry that produces cellulose and paper products such as paper and cardboard from plants such as wood, jute, and hemp. In recent years, the production of paper products from used paper has gained great importance. This industry covers the preparation of raw material, its conversion into cellulose, bleaching, and conversion into paper products.

In this study, CP practices realized in real scale in the processes of the facility examined were evaluated.

7.6.1 Processes in the facility

This study was carried out in all kinds of sanitary and cleaning paper production facilities with a paper production capacity of 385 837.26 tons/year. The facility consists of two main processes, the paper production section and the converting units.

Paper preparation unit; cellulose, which is a raw material and supplied from abroad, is loaded into the machine in bales in the feeding band section. According to the production plan and to meet the determined semi-product standards, the type and amount of strength, softener, and chemicals are added to the process according to their requirements. By systematically controlling the entire process, the paper is produced as a buffer in the wrapper part of the machine. Paper bobbins hung on the rubber bumper or axial bumper are directed to narrow dubbing or wide dubbing machines to be sized and wound according to their combination. The bobbins coming out of the dubbing machines are stretched several times and labeled and directed to the warehouse. In the warehouse, the coils are shipped to the coil customer according to the order, or they are forwarded to the converting department to be converted into products. The bobbins coming to the converting department are sent to the toilet paper/towel, napkin, and tissue machines in accordance with the production plan. The general flow chart of paper products production is given in Figure 7.7.

Cellulose bales are separated according to their types in the dough preparation section. Depending on the customer's demand, a mixture of long and short fibers is adjusted in accordance with the type and quality of the product to be produced. Cellulose bales are brought to the pulpers, which act as a special blade mixer, in order to open them and turn them into pulp in an aqueous medium. It is broken down with the help of white water system and chemicals that circulate in pulpers and save water. They are brought to a size that can pass through the holes under the pulpers. There are two pulp stock warehouses, long and short fibers, to be used in the production facility. It is passed through a coarse clean that cleans the foreign materials in the dough coming to the process from the dough stock tank. They are brought to a size that can pass through the holes under the pulpers. Thus, a more homogeneous and individual fibered suspension will be obtained. The homogenized fibers pass to the grinder, and the milled pulps coming out of the machine are taken to the mixing tank.

Converting unit: The coils kept in the warehouse are sent to the converting department in line with the order specifications. Coils, which are in various sizes in dubbing machines come to the converting department, and are directed to toilet paper/towel, napkin, or tissue machines in accordance with the production plan.

7.6.2 CP practices in the facility

The aim of the environmental engineers, environmental consultants and auxiliary technical personnel in the paper production facility is to develop environmentally friendly products and processes by using proactive methods, to minimize environmental impacts, and to continue production by minimizing environmental and energy losses in compliance with legal requirements.

7.6.2.1 Selection of production techniques that pollute the environment less

As a result of the use of the advantage air cap (AAC) system in the production process, the gap between the 'Hood' and 'Yankee Drying Cylinder' machines, which affect natural gas consumption,

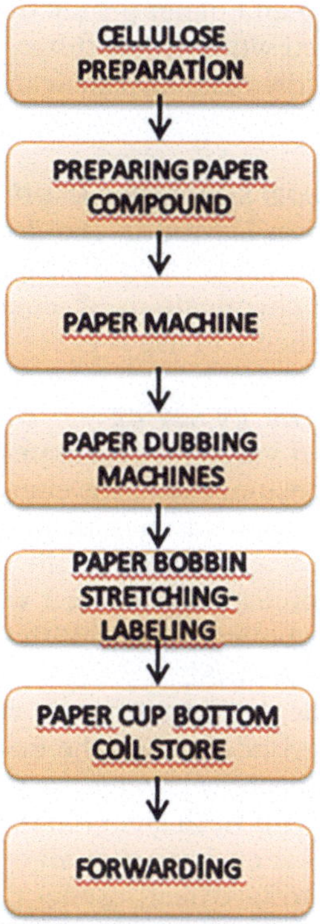

Figure 7.7 Pulp and paper manufacturing production processes.

can be reduced from 20 to 15 mm in order to save on natural gas consumption. Since the investigated facility is a pulp and paper production facility, the cellulose used as raw material is supplied from companies obtained from sustainable forests in accordance with the Forest Stewardship Council-Forest Management Council (FSC) and Program for the Endorsement of Forest Certification – Forest certification approval program (PEFC) standards. This is documented by the ECOLabel certificate, which states that the processes taking place in the production and use stages of the products cause less harm to the environment. There is an ECOLabel certificate stating that the processes that take place in the production and usage stages of the products are less harmful to the environment. There is a system called multiplayer in the paper production machines in the facility. There are long fibre and short fibre cellulose lines in the paper machines currently used in the facility. It is advantageous over a single type line. There are four paper production machines at the facility. The fourth machine in the facility uses completely pure cellulose, the pure cellulose used is supplied from forests grown for industrial purposes. Using pure cellulose requires less energy and water use than using recycled cellulose. Production at the facility continues with the product production plans created. While making the production plan, product productions of the same quality are planned one after the other. This reduces water and energy consumption during production.

7.6.2.2 Correct selection of the treatment system and reuse of waste water in paper processing

There is a wastewater treatment plant where the industrial wastewater from the factory and the domestic wastewater generated by the employees are treated. The pulp is squeezed and filtered between the sieve and the felt in the pulp and paper production machines. This filtered water returns to the White Water System. The so-called white water (filtered from the fibers) returns to the dough production process in the system. Thus, the need for raw water required for the production of pulp is reduced.

7.6.2.3 Reducing the amount of chemical substance used

Chemicals with Ecolabel Certificate are preferred in the processes. Ecolabel Certificate is an eco-label certification system that enables products to be distinguished as environmentally friendly and sustainable. To minimize the waste of chemicals used, chemicals are fed to the process with an automation system. Thus, unnecessary chemical use is prevented.

7.6.2.4 Evaluation of waste heat and energy saving

Equipment such as machinery, pipes and valves are insulated to prevent energy losses. Uninsulated valves are a constant source of energy consumption. This energy loss has been tried to be prevented by applying jackets to the valves. The jacket application is an isolation system installed on the valves and provides energy savings. An economizer was installed in the boiler in order to benefit from the waste flue gas heat of the natural gas steam boiler. 102° of feed water coming from the deaerator passes through the economizer and is fed to the boiler at 150°. With this application, waste heat transfer was made and natural gas usage amounts were saved.

In the paper production line, after the drying process is completed with the yankee cylinder, the saturated steam leaving its heat in the cylinder leaves the cylinder as condensate (condensed steam) and low-temperature steam. The condensate separated in the condensate tanks is fed back to the steam boiler. The rotten steam is mixed with fresh steam in the thermocompressor and fed back to the cylinder. Thanks to this heat recovery, it is ensured that all of the steam heat is used and turned into condensate. The method that enables us to save heating energy is: in the steam boiler fed by using coal, the air at atmospheric temperature coming from the primary and secondary fans with the air exchanger called requperator is heated up to 200° and pressed in the boiler. In the fourth paper production machine in the facility, the air inlet–outlet flow rates, temperature values, humidity values required for drying the paper are controlled by the air balance control (ABC) system, and the system is always operated at optimum values. In this way, 30–40 kWh of energy is saved for the electricity consumed per ton of paper. While processes are carried out using fresh steam in conventional paper production systems, the exhaust air from the process will be used thanks to the ReDry System implemented. Thanks to this system, 0.4 tons of steam will be saved per ton of paper and the waste heat is reused in the process and is not given back to the atmosphere. With the waste heat recovery system, the energy of the exhaust air is utilized, while the exhaust air is passed through a washer, reducing the emission values given to the environment and the fibers escaping from the process in this air are recovered. This system is environmentally friendly in terms of emission values and energy-friendly in terms of energy recovery. With this system, approximately 5.5 MWh of energy is recovered. Another environmentally friendly technology applied is the IQ Fiber Measurement system, which does not contain a radioactive source like conventional systems and measures the weight and moisture profile without carbon emission.

7.6.2.5 Wastes

The company has a Zero Waste Certificate at the basic level; within the scope of the Zero Waste Regulation applied in Turkey, all wastes generated as a result of the process are collected separately and sent to waste disposal facilities.

7.6.2.6 Emissions

The amount of dust emission generated in the machine is reduced with the help of Advantage Run System technology in the paper production machine. It is pulverized with the help of vacuuming technology and recycled to the system in order to reduce the volatile organic carbon (VOC) and dust emissions expected to occur in the drying section. In this way, emission reduction takes place. The advantage wet dust removal system in the facility is a wet-based system. A wet dust collector is integrated in the pulper under the machine, thus providing a clean, safe, and environmentally friendly tissue paper production environment. The system prevents dust from spreading by flying. Increases work environment safety and eliminates the risk of fire and clogging.

7.7 DISCUSSION AND CONCLUSION

As a result of eco-efficiency practices, the environmental issue for the industry is no longer just 'environmental legislation pressures' and 'additional costs for environmental protection,' it has evolved into a concept that includes opportunities for minimizing production and environmental costs by increasing efficiency in production and providing both environmental and economic benefits. The development and implementation of CP strategies for an organization can be beneficial in many ways.

One of the main goals of CP is to increase process efficiency. Increasing process efficiency for a facility will affect important cost items such as reduction in energy, natural resources and raw materials, and increase the profitability of the facilities. Increasing the efficiency of the process will also lead to results such as minimizing the use of energy, natural resources and raw materials, and increasing the efficiency and profitability of a facility directly due to the decrease in product or service costs. Technology applications used within the scope of CP, all production processes, machinery, raw materials, auxiliary chemicals and dyestuffs, water and energy consumption, all kinds of waste production, occupational and worker health as a result of examining and evaluating the negative effects of the facility on the environment, while minimizing the negative impact of the facility. At the same time, it will ensure quality production at the facility and the sustainability of environmentally friendly production by providing great financial gains.

CP strategies, due to their proactive nature, will ensure that the facility fulfills its legal obligations regarding environmental issues and that any incompatibilities are eliminated. Facilities that develop and implement clean technology strategies not only will gain the advantage of being in compliance with current regulations, they will also be prepared to comply with environmental regulations and international/national standards and regulations, which will be under even greater pressure in the future. Moreover, the resulting solid/liquid/gas wastes will be reduced at their source, if possible, and compliance with the relevant environmental legislation will be facilitated. The industries, on the one hand, while they contribute to economics by producing more and also providing development, on the other, they have to find a solution to reduce the wastes resulting in their production. In other words, they have to realize development with a sustainable approach in harmony with the environment.

In this chapter, it is seen that the industries (mostly small manufacture enterprises – SMEs) examined in the literature and the paper industry, which is examined as a case study, mostly focused on energy recovery and waste water recovery as CP practices. Therefore, management skills and institutional competencies of especially SMEs should be developed and studies should be carried out to increase the ability to comply with relevant national and international standards. 'Know-how' practices including a responsible environmental management should be developed. In other words, efficiency should be increased by applying new techniques and work programs to existing ones without applying a new technology or process, and by reviewing the policies of production, processes, and organizations. CP technology practices such as the manufacturing technology or manufacturing process, changing inputs, changing the final product, using non-product materials formed during production at the production site and in the process, or improving the existing ones, should be developed on a sectoral basis. The approach in the industry in the new millennium is not only to have an obligation to create a healthy society, but also to develop sustainable production systems that minimize environmental negatives. Minimizing environmental problems is perhaps the most important social responsibility of industries. In order to fulfill these responsibilities, industries should show sensitive approaches to environmental issues in decisions regarding production management and evaluate the production function and environmental issues together.

The future is focused on CP and consumption processes. While industries investing in this field with a sustainable society understanding, should consider not only the solution of short-term problems, but also to assess the contributions in medium- and/or long term to the environment and the society in general regarding these investments.

ACKNOWLEDGEMENT

The authors acknowledge the Pulp & Paper Facility task group who were responsible for applying the CP practices for their sharing networks.

Author's contributions

Conceive: G.Y.T.; design: G.Y.T.; supervision: G.Y.T.; literature review: G.Y.T. and M.C; writing: G.Y.T. and M.C.; and critical reviews: G.Y.T.

REFERENCES

Alkaya E., Böğürcü M., Ulutaş F. and Demirer G. N. (2011). Adaptation to climate change in industry: pilot practices for eco-efficiency approach and water saving. 9th National Environmental Engineering Congress, 05–08 October 2011, Samsun-Turkey.

Asia Pacific Economic Cooperation (APEC) (2006). Cleaner production and sustainable technology CAS studies from APEC economies with some economic analysis and interpretation, Industrial Technology Development, Industrial Science and Technology Working Group Institute Department of Science and Technology Republic of the Philippines.

Clark G. (2007). Evolution of the global sustainable consumption and production policy and the United Nations Environment Programme's (UNEP) supporting activities. *Journal of Cleaner Production*, **15**, 492–498, https://doi.org/10.1016/j.jclepro.2006.05.017

Das T. (2005). Toward Zero Discharge, Innovative Methodology and Technologies for Process Prevention. Wiley, New York, pp. 11–23.

Davarcıoğlu B. (2017). Adaptation to climate change and eco-efficiency (cleaner production) for more effective environmental management in industry. *International Journal of Recent Trends in Engineering and Research*, **3**(2), 100–112, https://doi.org/10.23883/IJRTER.2017.3011.GN6QY

Demirer G. N. (2003). Basic principles of pollution prevention approaches. *Environment and Engineer TMMOB*, (25), 13–20.

Demirer G. N. (2009). Cleaner Production and Eco-Efficiency in Industry: Best Practices. Unido Eco-Efficiency (Cleaner Production) Program Information Meeting, Dr. Fikret Yücel Conference Hall Cyberpark Cyberplaza, Bilkent-ANKARA.

Demirer G. and Mirata M. (1999). Industrial pollution prevention or cleaner production-1. *Industry & Automation*, **31**, 110–113.

Frondel M., Horbach J. and Rennings K. (2004). End-of-Pipe or Cleaner Production? An Empirical Comparison of Environmental Innovation Decisions Across OECD Countries, Center for European Economic Research, Discussion Paper No. 04-82.

Glavic P. and Lukman R. (2007). Review of sustainability terms and their definitions. *Journal of Cleaner Production*, **15**, 1875–1885.

Greco Initiative & Regional Activity Centre for Cleaner Production (CP/RAC) (2008). GRECO Med Clean Report Overview. Imprentaweb, Barcelona.

Kjaerheim G. (2005). Cleaner production and sustainability. *Journal of Cleaner Production*, **13**, 329–339, https://doi.org/10.1016/S0959-6526(03)00119-7

Kotan T. and Bakan G. (2007). Investigation of Cleaner Production Practices and Performance Studies in Various Industries. 7th National Environmental Engineering Congress, 24–27 October 2007 –İzmir-Turkey).

Luskin J. and Del Matto T. (2007). Introduction to the special issue on, sustainable production and consumption: making the connection. *Journal of Cleaner Production*, **15**, 489–491, https://doi.org/10.1016/j.jclepro.2006.05.009

MPM Publications (2007). Environmentally Sensitive Approaches in Industry: Cleaner Production and Eco-efficiency. Publication No: 693, Ankara.

Narayanaswamy V. and Stone L. (2007). From cleaner production to sustainable production and consumption in Australia and New Zealand: achievements, challenges, and opportunities. *Journal of Cleaner Production*, **15**, 711–715, https://doi.org/10.1016/j.jclepro.2006.06.011

Özbay A. (2005). Analysis of Cleaner Production Approach in Turkey and Institutional Restructuring Proposal for a Cleaner Production Center for SMEs. Specialization thesis, State Planning Organization, Ankara, p. 86.

REC (2011). Sustainable Production and Consumption Publications – II Cleaner Production. Regional Environment Center- REC Turkey, Ministry of Environment and Urbanization, Ankara.

Regional Activity Centre for Cleaner Production (CP/RAC) (2000). MOED: Minimization Opportunities Environmental Diagnosis. ALTES, SL, Barcelona.

TDFT (Turkey Technology Development Foundation) (2011). Eco-Efficiency in Industry (Clean Productıon) Guıde: Methods and Applıcatıons, Technology Development Foundation of Turkey (TDFT) UNIDO Eco-efficiency (Cleaner Production) Program. Date of access: https://www.ttgv.org.tr/tur/images/publications/6005ba5cf24cd.pdf

The UNEP Working Group for Cleaner Production in the Food Industry (2004). Eco-efficiency Toolkit for the Queensland Food Processing Industry. Australian Industry Group, Queensland.

UNEP (United Nations Environment Programme) (1996). Cleaner Production: A Training Resource Package, Industry and Environment.

UNEP (United Nations Environment Programme) (2002). Sustainable Consumption and Cleaner Production Global Status 2002, United Nations Environment Programme Division of Technology, Industry and Economics, ISBN: 92-807-2073-2, Cedex, France).

Vietnam Cleaner Production Center (VNCPC) (2000). Mini-guide to Cleaner Production. Vietnam Cleaner Production Center, Hanoi.

Zver L. (2005). Water minimization in process industries: case study in beet sugar plant. *Journal of Cleaner Production*, **43**, 133–145.

doi: 10.2166/9781789063783_0141

Chapter 8

Controlling organic micropollutants in urban (waste) water treatment by activated carbon adsorption and membrane technology

Maria João Rosa, Margarida Campinas, Catarina Silva, Elsa Mesquita and Rui M. C. Viegas

Water Quality and Treatment Laboratory, Urban Water Unit, Hydraulics and Environment Department, LNEC – National Laboratory for Civil Engineering, 1700-066 Lisbon, Portugal

Corresponding author: mjrosa@lnec.pt, mcampinas@lnec.pt, csilva@lnec.pt, emesquita@lnec.pt, rviegas@lnec.pt

ABSTRACT

Population growth and aging, as well as water scarcity driven by climate change call for an enhanced control of organic micropollutants and contaminants of emerging concern (CECs) in the urban water cycle, as progressively entailed in the EU and Portuguese legal framework on drinking water, urban wastewater treatment (UWWT) and water reuse. Multibarrier solutions, grounded on current barriers' improvement, prioritizing low-energy, physical barriers (to minimize byproducts, resources' use, and carbon footprint) and producing fit-for-purpose water(s) are needed. Activated carbon(AC)-based and hybrid powdered activated carbon/low-pressure membrane processes have proven a huge potential for controlling organic micropollutants in drinking water treatment and water reclamation, but there is room for improving their sustainability and cost-efficiency with process optimization. This chapter aims to introduce the CECs' problem in drinking water and UWWT and comprehensively explain the potential for controlling CECs using AC adsorption and hybrid adsorption/low-pressure membrane processes, based on our last decade's lab-, pilot-, and full-scale results herein summarized and discussed.

Keywords: activated carbon adsorption, contaminants of emerging concern, drinking water, hybrid adsorption/membrane process, ceramic microfiltration, organic micropollutants, urban wastewater treatment

ACRONYMS AND ABBREVIATIONS

A254	UV absorbance at 254 nm
AC	activated carbon
AMX	amoxicillin (PhC)
AOC	assimilable organic carbon
AOP	advanced oxidation process
APAP	acetaminophen (PhC)

ARB	antibiotic-resistant bacteria
ARG	antibiotic resistance gene
AS	activated sludge
ATN	atenolol (PhC)
BAC	biologically active carbon
BEI	Beirolas (WWTP)
BET	Brunauer–Emmet–Teller
BOD_5	biochemical oxygen demand at 5 days
BTZ	bentazone (pesticide)
BZ	1-H-benzotriazole (anticorrosive)
BZF	bezafibrate (PhC)
C	carbon
CAF	caffeine (PhC)
CAPEX	capital expenditures
CBZ	carbamazepine (PhC)
CEB	chemically enhanced bakwash
CEC	contaminant of emerging concern
CFA	clofibric acid (PhC)
C/F/S	coagulation/flocculation/sedimentation
CHL	chlortoluron (pesticide)
COD	chemical oxygen demand
CP	chlorophene (biocide)
CTS	cortisone (PhC)
CYP	cyclophosphamide (PhC)
d_{50}	median diameter
D_{pH}	(octanol–water) distribution coefficient at a given pH
DAF	dissolved air flotation
DCF	diclofenac (PhC)
DT	DEET (N,N-diethyl–3-methylbenzamide) (insect repellent)
DES	diethylstilbestrol (hormone)
DIU	diuron (pesticide)
DMT	dimethoate (pesticide)
DOC	dissolved organic carbon
DPR	direct potable reuse
DW	drinking water
DWT	drinking water treatment
DWTP	drinking water treatment plant
E1	estrone (hormone)
E2	17-β-estradiol (hormone)
EE2	17α-ethinylestradiol (hormone)
E3	estriol (hormone)

EDC	endocrine-disrupting compound
EfOM	effluent organic matter
EQS	environmental quality standard
EQSD	environmental quality standard directive
ERY	erythromycin (PhC)
EU	European Union
FLX	fluoxetine (PhC)
F/M	food to microorganisms ratio
FNW	Faro Noroeste (WWTP)
GAC	granular activated carbon
GTD	gestodene (hormone)
HAA	haloacetic acid
HSDM	homogeneous surface diffusion model
IBUP	ibuprofen (PhC)
IND	indomethacin (PhC)
IPR	indirect potable reuse
ISO	International Standardisation Organisation
JRC	Joint Research Centre
k_{bio}	biodegradability coefficient
K_d	sorption onto biomass coefficient
K_{ow}	octanol–water partition coefficient
LIN	linuron (pesticide)
LINC	lincomycin (PhC)
lmh	$L/(m^2.h)$
MBR	membrane bioreactor
MC-LF	microcystin-LF
MC-LR	microcystin-LR
$MC\text{-}LR_{eq.}$	microcystin-LR equivalent (overall concentration of all MC variants)
MC-LY	microcystin-LY
MC-LW	microcystin-LW
MF	microfiltration
μF	micro filter
MIB	2-methylisoborneol
ML	3-methylindole (fragrance)
MM	molar mass
MTPL	metoprolol (PhC)
MXL	mixed liquor
N	nitrogen
NF	nanofiltration
NH	nortriptyline.HCl (antidepressant)
NOM	natural organic matter

NPX	naproxen (PhC)
OPEX	operational expenditures
P	phosphorus
PAC	powdered activated carbon
PFAS	per- and poly-fluoroalkyl substance
pH_{pzc}	point of overall zero charge
PPNL	propranolol
PS	priority substance
PSA	polar surface area
PT	Portugal/Portuguese
RAN	ranitidine (PhC)
RO	reverse osmosis
SDG	sustainable development goal
SDZ	sulfadizine (PhC)
SF	sand filtration
SMX	sulfamethoxazole (PhC)
SPD	sulfapyridine (PhC)
SUVA	specific UV absorbance, given by A254/DOC
T254	transmittance at 254 nm
TCp	treatment capacity
TCZ	tebuconazole (pesticide)
THM	trihalomethane
THMFP	trihalomethane formation potential
TSS	total suspended solids
TTE	testosterone (hormone)
UF	ultrafiltration
UN	United Nations
UV	ultraviolet radiation
UWW	urban wastewater
UWWT	urban wastewater treatment
UWWTD	urban wastewater treatment directive
UWWTP	urban wastewater treatment plant, often referred simply as WWTP – wastewater treatment plant
VOC	volatile organic compound
WFD	Water Framework Directive
WHO	World Health Organisation
WL	Watch List
WTP	water treatment plant
WWTP	wastewater treatment plant

8.1 THE PROBLEM: ORGANIC MICROPOLLUTANTS IN WATER

Water is central to all human activities, to all components of the EU Green Deal and to several United Nations sustainable development goals (UN SDGs), starting with SDG 6 'Clean water and sanitation.' Furthermore, population growth and aging (SDGs 3, 11) and economic growth (SDG 8) are increasing the water demand and contaminants' discharge, while climate change (SDG 13) is decreasing source water availability (droughts and time concentrated rainfall) and increasing water demand (for irrigation, cleaning), thus promoting water exploitation and scarcity, and decreasing source water quality (increased organics, salts, toxic cyanobacterial blooms). These are reasons why we need to 'Take urgent action to combat climate change and its impacts' (SDG 13), 'Build resilient infrastructure' (SDG 9), promote 'Sustainable cities and communities' (SDG 11), carbon neutrality, energy self-sufficiency, resource efficiency, and circular economy, for example, water reuse [SDGs 6, 7 (energy), and 12 (responsible consumption)].

More than ever, effective, and efficient urban water systems are therefore crucial, and the drinking water treatment (DWT) and the urban wastewater treatment (UWWT) are, in addition to the pollution-source control, the key barriers for protecting human health, ecosystems, and their services against contaminants of health–environmental concern. These include regulated contaminants, but barriers must be prepared for addressing also contaminants of emerging concern (CECs).

As clearly proposed by Sauvé and Desrosiers (2014), CECs are naturally occurring or manufactured contaminants present or suspected to be present in various environmental compartments and whose toxicity or persistence are likely to significantly alter the metabolism of a living being; "should remain 'emerging' as long as there is a scarcity of information in the scientific literature or there are poorly documented issues about the associated potential problems they could cause"; 'are expected to be chemicals that show some potential to pose risks to human health or the environment and which are not yet subjected to regulatory criteria or norms for the protection of human health or the environment'; 'not all will actually prove to be evil and have some potential to cause tangible concerns'; 'an already regulated presumed well-known contaminant could certainly regain 'emerging' status as new scientific information becomes available and thus force regulatory agencies to re-evaluate their norms and guidelines'; 'will remain a moving target as new chemical compounds are continuously being produced and science continuously improves its understanding of current and past contaminants.'

CECs include synthetic chemicals, naturally produced metabolites and biological hazards. Among the latter, currently, antibiotic-resistant bacteria (ARB) and antibiotic resistance genes (ARGs), and viruses and virions are of top concern. Naturally occurring CECs include natural hormones excreted in water (Ternes & Joss, 2006) and cyanotoxins (hepato-, neuro-, and dermatotoxins) produced by toxic cyanobacterial blooms in surface waters. These blooms (in fresh and transitional waters) are generally triggered by phosphorus, temperature, and solar light conditions, and the most commonly occurring cyanotoxin is microcystin-LR (Chorus & Welker, 2021; Menaia & Rosa, 2006). MC-LR was therefore included in the Portuguese legislation relative to drinking water quality standards (Decree-law 306/2007, Decree-law 152/2017) following the World Health Organisation (WHO) guidelines (Chorus & Welker, 2021, the 2nd edition of the WHO Guide published in 1999) and some Portuguese or EU research projects developed in Portugal (e.g., TOXIC and CIANOTOX projects), and later included in the EU Drinking Water Directive (Directive (EU) 2020/2184).

Industrial chemicals, such as pharmaceutical compounds (PhCs) (human and veterinary drugs), synthetic hormones, cosmetics and personal care products, pesticides, flame-retardants, plasticizers, food additives, per- and poly-fluoroalkyl substances (PFAS), and nano- and microplastics, raise considerable toxicological concerns for the aquatic environment since they may be transported into waterbodies in some stage of their lifecycle (Mestre *et al.*, 2022a). Actually, most of these chemicals have been already detected in surface and groundwater in pg/L to μg/L worldwide (Geissen *et al.*, 2015), as illustrated for PhCs in Figure 8.1 (Aus der Beek *et al.*, 2016) and Figure 8.2 (Silva *et al.*, 2021), whereas in urban wastewaters they occur in ng/L to dozens in μg/L. Figure 8.2 compares the concentrations of 13 PhCs found by others in urban wastewaters in Portugal and beyond with those found in BEI and FNW WWTPs' influents in 2016–2019 (within LIFE Impetus project). We have statistically found the PhC occurrence seasonality is well-defined by the maximum air temperature, with a turning point of 20°C in Lisbon and 22°C in Faro. Some PhCs showed lower concentrations in colder (wet) months due to rainwater dilution, whereas others showed higher concentrations, reflecting an increased seasonal consumption and/or slower transformation due to lower air temperatures and/or shorter hydraulic retention times. Seasonal studies should therefore focus on temperature and rainfall rather than on calendar seasons, increasingly uncertain due to climate change (Silva *et al.*, 2021).

On the contrary, a review of pesticides' monitoring studies of surface waters worldwide (Souza *et al.*, 2020) showed a critical occurrence of atrazine and its metabolites, metalochlor, chlorpyrifos, and tebuconazole, and high concentrations and frequency of diuron (0.03–22,770 ng/L) and of the insecticide dimethoate (0.57–61,200 ng/L).

Domestic wastewater, hospital and industrial effluents, stormwater runoff, rural runoff, and manure media are the main sources of CECs entering the aquatic environment (Eggen *et al.*, 2014; Luo *et al.*, 2014). So far, urban wastewater treatment plants (UWWTPs) were essentially designed to remove total suspended solids (TSS) and reduce the chemical and the biochemical oxygen demands (COD and BOD_5, respectively) in the biological secondary treatment, and, when the receiving water requires so, also to remove nutrients (nitrogen and or phosphorus) to avoid eutrophication, or to disinfect to control water microbiological quality. CECs' control in such UWWTPs starts with a good quality secondary effluent since biodegradation and sorption onto particulate matter and biomass are the two major elimination mechanisms (Rosa *et al.*, 2019; Siegrist & Joss, 2012; Silva *et al.*, 2023; Ternes & Joss, 2006). In turn, the non-biodegradable and polar CECs can easily escape the conventional secondary treatment (Eggen *et al.*, 2014; Luo *et al.*, 2014; Rosa *et al.*, 2019; Siegrist & Joss, 2012; Silva *et al.*, 2022; Ternes & Joss, 2006), being released into the water bodies, where they may accumulate in biota (Cravo *et al.*, 2022). The same may happen with the CEC metabolites (e.g., PhCs produced and excreted by human and veterinary metabolism) that, although typically less studied (Pourchet *et al.*, 2020), may be as toxic or more than their parent compounds (Golovko *et al.*, 2021). The continuous discharge of CECs on ecosystems may affect their health, biodiversity, and

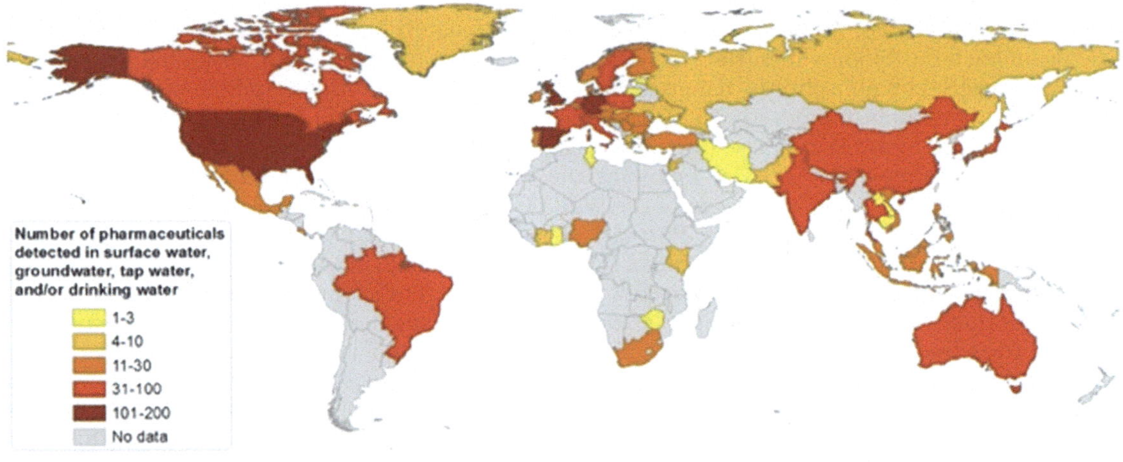

Figure 8.1 Country survey on the number of PhCs detected in surface waters, groundwater, or tap/drinking water. *Source*: From Aus der Beek *et al.* (2016).

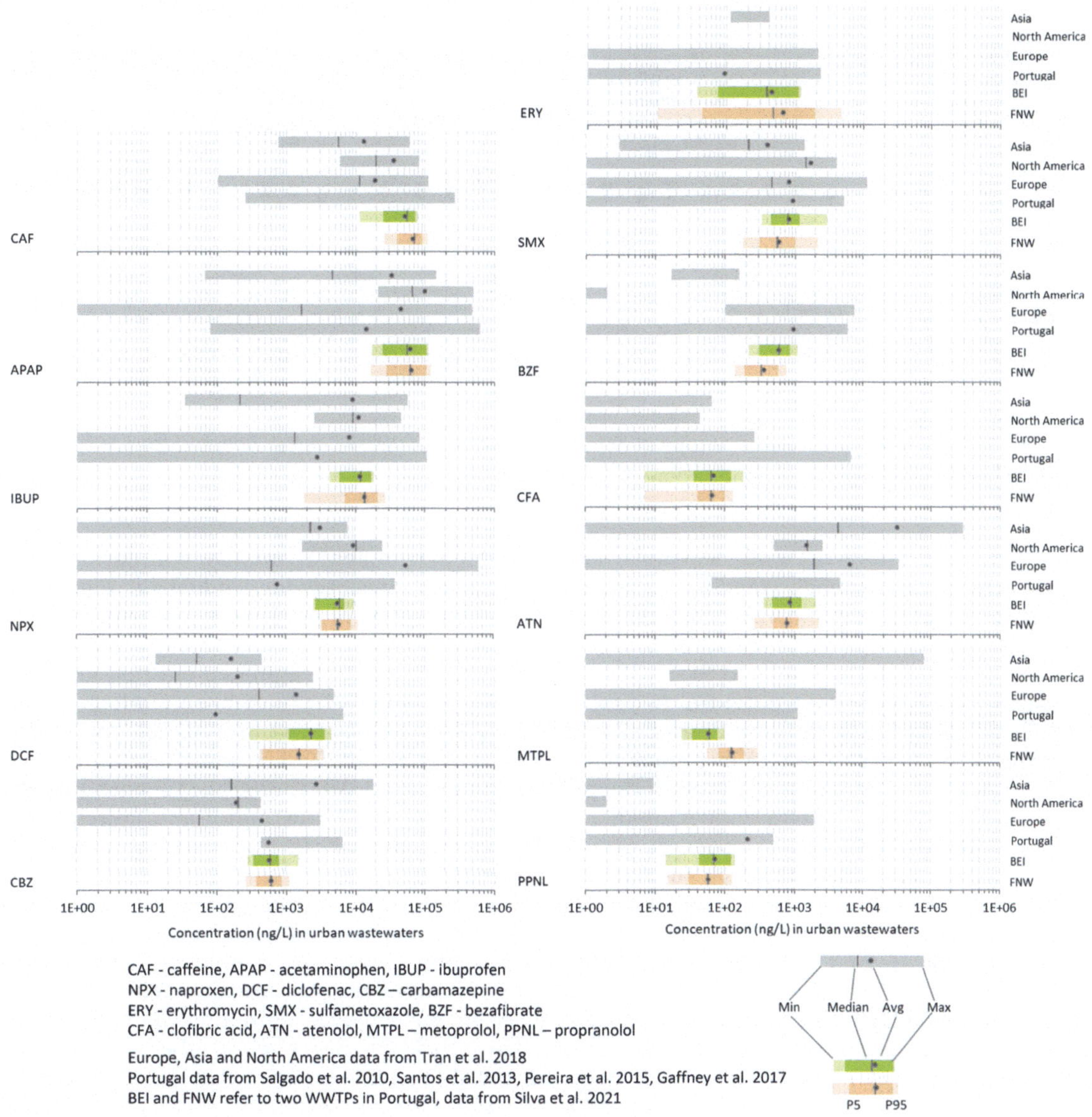

Figure 8.2 Examples of pharmaceuticals' concentration in urban wastewater. *Source*: Adapted from Silva *et al.* (2021).

eventually the quality of raw water for human consumption, particularly where highly populated and industrialized urban areas have closed water cycles and/or the dilution factors of rivers or receiving waters are not enough to disregard the risk for aquatic environment, as in some cities in Central Europe (Eggen *et al.*, 2014).

The EU water legislation evolution in the last two decades has been reflecting the CEC monitoring and mitigation needs, as illustrated in Figure 8.3, starting by monitoring and more recently addressing treatment. The Water Framework Directive (WFD) 2000/60/EC

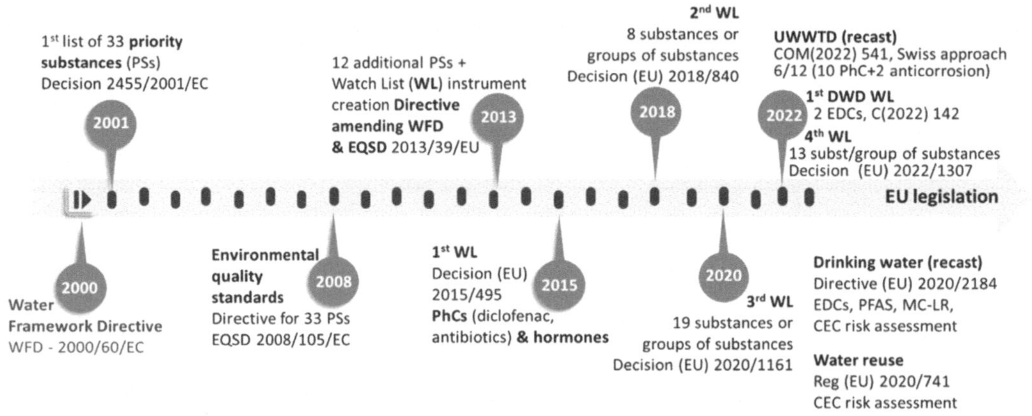

Figure 8.3 Evolution of water-related European legislation concerning CECs. *Source*: Adapted from Viegas *et al.* (2021b) and Mestre *et al.* (2022a and updated).

and subsequent directives and decisions established environmental quality standards (EQS) for 45 priority substances (specific herbicides, pesticides, insecticides, industrial chemicals, hydrocarbons, polyaromatic hydrocarbons, and heavy metals) and in 2013 (Directive 2013/39/EU) a new mechanism, the Watch List (WL), was introduced to support the identification of priority substances for regulation under WFD. A WL comprises CECs not yet regulated but that may pose significant risk due to their potential toxicological effects. Four WLs have been published since 2015, comprising PhCs, hormones, and pesticides/herbicides (Mestre *et al.*, 2022a), the 4th WL including seven new substances or groups (Gomez Cortes *et al.*, 2022). The same mechanism was recently adopted in drinking water, with the publication of the 1st WL (C(2022) 142) entailing that the potential presence of two endocrine-disrupting substances (17-β-estradiol and nonylphenol) will have to be monitored at relevant points throughout the whole water supply chain and indicating guidance values for each substance (CID, 2022).

In drinking water and in water reuse (namely for the highest quality reclaimed water for agriculture irrigation), there are no fixed limits for CECs, but additional requirements may have to be adopted for pharmaceuticals and other CECs depending on the outcome of the monitoring according to the 1st DWD WL (C(2022) 142) or of the mandatory risk assessment (Directive (EU) 2020/2184 and Regulation (EU) 2020/741). The same takes place in the national legislation, namely Decree-law 152/2017 (drinking water quality) and Decree-law 119/2019 (water reuse), the latter (as well as the Regulation (EU) 2020/741) fully aligned with the ISO standards on water reuse (ISO 16075 series – Guidelines for treated wastewater use for irrigation projects, ISO 16075 series (2015–2021) – Guidelines for treated wastewater use for irrigation projects, ISO 20426:2018 – Guidelines for health risk assessment and management for non-potable water reuse).

Regarding wastewater, the recently proposed revised text for the Urban Wastewater Treatment Directive (UWWTD recast, COM (2022)) adopts the Swiss approach, with whom the German Federal Centres for Trace Substances (KomS) in Baden-Württemberg, North Rhine-Westphalia, and Berlin (Central Europe) agree. This is, UWWTD (recast) introduces the obligation to apply additional treatment to urban wastewater (quaternary treatment) in order to eliminate the broadest possible spectrum of micropollutants, that is, 80% removal (contrasting with the typical WFD approach of limit concentrations) of 6 CECs of a list of 12 in all UWWTPs treating $\geq$100,000 p.e. by December 31, 2035 (at the latest) and, by 2040, in all facilities $\geq$10,000 p.e. in areas where the concentration or accumulation of micropollutants poses a risk to human health or the environment according to the specified criteria (areas to be identified by the member states). Furthermore, it introduces the extended producer (including importers) responsibility (EPR) to contribute to support the costs of the quaternary treatment, such financial contribution being established based on the quantities and toxicity of the products the producer (including importers) places on the market. In Germany, late 2019, more than 20 plants have been expanded with advanced treatment technologies and put into operation to provide 80% reduction of selected indicator substances, and further 27 facilities were planned or under construction (Metzger *et al.*, 2019).

Once a CEC is regulated it becomes a micropollutant, as the new UWWTD refers to the 12 PhCs from which six must be selected for minimum 80% removal in UWWTP and the EU & Portuguese drinking water standards refer to microcystin-LR and pesticides.

8.2 MICROPOLLUTANTS TARGETED AND KEY PROPERTIES FOR REMOVAL

This chapter aims to raise the awareness and the technical and scientific preparedness of the readers to support transforming the latest regulation in DWT and UWWT into practical impact, namely by comprehensively explaining a solution for controlling organic micropollutants of different molar mass (MM), charge, hydrophobicity/hydrophilicity, biodegradability, and sorption onto biomass – the key properties for removal by the targeted solutions.

Amongst the large spectrum of organic micropollutants, many may be controlled at source or its use limited or prohibited, while others will likely be naturally produced in surface waters (e.g., cyanotoxins) or inevitably (partially) discharged in urban wastewaters (e.g., pharmaceuticals). Therefore, for DWTPs, the selected organic micropollutants are MC-LR (and other microcystin variants naturally occurring with MC-LR, Table 8.1, often overall expressed as MC-LR_{eq}) and pesticides, regulated contaminants, as well as representative PhCs, considered via the risk assessment.

For UWWTPs, the selected organic micropollutants for advanced adsorption/membrane treatment are representative recalcitrant PhCs (i.e., of low biodegradability (k_{bio}) and sorption onto biomass (K_d), Figure 8.4), escaping the secondary (for C removal) or tertiary (for C, N and P removal) treatment, that is, poorly removed or exhibiting variable and unreliable removal by secondary/tertiary treatment. Considering their distinct physical–chemical properties, namely charge at pH 7.4 and hydrophobic/hydrophilic character (expressed by Log K_{ow} (the logarithm of the n-octanol/water partition coefficient) or Log D_{pH} (the logarithm of the distribution coefficient, which corrects the Log K_{ow} for the pK_a of each compound by quantifying the amount of both the ionized and non-ionized forms of the PhCs in octanol and water at the pH values under study; Viegas *et al.*, 2022), the higher the more hydrophobic), our studies have been often conducted with carbamazepine (CBZ), diclofenac (DCF), and sulfamethoxazole (SMX) (Table 8.2), though not limited to (other examples in Table 8.3 and Sections 8.6, 8.7, and 8.8).

8.3 AVAILABLE TECHNOLOGIES FOR CONTROLLING ORGANIC MICROPOLLUTANTS

As introduced earlier, conventional UWWTPs essentially target TSS, COD, and BODs reduction via physical unit operations and biological processes, as well as nitrogen and phosphorus reduction or disinfection (*Escherichia coli* inactivation) when needed, respectively, to avoid eutrophication or protect bathing waters/aquaculture. In turn, as described in Rosa *et al.* (2009), conventional DWTPs include (a) preoxidation, for controlling biological growth, taste, and odor compounds, ferrous and manganese ions, (b) coagulation/flocculation/sedimentation and rapid sand filtration for removing turbidity and reducing the natural organic matter (NOM), particularly the more hydrophobic fraction (expressed by SUVA $>$ 4–5 L/(mgC m), where SUVA represents the specific UV absorbance), the major precursor of organochlorinated by-products, and (c) final chlorination for water disinfection and for providing a disinfectant residual in the water distribution system (to

Table 8.1 Properties of microcystin variants.

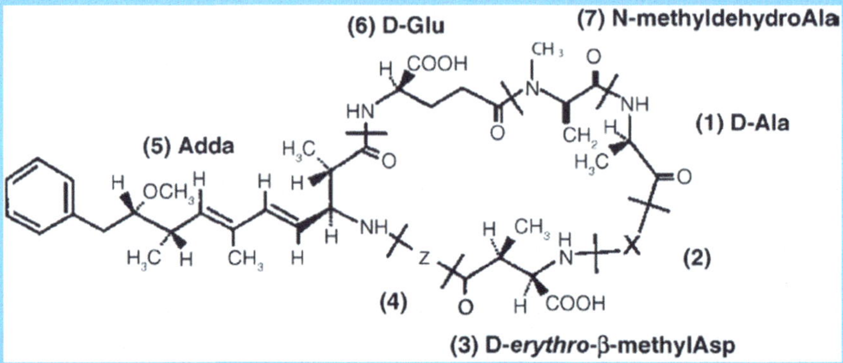

	MC-LR	MC-LY	MC-LW	MC-LF
Amino acids (X, Z)	Leucine, arginine	Leucine, tyrosine	Leucine, tryptophane	Leucine, phenylalanine
Molar mass (g/mol)	994	1001	1024	985
Net charge at pH 7	−1	−2	−2	−2

Increasing hydrophobicity →

Source: Adapted from Campinas and Rosa (2006).

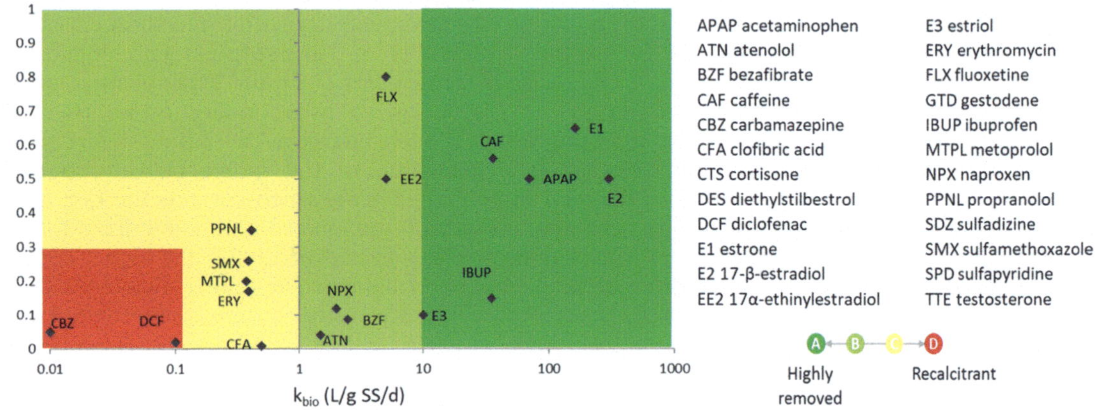

Figure 8.4 K_d vs k_{bio} for selected organic micropollutants and CECs. *Source*: From Rosa *et al.* (2019) and Silva *et al.* (2022).

prevent biological regrowth). Consequently, the conventional treatment plants were not designed as full barriers against organic micropollutants. The strategy to control them must therefore start by assessing and improving the performance of the existing barriers and then upgrade them with advanced treatment, as needed.

Operational, low-investment measures for improving current UWWTPs include providing conditions for the build-up of the slowly growing nitrifying bacteria, which can excrete enzymes

Table 8.2 Carbamazepine, diclofenac, and sulfamethoxazole key properties for removal by adsorption and adsorption/membrane filtration.

PhC Molecular Structure and Therapeutic Class	Physical–Chemical Properties	Optimized Geometries and Dimensions
Carbamazepine/CBZ 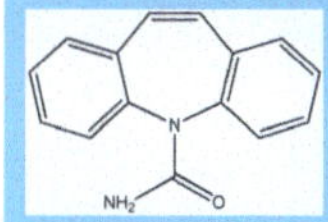 Anti-epileptic and psychiatric drug	MM = 236.3 g/mol Log K_{ow} = 2.67 Log $D_{7.4}$ = 2.28 pK_a = 13.9 Neutral and hydrophobic	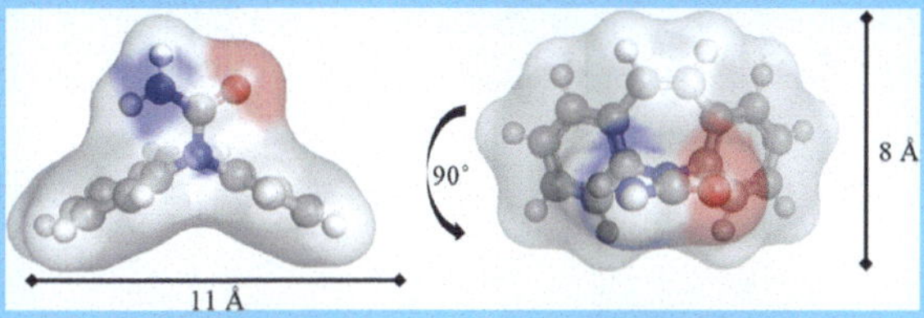
Diclofenac/DCF 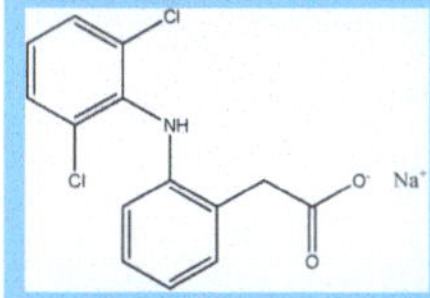 Non-steroidal analgesic and anti-inflammatory drug	MM = 318.1 g/mol Log K_{ow} = 4.06 Log $D_{7.4}$ = 1.37 pK_a = 4.0 Anionic and relatively hydrophobic	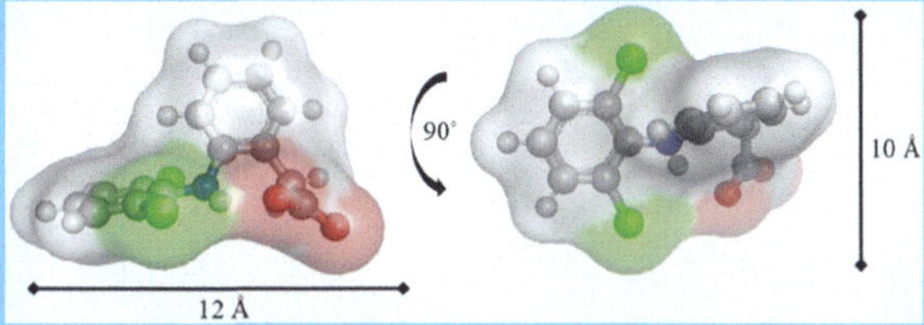
Sulfamethoxazole/SMX 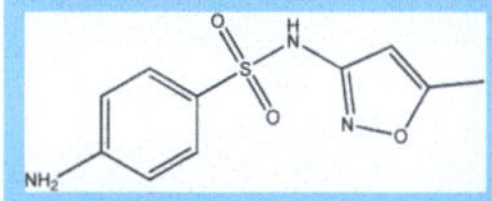 Antibiotic	MW = 253.3 g/mol Log K_{ow} = 0.89 Log $D_{7.4}$ = −0.56 pK_{a1} = 1.7; pK_{a2} = 5.6 Anionic and hydrophilic	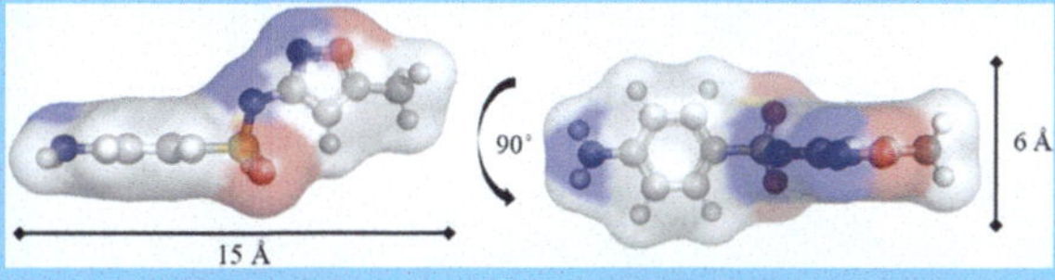

Source: From Mestre *et al.* (2022b) and Viegas *et al.* (2022).

and, by co-metabolism, may break down some low degradable molecules (Clara *et al.*, 2005; Ternes & Joss, 2006). These conditions include sludge retention times of, at least, 10–15 days (Clara *et al.*, 2005) or F/M (food to microorganisms) ratios below 0.08 d^{-1} that were associated with N-removal > 80% and alkalinity reduction > 40% (due to nitrification) (Rosa *et al.*, 2019; Silva *et al.*, 2022).

Improvement measures in conventional DWTP include adjusting preoxidation and C/F/S operating conditions for promoting NOM removal by enhanced coagulation and implementing C/F/S assisted with powdered activated carbon (PAC) addition. For instance, an enhanced control of microcystins (cyanobacteria and other cyanotoxins) may be achieved using low ozone doses in preoxidation (to avoid cell disruption prior to C/F/S and maximize intracellular $MC\text{-}LR_{eq}$ removal), adequate PAC and coagulant doses, dissolved air flotation (DAF) instead of sedimentation, smooth start-up/shut-down of sand filtration cycles, and multilayer filtration (Hall *et al.*, 2005; Raspati *et al.*, 2015; Ribau Teixeira *et al.*, 2020; Rosa *et al.*, 2009; Smeets *et al.*, 2015).

DWTP and UWWTP upgrade relies on implementing advanced treatment (quaternary treatment according to the proposed recasted UWWTD). Established advanced (waste)water treatment options for enhanced control of organic micropollutants and CECs include (Baresel *et al.*, 2017; Campinas *et al.*, 2017, 2022; Chorus & Welker, 2021; Eggen *et al.*, 2014; Hall *et al.*, 2005; Hillenbrand *et al.*, 2016; Luo *et al.*, 2014; Mestre *et al.*, 2022a, b; Metzger *et al.*, 2019; Mulder *et al.*, 2015; Pesqueira *et al.*, 2020; Rizzo *et al.*, 2019, 2020; Rosa *et al.*, 2009; Rosa *et al.*, 2019; Smeets

Table 8.3 Other selected organic micropollutants and their key properties for removal by AC adsorption and membrane filtration.

PPCP	Structure	Molecular Formula	MM, g/mol	V_{molar}, cm³ (1)	Log K_{ow}	Charge at pH 7	Log *D* at pH 7–7.5	Category
1-H-Benzotriazole (BZ)	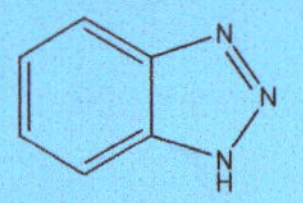	$C_6H_5N_3$	119.1	88.3	1.44	Neutral	1.42–1.39	Anticorrosive
DEET (DT)	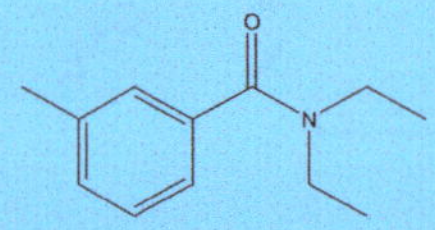	$C_{12}H_{17}NO$	191.3	194.0	2.18	Neutral	2.18	Insect repellent
3-Methylindole (ML)	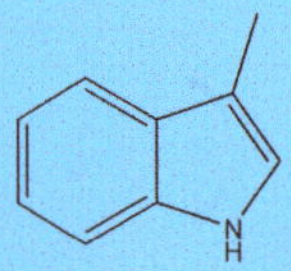	C_9H_9N	131.3	118.1	2.60	Neutral	2.60	Fragrance
Chlorophene (CP)	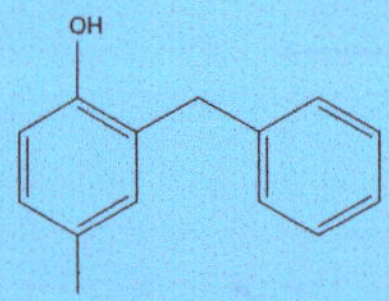	$C_{13}H_{11}ClO$	218.7	179.2	4.18	Neutral	4.18	Biocide
Nortriptyline. HCl (NH)	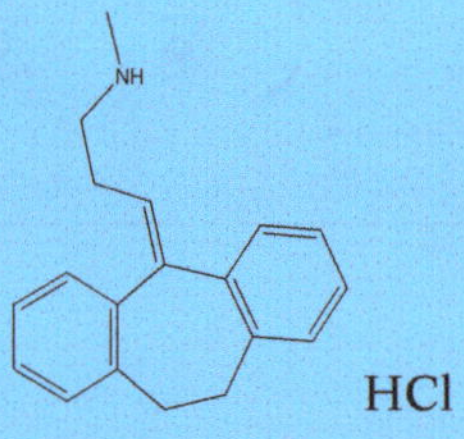	$C_{19}H_{21}NO{\cdot}HCl$	299.8	242.9	4.51	Positive	1.40–1.80	Antidepressant

(1) Calculated by ChemSketch

Source: From Rodriguez *et al.* (2016).

et al., 2015; Ternes & Joss, 2006; UMWELT.NRW, 2019; Viegas *et al.*, 2021b; Zietzschmann *et al.*, 2014; Zietzschmann, 2020):

- Ozonation and advanced oxidation processes (AOPs, e.g., O_3/H_2O_2, UV/ H_2O_2, Fenton).
- Adsorption-based systems, including PAC, granular activated carbon (GAC) filtration or biologically active GAC filters (BAC);
- Membrane filtration, that is, microfiltration (MF), ultrafiltration (UF), nanofiltration (NF), and reverse osmosis (RO);
- Adsorption/membrane hybrid processes, for example, PAC/MF, PAC/UF, PAC/NF.

The technology applicability depends on its separation mechanisms vs the target contaminants' properties (e.g., charge, hydrophobicity/hydrophilicity, solubility, volatility, chemical resistance, biodegradability, sedimentability/floatability, colloidal character, release of intracellular metabolites), as illustrated in Table 8.4 for macro and microcontaminants (Rosa *et al.*, 2009).

Each option has its pros and cons, as extensively compared in literature (e.g., in Luo *et al.*, 2014; Pesqueira *et al.*, 2020; Rizzo *et al.*, 2019, 2020). Regarding adsorption-based treatment, during the last decade, several projects addressed its introduction in an increasing number of WWTPs in

Table 8.4 Effectiveness of alternative and advanced processes for macro and microcontaminant control.

Contaminant	C/F + DAF	UV	GAC	BAC	MF	UF	PAC/UF	NF[a]	RO
Protozoa (cysts, oocysts)	+/-	+	+/-		+	+	+	+	+
Bacteria (vegetative forms)	-/+	+	-		+	+	+	+	+
Bacteria (endospores)	-/+	-	-		+	+	+	+	+
Helminth eggs	+/-	-	+/-		+	+	+	+	+
Cyanobacteria	+	- [b]	+/-		+	+	+	+	+
Enteroviruses	-	-	+/-		+/-	+	+	+	+
NOM_SUVA < 3 L/(mgC.m) [c]	-/+	+/-	-/+		+	-	-	+/-	+
NOM_SUVA > 4 L/(mgC.m) [c]	+/-		+/-		+/-	-/+	+/-	+	+
Assimable organic carbon (AOC)	-	+/-	-/+	+	-	-	-/+	+/-	+
Trihalomethanes (THMs)	-		+		-	-	+	+/-	+
Haloacetic acids (HAAs)	-		+		-	-	+	+/-	+
Bromate	-		+/-	+/-	-	-	-/+	+	+
Bromide	-				-	-	-	+/-	+
Chlorate	-				-	-	-	+	+
Chloride	-				-	-	-	+/-	+
Nitrate	-				-	-	-	+	+
Sodium	-				-	-	-	+/-	+
Sulfate	-				-	-	-	+	+
Microcystins	+ or -/+[d]		+ or +/- [e]		-	-	+ or +/- [e]	+	+
Taste and odour compounds (MIB, geosmin)	-/+		+		-	-	+	+	+
Volatile organic compounds (VOCs)	+[f]		+		-	-	+/-	-/+	+
EDCs and pharmaceuticals (hydrophobic and chemically resistant)	-/+		+		-	-	+	+	+
Pesticides (including chemically resistant)	-/+		+ or +/- [e]		-	-	+	+	+

- Not adequate.
-/+ Limited effectiveness.
+/- Partial control if adequate operation conditions are guaranteed.
+ Effective provided adequate operation conditions are guaranteed.
No information available.
[a] Considering 200 Da molecular cut-off.
[b] UV should not be used to control cyanobacteria, since it leads to cell rupture and cyanotoxin release.
[c] SUVA < 3 L/(mgC. m) indicates mainly hydrophilic NOM. SUVA > 4 L/(mgC.m) indicates mainly hydrophobic NOM and especially aromatic compounds.
[d] Effective removal of intracellular toxins; no significant removal of dissolved toxins.
[e] Depends on chemical characteristics of the target compound.
[f] There are volatilization conditions in C/F/DAF.
Source: From Rosa *et al.* (2009) and Smeets *et al.* (2015).

German-speaking countries, mainly because of Swiss and German governmental requirements or encouragement to accomplish CEC abatement from urban wastewater. As reviewed in Mestre *et al.* (2022a), results on the different technologies implemented have been published in reference journals, but the practical knowledge on the ACs' selection for pilot-, large-, and full-scale implementation can only be consulted in project reports and other documents written mostly in German. An exception is the Zietzschmann's (2020) book chapter, providing valuable information on the adsorptive removal of organic microcontaminants from wastewater, covering practical aspects of PAC and GAC processes and operation fundamentals, tools for their performance prediction and general considerations for an informed selection of an AC product.

This chapter focuses on PAC adsorption and on PAC/low-pressure membrane technology. These are two best available technologies, flexible and resilient to severe, climate change-driven variations in raw water composition, particularly suited as seasonal/sporadic (for episodes) barriers. Moreover, these technologies rely on physical separation, which allows minimizing the health and environmental risks associated with the potential formation of undesired by-products. Low-pressure-driven membranes are preferentially addressed to minimize the energy use and carbon footprint.

8.4 AC ADSORPTION

In water treatment, adsorption is the physical (or chemical) mechanism by which dissolved molecules (adsorbates or solutes) are captured on a solid interface (adsorbent). AC is the most used adsorbent in DWT and UWWT due to its nanoporous structure, responsible for an outstanding surface area, typically from hundreds to 2000 m^2/g (Mestre *et al.*, 2022a, b).

Most carbon-rich materials may be used as raw materials for AC production, but a large-scale production requires large amounts of precursors with reproducible features, controlled ash content, and cost. Commercially available products are mainly produced from coal, wood, or coconut shell; water treatment applications represent around one-third of the market share and are expected to remain the largest throughout 2020–2027, being the current driver for market growth (Figure 8.5, Mestre *et al.*, 2022a).

The precursors are first carbonized, and the chars are then physically or chemically activated (using high temperatures under controlled atmosphere, e.g., steam activation) to increase the porous network (of macro-, meso-, and micropores, Figure 8.6) yielding a high surface area. Given their relevance for PhC adsorption, the micropores' classification is further broken down into supermicropores (0.7–2 nm wide) and ultramicropores (<0.7 nm width). Manufacturers commonly report the iodine number and BET (Brunauer–Emmet–Teller) area values as indicators of the apparent surface area of a given AC, with both parameters usually presenting close values despite their distinct units (mg/g and m^2/g, respectively) (Mestre *et al.*, 2022a; Zietzschmann *et al.*, 2014).

ACs are industrially applied as PACs, average particle size $<100\,\mu m$ or as GAC filters (average particle size $>100\,\mu m$ and median $d_{50} > 1$ mm) (Figure 8.5), the lower the particle size, the lower the

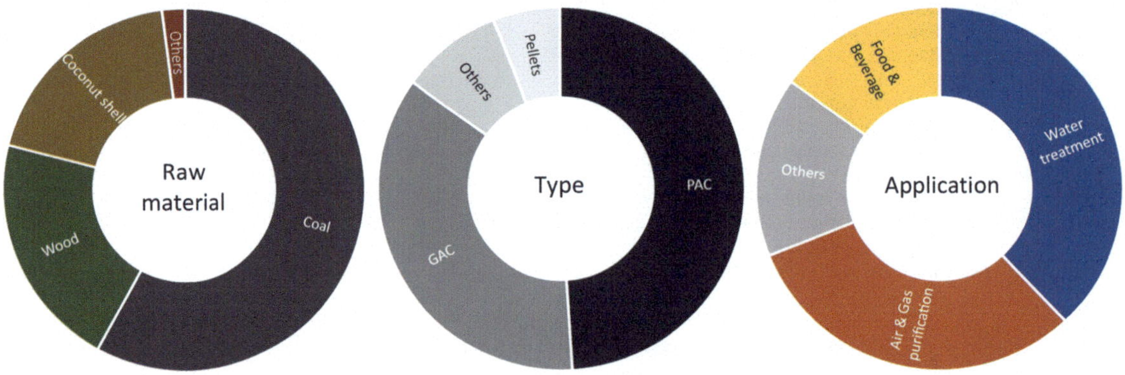

Figure 8.5 Global AC market share by raw material and type. *Source*: Data adapted from Inkwood Research (2019) and application (data adapted from Fortune Business Insights, 2019). From Mestre *et al.* (2022a).

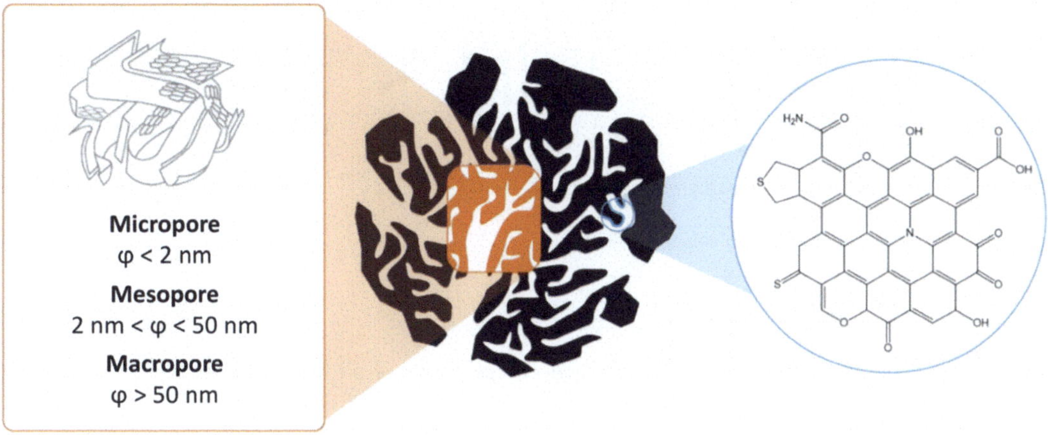

Figure 8.6 Representation of nanoporous structure and surface chemistry groups on AC. *Source*: From Mestre *et al.* (2022a).

adsorption path and therefore the faster the adsorption kinetics to the active sites for a similar porous structure (Figure 8.7). Super fine PACs (sPAC) with enhanced kinetics are emerging for applications not requiring particle settleability, that is, for PAC/membrane applications (Section 8.6). After exhaustion, PACs are usually discharged in landfill or incinerated, while GACs can be reactivated (or regenerated) and reused.

AC adsorption capacity (adsorbate/adsorbent mass ratio, often determined through Freundlich or Langmuir isotherms) and kinetics (HSDM (homogeneous surface diffusion model) vs Boyd's model, Viegas *et al.*, 2014) depend on (external and intraparticle) diffusion and mass transfer and affinity (Campinas *et al.*, 2013), that is, adsorbent–adsorbate hydrophobic and electrostatic interactions. These are determined by the AC textural properties (the closer the active-site pore size to the solute's size the better; Zietzschmann, 2020) and surface chemistry (e.g., hydrophobicity and net surface charge, expressed by pH_{pzc} – examples in Figure 8.8), as well as by the solutes' properties (e.g., size, hydrophobicity, and net charge; examples in Tables 8.2 and 8.3) and the water background organics and inorganics.

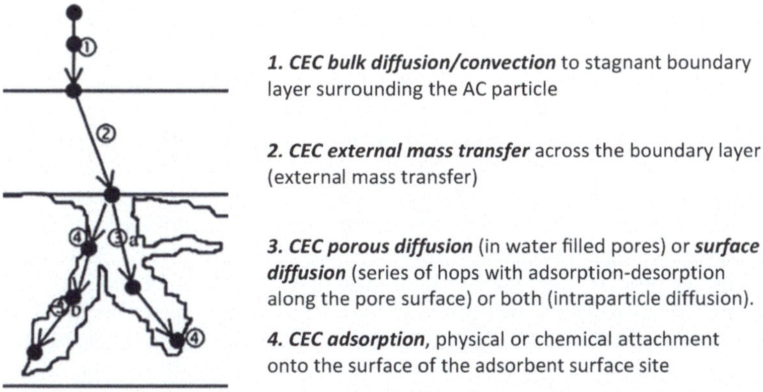

Figure 8.7 AC adsorption representation. *Source*: Adapted from Viegas *et al.* (2014).

BET surface area (m^2/g)	1112
Ash content (%)	8.17
Moisture (%)	2
Apparent density (g/cm^3)	0.16
Geometric mean diameter (μm)	6
Primary micropore (<7 Å) volume (cm^3/g)	0.343
Secondary micropore (7-20 Å) volume (cm^3/g)	0.194
Mesopore (20-500 Å) volume (cm^3/g)	0.357
Micropore surface area (m^2/g)	733
Mesopore surface area (m^2/g)	379

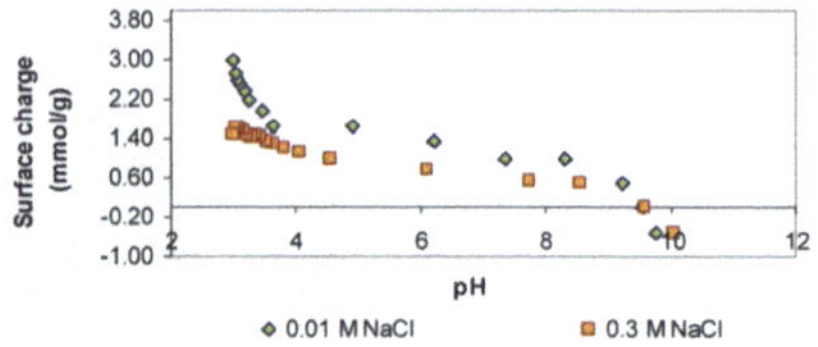

Figure 8.8 Key characteristics (textural, left; surface charge, right) of an AC illustrated for a high performing (mesoporous and microporous, alkaline, i.e., $pH_{pzc} \sim 9.5$) commercial PAC for CEC control (PAC Norit SA-UF). *Source*: Adapted from Campinas and Rosa (2006).

Water background organics determine the competitive adsorption by (a) direct competition for the adsorption sites (the closer the NOM/EfOM (effluent organic matter) size and hydrophobicity to the target solute the stronger the competition; Campinas *et al.*, 2013) or (b) pore blockage and are often characterized by DOC content and character (e.g., Rodriguez *et al.*, 2016). This is the reason why, in practical applications, one uses the DOC-normalized PAC dose (mg PAC/mg DOC). Background inorganic matrix may be easily characterized by two bulk parameters (i) the water pH, determining the AC and solute's ionization/net charge, and (ii) the electrical conductivity (expressing the ionic strength), determining the electrostatic repulsions' shielding effect and so the solute's hydrodynamic size (Campinas & Rosa, 2006).

The development of high-performing, environmental-friendly (biomass-derived, steam activated) PACs from local biomasses is very important for the sake of the technology's sustainability (Mestre *et al.*, 2022b; Viegas *et al.*, 2020b), as well as of fine sPAC and magnetic PACs allowing their recovery and regeneration (as ongoing within EMPOWER + project).

8.5 MEMBRANE FILTRATION

In membrane filtration for water treatment, an applied pressure forces water to pass through a semi-permeable membrane, which retains the solutes larger than the membrane pores, the narrower the pores, the higher the pressure, as illustrated in Figure 8.9 for MF, UF, NF, and RO (and also for sand filters, for comparison).

Membrane filtration covers a broad spectrum of options in terms of membrane pore/selectivity/applied pressure, but also in terms of (a) membrane materials and modules, (b) filtration mode, dead-end, or crossflow, in/out or out/in, pressurized or immersed, and (c) number of stages (Figure 8.10) (Pinho & Rosa, 1995). Common applications include (i) 1-stage cross-flow polymeric spiral-wound RO membranes for seawater desalination, (ii) polymeric spiral-wound NF membranes for DWT, that is, groundwater softening and NOM and CEC removal from surface water, (iii) pressurized in/out polymeric hollow-fibre UF membranes for DWT and water reclamation, (iv) immersed out/in polymeric hollow-fiber or flat sheet UF membranes for membrane bioreactors (MBRs) for UWWT/water reclamation, and (v) the emerging ceramic in/out dead-end tubular MF membranes for surface water treatment. All membranes (from MF to RO) are full physical barriers against particles (TSS and turbidity) and bacteria; UF membranes are also effective against viruses and macromolecules; NF membranes further remove small organics (usually > 300 Da), divalent salts, and partially monovalent salts, which are highly removed by RO. This is, all membranes are adequate for water (physical) disinfection, but the low-pressure MF and UF are not able to remove the organic micropollutants targeted.

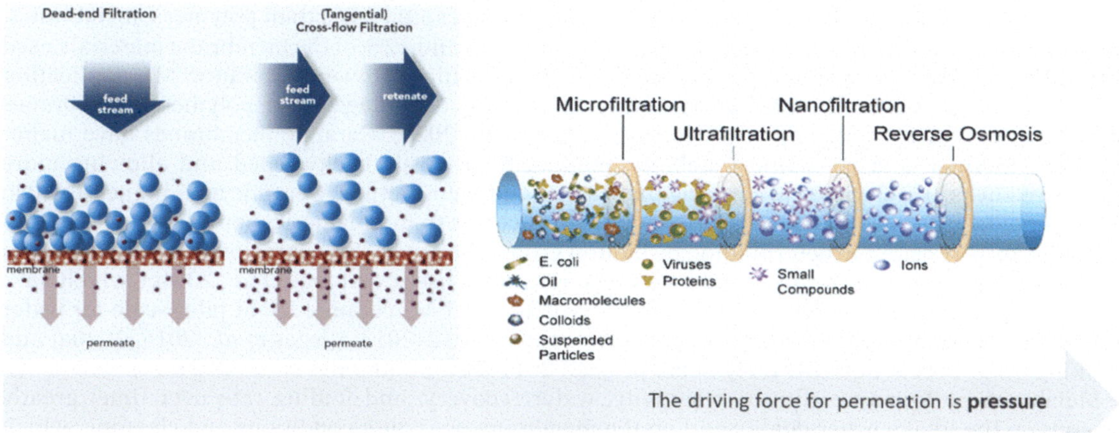

SAND FILTERS	MICROFILTRATION	ULTRAFILTRATION	NANOFILTRATION	REVERSE OSMOSIS
10 µm	0.1-2 µm	0.01-0.1 µm	1-10 nm	
	> 500.000 Da	1.000 – 500.000 Da	100-1.000 Da	< 100 Da
	0.1 - 1 bar	0.5 - 5 bar	5-40 bar	20-100 bar

Figure 8.9 Membrane filtration in water treatment – MF, UF, NF, and RO.

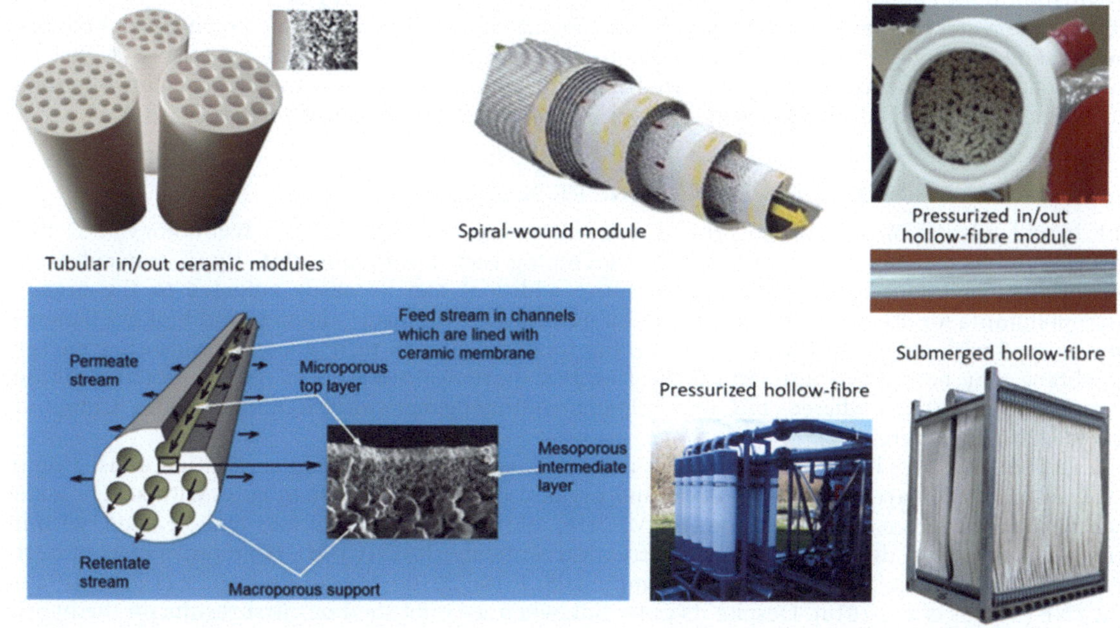

Figure 8.10 Membranes and modules.

As ceramic membranes have so far involved higher initial capital costs than polymeric membranes, they represented only 2–3% of the membrane market in 2015 and 12% of the membrane materials used until 2016 for DWT by membrane processes with backwashing. However, ceramic MF application is expected to grow over the next years due to their many advantages over polymeric membranes and also due to their significant cost decrease (Wise *et al.*, 2020). Ceramic membranes have higher chemical, thermal, and mechanical stability, making them easily backwashed and allowing more aggressive physical and chemical cleaning that can extend their lifetime. Ceramic membranes are also better candidates for hybrid adsorption/membrane processes than polymeric membranes as they can be significantly resistant to deterioration by biofilm growth and to surface abrasion by coarse particles circulation, two concerns mostly related with PAC long-term use (Campinas *et al.*, 2021c), which have driven our pioneering research on PAC/ceramic MF and/or PAC/ceramic UF at pilot scale for water reclamation (within EU-FP7 TRUST project developed in 2012–2013; Viegas *et al.*, 2015, 2020a) and drinking water production (LIFE Hymemb project, 2014–2016; Campinas *et al.*, 2021b, c).

Membrane performance (fluxes, selectivity, water recovery, and fouling rate over time) greatly depends on the intake water quality and on the membrane operation conditions and cleaning, which control the membrane (reversible and irreversible, bio- and chemical) fouling, scaling, and aging, ultimately determining the membrane productivity, lifespan and the energy costs (Figure 8.11). A new indicator was developed for performance assessment and optimization, the treatment capacity (TCp), that is, the design flowrate normalized to membrane area and intake pressure. TCp ($m^3/(m^2.d.bar)$) is the effective volume of permeate produced by unit time, membrane area, and intake feed pressure, deducting the permeate volume used for backwashing and chemically enhanced backwashing (CEB), and considering the productive time (filtration time) and the time a unit is off-line for cleaning procedures (backwashing and CEB). TCp integrates all key aspects of process productivity and therefore constitutes a useful indicator to balance flux, energy consumption, backwash frequency, and chemical cleaning frequency (Campinas *et al.*, 2021c).

The intake water background organics play a key role on membrane fouling, usually requiring coagulation pre-treatment for NOM control (Campinas *et al.*, 2021c; Ribau Teixeira & Rosa, 2006), whereas the water inorganics (pH, ions) affect the membrane scaling (inorganic precipitation) but also the flux and selectivity of NF membranes due to charge (pH) and shielding (ions) effects (higher pH corresponding to higher electrostatic repulsions, narrower pores, lower flux, and higher solute retention; deleterious effects partially balanced by the ions' shielding effect) (Ribau Teixeira *et al.*, 2005).

8.6 HYBRID AC ADSORPTION/MEMBRANE PROCESSES

The hybrid adsorption/membrane process aims at taking the best of each technology (i.e., PAC adsorption and membrane filtration) while overcoming their limitations. PAC (type, concentration, and dosing conditions, e.g., continuous or pulse addition) can be easily adjusted to the organic micropollutants (or other CECs) targeted and a finer PAC, with faster kinetics, and higher efficiency may be used with no risk of PAC release into the treated water. In turn, a low-pressure (low energy) membrane may be used to fully retain the fine PAC particles (unlike PAC conventional addition; Campinas *et al.*, 2021a), whereas PAC may help controlling the membrane fouling by NOM (Campinas & Rosa, 2010a) (Figure 8.12).

This combination ('hybridization') yields a flexible, compact, and modular solution consisting of a safe and reliable barrier for disinfection and organic CEC control, able to cope with severe water quality changes driven by climate change. MF, UF, and loose-NF membranes may be used, with or without coagulation, depending on the upstream treatment – PAC/(C)/MF (Campinas *et al.*, 2021b, c; Viegas *et al.*, 2020a, 2021a), PAC/UF (Campinas & Rosa, 2010a, b, c, d; Rodriguez *et al.*, 2016), or PAC/NF (Viegas *et al.*, 2018). Despite its potential, when we published our first results on the hybrid adsorption/membrane process for the removal of cyanotoxins from drinking water, back in 2010, only one published study was known on PAC/UF for such application (Campinas & Rosa, 2010b).

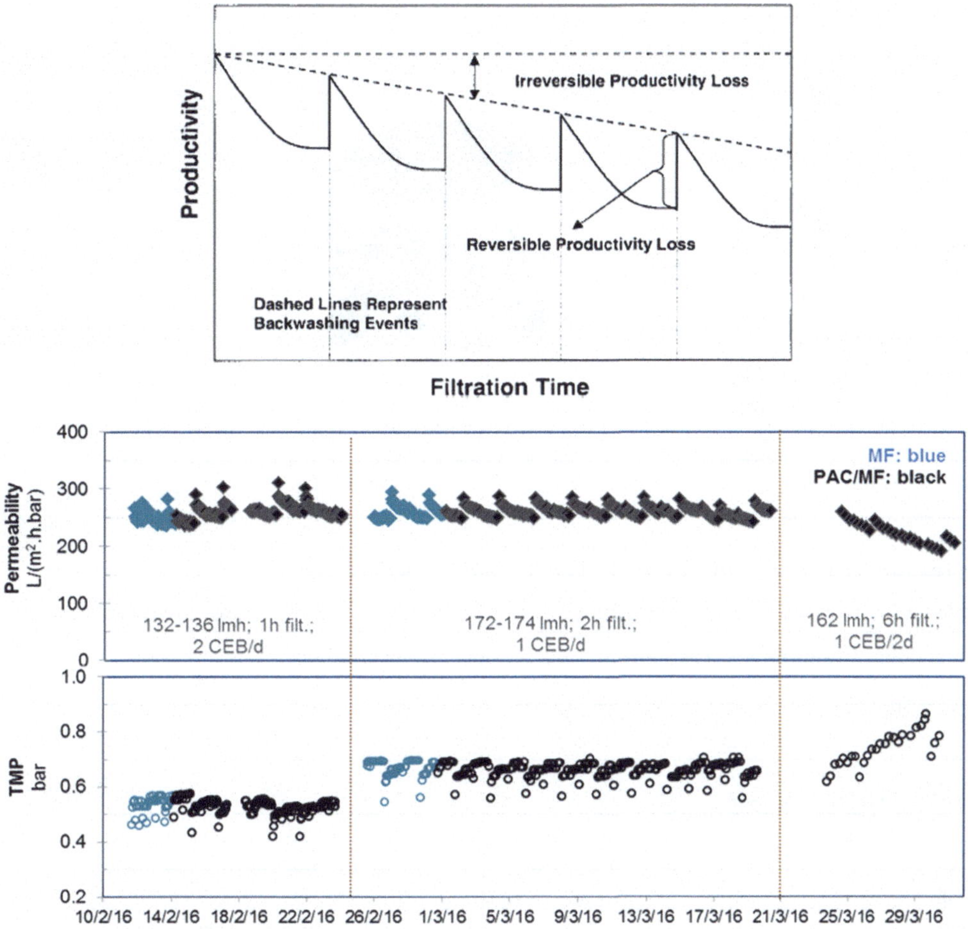

Figure 8.11 Membrane productivity over time/filtration cycles – conceptual scheme (left) and PAC/MF pilot results (right). *Source*: From Campinas *et al.* (2021c).

In addition, and as introduced in Section 8.5, ceramic membranes are particularly suited for hybrid adsorption/membrane processes due to their lower abrasion by PAC and subsequent longer lifetime (2–5 times the lifetime of polymeric membranes), their low energy consumption (<0.1 kWh/m^3) and higher water recovery rate (>95%) both in DWT (Campinas *et al.*, 2021c) and in water reclamation (Viegas *et al.*, 2020a). Again, in 2012, when our studies with pressurized hybrid low-pressure ceramic membrane processes started (we started investigating ceramic MF and PAC/MF for water reclamation in 2012 (FP7 TRUST, 2011–2014), the same year we submitted for funding LIFE Hymemb for benchmarking PAC/ceramic MF against PAC conventional addition in drinking water production (executed in 2014–2016, awarded as a best LIFE ENV project 2016–2017).), most research on PAC/MF was mainly dedicated to conventional polymeric membranes and submerged configuration. Moreover, many studies focused on membrane fouling and traditional water quality parameters conducted at the lab scale, sometimes with synthetic waters. Pilot studies of pressurized PAC/coagulation/ceramic MF envisaging organic microcontaminants, NOM, and microorganisms' removal under real scenarios, with real waters and quality variations were, and still are, scarce.

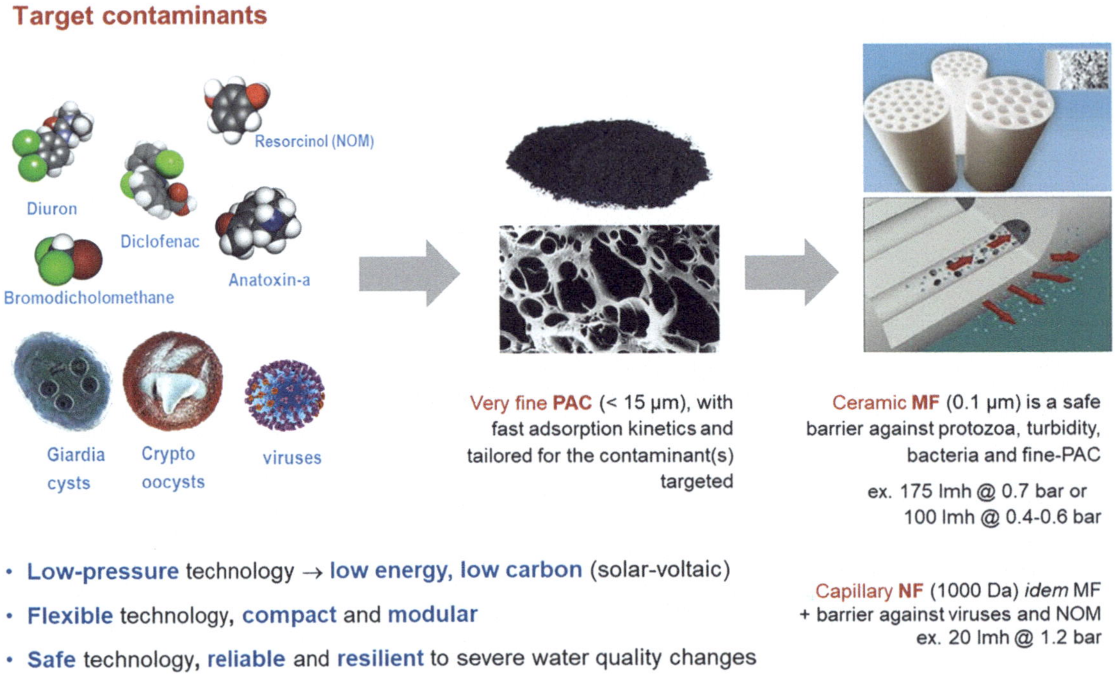

Figure 8.12 Hybrid adsorption/membrane process.

8.7 APPLICATIONS IN DWT

Complementing the former sections, which included a short critical state of the art on the problem and the technologies targeted, this and Sections 8.8 relative to the applications are focused on our last decade's results, presenting our main contributions to this research area. The applications studied are illustrated in Figure 8.13 and aimed at answering the following questions: 'to what extent & why?', 'where, how & at what cost?' PAC conventional application (PAC/C/F/S) and the innovative PAC/MF (or PAC/UF) can control organic micropollutants and CECs in surface water treatment for drinking water production. Our strategy was to start by optimizing the current barriers (option 1), then to find the best (tailored) solution to upgrade the treatment, if needed, by investigating the several alternative points where to apply the advanced treatment solution proposed (PAC/MF or UF, options 2–5). Such strategy involved studies at the lab-, but also at pilot- and full-scales (Figure 8.14), which allowed developing cost analysis, and were carried out within PhD (Campinas & Rosa, 2006, 2010a, b), postdoc (Campinas *et al.*, 2013), and demonstration projects (LIFE Hymemb; Campinas *et al.*, 2017, 2021a, b, c; Viegas *et al.*, 2021a).

Regarding the 'to what extent & why', our results allowed understanding and modelling the adsorption process. In brief:

- All options (1–5) achieve good removal of the pesticides, PhCs and hormones, and MC-LR_{eq} targeted (Campinas & Rosa, 2006, 2010b; Campinas *et al.*, 2017, 2021a, b).
- Charge, hydrophobicity (Log K_{ow}, Log D), molar mass (MM), and aromaticity are the adsorbate's key factors for adsorption (Figure 8.15, Campinas *et al.*, 2021a), with a turning point for Log K_{ow} 2.2, above which the compounds are very amenable to adsorption (Figure 8.16, Campinas *et al.*, 2021a), as found earlier in a UWWT study (Rodriguez *et al.*, 2016). Positively charged functional

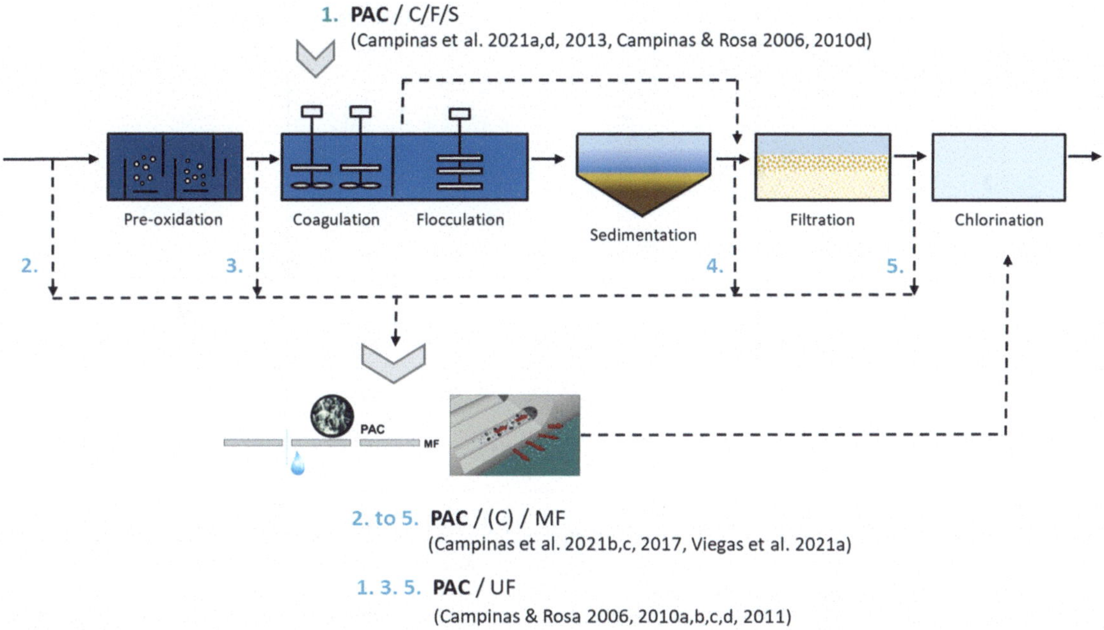

Figure 8.13 PAC application for controlling organic micropollutants in DWT (the conventional sequence represents Alcantarilha WTP).

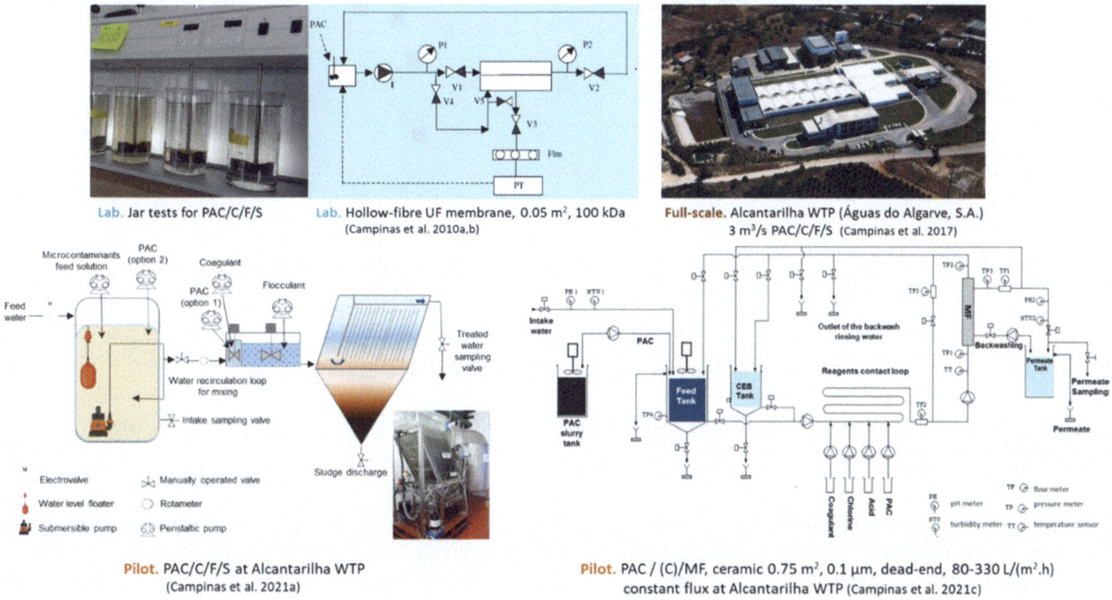

Figure 8.14 Lab, pilot, and full-scale tests of PAC application in DWT.

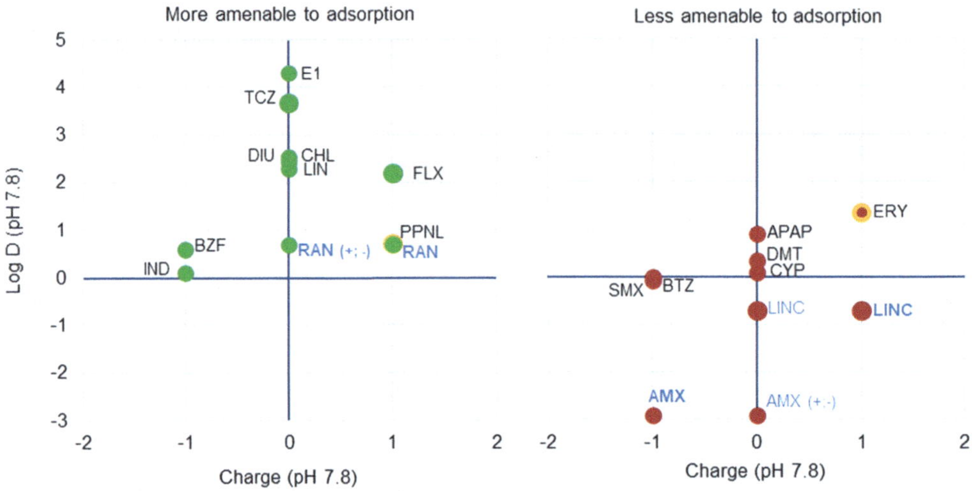

Figure 8.15 Log *D* vs charge for selected PhCs and pesticides. *Source*: From Campinas *et al.* (2021a) (very amenable to adsorption (green), relatively amenable (yellow), and less amenable (red). Compounds in blue indicate they have a distribution of species between two charges at the working pH, the dominant charge emphasized in bold letter. For neutral compounds, (+,-) highlights they have positive and negative functional groups, despite the overall neutral charge.

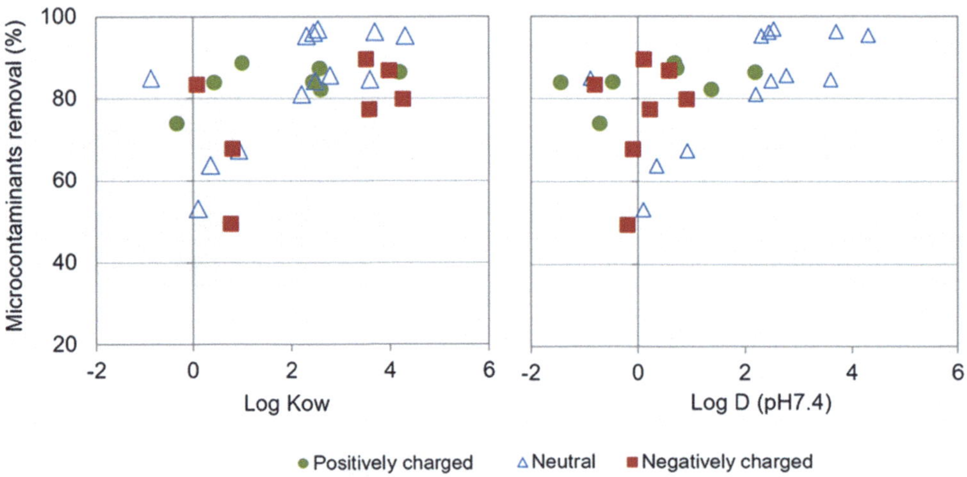

Figure 8.16 Microcontaminants' removal vs hydrophobicity measured by Log K_{ow} (left) or Log $D_{7.4}$ (right). *Source*: From Campinas *et al.* (2021a).

groups, low surface polar area (PSA) and/or high aromaticity acted as adsorption enhancers of low-hydrophobicity compounds (Log $K_{ow} < 2.2$) (Figure 8.17). Coherently with other studies, a better removal of the positively charged compounds was observed, the difference depending on the PAC net charge, that is, better with the negatively charged PAC (Campinas *et al.*, 2021a). A multiple linear regression allowed deriving correlations integrating the key variables of PhC

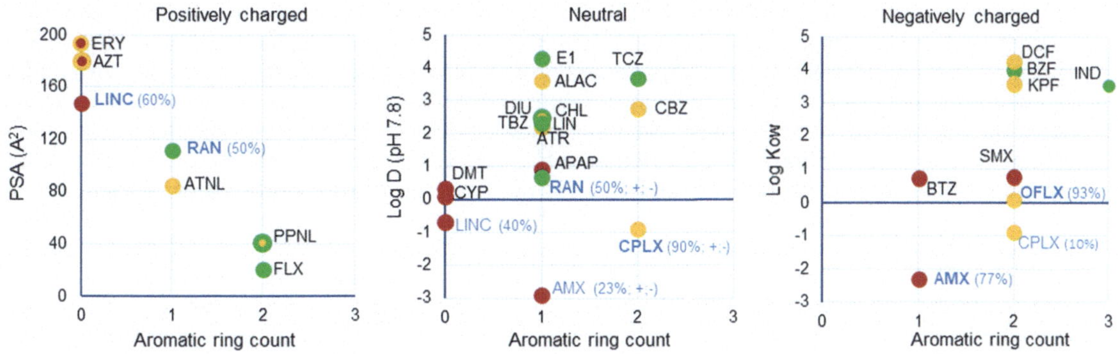

Figure 8.17 PSA, aromatic ring count, Log $D_{7.8}$ and Log K_{ow} role on the adsorption of positively charged (left), neutral (center) and negatively charged (right) PhCs and pesticides (compounds very amenable to removal in green, relatively amenable in yellow, and less amenable in red). *Source*: From Campinas *et al.* (2021a).

and pesticides' removal (Table 8.5, Campinas *et al.*, 2021a). Log *D* vs charge plot (Figure 8.15, Campinas *et al.*, 2021a) may be used to select the indicator CECs (surrogates) – an exhaustive monitoring is technically and economically unfeasible.

- Water background organics increase the PAC dose for balancing background NOM competition, which is stronger for the compounds less-amenable to adsorption and for similar-size NOM,

Table 8.5 Correlations between PAC/C/F/S removal efficiencies (Er) and the properties of 19 PhCs and nine pesticides.

19 PhCs + 9 pesticides studied	**Integrating variables using Multiple Linear Regression** (significance F should not exceed 0.05)
Set 2 (best set with MV118) 20-40 mg/L MV118, 12 min contact time	
All microcontaminants, disregarding charge	Er (%) = 66 + 6.6 **Log Kow** – 0.039 **MM** + 2.6 **Ar-rings** R^2 = 0.36 Significance F = 0.02
All microcontaminants, considering charge	Er (%) = 72 + 5.0 **Log Kow** + 8.1 **charge** – 0.0079 **MM** + 3.9 **Ar-rings** R^2 = 0.49 Significance F = 0.01
Set 4 (best set with MV121) 3-5 mg/L MV121, 72 min contact time	
All microcontaminants, disregarding charge	Er (%) = 71 + 3.7 **Log Kow** + 2.4 **Ar-rings** R^2 = 0.30 Significance F = 0.02
All microcontaminants, considering charge	Er (%) = 66 + 3.8 **Log Kow** + 7.6 **charge** + 6.0 **Ar-rings** R^2 = 0.49 Significance F = 0.001
Positively charged microcontaminants	Er (%) = 52 + 0.17 **PSA** + 15 **Ar-rings** R^2 = 0.77 Significance F = 0.05
Neutral microcontaminants *	Er (%) = 60 + 8.3 **Log Kow** + 3.7 **Ar-rings** R^2 = 0.73 Significance F = 0.003
Negatively charged microcontaminants	Er (%) = 13 + 1.9 **Log Kow** + 0.13 **MM** + 10 **Ar-rings** R^2 = 0.90 Significance F = 0.05

* except ciprofloxacin (charge 0 or -1)

Source: Adapted from Campinas *et al.* (2021a).

for example, tannic acid and MC-LR (Campinas *et al.*, 2013). Background inorganics affect the extent and mechanisms of NOM-target adsorption (via shielding mechanisms) and may have a positive effect on PAC adsorption depending on the PAC overall charge (Campinas & Rosa, 2006; Campinas *et al.*, 2013) – for example, for a solute and PAC with opposite charges, for low surface concentrations (low X/M), ions shield PAC-solute electrostatic attraction resulting in an ionic strength-driven adsorption reduction, whereas for high X/M ratios, ions shield solute–solute intra- and intermolecular repulsions resulting in an ionic strength-driven adsorption enhancement.

- PAC (10 mg/L in Campinas and Rosa, 2010b; 6–24 mg/L in Campinas *et al.*, 2021c) does not impair the membrane flux, rather it helps controlling the irreversible membrane fouling, minimizing the chemical cleaning frequency. It enhances NOM retention by UF/MF, including algogenic organic matter (AOM) (Campinas & Rosa, 2010a), though not significantly for the highly hydrophilic compounds whose membrane foulant behavior may increase driven by water divalent salts (Campinas & Rosa, 2010a).

Regarding the 'where, how & at what cost', the main results are:

- **All options** (1–5) allow an effective control of the target micropollutants and CECs as long as an adequate PAC type and dose is applied. The differences are more related with (a) the risk of PAC particles escaping the treatment (in PAC/C/F/S the dose appears to be limited to 10 mg/L) and (b) the energy consumption and the cost, for the upgrading solution or overall (Campinas *et al.*, 2021c).
- **Option 1. PAC/C/F/S** (Campinas *et al.*, 2021a) – removals of 65–79% for total-pharmaceuticals (19) and 73–83% for total-pesticides (9) were achieved with 3–9 mg/L of a mesoporous negatively charged PAC or with 20–24 mg/L of a microporous positively charged PAC. For hydrophilic low NOM waters, PAC textural properties and PAC surface chemistry were both important for adsorption. Half to one third of PAC dose would be enough as long as a PAC with adequate pore-size distribution is used, this means using a PAC with high secondary microporous volume but also with high volume of mesopores to avoid microcontaminants' size exclusion and minimize PAC pore blocking phenomenon by NOM. Similar PAC savings are possible providing an extra 1-h PAC contact time to the very short time observed in the regular operation of the WTP where the PAC/CFS pilot was installed (around 15 min), with benefits particularly for the compounds of higher molar mass and/or less amenable to adsorption. A high vulnerability of clarification effectiveness to PAC dosing was observed with the low-turbidity waters tested, with a higher risk of negative impact for PAC doses above 10 mg/L towards residual turbidity due to PAC fines (with particulate microcontaminants), residual aluminum and aerobic endospores, the latter used as surrogates of biological forms resistant to chemical disinfection, as *Cryptosporidium* oocysts and *Giardia* cysts. This result may compromise the microcontaminants' removal reliability unless a downstream filtration safely retains the PAC fines.
- **Options 2–5. PAC/MF** (Campinas *et al.*, 2021b) – PAC/(Alum)/MF (conditions below) achieved 75% to complete removal of total microcontaminants (pesticides, PhCs, or microcystins) with 4–18 mg/L of a mesoporous PAC and 2 h contact time (Figure 8.18), with a reliable particle separation (turbidity < 0.03 NTU) and low aluminum residuals. Microcontaminants showed different amenabilities to PAC adsorption, depending on their charge, hydrophobicity (Log K_{ow}), polar surface area and aromatic rings count. Compounds less amenable to adsorption showed higher vulnerability to NOM competition (higher A254 waters), greatly benefiting from DOC-normalized PAC dose increase (Figure 8.19, Campinas *et al.*, 2021b). PAC/Alum/MF also attained 29–47% NOM median removal as DOC, decreasing THMFP by 26%. PAC complemented NOM removal by coagulation (+15% to +19%, percentage points), though with no substantial improvement toward THMFP and membrane fouling. Furthermore, PAC/Alum/MF was a full barrier against aerobic endospores, and PAC dosing was crucial for ≥ 1.1-log reduction in bacteriophages.

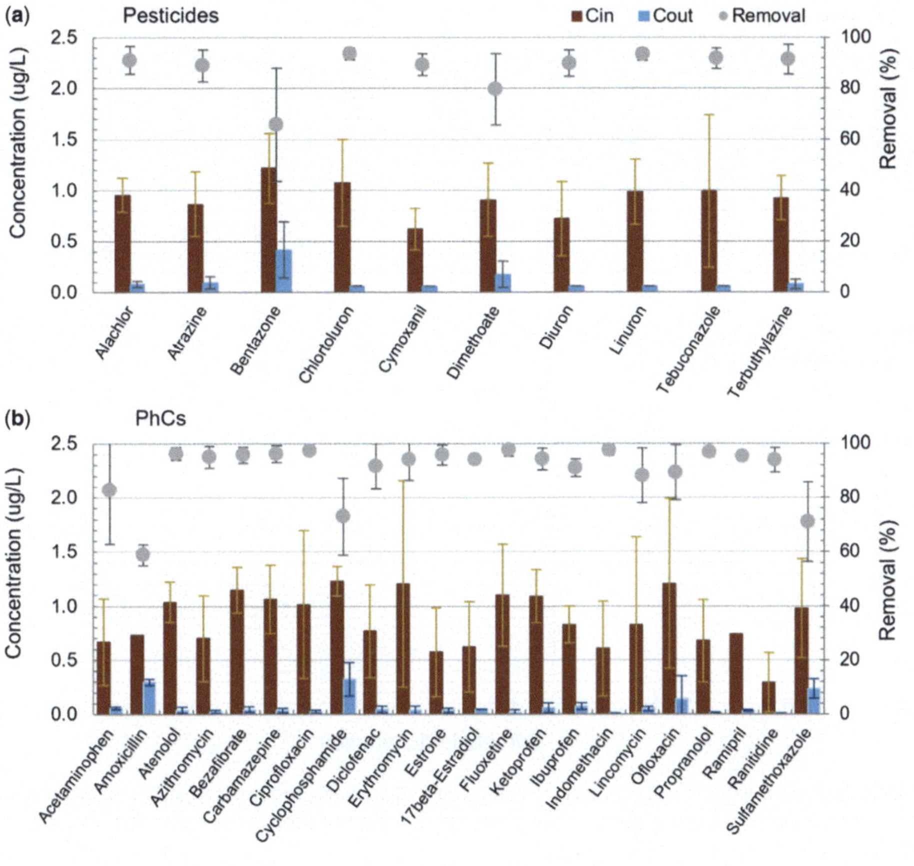

Figure 8.18 Average intake (brown bars) and permeate (blue bars) concentrations and removals (circles) of pesticides (a) and pharmaceuticals (b) after PAC/(Alum)/MF (average values of all spiking trials; error bars represent standard deviations between trials). *Source*: From Campinas *et al.* (2021b).

- **Options 2–5. PAC/MF** with 15 μm particle PAC, pH_{pzc} 10–11, 1126 m^2/g BET area and 0.83 cm^3/g total pore volume with 53% of mesopores (2–50 nm diameter); pressurized module, with three tubular MF (0.1 μm) ceramic (ZrO_2/TiO_2) membranes (0.75 m^2), operated in a continuous dead-end mode with constant permeate flowrate in the 80–330 L/(m^2.h) range (Campinas *et al.*, 2021c) – PAC addition (6–24 mg/L) did not promote membrane fouling and, for all conditions tested, TCp was kept constant or slightly increased with PAC dosing. Membrane fouling observed during the 1.5-year demonstration period was essentially reversible, since no significant change in clean membrane permeability was observed. As expected, a higher treatment capacity was obtained with filtered water (9.6 m^3/(m^2.d.bar), option 5), followed by ozonated/pre-coagulated water (5.6 m^3/(m^2.d.bar), option 4), with similar results for raw water and ozonated water with recirculated filter-backwash waters (4.7–4.9 m^3/(m^2.d.bar), options 2 and 3). The non-clarified waters tested (options 2 and 3) required pretreatment with in-line alum coagulation to minimize membrane fouling (PAC/C/MF). Water recovery was 97–99%. PAC/MF energy consumption was 0.045–0.053 kWh/m^3 (options 5–3, 0.02 kWh/m^3 for PAC dosing), inversely varying with TCp. The cost analysis showed MF total production costs

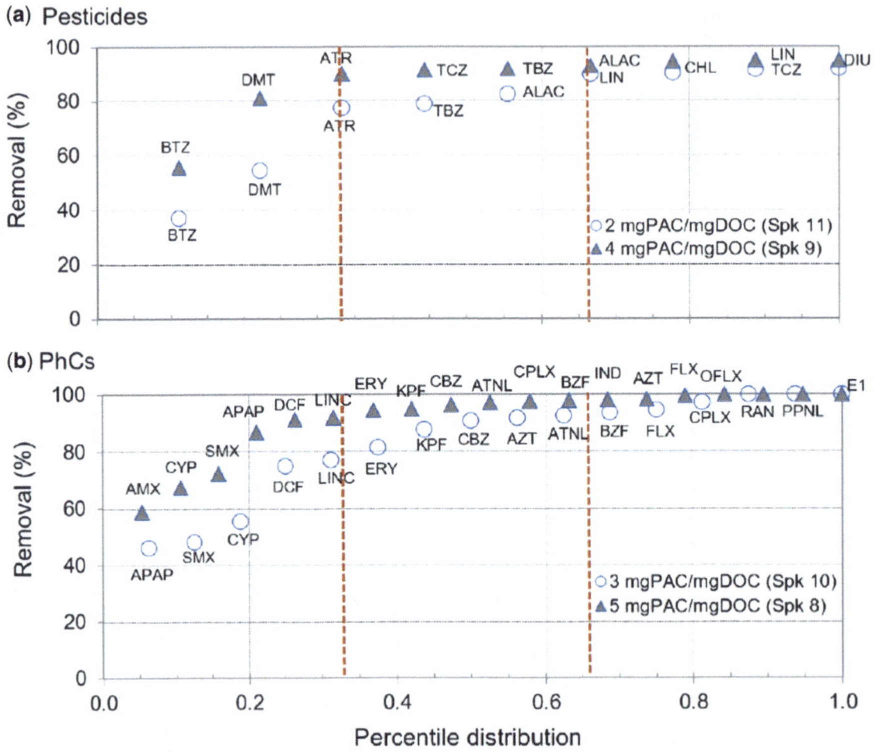

Figure 8.19 DOC-normalized PAC dose effect on the percentile distribution of removal efficiencies of nine pesticides (a) and 19 PhCs (b) during two spiking trials (those with the highest number of microcontaminants and the lowest PAC doses). *Source*: From Campinas *et al.* (2021b).

(CAPEX and OPEX) of 0.07 and 0.04€/m^3 for treating 1,00,000 m^3/d raw and filtered waters, respectively; PAC dosing increases costs 0.03€/m^3 for both waters. Figure 8.20 shows PAC/MF total costs' breakdown.

- **Options 2–5. PAC/MF** idem (Viegas *et al.*, 2021a) showed the way PAC is dosed matters – two pilot trials were conducted to compare continuous 10–12 mg/L PAC inline dosing with 8–10 mg/L dosing to a 2 h-contact tank. Two low turbidity/low natural organic matter (NOM <2 mgC/L) surface waters spiked with 7.2–10.3 μg/L total pesticides were tested. Removal differences between the two PAC dosing options depended on pesticides' amenability to adsorption and NOM characteristics (A254) (Figure 8.21). Waters with low A254-absorbing NOM and only pesticides amenable to adsorption showed very high removals (all pesticides ≥93%) and no significant differences between the two PAC dosing options. Waters containing higher A254-absorbing NOM and high loads of pesticides less amenable to adsorption (dimethoate, bentazone) required higher inline PAC dose. Those or more severe conditions may require PAC doses higher than tested to comply with the Drinking Water Directive limits for pesticides. Cost analysis showed PAC inline dosing is more cost-effective than PAC dosing to the contact tank when identical PAC dose is sufficient or when the doses are low, even if 50% higher for inline dosing, and the plant is small.

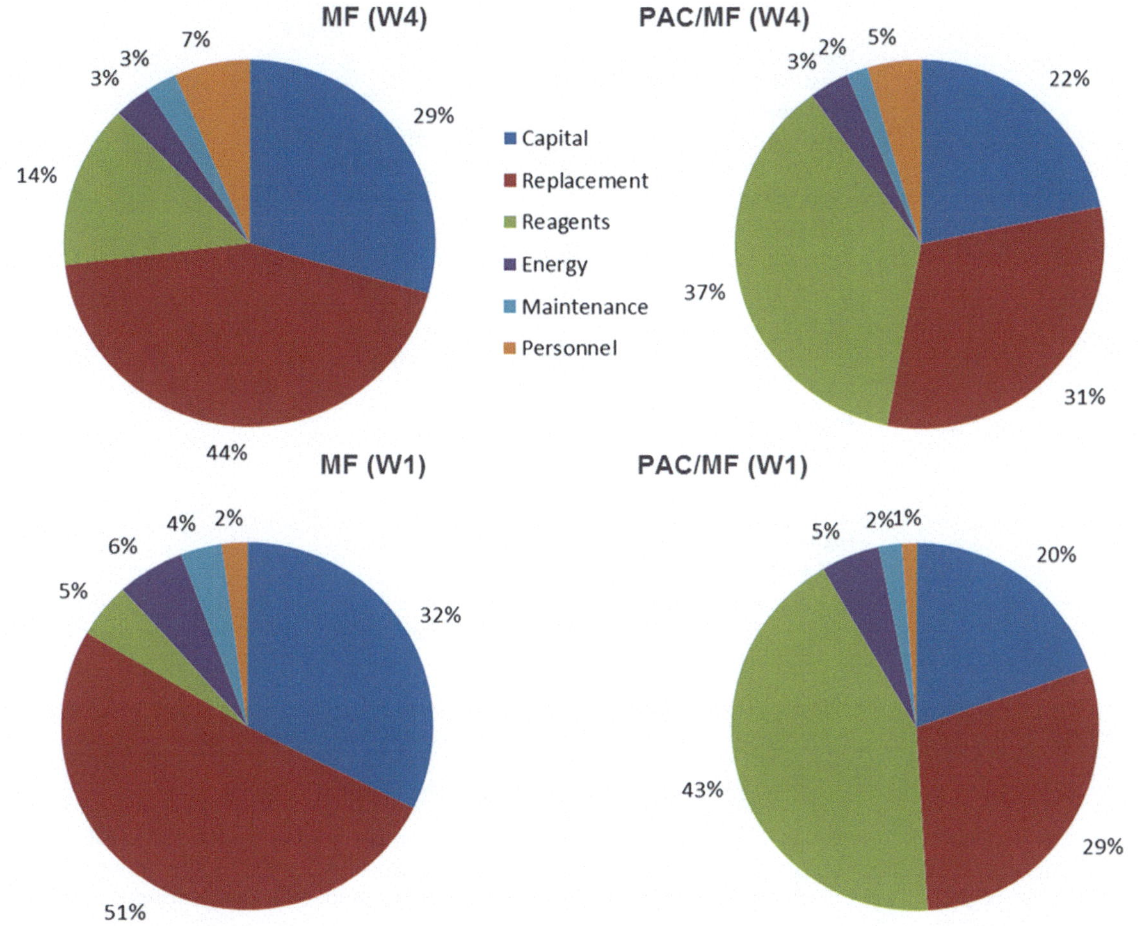

Figure 8.20 Cost structure of ceramic MF (left) and PAC/MF (right) for option 2 (W4) and option 5 (W1). *Source*: From Campinas *et al.* (2021c).

- **Option 5. PAC/UF** with mesoporous and microporous positively charged (alkaline) fine PAC and hydrophilic hollow-fiber UF membrane, 100 kDa cut-off, in/out cross-flow filtration (Campinas & Rosa, 2010a, b) – in the absence of background NOM, PAC/UF with 10 mg/L PAC and up to 20 μg/L MC-LR_{eq} feed concentration achieved 93–98% MC-LR_{eq} removal and a cycle-averaged permeate concentration below the WHO drinking water guideline-value for microcystin-LR variant. NOM type and concentration and MC initial concentration determined the PAC dose to be used. While 10 mg/L PAC effectively controlled ca. 5 μg/L MC-LR_{eq} in a model water with 2.5 mg/L NOM or with *M. aeruginosa* culture (cells and algogenic organic matter), 15 mg/L PAC were unable to achieve the WHO quality with a water containing higher concentrations of NOM (5 mg/L) and microcystins (ca. 20 μg/L MC-LR_{eq}) (Campinas & Rosa, 2010c, 2011). UF is a safe barrier against cyanobacteria, ensuring an absolute removal of *M. aeruginosa* single cells, the smallest cyanobacterial cells and hence the most difficult to remove. An increased cell lysis was observed with cell aging, although it did not necessarily degrade permeate quality, as in parallel to cell damage an enhancement of microcystin rejection by the

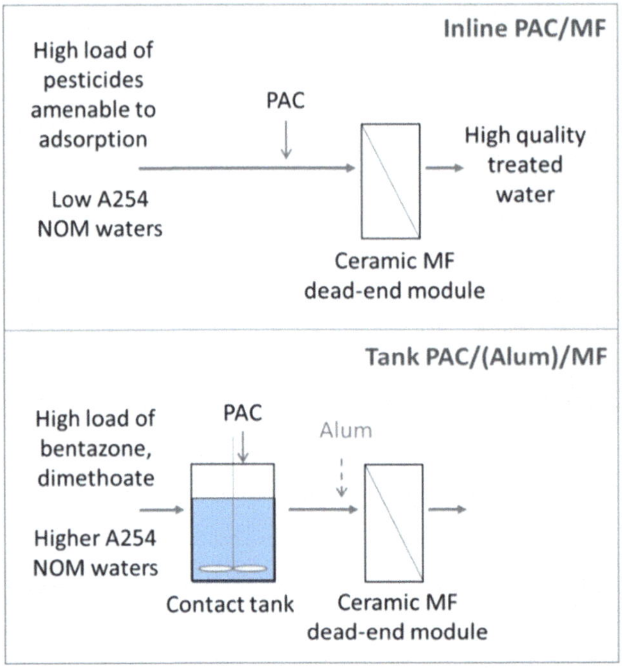

Figure 8.21 Inline vs tank PAC dosing in PAC/MF for pesticides and NOM removal in DWT. *Source*: From Viegas *et al.* (2021a).

UF hydrophilic membrane was observed with cell aging, most probably due to AOM-driven microcystin adsorption on the membrane, connected to the greater content of the older cultures in segregated AOM (mucopolysaccharides) and/or protein lysed AOM (Campinas & Rosa, 2010c, 2011).

8.8 APPLICATIONS IN UWWT AND WATER RECLAMATION

The applications we have been studying over the last decade are illustrated in Figure 8.22 and aimed at answering the following questions: 'to what extent and why?', 'where, how and at what cost?' PAC different set-ups, PAC dosing directly into activated sludge reactor (PAC-AS) and PAC dosing downstream the biological treatment in a hybrid adsorption/membrane process (PAC/MF, or PAC/UF, or PAC/loose NF) can control organic micropollutants and CECs in UWWT/resource recovery or water reclamation. Our strategy is to start by optimizing the current barriers (option 1), then to find the best (tailored) solution to upgrade the treatment if needed by investigating several alternatives (options 2–5), and finally to look ahead to more demanding needs likely to occur in the (near) future, namely the direct potable reuse (DPR). Such strategy involved studies at lab-, pilot-, and full-scales (Figure 8.23), which allowed developing cost analysis, and were conducted within PhD (Rodriguez *et al.*, 2016), postdoc (Viegas *et al.*, 2018), and R&I projects – LIFE Impetus (Campinas *et al.*, 2022; Mestre *et al.*, 2022a; Rosa *et al.*, 2019; Silva *et al.*, 2022; Viegas *et al.*, 2022), FP7 TRUST (Viegas *et al.*, 2020a), LIFE aWARE (Viegas *et al.*, 2018), and H2020 B-WaterSmart (Galego *et al.*, 2022). We also investigated the development of new PACs, framed within a circular economy approach (Mestre *et al.*, 2022a, b; Viegas *et al.*, 2020b).

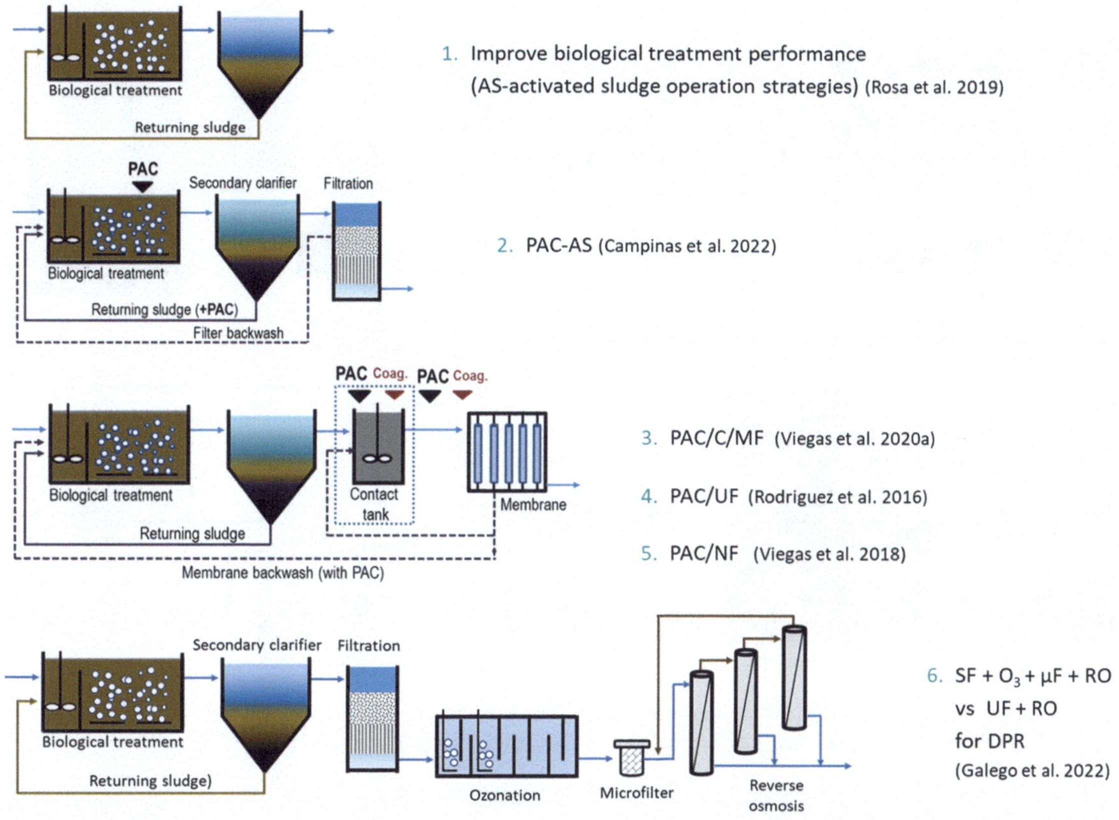

Figure 8.22 PAC application for controlling organic micropollutants in UWWT/water reclamation.

Regarding 'to what extent & why' and 'where, how & at what cost', briefly, our results showed:

- **All options**, including option 1, achieve some or good removal of the organic micropollutants targeted. So, if possible, it is worth starting by implementing option 1 (no/low investment) and then upgrade as needed.
- **Option 1. AS – activated sludge performance improvement** (Rosa *et al.*, 2019; Silva *et al.*, 2022) – the control of a representative range of PhCs (24) was studied in two activated sludge UWWTPs with nitrification/ denitrification (oxidation ditch in FNW, anaerobic/anoxic/aerobic system in BEI), involving 55 sampling campaigns during 2.5 years. The results show similar removals in both plants and, as earlier found, some of the most abundant PhCs at the WWTP inlet were highly removed during treatment (caffeine, acetaminophen, ibuprofen, and naproxen), others presented intermediate and variable removals (the antibiotics erythromycin, sulfamethoxazole, sulfapyridine, and sulfadiazine; the beta-blockers metoprolol and propranolol) and some compounds were recalcitrant to treatment, as the antiepileptic/ anticonvulsant carbamazepine and the anti-inflammatory diclofenac. The PhC removals in both plants highly correlated with the PhCs' biodegradation constants (k_{bio}), with a turning point at 1 L/(g SS.d), above which the median removals were above 86% in BEI and above 74% in FNW (Figure 8.24). A 4-class (from A – easily removed to D – recalcitrant) biodegradation/sorption framework is proposed for interpreting and predicting PhC control in urban AS WWTPs (Figure 8.4), as well as for

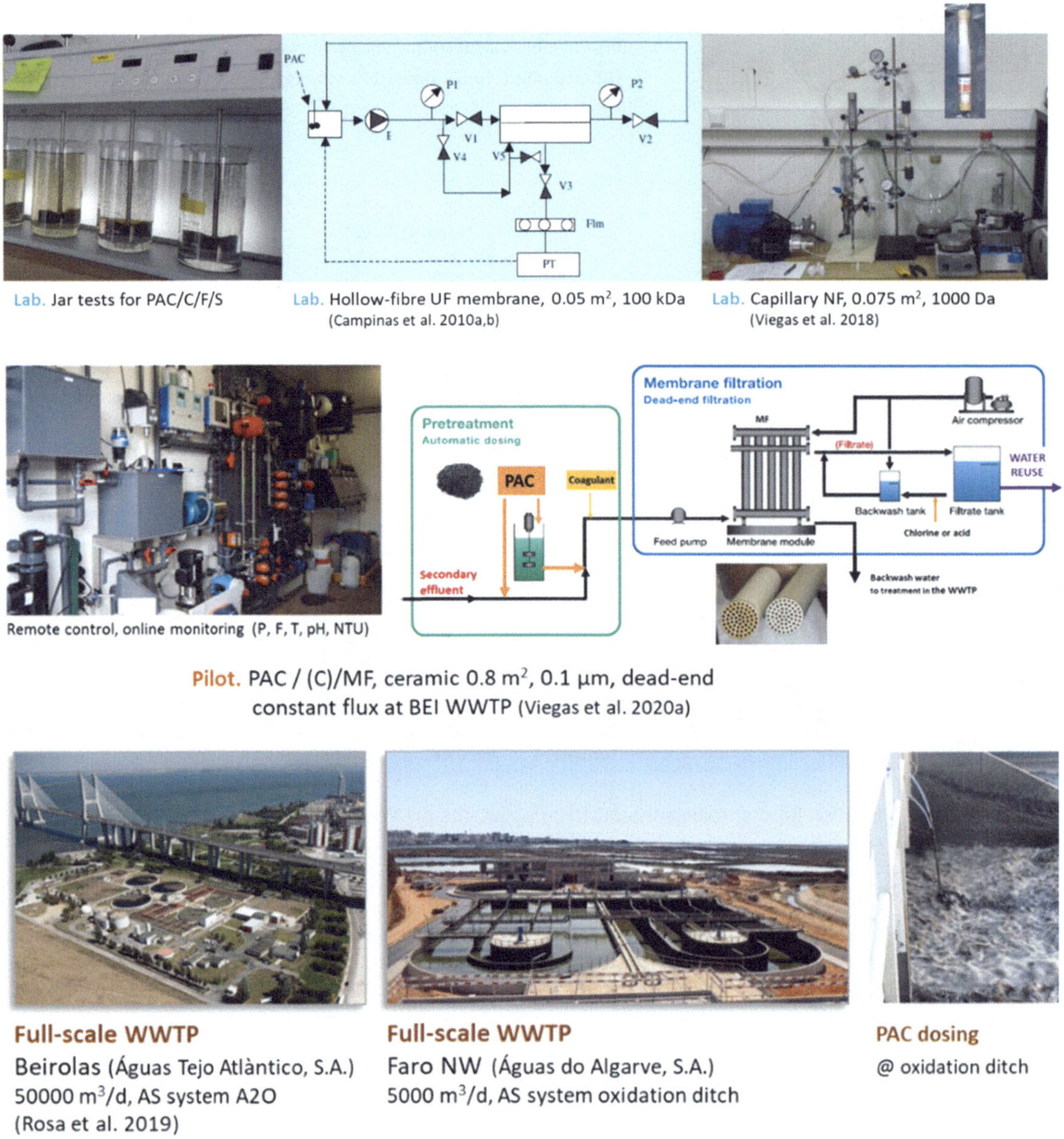

Figure 8.23 Lab, pilot, and full-scale tests of PAC application in UWWT.

selecting the indicator CECs (surrogates, as proposed in UWWTD recast). A statistical analysis indicated significantly higher removals in BEI associated with F/M values below 0.08 d^{-1}, and in FNW associated with N-removals > 80% and alkalinity reductions > 40% (indirect effect of nitrification) and were associated with an effluent transmittance (T254) of 67% (coherent with Zietzschmann *et al.* (2014) and Zietzschmann (2020), who have found A254 to be a reasonable/good surrogate parameter). Though expressed by different proxys, these results are

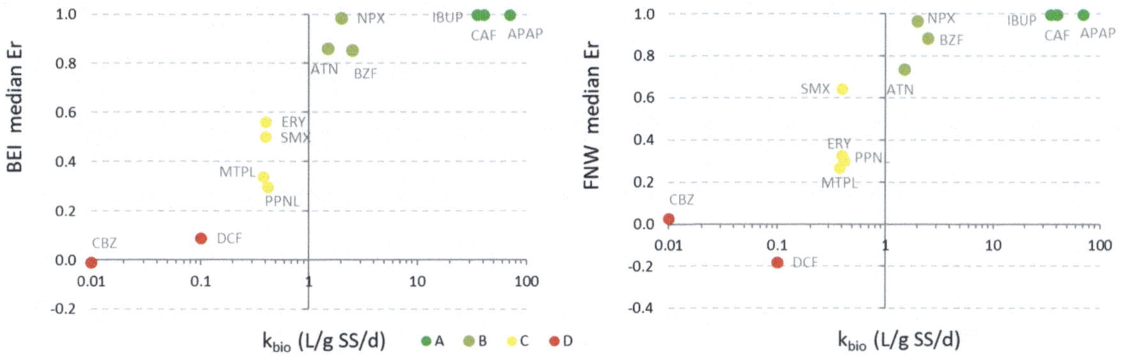

Figure 8.24 PhC median removal vs kbio of each PhC in BEI and FNW UWWTPs (the colours correspond to the A–D class of the PhC in Figure 8.4). *Source*: From Silva *et al.* (2022).

consistent with an enhanced elimination of some low biodegradable PhCs when conditions for the build-up of the slowly growing nitrifying bacteria (which can excrete enzymes and, by co-metabolism, break down some low degradable molecules) are provided, as found by Clara *et al.* (2005) for sludge retention times of, at least, 10–15 days. Promoting nitrification is therefore an operational low-investment measure for improving current UWWTPs towards PhC elimination.

- **Option 2. PAC-AS** (Campinas *et al.*, 2022) – PAC dosing to a conventional AS reactor is a low-investment option for controlling PhCs in UWWT, but its advantages and limitations in real operating environments are not fully assessed. A 3-week PAC-AS full-scale test was conducted in an oxidation ditch plant to assess PAC impact on effluent quality (PhCs, DOC, and other parameters), energy consumption, sludge production, and direct costs. DOC-normalized PAC doses of 0.7–2.6 mgPAC/mgDOC significantly reduced recalcitrant PhC discharged (e.g., by 63–84% for carbamazepine and 63–70% for diclofenac), the higher dose yielding a more reliable effluent quality (Figure 8.25). Effluent quality for total phosphorus, color, organic matter and transmittance was also enhanced and no interference with nitrification, oxidation-reduction potential or dissolved oxygen in the oxidation ditch was observed, resulting in no energy consumption increase. PAC had no impact on effluent turbidity and mixed liquor (MXL) suspended solids settleability, showing a positive effect on dewatered sludge dry weight and ultimately a 7–9% increase on final sludge production. After stopping PAC dosing, PAC remaining in the recirculated sludge presented adsorption capacity for some PhCs until it was completely out of system. Estimated direct costs (CAPEX + OPEX) for PAC addition to AS-reactor [0.064–0.055 €/m^3 for 5000–50,000 m^3/d, for 20 mg/L PAC dosed 12 months/year] are almost plant-size independent and compare favorably with literature data for PAC and GAC post-treatment (Baresel *et al.*, 2017; Rizzo *et al.*, 2019) and similarly with post-ozonation (Abegglen & Siegrist, 2012; Baresel *et al.*, 2017; Hillenbrand *et al.*, 2016; Mulder *et al.*, 2015).
- **AC adsorption UWWT vs DWT** – UWWT results are very consistent with the analogous ones found for DWT, though requiring more demanding conditions (higher PAC doses, lower membrane productivity) to balance the more concentrated water composition, so the removal mechanisms are essentially the same in both types of water. Charge, hydrophobicity, and aromaticity are again the adsorbate's key factors for PAC adsorption, as well as the water background organics and inorganics, the former with a stronger competing effect on the compounds less amenable to adsorption and the latter via the same shielding effect, as thoroughly analyzed in option 4 investigation (conducted at lab scale). A partial least-squares (PLS) regression with 25 descriptors

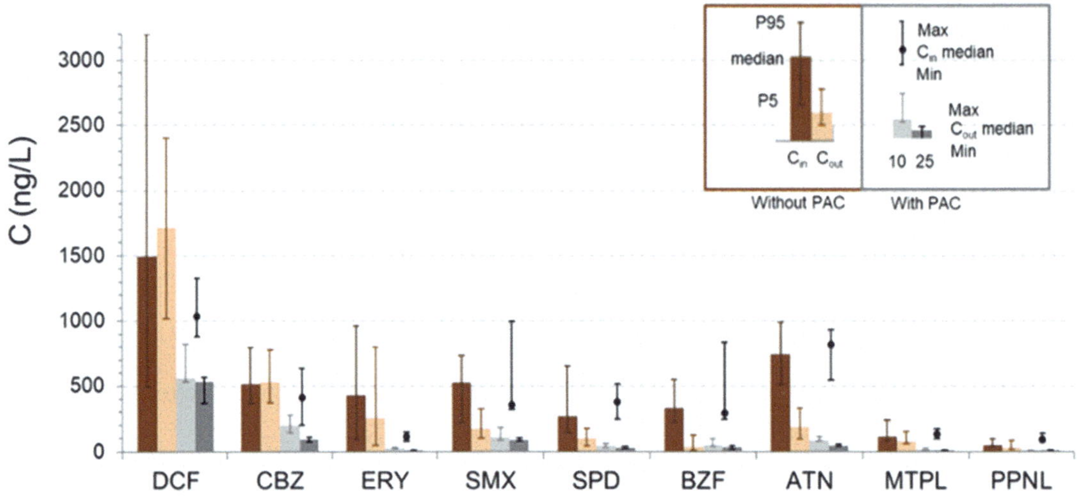

Parameter	No PAC	PAC-AS trial 1 (10 mg/L)		PAC-AS trial 2 (25 mg/L)	
	Median effluent conc. or value	Median effluent conc. or value	Reduction (%)	Median effluent conc. or value	Reduction (%)
DOC (mg C/L)	12.5	8.7	31	8.7	30
A254 (cm^{-1})	0.21	0.14	35	0.12	44
SUVA (L/(mg.m))	1.7	1.6	6	1.3	20
COD-Total (mg O_2/L)	42	35	17	30	29
COD-soluble (mg O_2/L)	70	29	59	26	63
A436 (cm^{-1})	0.019	0.011	42	0.009	53

Figure 8.25 Example of PAC-AS performance in PhC (left) and organic matter (right) control. *Source*: Adapted from Campinas *et al.* (2022).

(10 water-related (8 OM, 2 ion matrix), 7 PhC-related, and 8 PAC-related descriptors) yielded the following correlations of PhC removal from urban wastewater (Viegas *et al.*, 2022), which are overall coherent with those found in DWT (Table 8.5):

- Describing the adsorption capacity

$$\%\,removal\ (21\ h) =$$

$$-0.147 \cdot sHB\ \left(mg\frac{C}{L}\right) - 1.45x10^{-4} \cdot EC\ \left(\frac{uS}{cm}\right) + 6.73x10^{-3} \cdot Solv.energy\ \left(\frac{kJ}{mol}\right)$$

$$+0.0978 \cdot \log Kow + 2.04x10^{-4} \cdot A_{BET}\ \left(\frac{m^2}{g}\right) + 0.709$$

- Describing the short-term adsorption

$$\%\,removal\ (1\ h) = -0.0469 \cdot cHF\ \left(mg\frac{C}{L}\right) + 4.49x10^{-3} \cdot Solv.energy\ \left(\frac{kJ}{mol}\right) +$$

$$0.0190 \cdot \log P + 0.190 \cdot V_{super}\left(\frac{cm^3}{g}\right) - 1.24x10^{-3} \cdot PAC_{dens}\ \left(\frac{kg}{cm^3}\right) + 0.928$$

These correlations show the existence of complex interactions that a univariate analysis is not sufficient to describe and a good descriptive capability. Namely, it shows (a) the hydrophobic PhC–PAC interactions play a major role in the adsorption process, with the solvation energy and log K_{ow} being the most suitable descriptors, (b) the slightly hydrophobic fraction of the water DOM impact the adsorption capacity whereas its charged hydrophilic compounds impact the short-term adsorption, (c) the water inorganic matrix appears to impact only the PAC adsorption capacity and not the short-term adsorption, (d) for the pool of PACs tested, the results point to the BET area as a good descriptor of the PAC capacity, while the short-term adsorption kinetics appears to be better related to its supermicropore volume and density, the latter inversely related with the particle size, that is, the solute's intraparticle path.

- **Option 4. PAC/UF** (Rodriguez *et al.*, 2016) – for the compounds in Table 8.3, hydrophobicity is the adsorption key property for neutral adsorbates, with a turning point at Log K_{ow} 2.2–2.6, above which the compounds are very amenable to adsorption, that is, present low residual normalized concentrations (Figure 8.26). The uptake of the positively charged nortriptyline by the positively charged, meso- and microporous PACs exceeded the expected from Log *D* due to its high aromaticity and the background ions, which partially shielded PAC-nortriptyline electrostatic repulsions. Adsorption capacity depended on the solute's hydrophobicity, whereas the kinetics further depended on its charge. Hydrophobic EfOM was preferentially adsorbed and a stronger competitor, particularly for microcontaminants with Log $K_{ow} < 2.6$. The highly microporous PAC better adsorbed these target compounds and the hydrophobic EfOM, and it attenuated the EfOM competition. So, in real applications, PACs with higher volume of secondary micropores or small mesopores (SA vs SAE) might be a strategy for attenuating EfOM competition. For all waters, PAC had no effect on UF-flux, and it significantly improved the microcontaminants' and EfOM removal by PAC/UF over standalone PAC and UF; PAC/UF was more effective and efficient than PAC/sedimentation; PAC dose should target the contaminants with Log $K_{ow} < 2.6$, the weaker adsorbates, as found by others (e.g., Zietzschmann, 2020).

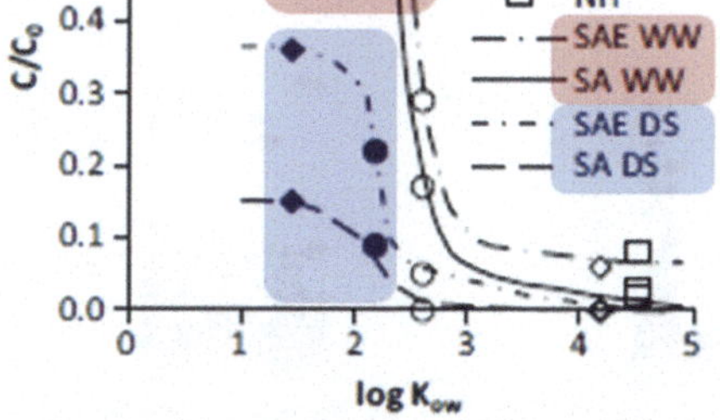

	WW	DS
pH	7.4 ± 0.4	7.1 ± 0.2
DOC, mg CL^{-1}	3.0 ± 0.6	–
Electrical conductivity, µS cm^{-1}	1432 ± 116	1523 ± 67
Alkalinity, mg $CaCO_3$ L^{-1}	117	117
Hardness, mg $CaCO_3$ L^{-1}	370	370
$UV_{254\ nm}$, m^{-1}	11.61 ± 0.27	–
$A_{436\ nm}$, m^{-1}	1.004 ± 0.035	–
SUVA, L m^{-1} mg^{-1}C	3.8 ± 0.4	–
EfOM composition, %		
Strongly hydrophobic	71.2	–
Moderately hydrophobic	14.0	–
Charged hydrophilic	10.5	–
Neutral hydrophilic	4.3	–

Figure 8.26 Log K_{ow} role on organic microcontaminants' control by PAC (SA and SAE) adsorption and the effect of water background organics (wastewater – WW and deionized water with equivalent inorganic matrix – DS) (compounds in Table 8.3). *Source*: Adapted from Rodriguez *et al.* (2016).

- **Option 3. PAC/C/MF** (Viegas *et al.*, 2020a) – pilot assays (100 L/(m^2h), 10 mg Fe/L) were conducted with sand-filtered secondary effluent spiked with four chemically diverse PhCs (ibuprofen, carbamazepine, sulfamethoxazole, and atenolol; Table 8.2, Figure 8.4, Figure 8.17) and 15 mg/L PAC dosed inline or to a 15-min contactor. The results showed no PAC-driven membrane fouling and +15 to +18% added removal with PAC contactor, reaching significant removals of CBZ and ATN (59–60%), SMX (50%), color (48%), A254 (35%), and DOC (28%). Earlier long-term demo tests with the same pilot proved PAC/Fe/ceramicMF to consistently produce highly clarified (monthly median < 0.1 NTU) and bacteria-free water, regardless of the severe variations in its intake (Figure 8.27, Viegas *et al.*, 2015). A detailed cost analysis points to total production costs of 0.21 €/m^3 for 50,000 m^3/day and 20 years membrane lifespan, mainly associated with equipment/membranes replacement, capital and reagents, the energy having the smallest share (for a specific flux of 261 L/(m^2.h.bar) and an energy consumption of 0.026 kWh/m^3).
- **Option 5. PAC/NF** (Viegas *et al.*, 2020a) – bench-scale results with a loose-NF membrane (capillary NF, 1000 Da cut-off) showed high removal efficiencies of four target PhCs (ATN, CBZ, DCF, SMX; 100 μg/L each spiked in secondary sand-filtered effluent) and the PAC continuous dosing to be more efficient than the pulse dosing, 50 mg/L PAC achieving a total-PhC removal of 68% (ranging from 58% to 89%), 58% DOC removal and 90% color removal. NF fluxes of 20 L/(m^2.h) were achieved with 1.2 bar transmembrane pressure, 1 m/s crossflow velocity, with no pressure increase up to 100 mg/L PAC. The results were successfully used in the design and operation of the PAC/NF pilot for the technology demonstration in El Prat WWTP (Barcelona). Compared to UF/RO (50/50), PAC/NF at pilot scale yielded comparable EfOM and PhC removals, PAC costs similar to UF/RO reagent costs, 40% less energy. Furthermore, PAC/NF concentrate recirculation to MBR showed synergic effects (not possible for UF/RO concentrate) with economic (lower cost) and operational advantages – PAC in the MBR improved the removal of refractory microcontaminants (diuron and carbamazepine) and metals (Cu, Ni, Zn, Pb) (Martin *et al.*, 2016).
- **PAC adsorption modelling** (integrating adsorption kinetics and isotherms) has the predictive ability to forecast the contaminant removal as a function of PAC dose, contact time and adsorbate concentration (Figure 8.28, Rosa *et al.*, 2019); it is therefore a powerful tool to support the design and operation of real systems. Under real scenarios with multisolute competitive adsorption, sound modelling involves integrating models describing the competitive adsorption

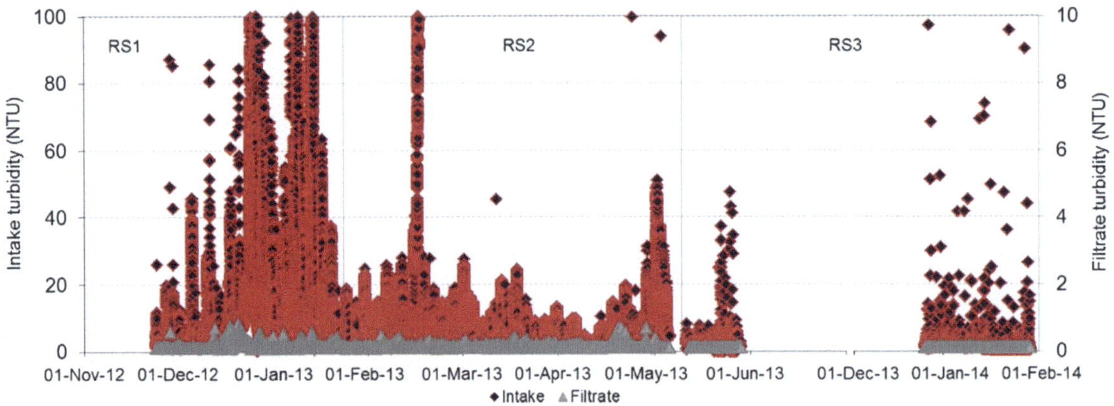

Figure 8.27 In-line records of water turbidity in PAC/MF intake (left axis) and filtrate (right axis). *Source*: From Viegas *et al.* (2015).

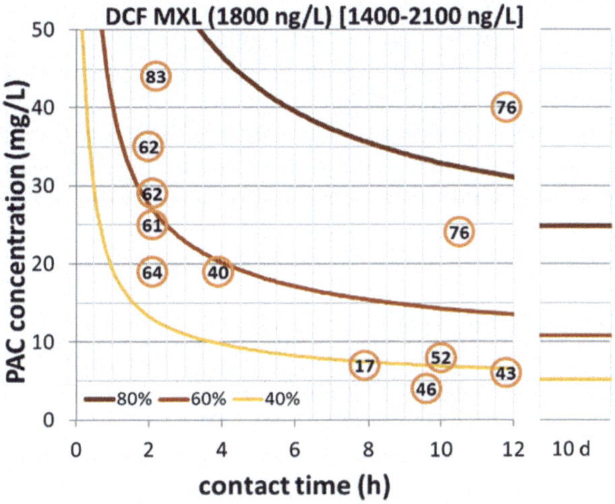

Figure 8.28 Model (IAST/EBCM, HSDM) predicted PAC adsorption removals (solid lines for 80%, 60%, 40%) of diclofenac from FNW WWTP as a function of PAC dose and contact time compared to field results from pilot-scale trials (circles). *Source*: From Rosa *et al.* (2019) (the 10-day prediction corresponds to the PAC full capacity use due to PAC recirculation; the values inscribed in the circles are the removals obtained in the pilot trials).

isotherms (e.g., the IAST/EBCM – ideal adsorption solution theory/equivalent background compound model) and kinetics (e.g., HSDM) (Campinas *et al.*, 2013; Rosa *et al.*, 2019; Viegas *et al.*, 2014; Worch, 2012).

- **Options 3–5.** As in DWT, these options with MF, UF, or NF membranes ensure a reliable water **disinfection** up to class A of reclaimed water (the highest water quality in Reg (EU) 741/2020, DL 119/2019), which is adequate for unrestricted water reuse, an extra feature which should be considered while comparing alternative options for PhC control. The pressure increase from MF to UF and NF [e.g., from 261 L/(m^2.h.bar), 0.40 bar for ceramic MF (Viegas *et al.*, 2020a) to 17 L/(m^2.h.bar), 1.2 bar for NF (Viegas *et al.*, 2018)] corresponds to an energy-intensity increase.
- **Option 6. direct potable reuse** – building on an earlier experience on batch production (Galego *et al.*, 2022), tests with a 24/7 automated pilot are starting to demonstrate the safety of potable water production by post-ozonation and or three-stage reverse osmosis, after sand-filtration or UF, and to develop the protocol to produce reclaimed water for artisanal beer production.

8.9 CLOSING REMARKS

Population growth and aging, as well as water scarcity driven by climate change call for an enhanced control of organic micropollutants and CECs in the urban water cycle, as progressively entailed in the EU and Portuguese legal framework on drinking water, UWWT and water reuse. In this context, there is a crucial need for multibarrier solutions, grounded on current barriers' improvement, prioritizing low-energy, physical barriers (to minimize by-products, resources' use, and carbon footprint) and producing fit-for-purpose water(s).

AC-based and hybrid PAC/membrane processes have proven a huge potential for controlling CECs in water treatment and water reclamation, but there is room for improving their sustainability and cost-efficiency with PAC development and process optimization. For instance, new high-performing environmental-friendly (biomass-derived, steam-activated) PACs from local biomasses (circular economy approach), as well as of more efficient finer sPAC and magnetic PACs allowing their recovery and regeneration are very important. Each application is water and target contaminant-specific and requires methodologies for selecting representative contaminants, the characterization of the inorganic and organic matrices of the waters, and to understand and anticipate/overcome competing factors, namely through modelling. Pilot demonstration and mathematical modelling are therefore success factors for process design and operation.

Strategic planning, multiobjective-driven (e.g., organic micropollutant control, ARB&ARGs control, and disinfection for water reuse) and performance-based decisions taken by skilled and competent human resources are required to successfully address the moving targets that the CECs constitute.

FUNDING AND ACKNOWLEDGMENTS

This work has received financial support from the European Union LIFE ENV Programme under Grant Agreements LIFE12 ENV/PT/001154 (www.life-hymemb.eu), LIFE14 ENV/PT/000739 (www.life-impetus.eu), and LIFE11 ENV/ES/000606 (LIFE aWARE), from EU FP7 contract 265122 (TRUST) and H2020 contract 869171 (https://b-watersmart.eu/), as well as from the Portuguese Fundação para a Ciência e Tecnologia (FCT) through PTDC/QUE-QUE/6024/2020 (EMPOWER+). This publication reflects only the authors' views and the European Union and FCT are not liable for any use that may be made of the information contained therein. We also acknowledge Metawater, Pentair, and Salmon for kindly providing membranes and ACs.

REFERENCES

Abegglen C. and Siegrist H. (2012). Mikroverunreinigungen aus kommunalem Abwasser. Verfahren zur weitergehenden elimination auf Kläranlagen. (Micropollutants from municipal wastewater. Process for further elimination in sewage treatment plants.) (In German). Bundesamt für Umwelt, Bern, Umwelt-Wissen. *Nr*, **1214**, 210 S.

Aus der Beek T., Weber F.-A., Bergmann A., Hickmann S., Ebert I., Hein A. and Küster A. (2016). Pharmaceuticals in the environment – global occurrences and perspectives. *Environmental Toxicology and Chemistry*, **35**(4), 823–835, https://doi.org/10.1002/etc.3339

Baresel C., Magnér J., Magnusson K. and Olshammar M. (2017). Tekniska lösningar för Avancerad Rening av Avloppsvatten. (Technical Solutions for Advanced Wastewater Treatment.) (In Swedish). IVL Svenska Miljöinstitutet AB, Stockholm

Campinas M. and Rosa M. J. (2006). The ionic strength effect on microcystin and natural organic matter surrogate adsorption onto PAC. *Journal of Colloid and Interface Science*, **299**, 520–529, https://doi.org/10.1016/j.jcis.2006.02.042

Campinas M. and Rosa M. J. (2010a). Assessing PAC contribution to the NOM fouling control in PAC/UF systems. *Water Research*, **44**(5), 1636–1644, https://doi.org/10.1016/j.watres.2009.11.012

Campinas M. and Rosa M. J. (2010b). Removal of microcystins by PAC/UF. *Separation and Purification Technology*, **71**(1), 114–120, https://doi.org/10.1016/j.seppur.2009.11.010

Campinas M. and Rosa M. J. (2010c). Evaluation of cyanobacterial cells removal and lysis by ultrafiltration. *Separation and Purification Technology*, **70**, 345–353, https://doi.org/10.1016/j.seppur.2009.10.021

Campinas M. and Rosa M. J. (2010d). Comparing PAC/UF and conventional clarification with PAC for removing microcystins from natural waters. *Desalination and Water Treatment*, **16**, 120–128, https://doi.org/10.5004/dwt.2010.1092

Campinas M. and Rosa M. J. (2011). PAC/UF for removing cyanobacterial cells and toxins from drinking water. Chap 12 In: Expanding Issues in Desalination, R. Y. Ning (ed.), InTech. Rijeka, Croatia, pp. 233–252. ISBN: 978-953-307-624-9.

Campinas M., Viegas R. and Rosa M. J. (2013). Modelling and understanding the competitive adsorption of microcystins and tannic acid. *Water Research*, **47**, 5690–5699, https://doi.org/10.1016/j.watres.2013.06.048

Campinas M., Viegas R. M. C., Silva C. and Rosa M. J. (2017). Technical Guidelines of PAC/MF – Powdered Activated Carbon/Ceramic Microfiltration for Drinking Water Production. Deliverable of LIFE Hymemb Project (LIFE12 ENV/PT/001154). LNEC, Lisboa, ISBN: 978-972-49-2299-7. 32 pp.

Campinas M., Silva C., Viegas R. M. C., Coelho R., Lucas H. and Rosa M. J. (2021a). To what extent may pharmaceuticals and pesticides be removed by PAC conventional addition to low-turbidity surface waters and what are the potential bottlenecks? *Journal of Water Process Engineering*, **40**, 101833, https://doi.org/10.1016/j.jwpe.2020.101833

Campinas M., Viegas R. M. C., Coelho R., Lucas H. and Rosa M. J. (2021b). Adsorption/coagulation/ceramic microfiltration for treating challenging waters for drinking water production. *Membranes*, **11**(2), 91, https://doi.org/10.3390/membranes11020091

Campinas M., Viegas R. M. C., Silva C. and Rosa M. J. (2021c). Operational performance and cost analysis of PAC/ceramic MF for drinking water production: exploring treatment capacity as a new indicator for performance assessment and optimization. *Separation and Purification Technology*, **255**, 117443, https://doi.org/10.1016/j.seppur.2020.117443

Campinas M., Viegas R. M. C., Almeida C. M. M., Martins A., Silva C., Mesquita E., Coelho M. R., Silva S., Cardoso V. V., Benoliel M. J. and Rosa M. J. (2022). Powdered activated carbon full-scale addition to the activated sludge reactor of a municipal wastewater treatment plant: pharmaceutical compounds control and overall impact on the process. *Journal of Water Process Engineering*, **49**, 102975, https://doi.org/10.1016/j.jwpe.2022.102975

Chorus I. and Welker M. (eds.) (2021). Toxic Cyanobacteria in Water. A Guide to Their Public Health Consequences, Monitoring and Management. 2nd ed. World Health Organisation, CRC Press, London. eBook ISBN 9781003081449. 859 pp.

CID (2022). Commission Implementing Decision of 19.1.2022 establishing a watch list of substances and compounds of concern for water intended for human consumption as provided for in Directive (EU) 2020/2184 of the European Parliament and of the Council. C(2022) 142 final.

Clara M., Kreuzinger N., Strenn B., Gans O. and Kroiss H. (2005). The solids retention time – a suitable design parameter to evaluate the capacity of wastewater treatment plants to remove micropollutants. *Water Research*, **39**(1), 97–106, https://doi.org/10.1016/j.watres.2004.08.036

COM (2022). Proposal for a Directive of the European Parliament and of the Council concerning urban wastewater treatment (recast). 2022/0345 (COD). COM(2022) 541 final.

Cravo A., Silva S., Rodrigues J., Cardoso V. V., Benoliel M. J., Correia C., Coelho M. R., Rosa M. J. and Almeida C. M. M. (2022). Understanding the bioaccumulation of pharmaceutical active compounds by clams *Ruditapes decussatus* exposed to a UWWTP discharge. *Environmental Research*, **208**, 112632, https://doi.org/10.1016/j.envres.2021.112632

Eggen R. I. L., Hollender J., Joss A., Schärer M. and Stamm C. (2014). Reducing the discharge of micropollutants in the aquatic environment: the benefits of upgrading wastewater treatment plants. *Environmental Science & Technology*, **48**, 7683–7689, https://doi.org/10.1021/es500907n

Fortune Business Insights (2019). Activated carbon market size, trends, growth & report [2020–2027]. Retrieved from https://www.fortunebusinessinsights.com/activated-carbon-market-102175 (accessed 4 May 2021)

Galego J., Lourinho R., Figueiredo D., Costa C., Viegas R. and Rosa M. J. (2022). Protocol for water reclamation by ozone/reverse osmosis for safe direct potable reuse in beer production. IWA World Water Congress & Exhibition, September 11–15, Copenhagen, Denmark. 4 pp. (poster)

Geissen V., Mol H., Klumpp E., Umlauf G., Nadal M., van der Ploeg M., van de Zee S. E. A. T. M. and Ritsema C. J. (2015). Emerging pollutants in the environment: A challenge for water resource management. *International Soil and Water Conservation Research*, **3**, 57–65, https://doi.org/10.1016/j.iswcr.2015.03.002

Golovko O., Örn S., Sörengård M., Frieberg K., Nassazzi W., Lai F. Y. and Ahrens L. (2021). Occurrence and removal of chemicals of emerging concern in wastewater treatment plants and their impact on receiving water systems. *Science of the Total Environment*, **754**, 142122, https://doi.org/10.1016/j.scitotenv.2020.142122

Gomez Cortes L., Marinov D., Sanseverino I., Navarro Cuenca A., Niegowska Conforti M., Porcel Rodriguez E., Stefanelli F. and Lettieri T. (2022). Selection of Substances for the 4th Watch List under the Water Framework Directive. EUR 31148 EN, Publications Office of the European Union, Luxembourg. ISBN 978-92-76-55021-1. JRC130252. https://doi.org/10.2760/01939

Hall T., Schmidt W., Codd G. A., von Gunten U., Kaas H., Acero J. L., Heijman B., Meriluoto J., Rosa M. J., Manckiewicz J. (2005). Best Practice Guidance for Management of Cyanotoxins in Water Supplies. TOXIC 5FP project (EVK1-CT-2002-00107). Abo Akademi University (Finland), DHI (Denmark), DVGW-TZW (Germany), EAWAG (Switzerland), KIWA (The Netherlands), University of Algarve (Portugal), University of Dundee (UK), University of Extremadura (Spain), University of Lodz (Poland), WRc plc (UK). Water Research Centre, UK. Sept., 136 pp.

Hillenbrand T., Tettenborn F., Fuchs S., Toshovski S., Metzger S., Tjoeng I., Wermter P., Hecht D., Kersting M., Werbeck N. and Wunderlin P. (2016). Maßnahmen zur Verminderung des Eintrages von Mikroschadstoffen in die Gewässer – Phase 2 (Measures to Reduce the Input of Micropollutants Into Water Bodies – Phase 2.) (In German), Texte 60/2016, Umweltsbundesambt, Dessau-Roßlau, Germany, 235 pp. ISSN 1862-4804. Retrieved from https://www.umweltbundesamt.de/sites/default/files/medien/377/publikationen/mikroschadstoffen_in_die_gewasser-phase_2.pdf (accessed 16 May 2019)

Inkwood Research (2019). Global activated carbon market forecast 2019–2028: growth, size, opportunity, share. Retrieved from https://inkwoodresearch.com/reports/activated-carbon-market/ (accessed 4 May 2021).

Luo Y., Guo W., Ngo H. H., Nghiem L. D., Hai F. I., Zhang J., Liang S. and Wang X. C. (2014). A review on the occurrence of micropollutants in the aquatic environment and their fate and removal during wastewater treatment. *Science of the Total Environment*, **473–474**, 619–641, https://doi.org/10.1016/j.scitotenv.2013.12.065

Martin I., Echevarría C., Mesa C., Lefèvre B., Bernat X., Viegas R. M. C., Mesquita E., Rosa M. J., Borràs J., Martínez B., Vega A., Aguiló P. and Castellví E. (2016). LIFE aWARE project: development of hybrid wastewater treatment and reclamation schemes to enhance removal of micro pollutants and promote water reuse. LIFE aWARE Lisbon Workshop, June 6, Lisbon, LNEC.

Menaia J. and Rosa MJ. (2006). Cianobactérias e Água Para Abastecimento Público. Manual de Apoio ao Curso AdP/LNEC 'Florescências de algas e cianobactérias nas águas de Captação' Carvoeiro, Algarve, nov. (Cyanobacteria and Public Water Supply. Course Support Manual AdP/LNEC 'Algal blooms and cyanobacteria in catchment water') (In Portuguese). Laboratório Nacional de Engenharia Civil, Portugal. 14 pp.

Mestre A. S., Campinas M., Viegas R. M. C., Mesquita E., Carvalho A. P. and Rosa M. J. (2022a). Activated carbons in full-scale advanced wastewater treatment. Chap. 17 In: Advanced Materials for Sustainable Environmental Remediation: Terrestrial and Aquatic Environments, D. A. Giannakoudakis, L. Meili and I. Anastopoulos (eds.), Elsevier, MA, USA, pp. 433–474. eBook ISBN: 9780323904865. https://doi.org/10.1016/B978-0-323-90485-8.00001-1

Mestre A. S., Viegas R. M. C., Mesquita E., Rosa M. J. and Carvalho A. P. (2022b). Engineered pine nut shell derived activated carbons for improved removal of recalcitrant pharmaceuticals in urban wastewater treatment. *Journal of Hazardous Materials*, **437**, 129319, https://doi.org/10.1016/j.jhazmat.2022.129319

Metzger S., Alt K., Benstoem F., Biebersdorf N., Boehler M., Bornemann C., Christian H., Jedele K., Jekel M., Lyko S., Andreas N., Wintgens T., Krahnstöver T., Mätzig H., Remy C., Rößler A., Zietzschmann F. and Wilhelm C. (2019). Aktivkohleeinsatz auf Kommunalen Kläranlagen zur Spurenstoffentfernung – Verfahrensvarianten, Reinigungsleistung und Betriebliche Aspekte (Activated Carbon Application on Municipal Wastewater Treatment Plants for Trace Compound Removal – Process Variants, Treatment Performance and Operational Aspects.) (In German). Deutsche Vereinigung für Wasserwirtschaft, Abwasser und Abfall e.V. (DWA), Hennef (Germany).

Mulder M., Antakyali D. and Ante S. (2015). Costs of Removal of Micropollutants From Effluents of Municipal Wastewater Treatment Plants – General Cost Estimates for The Netherlands Based on Implemented Full Scale Post Treatments of Effluents of Wastewater Treatment Plants in Germany and Switzerland. STOWA and Waterboard the Dommel, The Netherlands. 55 pp.

Pesqueira J. F. J. R., Pereira M. F. R. and Silva A. M. T. (2020). Environmental impact assessment of advanced urban wastewater treatment technologies for the removal of priority substances and contaminants of emerging concern: a review. *Journal of Cleaner Production*, **261**, 121078, https://doi.org/10.1016/j.jclepro.2020.121078

Pinho M. N. N. C. and Rosa M. J. F. (1995). Membranas e processos de Separação com Membranas. Instituto Superior Técnico, Lisboa, dezembro, 56 pp.

Pourchet M., Debrauwer L., Klanova J., Price E. J., Covaci A., Caballero-Casero N., Oberacher H., Lamoree M., Damont A., Fenaille F., Vlaanderen J., Meijer J., Krauss M., Sarigiannis D., Barouki R., Le Bizec B. and Antignac J.-P. (2020). Suspect and non-targeted screening of chemicals of emerging concern for human biomonitoring, environmental health studies and support to risk assessment: from promises to challenges and harmonisation issues. *Environment International* **139**, 105545, https://doi.org/10.1016/j.envint.2020.105545

Raspati G., Menaia J., Raat K., Rosa M. J., Gonzalez B. L. and Sivertsen E. (2015). Assessment of current treatment works to handle climate change related pollutants and options to make current multi-barrier systems climate change proof. Chap. 4.7 In: Climate Change, Water Supply and Sanitation: Risk Assessment, Management, Mitigation and Reduction, A. Hulsmann *et al.* (eds.), IWA Publishing, London, pp. 284–287. ISBN: 9781780405001 (eBook). https://doi.org/10.2166/9781780405001

Ribau Teixeira M. and Rosa M. J. (2006). The impact of the water background inorganic matrix on the natural organic matter removal by nanofiltration. *Journal of Membrane Science*, **279**, 513–520, https://doi.org/10.1016/j.memsci.2005.12.045

Ribau Teixeira M., Rosa M. J. and Nystrom M. (2005). The role of membrane charge on nanofiltration performance. *Journal of Membrane Science*, **265**, 160–166, https://doi.org/10.1016/j.memsci.2005.04.046

Ribau Teixeira M., Rosa M. J., Sorlini S., Biasibetti M., Christophoridis C. and Edwards C. (2020). Removal of cyanobacteria and cyanotoxins by conventional physical-chemical treatment. Chap. 3 In: Water Treatment for Purification From Cyanobacteria and Cyanotoxins, A. E. Hiskia, T. M. Triantis, M. G. Antoniou, T. Kaloudis and D. D. Dionysiou (eds.), 1st edn, Wiley, Hoboken, NJ, USA, pp. 69–97. Online ISBN: 9781118928677. https://doi.org/10.1002/9781118928677.ch3

Rizzo L., Malato S., Antakyali D., Beretsou V. G., Dolić M. B., Gernjak W., Heath E., Ivancev-Tumbas I., Karaolia P., Lado Ribeiro A. R., Mascolo G., McArdell C. S., Schaar H., Silva A. M. T. and Fatta-Kassinos D. (2019). Consolidated vs new advanced treatment methods for the removal of contaminants of emerging concern from urban wastewater. *Science of the Total Environment*, **655**, 986–1008, https://doi.org/10.1016/j.scitotenv.2018.11.265

Rizzo L., Gernjak W., Krzeminski P., Malato S., McArdell C. S., Perez J. A. S., Schaar H. and Fatta-Kassinos D. (2020). Best available technologies and treatment trains to address current challenges in urban wastewater reuse for irrigation of crops in EU countries. *Science of the Total Environment*, **710**, 136312, https://doi.org/10.1016/j.scitotenv.2019.136312

Rodriguez E., Campinas M., Acero J. L. and Rosa M. J. (2016). Investigating PPCP removal from wastewater by powdered activated carbon/ultrafiltration. *Water Air & Soil Pollution*, **227**(6), 1, https://doi.org/10.1007/s11270-016-2870-7

Rosa M. J., Vieira P. and Menaia J. (2009). O Tratamento de água Para Consumo Humano Face à Qualidade da água de Origem. Guia Técnico 13. IRAR, LNEC, Lisboa. ISBN 978-989-95392-7-3. 82 pp.

Rosa M. J., Campinas M., Viegas R. M. C. and Silva C. (2019). LIFE IMPETUS Technical Guidelines for Improved Control of Pharmaceutical Compounds in Urban Activated Sludge WWTPs. Deliverable of LIFE Impetus project (LIFE14 ENV/PT/000739). LNEC, Lisboa. 49 pp.

Sauvé S. and Desrosiers M. (2014). A review of what is an emerging contaminant. *Chemistry Central Journal*, **8**(1), 15, https://doi.org/10.1186/1752-153X-8-15

Siegrist H. and Joss A. (2012). Review on the fate of organic micropollutants in wastewater treatment and water reuse with membranes. *Water Science and Technology*, **66**(6), 1369–1376, https://doi.org/10.2166/wst.2012.285

Silva C., Almeida C. M. M., Rodrigues J. A., Silva S., Coelho M. R., Martins A., Alves R., Cardoso E., Cardoso V. V., Benoliel M. J. and Rosa M. J. (2021). Occurrence and seasonality of pharmaceutical compounds in urban wastewaters in two Portuguese regions. *Urban Water Journal*, **18**(6), 465–478, https://doi.org/10.1080/1573062X.2021.1893365

Silva C., Almeida C. M. M., Rodrigues J. A., Silva S., Coelho M. R., Martins A., Lourinho R., Cardoso E., Cardoso V. V., Benoliel M. J., Mesquita E., Ribeiro R. and Rosa M. J. (2023). Improving the control of pharmaceutical compounds in activated sludge wastewater treatment plants: Key operating conditions and monitoring parameters. *Journal of Water Process Engineering*, **54**, 103985, 214–7144, https://doi.org/10.1016/j.jwpe.2023.103985

Smeets P., Kardinaal E., Eikebrokk B., Menaia J., Mesquita E. and Rosa M. J. (2015). Optimised operation of drinking water systems. Chap. 4.14 In: Climate Change, Water Supply and Sanitation: Risk Assessment, Management, Mitigation and Reduction. A. Hulsmann G. Grützmacher, G. van den Berg, W. Rauch, A. L. Jensen, V. Popovych, M. R. Mazzola, L. S. Vamvakeridou-Lyroudia and D. A. Savic (eds.), IWA Publishing, London, pp. 332–337. ISBN: 9781780405001 (eBook). https://doi.org/10.2166/9781780405001

Souza R. M., Seibert D., Quesada H. B., Bassetti F. J., Fagundes-Klen M. R. and Bergamasco R. (2020). Occurrence, impacts and general aspects of pesticides in surface water – a review. *Process Safety and Environmental Protection*, **135**, 22–37. https://doi.org/10.1016/j.psep.2019.12.035

Ternes T. A. and Joss A. (eds.) (2006). Human Pharmaceuticals, Hormones and Fragrances. The Challenge of Micropollutants in Urban Water Management. IWA Publishing, Cornwall, UK. ISBN 10 1843390930. 453 pp.

UMWELT.NRW (2019). Entwicklung und Stander Abwasserbeseitigungin Nordrhein-Westfalen (Development and Standing of Wastewater Disposal in North Rhine-Westphalia.) (In German). Report of the Ministry for Environment, Agriculture, Conservation and Consumer Protection of the State of North Rhine-Westphalia, Germany.

Viegas R. M. C., Campinas M., Costa H. and Rosa M. J. (2014). How do the HSDM and Boyd's model compare for estimating intraparticle diffusion coefficients in adsorption processes. *Adsorption*, **20**(5–6), 737–746, https://doi.org/10.1007/s10450-014-9617-9

Viegas R. M. C., Mesquita E., Inocêncio P., Teixeira A. P., Martins J. and Rosa M. J. (2015). Water reclamation with hybrid coagulation–ceramic microfiltration: first part of a long-term pilot study in Portugal. *Journal of Water Reuse and Desalination*, **5**(4), 550–556, https://doi.org/10.2166/wrd.2015.122

Viegas R. M. C., Mesquita E., Campinas M., Almeida C. M. M. and Rosa M. J. (2018). Designing a PAC/NF advanced process for controlling pharmaceutical compounds in reclaimed water. In: Sanitation Approaches and Solutions and the Sustainable Development Goals, J. S. Matos and M. J. Rosa (eds.), Technical course 08. ERSAR, EWA, APESB, Lisbon, pp. 97–108. ISBN: 978-972-98996-8-3.

Viegas R. M. C., Mesquita E., Campinas M. and Rosa M. J. (2020a). Pilot studies and cost analysis of hybrid powdered activated carbon/ceramic microfiltration for controlling pharmaceutical compounds and organic matter in water reclamation. *Water*, **12**(1), 33, https://doi.org/10.3390/w12010033

Viegas R. M. C., Mestre A. S., Mesquita E., Campinas M., Andrade M., Carvalho A. P. and Rosa M. J. (2020b). Assessing the applicability of a new carob waste-derived powdered activated carbon to control pharmaceutical compounds in wastewater treatment. *Science of the Total Environment*, **743**, 140791, https://doi.org/10.1016/j.scitotenv.2020.140791

Viegas R. M. C., Campinas M., Coelho R., Lucas H. and Rosa M. J. (2021a). Hybrid process of adsorption/coagulation/ceramic MF for removing pesticides in drinking water treatment – inline vs. contact tank PAC dosing. *Membranes*, **11**(2), 72, https://doi.org/10.3390/membranes11020072

Viegas R. M. C., Ribeiro R., Rosa M. J., Campinas M., Mesquita E. and Silva C. (2021b). Controlo de contaminantes de interesse emergente em ETA e ETAR. Dossier 'Contaminantes emergentes na Água'. *Indústria e Ambiente*, **126**, 16–19.

Viegas R. M. C., Mestre A. S., Mesquita E., Machuqueiro M., Andrade M. A., Carvalho A. P. and Rosa M. J. (2022). Key factors for activated carbon adsorption of pharmaceutical compounds from wastewaters: a multivariate modelling approach. *Water*, **14**(2), 166, https://doi.org/10.3390/w14020166

Wise B. L., Shih W. and Smith S. (2020). Ceramic Ultrafiltration Membranes with Improved Economics, Operability, and Process Design Flexibility. Nanostone Water Inc., Minnesota, USA, IWC 17–34.

Worch E. (2012). Adsorption Technology in Water Treatment. Fundamentals, Processes, and Modeling. Walter de Gruyter GmbH & Co. KG, Berlin/Boston. ISBN 978-3-11-024022-1

Zietzschmann F. (2020). Adsorptive removal of pharmaceutically active compounds from wastewater. In: Removal and Degradation of Pharmaceutically Active Compounds in Wastewater Treatment, S. Rodriguez-Mozaz, P. B. Cano and M. S. Adroguer (eds.), Springer Berlin Heidelberg, Berlin, Heidelberg, Vol. **2020**, pp. 1–29. https://doi.org/10.1007/698_2020_687

Zietzschmann F., Altmann J., Ruhl A. S., Dünnbier U., Dommisch I., Sperlich A., Meinel F. and Jekel M. (2014). Estimating organic micro-pollutant removal potential of activated carbons using UV absorption and carbon characteristics. *Water Research*, **56**, 48–55, https://doi.org/10.1016/j.watres.2014.02.044

doi: 10.2166/9781789063783_0181

Chapter 9

Constructed wetlands: an approach toward phytoremediation for wastewater treatment

Sahail Alvi[1], Sai Shankar Sahu[1], Vivek Rana[2] and Subodh Kumar Maiti[1,]*

[1]Department of Environmental Science and Engineering, Indian Institute of Technology (Indian School of Mines), Dhanbad, 826004, Jharkhand, India
[2]Central Pollution Control Board, New Delhi, India
**Corresponding author: subodh@iitism.ac.in*

ABSTRACT

Constructed wetlands (CWs) have been used as a proven efficient technology for wastewater treatment since the early 1950s. However, the first full-scale operation was conducted in the late 1960s. CW is an engineered system that utilizes various macrophytes to treat wastewater. It is cost effective, easy to operate, and maintain as compared to other conventional wastewater treatment technologies. It works with the mechanisms of different phytoremediation processes (i.e., phytoextraction, phytostabilization, phytovolatization, phytodegradation) and sedimentation, agglomeration and biodegradation of wastewater. CW can be classified on the basis of vegetation types (submerged, emergent, free-floating, and floating leaved), hydrological properties (free water, surface flow, and subsurface flow). Subsurface flow wetlands can be further classified as vertical or horizontal depending on the flow direction. Hybrid systems such as CWs clubbed with tube settler for hospital wastewater, biochar-based CW for manure wastewater are reported to have better treatment efficiency. However, the performance of CWs depends on several factors like vegetation type, type of wetland, type of wastewater treated, design specification of CWs, substrate used, microbiology of CW, climate and hydrological factors, and so on. Therefore, this chapter mainly focuses on all these aspects to fill the knowledge gaps and gives an insight to future research challenges and opportunities to achieve sustainability in the field of wastewater treatment.

Keywords: constructed wetland, wastewater treatment, phytoremediation, sustainability

9.1 INTRODUCTION

Constructed wetland (CW) is an artificial wetland, mainly dominated by emergent plants (*Typha*, *Sagittaria*, *Canna indica*, etc.), which is a low-cost natural treatment used commonly for sewage and graywater treatment but currently industrial wastewater is also treated. The CW uses natural functions of vegetation, soil, organisms, and mechanisms of removal of pollutants through combination of physical, chemical, and biological actions. In many cases, CWs are also used as secondary treatment to treat wastewater. Based on type and characteristics of wastewater to be treated, the design of CW is varied. Research studies recommended that CWs can also be used as primary treatment, when concentration of suspended solids (SS), biochemical oxygen demand (BOD) and chemical oxygen demand (COD) is higher.

9.2 PHYTOREMEDIATION

Phytoremediation is a process to restore the ecosystem by using plants and related microorganisms. They are used to reduce the concentration of contaminants and their impacts on the system. Some of the popular accumulator plants for cleaning up the contaminated water are duckweed (*Lemna minor*), water lettuce (*Pistia stratiotes*), water hyacinth (*Eichhornia crassipes*), and so on. In addition to this, various novel technologies such as microbial intervention, use of genetic innovation and pharmaceutical additives, and so on are applied for enhancing the remediation potential. Different mechanisms of phytoremediation are shown in Figure 9.1.

9.2.1 Mechanisms of phytoremediation

Phytoremediation technology can be applied on a range of contaminants by using different habitats. They entail various mechanisms including in-situ stabilization, degradation and retrieval of pollutants.

9.3 PHYTOEXTRACTION

Phytoextraction is the removal of heavy metals or inorganic contaminants from water by using metal-tolerant plants. The plants uptake the contaminants by their roots and concentrate them in their above-ground biomass. They can further be harvested, recovered, disposed, or recycled. Plants that have a high metal accumulating potential are called hyperaccumulators.

9.4 PHYTODEGRADATION

Phytodegradation is the breakdown of organic pollutants either directly, by the release of enzymes from the roots, or by biochemical processes occurring within plant cells. Organic pollutants are absorbed by roots and converted into less toxic compounds in plant cells during phytodegradation. Organic pollutants that are hydrophobic have been especially effective at being biodegraded by plants. A few research studies show that successful phytodegradation of hazardous and resistant organic substances has been achieved using poplar plants (*Populus* spp.).

9.5 PHYTOVOLATIZATION

Phytovolatization is the process by which pollutants are uptaken by plant roots, transformed into gaseous form and released into the environment. Evapotranspiration process of plants enhances phytovolatization. So, the plants having strong evapotranspiration rates are desired. In this process, plants actively take up volatile organic pollutants, particularly, volatile organic compounds (VOCs).

9.6 PHYTOSTABILIZATION

Phytostabilization is the phenomenon by which metals or contaminants are immobilized within the profile of soil, so that the pollutant has minimum pollutant escape and biological exposure. Long-term stabilization and containment of pollutants are desired.

9.7 RHIZODEGRADATION

Rhizodegradation is the biodegradation of organic pollutants by soil microorganisms into less toxic compounds within the rhizosphere. Root exudates, which are organic compounds that support soil

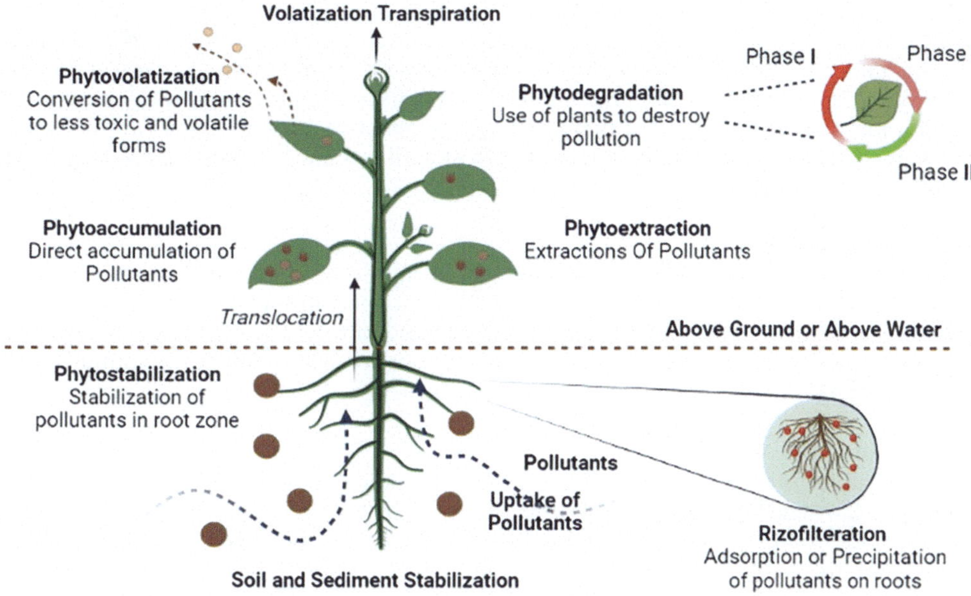

Figure 9.1 Mechanisms of phytoremediation.

microbial communities, aid in this process. The inocula of a particular bacterium can be introduced into contaminated soils to speed up this process. According to Campbell and Greaves (1990), plants discharge about 20% of the total photosynthates, including sugars, organic acids, and amino acids, to the rhizosphere, promoting the development of microorganisms. The area of about 1–3 mm from the root surface is known as the rhizosphere, and it is here that soil microbes can proliferate up to three or four orders of magnitude more than in nonvegetated soils (Schnoor & Huang, 1993). Some soil microbes co-metabolistically break down or mineralize organic pollutants such as PCBs and PAHs (Donnelly & Fletcher, 1994).

9.8 CONSTRUCTED WETLAND

CWs comprise of engineered systems that use biological processes to treat a wide array of wastewaters emanating from municipal and industrial sources. CWs utilize natural sources such as plants, soils, and associated microbial assemblages, which result in wastewater treatment through various physical, chemical, and biological processes that work individually and with each other. Helophytes, which are used in CWs, play a vital role in the treatment of wastewater. The CW is composed of one or more substrate-filled beds with a specific flow regime that promote the development of plants and microorganisms. The bed is reinforced by an impermeable layer at the bottom. The substrate actively removes contaminants by physio-chemical processes such as filtration and sedimentation of suspended particles, filtration of pathogenic organisms, and sorption of organic materials, nitrogen, phosphorus, and heavy metals.

In developing countries, there is a huge gap between wastewater generation and treatment due to which untreated wastewater is discharged into natural water bodies and lead to their contamination. To meet the increasing gap between wastewater generation and treatment, there is a strong need to reduce the load on existing wastewater treatment infrastructures. CWs, as a green technology, may be employed to treat wastewater. CWs are an inexpensive technology than conventional and advanced wastewater treatment systems, which require less operation and maintenance cost. In the past few decades, CWs, a green wastewater treatment method that mimics natural wetlands, has been used extensively to treat a variety of wastewaters, including sewage, agrarian effluent, industrial effluents, mine drainage, landfill leachate, stormwater runoff, polluted river water, and urban runoff (Vymazal, 2014). The design, development, and performance of CWs have been the subject of numerous studies recently, and it has also been suggested that CWs may be effective at removing a variety of pollutants from wastewater, including pathogens, pharmaceutical contaminants, organic matter, nutrients, and trace elements (Vymazal, 2014).

9.8.1 Merits of CW systems

It is only reliant on solar energy and is visually appealing. Environmental disturbance are kept to a minimum, and in situ treatment protects topsoil. It works best at locations with modest, minimal contamination levels but with specialized plants and process modifications its use can even be applied to industrial wastewaters. It is effective for treating a variety of environmental pollutants and is less pricey than traditional physico-chemical techniques by 60–80% or even more.

9.8.2 Demerits of CW systems

Cleaning up a location is a time-intensive procedure that could take at least several growing seasons. Animals may be harmed by plants that extract harmful heavy metals or persistent toxins, which can also pollute the food chain. Additionally, those organic and inorganic pollutants may produce intermediates that are cytotoxic to people, plants, and other creatures. The key to demonstrating the viability of phytoremediation is to conduct analyses on the mass balance and metabolic destiny of contaminants in plants.

9.8.3 Types of CW systems

CW systems may be categorized on the basis of hydrology (surface flow and subsurface flow in open water), macrophytic growth types (emergent, submerged, free-floating, and floating-leaved wetlands), and flow channel (horizontal and vertical). Several types of CWs can be merged to take use of the distinct advantages of the various systems (hybrid or mixed systems).

9.8.3.1 CW with horizontal subsurface flow

In CWs with horizontal subsurface flow (HSSF), wastewater is fed into the inlet and slowly moves through the porous medium beneath the bed in an estimated horizontal path till it approaches the outlet zone, where it is collected before being expelled via level control device at the outlet. Throughout this process, the wastewater will encounter a network of aerobic, anoxic, and anaerobic zones. The aerobic zones are located around roots and rhizomes, which add oxygen to the substrate. Both aerobic and anaerobic microbial activities, as well as sedimentation and filtering of particle organic matter, all contribute to the decomposition of organic matter in horizontal flow constructed wetlands (HF CWs). Due to severe loading and ongoing saturation of the filter bed, anoxic/anaerobic activities predominate, with aerobic processes limited to thin surface layers where diffusion of oxygen from the environment may occur and tiny zones next to rhizomes and roots (radial oxygen loss). Dissolved oxygen may also be transported by inflowing wastewater in systems with modest loads. The flocculation and settling of colloidal and super colloidal particles are one of the main removal/retention processes for SS in HF CWs. Gravity sedimentation, straining, physical capture, and adsorption on biomass film linked to gravel and root systems are additional efficient removal processes in HF systems. In HF CWS, nitrification and denitrification are the main methods of nitrogen removal. However, nitrification is restricted by the low level of oxygen in the filtration bed as a result of the filtration bed's continual submersion, hence, HF CWs are ineffective at removing ammonia. On the contrary, denitrification can occur in anoxic or anaerobic circumstances. Through sorption and precipitation, phosphorus can be eliminated. However, popular filtering materials like crushed rock or gravel do not have a large sorption capacity. The selection of materials with a superior P adsorption capacity, which depends on chemical and physical qualities, is required to improve phosphorus removal. These materials may be calcareous substances that might encourage the precipitation of Ca-phosphate or minerals having reactive Fe or Al hydroxide or oxide groups on their surfaces. Industrial waste and by-products, such as steel slags from electric arc and blast furnaces, fly ash, broken concrete, iron ore, and treated wood chips, have recently found value (Vymazal, 2005).

9.8.3.2 CWs with vertical subsurface flow

CWs with vertical flow (VSSF) consist of a flat bed of graded gravel covered in sand containing macrophytes. A big batch is intermittently supplied to VF CWs, saturating the top. After that, wastewater slowly seeps through the bed and is gathered at the base by a drainage system. The bed drains entirely without obstruction, allowing air to fill the bed. This type of dosage promotes efficient oxygen transport, which enhances the capacity for nitrification. Contrary to transfer of oxygen through the aerenchyma system of plants, airborne oxygen diffusion significantly increases the oxygenation of the filter bed. The fundamental function of vegetation in VF CWs is to support the bed's hydraulic conductivity (Vymazal, 2018).

9.8.3.3 CWs with free water surface

A typical FWS CW with sprouting macrophytes is a shallow isolated basin or series of basins with a liquid depth of 20–40 and 20–30 cm of rooting soil. A considerable portion of the surface is often covered with dense emergent vegetation, usually more than 50%. There may be naturally occurring species in addition to cultivated macrophytes. The litter provides the organic carbon required for denitrification, which can take place in anaerobic pockets inside this litter layer as plants are typically not harvested. FWS CWs are effective in removing organic materials owing to microbial decomposition and colloidal particle settling. By settling and filtering through the thick vegetation, suspended particles are successfully eliminated. Nitrification (in the water column) and denitrification (in the litter layer) are the main processes employed to remove nitrogen, as well as ammonia volatilization at higher pH levels because of algal photosynthesis. Due to the little water interaction with soil particles that adsorb and/or precipitate phosphorus, its retention is often poor. Because nutrients are returned to water during plant breakdown, plant intake only reflects temporary storage.

9.8.3.4 Hybrid CWs

Hybrid CWs with both horizontal and vertical flow both have distinct disadvantages (Cooper, 1999). BOD_5 and total suspended particles are effectively removed by horizontal flow-built wetlands, however, nitrification is not successfully removed due to a restricted oxygen transfer capacity. Since they have a significantly higher capacity for oxygen transfer and need less space to build, vertical flow-built wetland is of more importance. Yet, there are some drawbacks to vertical flow-built wetlands as well. They are less effective at removing sediments and can potentially produce clogging issues. Hybrid systems, a novel technology designed to address these issues, have been developed to lessen the issues associated with just horizontal and vertical flow.

9.8.4 Macrophyte used for CWs

There are three types of wetland plants that are used in CWs: free floating, submerged, and emergent. Researchers have reported different types of macrophytes used in CWs and their properties are presented in Table 9.1 (Kataki *et al.*, 2021).

9.8.4.1 Floating-leaved macrophytes

Plant species with roots in the substrate and leaves floating above are referred to as floating-leaved macrophytes. The typical members of this group are the waterlilies (*Nymphaea* spp.), the Indian lotus (*Nelumbo nucifera*), and the yellow waterlily (*Nuphar lutea*). Floating leafed macrophytes in CWs have a strong chance of removing suspended particles because leaves on the surface lessen the

Table 9.1 Concentrations of major pollutants removal reported in the literature for various industrial wastewater.

Location	Type of Macrophyte	Species of Macrophyte	Properties
India	Emergent, evergreen	*Cymbopogon citratus*	Antimicrobial characteristics, high COD, TSS, TN, and TP elimination.
India, Pakistan, Cameroon Nigeria	Emergent, submerged	*Pistia stratiotes*	Heavy metal accumulation, effective TN and ammonia-N removal, enhanced microbial development.
India, Colombia	Emergent perennial	*Colocasia esculenta*	Metal uptake.
India, Malaysia, Italy, Greece, Morocco, Portugal	Emergent	*Arundo donax*	High energy output, high methane generation, salt tolerance.
India, China, Thailand	Emergent, perennial	*Pennisetum purpureum*	Very high yield, low nutrient demand, high COD, TSS, TN, and TP removal.
India, China, Denmark, Vietnam	Emergent	*Canna indica*	High rate of loading, increased Fe plaque formation, high pollutant load tolerance, better TN, TP, and COD removal, as well as high capacity to absorb nutrients and high root activity in saline water.
India, Argentina, Brazil	Emergent, perennial	*Typha domingensis*	High productivity, high metal (Hg) accumulation and tolerance.
Malaysia	Perennial, evergreen	*Lepironia articulata*	High removal efficiency for SS, turbidity, available N, SS, BOD, and COD.
China	Emergent	*Acorus calamus*	Suggested for wastewater containing petroleum, good ability for bacterial enrichment, grows well in water with a high (N) load, extended duration of growth, high capacity for root aeration.
Thailand, China, Iran	Emergent	*Cyperus flabelliformis*	Excellent nutrient removal capability and rate of loading capacity.
China	Submerged, perennial	*Potamogeton crispus*	Increase the appropriate bacterial percentage (ammonia oxidizing bacteria), the ability for high oxygenation, and the efficiency of COD removal.
China	Emergent, perennial	*Zizania latifolia*	Low nutrient removal and intrusive.
China	Submerged	*Cladophora*	Removal of Mn.
China	Emergent, perennial	*Veronica serpyllifolia*	High rate of loading, good Zn, Pb, and Fe plaque formation.
China	Emergent, submerged	*Myriophyllum aquaticum*	Effective reduction of N, high discharge of organic exudates, metal tolerant.

Pakistan	Emergent, perennial	*Leptochloa fusca*	High growth rate and metal removal.
Pakistan, Vietnam	Emergent	*Brachiaria mutica*	High growth rate under high N stress make it good for COD, BOD, and Cr removal.
Thailand, China	Emergent	*Acorus calamus*	Increased O_2 release by roots and effective pollutant removal.
China, Thailand, Brazil	Emergent	*Pennisetum americanum*	High BOD, TN, and COD removal.
India, Thailand, South Africa	Emergent	*Vetiveria zizanioides*	Rapid growth and effective removal of ammonia, TN, TSS, COD, and chromium.
India, Sri Lanka, Argentina, China, Brazil	Free floating	*Eichhornia crassipes*	Excellent heavy metal accumulator (Cd, Cu, Al, productive, and highly invasive.
Throughout temperate and tropical regions	Perennial, floating	*Phragmites australis*	Highly invasive, metal tolerant, effective at removing N, P, BOD, and COD
USA	Emergent	*Scirpus pungens/ validus*	Quick colonizer, drought-tolerant, efficient pollutant removal ability
USA	Submerged	*Najas guadalupensis*	Favorable for P uptake
Portugal	Emergent	*Sarcocornia fruticosa*	Salinity-tolerant and appropriate for highly salinized effluent, such as that from tanneries
Germany, Italy	Perennial, terrestrial	*Miscanthus × giganteus*	High biomass and energy yield, low nutrient requirement, low evapo-transpiration and absorption of Fe, Cu, and Zn.
Czech Republic, New York	Perennial, emergent	*Phalaris arundinacea*	High biomass production, and effective removal of metals (Cr, Cd, Zn, Ni)

impact of wind affecting possible dispersion and water movement. The main methods for removing organics are microbial decomposition and particulate matter sedimentation. Although there is very little information available on these systems, it is probable that most of the activities take place in the water column aerobically since algae produce oxygen.

The removal of nitrogen from the water column can be done by plankton and periphytic communities clinging to the leaf peduncles and blades, which provide higher pH levels that allow for volatilization of nitrogen. The presence of algae within these systems is quite probable since the leaves of species with floating leaves frequently do not fully cover the water surface. While denitrification is unlikely to take place in the water column under aerobic circumstances, it may occur at the sediment–water interface if anoxic conditions exist. Algae may absorb phosphorus from the water column, however, this removal compartment is only temporary, as nutrients are soon washed out of decomposing algal tissue.

Roots and rhizomes of plants are capable of absorbing both nitrogen and phosphorus. According to literature, *Nuphar lutea* leaves that are floating and submerged may also absorb phosphorus. According to the season, this absorption through leaves, however, is 10–20 times less than root adsorption. The highest levels of phosphorus absorption by leaves are seen during summer.

9.8.4.2 Submerged macrophytes

Aquatic macrophytes that are submerged have all their photosynthetic tissue submerged. Nevertheless, submerged plants cannot be employed in wastewater that contains a lot of easily biodegradable organic matter since the microbial process that breaks down the organic matter would result in anoxic circumstances (Brix, 1997). Subsurface vegetation only thrives in oxygenated water. There is no concrete proof that the distribution of submerged plants is impacted by sudden reduction in oxygen content. But prolonged exposure to low oxygen levels can retard the development of submerged macrophytes (Shelef *et al.*, 2013).

9.8.4.3 Emergent macrophytes

Aquatic plants with growing sections that emerge above the water's surface are known as emergent aquatic macrophytes. Aquatic macrophytes are believed to be especially fruitful, with their roots in sediments beneath water and their photosynthetic components in the air, making the greatest use of all three potential states (Westlake, 1963). Complicated development usually takes place above the water's surface and is inaccessible to many aquatic creatures, emergent macrophytes are generally thought to be less structurally complex. It is hypothesized that all emergent macrophytes provide homogenous habitat and perform a comparable role for many trophic levels due to the basic structural intricacy of these plants.

9.8.5 Microbiology in CWs

When organics and nitrogen are removed from wastewater in CW's, microorganisms are very important. The microbial assemblages have not been thoroughly investigated, despite playing crucial roles. Yet, the development of new, cutting-edge methodologies has made it possible to identify the microbial communities that are in charge of the biodegradation of particular contaminants. Bernardes *et al.* (2019) found an incredibly high variety of bacteria and archaea. The makeup of wastewater also affects the microbial community. In a system made up of a moving-bed biofilm reactor and a built wetland on top, Lai *et al.* (2021) studied microbial assemblages. The predominant genera detected in a created wetland, were *Gemnobacter* and *Parasegetibacter*, which are heterotrophic bacteria engaged in the breakdown of organic materials. The most common genera of denitrifiers included *Hydrogenophaga*, *Inhella*, *Methyloversatilis*, *Azo-spira*, and *Dechloromonas*, whereas *Hyphomicrobium* was associated with the nitrification process.

Many studies have established that the presence of substances can influence the existence of particular microorganisms. Organo-phosphate flame retardants (OPFRs) were shown to have a noticeable impact on the richness, variety, and nitrogen-related microorganisms in the rhizosphere (Liu *et al.*, 2020). Importantly, *Pseudomonas* and *Sphingobium* genera may serve as the necessary microorganisms for the biodegradation of OPFRs. Dibutyl phthalate, reduced the variety of the bacterial and fungal population. Liu *et al.* (2020) suggested that the addition of biochar boosted the abundance of *Gemnobacter* and denitrifiers like *Hydrogenophaga*.

The filtering material is another element influencing the bacterial diversity in built wetlands. When intertidal wetland sediments were added to VF CW, anaerobic decomposition primarily by the taxa Anaero-lineae, Desulfobacterales, and Desulfuromonadales improved the treatment efficacy. Also, it has been discovered that ethylene production from *Dehalococcoides* triggered by intertidal wetland sediment enhanced plant stress tolerance (Wang *et al.*, 2019). Lu *et al.* (2019) reported a significant change in the composition of the bacterial population during the studies on the arsenic removal and fluoride in a hybrid system made up of a sorption unit filled with pebbles and activated alumina and VF CW filled with coal cinder.

Currently, there are several studies conducted on the establishment of microbial assemblages in CWs, and it appears to be very difficult to pinpoint the single most crucial feature. The findings showed that there are numerous factors, including wastewater composition, the presence of specific compounds in the wastewater, the composition of the filtration material, the impact of different macrophyte species, the type of CW, oxygenation level, including artificial aeration, and seasonal variation. Nevertheless, almost all the research carried out so far have used artificial wastewater and have been conducted in laboratories or small-scale outdoor systems. Very little research has been conducted on fully created wetlands, and there is little information available regarding how microbial communities have changed over the years.

9.8.6 Design of CWs

The principal design parameters of CWs include the topography, soil permeability, substrate, hydraulic retention time, basin geometry, BOD_5 loading rate, hydraulic loading rate, vegetation, and so on.

One of the most crucial elements in the design of a wetland is topography (Lee *et al.*, 2009; Liu *et al.*, 2015). The best place to build a wetland is a low-level region where water flows naturally due to gravity. The landform concerns, such as shape, size, and orientation, are determined by the direction of the prevailing wind. While a gentle slope at the design site is preferred, artificial wetlands can be erected anyplace that satisfies the design requirements. With the approval of the undertaking authority, converted lands like agricultural sites could also be utilized for the development of wetland since they naturally hold all the designing requirements needed for wetland construction.

Another crucial parameter is the soil permeability and kind of substrate present. They determine the effectiveness of a CW. The accompanying soil permeability must be taken into account when choosing any site for the development of wetlands. For design purposes, a soil permeability of between 7 and 10 m/s is ideal. The wetlands' permeability can be maximized by adding a restrictive layer to the soil profile with reduced permeability since sandy soils are too permeable to sustain wetland plants. Small wastewater facilities may also be built on high-permeable soils by digging narrow trenches, but this requires that the ditches first be lined with clay. Small wastewater treatment facilities can also be built on highly permeable soils by digging shallow trenches first and filling them with clay or another synthetic liner.

The purpose of substrates in a wetland is to support the vegetation and to give room for all biochemical and chemical processes necessary for effective and smooth operation (Rana & Maiti, 2018). Substrates also serve as a place to store pollutants throughout the treatment process and to remove them. Natural substances including soil, sand, gravel, and broken boulders are frequently utilized as substrates for CWs. Organic compounds are another option. Natural substrate materials such as limestone, fly ash, slag, and coal cinder can be replaced with artificial substrates or industrial by-products. Natural substrates for subsurface flow CWs may be substituted in developing nations or regions with warm climates by lightweight aggregate, activated carbon, calcium silicate, or compost.

As it provides optimal permeability, hydraulic loading, and organic loading, all of which are crucial for the construction of wetlands. Coarse sand is chosen as the most ideal substrate.

One of the crucial design factors affecting how effectively pollutants and nutrients are removed from CWs is hydraulic retention time (HRT). Calculating nominal HRT, which is defined as the proportion of notional wetland volume to flow rate at outflow, is the simplest approach for estimating HRT (Kadlec & Wallace, 2008). The nominal HRT is predicated on 'plug-flow' reactors (PFR) as not all wetland volumes are included in the flow route. In fact, several researchers discovered that wetlands have three distinct hydraulic volumes or zones. The dynamic primary channel is the first zone; the temporary storage zone, is the second zone, where water and other elements are interchanged with the main flow channel; and the entirely isolated, 'dead' water is the third zone (Aleissa & Bakshi, 2021; Zahraeifard & Deng, 2011). Field tracer experiments and tracer test data are frequently used in conjunction with the moment technique to estimate HRT in conjunction to the empirical approach (Jadhav & Buchberger, 1995). The benefit of this method is that it enables the production of a residence time distribution (RTD), which in turn enables the calculation of a wetland's mean HRT. The main drawback of this strategy is that HRT is required during the design phase of a built wetland, whereas tracer tests can only be carried out in an existing wetland.

$$t = \frac{LWnd}{Q}$$

where L is the length of the CW in meters; W is the width of the CW in meters; Q is the average flowrate through the system in m^3/d; n is the fraction of cross-sectional area not occupied by plants and d is the depth of the basin in meters.

The required cross-sectional area is calculated using the following equation for subsurface flow:

$$A_c = \frac{Q}{K_s * S}$$

$K_s * S$ is the flow velocity and is to be limited to 6.8 m/d to reduce localized shearing of bacterial films.

The required width of the system is the function of the cross-sectional area and design depth and is calculated as

$$W = \frac{A_c}{d}$$

The BOD removal in CWs follows the first order, plug-flow kinetics (Vymazal, 2018) and hence can be computed as

$$C_t = C_0 e^{-kt}$$

Since t is a function of bed area (A), we can also write

$$A = \frac{Q * \left(ln\left(C_0 - C_t\right)\right)}{K_{BOD}}$$

where A is the bed area in m^2; Q is the average flow in m^3/day; C_0 is the inlet 5-day BOD_5 in mg/L; C_t is the outlet BOD_5 in mg/L; K_{BOD} is the BOD_5 reaction constant per day.

9.8.7 Use of CWs for the treatment of various industrial wastewater

In perspective of the urgent need to concentrate on the water–energy–environment via environmentally friendly and cost-effective treatment technologies, CW applications have recently been shown to have relatively high pollutant removal percentage and ecological advantages in both lab and field settings.

CWs may be used in a number of environments, from dry to tropical to cold; for water with a variety of contaminants in it; and for a broad range of pollutants.

Recently, treatment of industrial wastewater using CWs has received more attention. Treatment of effluents from various industries, including distilleries, cork-boiling operations, petro-chemical plants, coffee, potato, and seafood processors, as well as wastewater from dairies, slaughter houses, steel mills, coke-ovens, pulp and paper mills, oil fields, floriculture, textile, and agro-food effluents, using CW has been found to be satisfactory. Because of the different inherent underlying mechanisms, such as phytodegradation, rhizofiltration, microbial degradation, sorption, and so on, CWs could successfully remove the recalcitrant organic fraction of the wastewater (Jain *et al.*, 2020). Chang *et al.* (2022) worked on the treatment of high metal-bearing and acidic wastewater, also known as 'acid mine drainage' and found that the biochar substrate had the greatest proportional concentration of sulfate-reducing bacteria (SRB), followed by walnut shells. The biotic acid mine drainage therapy in the CW was made possible by the collaboration of organic-degrading bacteria, primarily *Cellulomonas*, *Clostridium*, and *Bacteroides*, and SRB groups. For CW systems addressing acid mine drainage, a combination of organic solid refuse and biochar is a better filler option; however, more research is needed to determine its capability under various conditions (Chang *et al.*, 2022).

Industries have different discharge of various pollutants in their effluents depending on the material inflow and outflow, the adopted unit operations, treatment efficiency, and so on. Table 9.2 gives a brief overview for some of the pollutants being discharged in the effluents by different industries in their waste water.

CWs treatment strategies are utilized effectively to treat a variety of wastewaters with varied organic or inorganic loadings. Many countries around the world, including China, India, Australia, Finland, United States, and so on use CW's technology on a large scale. In recent years, the use of CWs has been expanding to include additional types of wastewaters from industrial, agricultural, and agro-industrial sources (Feng *et al.*, 2020; Gomes *et al.*, 2018; Saeed *et al.*, 2018; Tunçsiper *et al.*, 2015). However, the efficiency of treatment may vary at times with respect to the type of wastewater and type of plant species used in CW system. Table 9.3 gives the basic outline of the removal efficiency observed along with the treatment mechanism employed for the CW system.

Table 9.2 Concentrations of major pollutants removal reported in the literature for various industrial wastewater.

Type of Industrial Wastewater	TSS (mg/L)	TKN (mg/L)	COD (mg/L)	BOD_5 (mg/L)	Phenol (mg/L)	TP (mg/L)	Reference(s)
Pulp and paper	100–23 300	11–600	500–40 000	100–13 300	–	0.02–36	Vymazal (2014)
Distillery	1000–17 000	1000–17 000	4000–212 000	6000–65 000	35–10 000	740–2500	Chandra *et al.* (2008), Mohana *et al.* (2009)
Tannery effluents	1500–42 500	–	2000–23 000	800–4000	–	8–40	Espinoza-Quiñones *et al.* (2009), Mannucci *et al.* (2010)
Sugar mill	350	53	3500–10 000	4000–7000	–	4.8	Güven *et al.* (2009)
Soft drink	140–5000	54	1400–33 000	770	–	2.5	Oktay *et al.* (2007)
Pharmaceuticals	2500	–	5000–12 000	1300–6400	–	–	Vymazal (2014)
Textile	25–24 500	–	220–31 300	75–6300	–	–	dos Santos *et al.* (2007), Sharma *et al.* (2007)

Table 9.3 Efficiency of CW system treating different sources wastewater including species of macrophyte and treatment system adopted.

Type of Wastewater	Plant Species	Type of CW	Pollutant Removal Efficiency	Reference
Activated sludge effluent	*Phragmites australius*	Horizontal subsurface flow	BOD: 89%, TSS:81%, FC: 99%, TC: 97%,	Ahmed *et al.* (2008)
Domestic	*Phragmites carca* *Typha latifolia*	Horizontal subsurface flow	N: 58–63%, BOD: 67–90% Higher N was removed by *P. carca*	Juwarkar *et al.* (1995)
Municipal	*Canna, Phragmites,* and *Cyperus* sp.	Horizontal subsurface flow Vertical subsurface flow	BOD: 93%, TSS: 92%, COD: 91%, BOD: 94%, TSS: 94%, COD: 93%,	Abou-Elela *et al.* (2013)
	Phragmites australis	Hybrid	COD:91%, BOD5: 93%, TSS: 95%, TN: 67% and TP:62%	Elfanssi *et al.* (2018)
	Phragmites australis and Cyperus papyrus	Vertical subsurface flow	RE of *C. papyrus* higher (COD:70%, BOD: 81%, TP: 50%, ammonia-N: 70%, FC: 95%, TC: 98%,	García-Ávila *et al.* (2019)
Paper and pulp	*Typha domingensis, Phragmites mauritianus, Cyperus immensus, and Cyperus papyrus*	Subsurface flow	N: 49–75% in planted CW and 42–49% in unplanted CW; phenol: 60–77% (higher at lower HRT); P: 4–38% in the unplanted CW and 30–60% in planted CW	Abira *et al.* (2005)
Distillery	*Typha latifolia, Phragmites karka*	Horizontal	BOD: 84%, COD: 64%, TP: 79%, TN: 59%	Billore *et al.* (2001)
Tannery	*Phragmites* sp.	Horizontal subsurface flow	COD: 82%, NH_4^+-N: 96%	Ramírez *et al.* (2019)
Textile	*Brachiaria mutica*	Vertical	Color: 74%, TN: 84%, BOD: 72%, COD: 81%, TSS: 32%, TP: 79% Cd: 72%, Cr: 97%, Ni: 88%, Fe: 89%	Hussain *et al.* (2018)

9.9 CONCLUSION

As an alternative to current chemical-intensive treatment systems consuming surplus amounts of energy and resources, CW is a sustainable and economically advantageous technology. In virtue to being a potential net energy producer with remarkably low operational costs, CW is also capable of providing a range of ecosystem services and lowering its carbon impact. The development of several applications has made it clear that more work has to be conducted to increase the CW system's efficiency through a broader set of design criteria, including component change or upgrade. This review summarizes some relatively unexplored issues relating to improving vital wetland elements (macrophytes, substrates, and biofilm), which would improve system efficiency for removing various contaminants and the potential for expanding the technology's application to effluent emanating from various industries.

The goal of this study is to explain how the interface structure and synergetic effects of these components may be functionalized and developed logically to aid advance CW efficiency. Additionally, the main difficulties, possibilities, and options for future growth of these elements are highlighted. It is well known that the growth of rhizosphere manipulation/management methods would enable the interaction of macrophyte–biofilm–substrate, which in turn would enhance overall efficiency and implementations of the technology to a wide range of wastewater, by appropriately designing and tying in every individual component.

Owing to the future of the CW's there can be various aspects on which further thought can be put into for macrophyte utilization of genetically modified plants, using hyperaccumulators, phyto mining. For the influent wastewater that has already been given a pre-treatment can be used. Substrate modifications can also be done as in using substrate from organic carbon source, utilization of hybrid substrate that can target multiple pollutants and focus on reusability and recyclability of substrate. Biofilm development can be augmented furthermore and designer microbes developed in the laboratory can also be utilized.

REFERENCES

Abira M. A., van Bruggen J. J. A. and Denny P. (2005). Potential of a tropical subsurface constructed wetland to remove phenol from pre-treated pulp and papermill wastewater. *Water Science and Technology*, **51**(9), 173–176, https://doi.org/10.2166/wst.2005.0312

Abou-Elela S. I., Golinielli G., Abou-Taleb E. M. and Hellal M. S. (2013). Municipal wastewater treatment in horizontal and vertical flows constructed wetlands. *Ecological Engineering*, **61**, 460–468, https://doi.org/10.1016/j.ecoleng.2013.10.010

Ahmed S., Popov V. and Trevedi R. C. (2008). Constructed wetland as tertiary treatment for municipal wastewater. *Proceedings of Institution of Civil Engineers: Waste and Resource Management*, **161**(2), 77–84, https://doi.org/10.1680/warm.2008.161.2.77

Aleissa Y. M. and Bakshi B. R. (2021). Constructed wetlands as unit operations in chemical process design: benefits and simulation. *Computers and Chemical Engineering*, **153**, 107454, https://doi.org/10.1016/j.compchemeng.2021.107454

Bernardes F. S., Herrera P. G., Chiquito G. M., Morales M. F., Castro A. P. and Paulo P. L. (2019). Relationship between microbial community and environmental conditions in a constructed wetland system treating greywater. *Ecological Engineering*, **139**, 105581, https://doi.org/10.1016/j.ecoleng.2019.105581

Billore S. K., Singh N., Ram H. K., Sharma J. K., Singh V. P., Nelson R. M. and Dass P. (2001). Treatment of a molasses based distillery effluent in a constructed wetland in central India. *Water Science and Technology*, **44**(11–12), 441–448, https://doi.org/10.2166/wst.2001.0864

Brix H. (1997). Do macrophytes play a role in constructed treatment wetlands? *Water Science and Technology*, **35**(5), 11–17, https://doi.org/10.1016/S0273-1223(97)00047-4

Campbell R. and Greaves M. P. (1990). Anatomy and community structure of the rhizosphere. In: The rhizosphere, Wiley, Lynch, J.M. (ed.), Chichester, United Kingdom, pp. 11–34.

Chandra R., Yadav S., Bharagava R. N. and Murthy R. C. (2008). Bacterial pretreatment enhances removal of heavy metals during treatment of post-methanated distillery effluent by *Typha angustata* L. *Journal of Environmental Management*, **88**(4), 1016–1024, https://doi.org/10.1016/j.jenvman.2007.05.001

Chang J., Deng S., Li X., Li Y., Chen J. and Duan C. (2022). Effective treatment of acid mine drainage by constructed wetland column: coupling walnut shell and its biochar product as the substrates. *Journal of Water Process Engineering*, **49**(August), 103116, https://doi.org/10.1016/j.jwpe.2022.103116

Cooper P. (1999). A review of the design and performance of vertical-flow and hybrid reed bed treatment systems. *Water Science and Technology*, **40**(3), 1–9, https://doi.org/10.1016/S0273-1223(99)00414-X

Donnelly P.K. and Fletcher J. S. (1994). Potential use of mycorrhizal fungi as bioremediation agents. In: Bioremediation Through Rhizosphere Technology, Anderson, T.A. (ed.), pp. 93–99. American Chemical Society. Symposium Series 563. ACS Chicago, Illinois, USA, https://doi.org/10.1021/bk-1994-0563.ch008

dos Santos A. B., Cervantes F. J. and van Lier J. B. (2007). Review paper on current technologies for decolourisation of textile wastewaters: perspectives for anaerobic biotechnology. *Bioresource Technology*, **98**(12), 2369–2385, https://doi.org/10.1016/j.biortech.2006.11.013

Elfanssi S., Ouazzani N., Latrach L., Hejjaj A. and Mandi L. (2018). Phytoremediation of domestic wastewater using a hybrid constructed wetland in mountainous rural area. *International Journal of Phytoremediation*, **20**(1), 75–87, https://doi.org/10.1080/15226514.2017.1337067

Espinoza-Quiñones F. R., Fornari M. M. T., Módenes A. N., Palácio S. M., da Silva F. G., Szymanski N., Kroumov A. D. and Trigueros D. E. G. (2009). Pollutant removal from tannery effluent by electrocoagulation. *Chemical Engineering Journal*, **151**(1–3), 59–65, https://doi.org/10.1016/j.cej.2009.01.043

Feng L., Wang R., Jia L. and Wu H. (2020). Can biochar application improve nitrogen removal in constructed wetlands for treating anaerobically-digested swine wastewater? *Chemical Engineering Journal*, **379**, 122273, https://doi.org/10.1016/j.cej.2019.122273

García-Ávila F., Patiño-Chávez J., Zhinín-Chimbo F., Donoso-Moscoso S., Flores del Pino L. and Avilés-Añazco A. (2019). Performance of *Phragmites australis* and *Cyperus papyrus* in the treatment of municipal wastewater by vertical flow subsurface constructed wetlands. *International Soil and Water Conservation Research*, **7**(3), 286–296, https://doi.org/10.1016/j.iswcr.2019.04.001

Gomes A. C., Silva L., Albuquerque A., Simões R. and Stefanakis A. I. (2018). Investigation of lab-scale horizontal subsurface flow constructed wetlands treating industrial cork boiling wastewater. *Chemosphere*, **207**, 430–439, https://doi.org/10.1016/j.chemosphere.2018.05.123

Güven G., Perendeci A. and Tanyolaç A. (2009). Electrochemical treatment of simulated beet sugar factory wastewater. *Chemical Engineering Journal*, **151**(1–3), 149–159, https://doi.org/10.1016/j.cej.2009.02.008

Hussain Z., Arslan M., Malik M. H., Mohsin M., Iqbal S. and Afzal M. (2018). Integrated perspectives on the use of bacterial endophytes in horizontal flow constructed wetlands for the treatment of liquid textile effluent: phytoremediation advances in the field. *Journal of Environmental Management*, **224**(May), 387–395, https://doi.org/10.1016/j.jenvman.2018.07.057

Jadhav R. S. and Buchberger S. G. (1995). Effects of vegetation on flow through free water surface wetlands. *Ecological Engineering*, **5**(4), 481–496, https://doi.org/10.1016/0925-8574(95)00039-9

Jain M., Majumder A., Ghosal P. S. and Gupta A. K. (2020). A review on treatment of petroleum refinery and petrochemical plant wastewater: a special emphasis on constructed wetlands. *Journal of Environmental Management*, **272**(May), 111057, https://doi.org/10.1016/j.jenvman.2020.111057

Juwarkar A. S., Oke B., Juwarkar A. and Patnaik S. M. (1995). Domestic wastewater treatment through constructed wetland in India. *Water Science and Technology*, **32**(3), 291–294, https://doi.org/10.1016/0273-1223(95)00637-0

Kadlec R. H. and Wallace S. (2008). Treatment Wetlands, https://doi.org/10.1201/9781420012514

Kataki S., Chatterjee S., Vairale M. G., Dwivedi S. K. and Gupta D. K. (2021). Constructed wetland, an eco-technology for wastewater treatment: a review on types of wastewater treated and components of the technology (macrophyte, biolfilm and substrate). *Journal of Environmental Management*, **283**, 111986, https://doi.org/10.1016/j.jenvman.2021.111986

Lai C., Sun Y., Guo Y., Cai Q. and Yang P. (2021). A novel integrated bio-reactor of moving bed and constructed wetland (MBCW) for domestic wastewater treatment and its microbial community diversity. *Environmental Technology (United Kingdom)*, **42**(17), 2653–2668, https://doi.org/10.1080/09593330.2019.1709904

Lee C. G., Fletcher T. D. and Sun G. (2009). Nitrogen removal in constructed wetland systems. *Engineering in Life Sciences*, **9**(1), 11–22, https://doi.org/10.1002/elsc.200800049

Liu R., Zhao Y., Doherty L., Hu Y. and Hao X. (2015). A review of incorporation of constructed wetland with other treatment processes. *Chemical Engineering Journal*, **279**, 220–230, https://doi.org/10.1016/j.cej.2015.05.023

Liu T., Lu S., Wang R., Xu S., Qin P. and Gao Y. (2020). Behavior of selected organophosphate flame retardants (OPFRs) and their influence on rhizospheric microorganisms after short-term exposure in integrated vertical-flow constructed wetlands (IVCWs). *Science of the Total Environment*, **710**, 136403, https://doi.org/10.1016/j.scitotenv.2019.136403

Lu H., Li J., Liu X., Yu Z. and Liu R. (2019). Removal of fluoride and arsenic by a hybrid constructed wetland system. *Chemistry and Biodiversity*, **16**(7), 1–8, https://doi.org/10.1002/cbdv.201900078

Mannucci A., Munz G., Mori G. and Lubello C. (2010). Anaerobic treatment of vegetable tannery wastewaters: a review. *Desalination*, **264**(1–2), 1–8, https://doi.org/10.1016/j.desal.2010.07.021

Mohana S., Acharya B. K. and Madamwar D. (2009). Distillery spent wash: treatment technologies and potential applications. *Journal of Hazardous Materials*, **163**(1), 12–25, https://doi.org/10.1016/j.jhazmat.2008.06.079

Oktay S., Iskender G., Germirli Babuna F., Kutluay G. and Orhon D. (2007). Improving the wastewater management for a beverage industry with in-plant control. *Desalination*, **211**(1–3), 138–143, https://doi.org/10.1016/j.desal.2006.02.088

Ramírez S., Torrealba G., Lameda-Cuicas E., Molina-Quintero L., Stefanakis A. I. and Pire-Sierra M. C. (2019). Investigation of pilot-scale constructed wetlands treating simulated pre-treated tannery wastewater under tropical climate. *Chemosphere*, **234**, 496–504, https://doi.org/10.1016/j.chemosphere.2019.06.081

Rana V. and Maiti S. K. (2018). Municipal wastewater treatment potential and metal accumulation strategies of *Colocasia esculenta* (L.) Schott and *Typha latifolia* L. in a constructed wetland. *Environmental Monitoring and Assessment*, **190**(6), 328, https://doi.org/10.1007/s10661-018-6705-4

Saeed T., Muntaha S., Rashid M., Sun G. and Hasnat A. (2018). Industrial wastewater treatment in constructed wetlands packed with construction materials and agricultural by-products. *Journal of Cleaner Production*, **189**, 442–453, https://doi.org/10.1016/j.jclepro.2018.04.115

Schnoor J. L. and Huang W. (1993). Beneficial effects of plants in the remediation of soil and groundwater contaminated with organic materials. *Critical Reviews in Environmental Science and Technology*, **23**(1), 41–77, https://doi.org/10.1080/10643389309388441

Sharma K. P., Sharma S., Sharma S., Singh P. K., Kumar S., Grover R. and Sharma P. K. (2007). A comparative study on characterization of textile wastewaters (untreated and treated) toxicity by chemical and biological tests. *Chemosphere*, **69**(1), 48–54, https://doi.org/10.1016/j.chemosphere.2007.04.086

Shelef O., Gross A. and Rachmilevitch S. (2013). Role of plants in a constructed wetland: current and new perspectives. *Water (Switzerland)*, **5**(2), 405–419, https://doi.org/10.3390/w5020405

Tunçsiper B., Drizo A. and Twohig E. (2015). Constructed wetlands as a potential management practice for cold climate dairy effluent treatment – VT, USA. *Catena*, **135**, 184–192, https://doi.org/10.1016/j.catena.2015.07.028

Vymazal J. (2005). Horizontal sub-surface flow and hybrid constructed wetlands systems for wastewater treatment. *Ecological Engineering*, **25**(5), 478–490, https://doi.org/10.1016/j.ecoleng.2005.07.010

Vymazal J. (2014). Constructed wetlands for treatment of industrial wastewaters: a review. *Ecological Engineering*, **73**, 724–751, https://doi.org/10.1016/j.ecoleng.2014.09.034

Vymazal J. (2018). Constructed Wetlands for Wastewater Treatment. Encyclopedia of Ecology (2nd Edition), **45**, 14–21, https://doi.org/10.1016/B978-0-12-409548-9.11238-2

Wang Q., Cao Z., Liu Q., Zhang J., Hu Y., Zhang J., Xu W., Kong Q., Yuan X. and Chen Q. F. (2019). Enhancement of COD removal in constructed wetlands treating saline wastewater: intertidal wetland sediment as a novel inoculation. *Journal of Environmental Management*, **249**(August), 109398, https://doi.org/10.1016/j.jenvman.2019.109398

Westlake D. F. (1963). Comparisons of plants productivity. *Biological Reviews*, **38**, 385–425, https://doi.org/10.1111/j.1469-185X.1963.tb00788.x

Zahraeifard V. and Deng Z. (2011). Hydraulic residence time computation for constructed wetland design. *Ecological Engineering*, **37**(12), 2087–2091, https://doi.org/10.1016/j.ecoleng.2011.08.011

doi: 10.2166/9781789063783_0197

Chapter 10

Myconanoremediation of various environmental toxicants: challenges and future perspectives

N. Prabhu[1]*, A. Archana*[1]*, G. Someshwari*[1]*, G. Priyanka*[1]*, S. Livithra*[1]*, S. Chozhavendhan*[1] *and T. Gajendran*[2,*]

[1]*Department of Biotechnology, Vivekanandha College of Engineering for Women (Autonomous), Tiruchengode 637-207, Namakkal, Tamilnadu, India*
[2]*Department of Biotechnology, Bharathidasan Institute of Technology, Anna University, Tiruchirappalli 620024, Tamilnadu, India*
**Corresponding author: gajendran01@gmail.com*

ABSTRACT

The application of nanotechnology in the area of environment has been studied extensively. With the rise in environmental challenges, one of biotechnology's primary concerns is the remediation of the environment. This technique can be called bioremediation and it works to decontaminate the environment by using microorganisms like fungi or their enzymes to break down hazardous compounds into non-hazardous ones. The combination of various particle and organism types, including fungi along with nanotechnology, can offer ecologically responsible alternatives for bioremediation. Nanoparticles are one of the varieties of nanomaterials and have found extensive use in a wide range of applications. Iron oxide nanoparticles were considered a good candidate for nanoparticles that is appropriate for usage in biosystems because of their chemical stability and low toxicity. As a result, components like bioactive nanoparticles, nanocomposites, and nanocatalysts have been used more in the creation of procedures and techniques for cleaning up an environment contaminated with chemical waste.

Keywords: nanosensor, nanoremediation, fungi, nanoencapsulation, nanocatalyst

10.1 INTRODUCTION

Environmental pollution and natural contamination are key issues in society. Innovations are reconnoitered steadily for the remediation of pollutants in the air, water, and soil (Masciangoli *et al.*, 2003). Particulate matter, herbicides, weight metals, pesticides, manures, oil slicks, harmful gases, sewage effluents, and natural mixtures are some of the instances of the many concerning impurities (Vaseashta *et al.*, 2007). As the altered form of environmental contamination can be trying because of the difficulty of the combination of various mixtures, high instability, and low reactivity, late investigations have zeroed in on the utilization of nanomaterial and the advancement of new natural remediation innovation.

Numerous studies and applications of nanotechnology are being made in the field of the environment. Together with other fields of knowledge, this 'new' scientific field can significantly advance environmental research and application. The advancement and significant growth of nanotechnology will enhance human well-being and contribute to environmental preservation (Kango *et al.*, 2013). The basic sciences, engineering, and their corresponding applications have been revolutionized by this science, which has developed into a multidisciplinary discipline (Lodhia *et al.*, 2010). The decontamination of environments is one of biotechnology's primary concerns in light of the growing environmental issues. This process is known as bioremediation and this process uses microorganisms and their enzymes to break down toxic substances into non-toxic ones to decontaminate the environment.

Nanoparticles are created due to their normal systems for the detoxification of metallic particles through biosystems (Vijay Pradhap Singh *et al.*, 2021). In any case, such cycles are frequently not achievable because of the cost of production, harmful muck synthesis, and the requirement for a persistent contribution of synthetic compounds (Han *et al.*, 2006). In this circumstance, metal-poisoning bioremediations have arisen as an innovation that welcomes nature. A few microorganisms are powerful that eliminate metal particles through biosorption instruments (Volesky, 2022). As a result, they can adapt to grow in abnormal pH, temperature, and nutrient availability conditions and at high heavy metal fixation levels (Anand *et al.*, 2006). Additionally, important to reduction are the infectious microbial structures. A practical strategy for both the biosynthesis of nanoparticles and contemporary bioremediation is bioreduction, one of the crucial cycles of biosynthesis (Malik, 2004; Srivastava *et al.*, 2013).

10.2 SYNTHESIS OF MYCONANOPARTICLES

Each extracellular active organism produces metal nanoparticles. There are numerous studies on fungi that can produce nanoparticles both intracellularly and extracellularly. In intracellular active organisms, the fungal biomass reacts with metals intracellularly and the desired nanoparticles are produced. Whereas in extracellular synthesis the enzyme in the cell wall of the microbes reduces the metal ions and aggregates to produce the nanoparticles. Therefore, extracellular synthesis is mostly preferred (Yadav *et al.*, 2015). When the fungi were exposed to metal ions during extracellular nanoparticle production, the result has been unused enzymes and the generation of very stable myconanoparticles (Kashyap *et al.*, 2013). The fungus species was made to treat extracellularly and the gold nanoparticles have been produced (Du *et al.*, 2011). The extracellular synthesis started after the resolution of $AuCl_4$ was incubated with fungal biomass for 8 h, whereas the golden nanoparticles were obtained when $AuCl_4$ ions reacted with the fungal cell deposit over a period of time. A living thing's biosynthesis may be a time-limited problem that depends on in vivo cell synthesis because there are two fundamentally different extracellular syntheses, namely fast synthesis and slow synthesis (Moghaddam, 2010). Apart from the two processes, an additional process is added to produce the uncontaminated nanoparticles. Hence, the extracellular production process is significantly more favorable than the intracellular method (Kuber *et al.*, 2006).

The extracellular and intracellular fungal nanoparticle synthases have been reported by various researchers (Table 10.1). Fatima *et al.* (2016) stated that the 50 nm silver nanoparticle produced from *Aspergillus flavus* has significant antibacterial activity against pathogenic bacteria and fungi. Myconanoparticles are said to be microbiological agents on agricultural grounds. Ingle *et al.* (2009) reported that *Fusarium solani* is a unique biohazard that has been used in the fusion of silver nanoparticles in a separate investigation.

10.3 BIOSYNTHESIS OF NANOPARTICLES USING FILAMENTOUS FUNGI

Fungi are heterotrophic, eukaryotic cells that could not produce chlorophyll but can produce spores (Veglio & Beolchini, 1997). The fungus is capable of absorbing and accumulating the metals through

Table 10.1. List of fungal synthesis of nanoparticles.

Fungal Species	Nano Particles	Synthesis	Reference
Mucorhils	Silver	Intracellular	Aziz *et al.* (2016)
Aspergills KF934407)	Silver	Extracellular	Fatima *et al.* (2016)
Coriolusr	Silver	Intracellular Extracellular	Sanghi and Verma (2009)
Fusarium oxysporm	Platinum	Extracellular	Syed and Ahmad (2012)
Aspergillus terreus	Zinc oxide	Extracellular	Baskar *et al.* (2013)
Fusarium solani	Silver	Extracellular	Ingle *et al.* (2009)
Saccharomyces cerevisiae	Cds	Extracellular	Prasad and Jha (2010)
Alternaria alternata	Silver	Extracellular	Sarkar *et al.* (2011)
Aspergillus clavatus	Silver	Extracellular	Verma *et al.* (2010)
Aspergillus flavus	Silver	Intracellular	Vala *et al.* (2014)
Aspergillus fumigatus	Silver	Extracellular	Ranjbar-Navazi *et al.* (2010)
Aspergillus fumigatus	Silver	Extracellular	Bhainsa and D'souza (2006)
Aspergillus niger	Silver	Extracellular	Gade *et al.* (2008)
Aspergillus niger	Silver	Extracellular	Kathiresan *et al.* (2010)
Aspergillus tamarii	Silver	Extracellular	Kumar *et al.* (2012)
Aspergillus terreus	Silver	Extracellular	Li *et al.* (2011)
Aspergillus species	Zinc	Extracellular	Pavani *et al.* (2012)
Cladosporium cladosporioides	Silver	Extracellular	Balaji *et al.* (2009)
Coriolus versicolor	Cadmium sulfide	Extracellular	Sanghi and Verma (2009)
Fusarium acuminatum	Silver	Extracellular	Ingle *et al.* (2008)
Fusarium semitectum	Silver	Extracellular	Basavaraja *et al.* (2008)
Fusarium solani	Silver	Extracellular	Ingle *et al.* (2009)
Fusarium oxysporum	Silver	Extracellular	Ahmad *et al.* (2003)
Fusarium oxysporum	Zirconia	Extracellular	Bansal *et al.* (2004)
Fusarium oxysporum	Silica/titanium	Extracellular	Bansal *et al.* (2005)
Fusarium oxysporum	Magnetite	Extracellular	Bharde *et al.* (2006)
Fusarium oxysporum	Gold	Extracellular	Mukherjee *et al.* (2002)
Fusarium oxysporum f.sp. *lycopersici*	Platinum	Intra-/extracellular	Riddin *et al.* (2006)
Hormoconis resinae	Gold	Extracellular	Varshney *et al.* (2009)
Helminthosporium tetramera	Silver	Extracellular	Shelar and Chavan (2014)
Mucor hiemalis	Silver	Extracellular	Aziz *et al.* (2016)
Neurospora crassa	Gold, silver	Extracellular	Castro-Longoria *et al.* (2011)
Nigrospora oryzae	Gold	Extracellular	Kar *et al.* (2014)
Penicillium brevicompactum	Gold	Extracellular	Mishra *et al.* (2011)
Penicillium citrinum	Silver	Extracellular	Honary *et al.* (2013)
Penicillium citrinum	Copper oxide	Extracellular	Honary *et al.* (2012)
Penicillium fellutanum	Silver	Extracellular	Kathiresan *et al.* (2009)
Penicillium nalgiovense	Silver	Extracellular	Maliszewska *et al.* (2014)
Penicillium purpurogenum NPMF	Silver	Extracellular	Nayak *et al.* (2011)
Penicillium glabrum	Silver	Extracellular	Nanda and Majeed (2014)

(*Continued*)

Table 10.1. List of fungal synthesis of nanoparticles. *(Continued)*

Fungal Species	Nano Particles	Synthesis	Reference
Phanerochaete chrysosporium	Silver	Extracellular	Vigneshwaran *et al.* (2006)
Phoma glomerata	Silver	Extracellular	Birla *et al.* (2009)
Phoma gardenia	Silver	Extracellular	Rai *et al.* (2014)
Phoma sp.3.2883	Silver	Extracellular	Chen *et al.* (2003)
Pleurotus sajor-caju	Silver	Extracellular	Nithya and Ragunathan (2009)
Thermomonospora sp.	Gold	Extracellular	Ahmad *et al.* (2003)
Trichoderma asperellum	Silver	Extracellular	Mukherjee *et al.* (2008)
Trichoderma viride	Silver	Extracellular	Fayaz *et al.* (2010)
Trichothecium sp.	Gold	Intra-/extracellular	Ahmad *et al.* (2005)
Verticillium sp.	Magnetite	Extracellular	Bharde *et al.* (2006)
Yeast cells	Cadmium telluride	Extracellular	Bao *et al.* (2010)

its metal-binding functional group present in it via a biological metabolism-based mechanism (Sastry *et al.*, 2003). The characteristic of microbes as potent secretors of oxidative enzymes being utilized in metallic nanoparticle synthesis and an additional evaluation of distinct microorganisms makes the fungi seen as a promising tool for large-scale nanoparticle manufacturing (Kitching *et al.*, 2015; Prasad, 2016, 2017). This technique uses fungi to synthesize clean and economical fungal nanoparticles produced either in the laboratory or on a commercial scale (Moghaddam *et al.*, 2015). Through the use of biomimetic mineralization and lowering enzymes either intracellularly or extracellularly, fungi can produce metal nanoparticles and nanostructures (Ahmad *et al.*, 2003).

Due to its operational flexibility and affordability, using fungal dead biomass in the design of nanomaterials is particularly advantageous (Salvadori *et al.*, 2013). The filamentous fungus *Verticillium* sp. mediates the biosynthesis of silver nanoparticles. It has been noted by one of the earliest reviews on the synthesis of nanoparticles by a fungus that was illustrated by Mukherjee *et al.* (2001). *Rhizopus oryzae* has produced other noble metallic nanoparticles as nano gold-bio coupled within the side of the floor cell, with the length of 10 nm fungal lines isolated from wastewater present in a copper mine located in the Amazon region of Brazil as *Hypocrealixii* and *Trichoderma*. Steel or copper nanoparticles are not produced by yeasts (Salvadori *et al.*, 2013, 2014a). So, many fungi revealed a plethora of potential metal nano adsorbents for bioremediation.

10.4 FACTORS THAT AFFECT SYNTHESIS OF MYCO NANOPARTICLE

The factors such as temperature, biomass, concentration, substrate utilization time, pH, and the presence of a specific catalyst all have an impact on the biosynthesis of fungus nanoparticles. These are widely acknowledged as the primary factors that are influencing the shape and size of the nanoparticles (Kashyap *et al.*, 2013). Khan *et al.* (2016) investigated different parameters such as pH, quantity of fungal biomass, temperature, and concentration of nitrate, which could be improved to produce silver nanoparticles produced by *Aspergillus niger*. It has been stated that an increase in the above-mentioned factors can improve both the yield and synthesis of silver nanoparticles. Birla *et al.* (2013) looked into ways to improve the social and environmental conditions that were involved in the synthesis of the nanoparticles from *Fusarium oxysporum* and the produced nanoparticle was silver.

The pH is considered as one of the important factors that help in the improvement of fungal nanoparticles production. The properties and size of nanoparticles that have been synthesized are strongly influenced by the pH (Armendariz *et al.*, 2004; Gardea-Torresdey *et al.*, 1999). The nanoparticles that are synthesized from the species *Aspergillus terreus* and their efficacy against *E. coli* have been discussed by Priyadarshini *et al.* (2014) and the type of nanoparticle produced is gold. The nanoparticles have a size range between 10 and 19 nm. Fourier transform infrared spectroscopy (FTIR) has been used for the characterization of the nanoparticles that were synthesized. The regulation of biological activity and the transport of ions were greatly influenced by temperature (Dhillon *et al.*, 2012). Temperatures less than or near 100°C are required for the environmentally friendly method of synthesizing nanoparticles. Fayaz *et al.* (2009) investigated the effect of temperature on the production of plant-based nanoparticles. They discovered that raising the temperature of the reaction can cause a reduction in the size of the nanoparticles but an increase in monodispersity. Another crucial thing influencing the advent of fungal nanoparticles is the incubation period. The response medium's simple composition in large part impacts the type of nanoparticle that is produced and consequently, their quality (Darroudi *et al.*, 2011). Due to combining the debris, the nanoparticle can be stored for an extended period of time, and the quality can be modified by making changes in the incubation time (Baer, 2011).

For the biosynthesis of nanoparticles, there are extensive strategies. Mistreatment of the upstream and downstream processes is regularly combined. Physical strategies, including diffusion, thermal breakdown, irradiation, and so on, are used in the upstream method. In the downstream method, nanoparticles have created the usage of organic strategies which include organic entities and chemical strategies like a chemical discount, seeded growth, chemical technological know-how synthesis, and polyol synthesis (Tikariha *et al.*, 2012). Etching and machining tactics are used in the upstream method, where huge substances slowly decrease to nano-sized substances (Lengke *et al.*, 2011). Fungal nanoparticle synthesis has probably finished the usage of a downstream strategy. The majority of mycosynthesis reactions that bring about nanoparticles resemble backside-up processes due to the fact they encompass substrate oxidation or discount that ends in combined structures. The downstream method has more benefits over different processes due to the fact of a better probability of the usage of structures in nanoscale such as nanotubes, nanorods, nanowires, nanocubes, nanosheets, and so forth with better chemical composition homogeneity. This is regularly carried out because of the method, which is guided especially via lower chemist-unfastened power that moves the synthesized nanomaterials or nanostructures toward the nation of thermodynamic equilibrium (Kashyap *et al.*, 2013).

10.5 FUNGAL DEAD BIOMASS IN THE BIOREMEDIATION PROCESS

Microscopic fungi-like yeast and fungi that are single cellular can grow facultatively anaerobically and are frequently spherical or oval. The last electron acceptor allows yeast to utilize oxygen or a natural substance. This is a priceless trait that enables the fungi to survive and thrive in a variety of situations. It was one of the first attempts to employ organic materials for this purpose, producing CdS nanoparticles intracellularly using *Schizosaccharomyces pombe* and *Candida glabrata* (Dameron *et al.*, 1989). Silver NPs were created using yeast pressure MKY3, while gold nanoparticles were created using the humid marine yeast *Yarrowia lipolytica* NCIM 3589 (Agnihotri *et al.*, 2009). Kowshik *et al.* (2003) extracted Au and Ag's nanoparticles using *Candida guilliermondii* and *Saccharomyces cerevisiae* (Lim *et al.*, 2011; Mishra *et al.*, 2011). As part of the bioremediation process, the yeast develops the ability to synthesize intracellular metal copper nanoparticles with a length of 10.5 nm and round shaped, simultaneously for the metal ions in wastewater (Salvadori *et al.*, 2014b).

There have been few studies on the formation of nanoparticles by contagious dead biomass, such as the growth of fungi. *Hypocrealixii* can deliver extracellular metallic copper nanoparticles as well as intra and extracellular NiO_2 nanoparticles (Salvadori *et al.*, 2015). Extracellular metallic copper nanoparticles were produced by *Trichoderma koningiopsis*. *Aspergillus aculeatus* could produce extracellular nickel oxide nanoparticles in the form of film structures. *Rhodotorula mucilaginosa* could approve intracellular, metallic copper nanoparticles, and appealing Ni/NiO_2 center shell nanostructured nanoparticles in film structures (Salvadori *et al.*, 2016). Because the metal ions cannot biodegrade, substance or real treatment is largely responsible for removing them from contaminated soil and water. The readily available treatment methods are precipitation, particle exchange, electrolysis, and converse membrane partition.

Today, progressively more research is being done on the bioremediation process and the synthesis of metallic nanoparticles (Gad, 2010). The primary method commonly used for each nanoparticle production and bioremediation using fungal waste biomass includes a process known as biosorption that involves floor-level adherence of ions that are metallic to the wall of the molecular surface (Bishnoi, 2005; Dhankhar & Guriyan, 2011). The chitin, chitosan, glucan, lipid, and phospholipid walls of the fungal biomass are composed of COH groups, amino organization phosphates, lipids, melanin, sulfates, and hydroxides. The great advantage of waste biomass is attributed to its superior ability to bind metals compared to living cells, mathematical modulation of the metallic elimination reactors, lack of need for media and supplements present in solution feed, reusability of the biomass regenerated, and low toxicity (Kapoor & Viraraghavan, 1996; Kogej & Pavko, 2001; Merrin *et al.*, 1998; Salvadori *et al.*, 2013).

Physicochemical factors such as concentration of biomass, pH, temperature, agitation rate, contact time, and metallic ion concentration, on biosorption must be highlighted to improve a green nanoparticle biosynthesis process. The amount of biosorbents is a crucial factor in determining the system's sorbents equilibrium (Hanif *et al.*, 2007; Sari *et al.*, 2007). The pH of the solution indicates how well the organizations are ionized at the molecular floor of the biomass wall (Ozer & Ozer, 2003). The extracellular nanoparticles within the molecular wall of the fungi *Trichoderma koningiopsis* impregnated with copper were confirmed by the transmission electron microscope (TEM) as having a mean diameter of 87.5 nm, with a majority of them being round shaped, and few aggregates of NPs having a mean diameter of 328.27 nm. The bottom of the copper nanoparticle was covered with a capping fabric made of proteins, as revealed by the energy-dispersive and x-ray photoelectron spectra.

10.5.1 Microorganisms used in bioremediation

Microorganisms play a vital part in the dietary process and are a dominant part of life's organic stability. Bioremediation is the process of removing contaminated substances with the service of multicellular organisms such as yeast, fungi, and algae. Microbes can expand in subzero temperatures and maximum conditions of heat in the presence of hazardous compounds or streams of waste. The changeability of microbes and properties that are machinable make them fit for remediation. The activity of microorganisms is determined based on the amount of carbon. In exceptional environments, the monopoly of microorganisms can also be used for bioremediation. *Achromobacter, Arthrobacter, Alcaligenes, Bacillus, Corynebacterium, Pseudomonas, Flavobacterium, Mycobacterium, Nitrosomonas, Xanthobacter,* and other microorganisms are among these (Jaiswal, 2016). Aerobic microorganisms have degradative properties to degrade complex compounds along with *Pseudomonas, Acinetobacter, Sphingomonas, Nocardia, Flavobacterium, Rhodococcus,* and *Mycobacterium*. The above-mentioned microorganisms are used to degrade pesticides, hydrocarbons, alkanes, and polyaromatic compounds. Many microorganisms use carbon from pollutants as a sole representation of strength.

Cardio microorganisms are not often used as anaerobic microorganisms. A growing hobby in cardio microorganisms used for bioremediation of fragmented compounds that are chlorinated, polychlorinated biphenyls, and de-chlorination belong to the solvents trichloroethylene and chloroform, breakdown pollutants to reduce them further.

10.6 SOIL AND GROUNDWATER BIOREMEDIATION WITH METAL NANOPARTICLES

An important natural resource that significantly helps the ecosystem is the soil. Agriculture is a major part of the economy in many countries. The area is polluted by industrial effluents, metals from miners, and home wastes as a result of anthropogenic activities. Excavation and landfilling with incineration have been used to remediate soil, but this process takes a long time and uses items to create toxic gases as well as involves months to complete (Theng & Yuan, 2008). The Green era method of bioremediation currently reduces environmental pollution and the utilization of electricity consumption (Chozhavendhan *et al.*, 2021). Pollutants are continuously released by households and businesses, mixing with the surrounding soil. Their greatest attributes are agricultural damage, chemical harm, and landfill leakage. Ex-situ soil cleaning, pump-and-deal operations, chemical oxidation, in-situ thermal treatment, reactive boundary use, and iron treatment are all examples of traditional remediation technology (Hodson, 2010). The cost of soil and groundwater cleanup is frequently quite high, and conventional methods either fail or take a very long time to be effective. Pump-and-deal techniques, according to Bezbaruah *et al.* (2009), require several years of preparation than 1–2 years of actual use.

The containment of the groundwater is indirectly impacted by methods aimed at reducing soil contamination and vice versa. Metallic nanoparticles are being used to complete small paintings on topsoil remediation. Globally, natural or metallic substances that are carcinogenic can contaminate soil and groundwater. Halogenated substances or herbicides are also frequently found in groundwater resources. Chemical pesticides and metals in wastewater can also be purified using nanoparticles. Bootharaju and Pradeep (2010) discussed that the relationship between Hg^{2+} ions and silver nanoparticles is identified with the unification of mercury into the silver nanoparticle. As a way to create nanoparticle alloys, silver nanoparticles can decrease other metals that are apparent. For instance, the development of Pd–Ag and Pt–Ag nanoparticles using the electroplating method serves as an illustration of the development of alloy nanoparticles. Through the use of nanomaterials and colorimetric techniques, several heavy metals, including lead, arsenic, and mercury have been identified. Ono and Togashi (2004) stated that the exhaustion of steel ions mixture and nucleotides in which the Hg^{2+} facilitated the creation of thymine–thymine base pairs.

Using nanomaterials to eliminate insecticides play a favorable solution in significant surroundings because pesticides are currently heavily used in agricultural operations as organophosphorus groups, leaving toxic residues in potable water. Numerous techniques have been improved with nanotechnology in recent decades to reduce the impact of insecticides on the environment. The adsorption of organophosphorus insecticides was discovered by Das *et al.* (2009) using gold nanoparticles produced by the fungus *Rhizopus oryzae*.

10.6.1 Application in environmental areas

10.6.1.1 Heavy metals removal by mycoremediation

Heavy metals are a major problem for the ecosystem because they are frequently released into the earth and groundwater (Gupta *et al.*, 2010). The finding of metals, effective treatment, and heavy metal removal from the water bodies are the more pressing environmental problems. Due to their increased discharge, acute toxicity, non-biodegradable, harm to the environment, and propensity for bioaccumulation, these metals have been a major worry for decades (Feng *et al.*, 2010). These factors could be caused by a variety of activities, including agricultural and manufacturing activities and mineral extraction (Xu *et al.*, 2012). Prominent heavy metals and lead contaminants cause neurological effects on the central nervous system, the brain, and kidneys (Southichak *et al.*, 2006; Xu *et al.*, 2012). The abundant presence of a beneficial group of fungi has an upper hand in binding heavy metals and the elimination system (Cheng *et al.*, 2015).

10.6.1.2 Textile industries effluent treatment by myconanoparticles

The textile industry's effluent water causes major contamination of the environment because of the intake of huge quantities of water consumed in complex processes discarding the difficult ones. The effluent water from the textile industries contains a dye and its by-product which causes deleterious effects like carcer (Saharan *et al.*, 2014). Around 15% of the dye used for dyeing is directly released as effluents into the water and makes them unusable (Mollea *et al.*, 2005).

10.6.1.3 Phenol removal by mycoremediation

The phenol substance, which is harmful even in low quantities, can be ingested, touched, or inhaled, is another significant environmental pollution issue (Huang *et al.*, 2015). Acute phenol exposure disrupts the central nervous system, resulting in coma and collapse (Nair *et al.*, 2008). This contaminant is typically found in contaminated rivers, lakes, and industrial effluents. This harmful pollutant is removed from these environments using microbes and nanoparticles (Martnkova *et al.*, 2009). Large concentrations of this contaminant have been discovered in effluent from coal gasification facilities, phenol polymer manufacturing facilities, and petroleum refineries.

Magnetic nanoparticles are being employed frequently in the bioremediation process to create distinct systems that are removed using a peripheral system of magnets. A biological material possesses diamagnetic properties and can enable the efficient and selective separation of materials with magnetic modifications, this capability is crucial for biological applications.

Huang *et al.* (2015) revealed removal of phenol from effluents under the influence of Fe_3O_4 nanoparticles and Pc fungus (BKMF-1767, *P. chrysosporium*). The experiment was carried out at various concentrations of Fe_3O_4 nanoparticles, phenol concentrations, and the oxalate generation process. The binding of two important ligninolytic enzymes, MnP and LiP, released by Pc (BKMF-1767), with the phenol degradation was recorded.

10.6.1.4 Treatment of radioactive pollutants

Li *et al.* (2015) investigated *Aspergillus* with Fe_3O_4 nanoparticles as a successful application in nanobiocomposites to eliminate radioactive uranium oxide ions from uranium $(UO_2)^{2+}$ polluted water. The fungus-derived iron oxide bio-nanocomposite had a better capacity for adsorbing radioactive $(UO_2)^{2+}$ ions, with a 55% increase in comparison to the fungus and a 170% increase in comparison to the Fe_3O_4 nanoparticles. The biological system can create a magnetic adsorbent with the magnetic nanoparticles attached to the fungus and allow for the separation of the fungus from water using a magnet (Li *et al.*, 2015).

Ding *et al.* (2015) carried out a similar experiment with nanocomposites that are bio-based like fungus like *Penicillium* sp. and particles of nano-size Fe_3O_4 (VI) for the elimination of the radionuclides such as U, Th (IV), and Sr (II). This technique was low-cost and during *Penicillium* sp. growth, nano- Fe_3O_4 particles are trapped on the core of the pellet mycelia. The new bio-nanocomposites of fungus- Fe_3O_4 are more economical and environmentally friendly than the other mentioned templates because they may expand evenly on the fungal wall without aggregation. According to their studies, the fungus–Fe_3O_4 has a higher adsorption capacity than *Penicillium* sp., nano- Fe_3O_4, and other adsorbents. Sr (II), U (VI), and Th (IV) had the highest sorption capacities of the fungus–Fe_3O_4 at pH's of 5.0 and 3.0, respectively.

As a result of their capacity to alter or degrade substances, as well as their role as decomposing organisms and recyclers, fungi are extensively studied. However, when combined with magnetic nanoparticles, they create novel systems that might provide exceptional processes that are often absent when simply fungus or nanoparticles are employed. These nanobiocomposites have degraded from the polluted environment using a magnetic separation technique, which makes them potentially effective in the remediation of habitats contaminated with various pollutants, as we have already seen in multiple situations. For usage in remediation contexts, inexpensive technology, quick, and large-scale approaches are crucial.

10.6.1.5 Other pollution removal

Xenobiotic compounds are persistent organic pollutants that are manmade chemicals released from various industrial and agricultural activities. The release of xenobiotic compounds creates an impact on the environment and the ecosystem through the outbreak of adverse diseases (Karthigadevi *et al.*, 2021). Activated mono-adsorbents interact with contaminants over other things at various adsorption sites to absorb the contaminants by flexible chemistry (Kurwadkar *et al.*, 2019).

10.7 CONCLUSION

The amazing potential of using fungal dead biomass as a remediation method in the synthesis of nanoparticles in contaminated areas using hazardous metals left over from industry. Biomass is utilized (biosorption based on bioremediation) through sustainable management and reduced cost of operation, which may be employed both on small and bioremediation processes expands, contributing to sustainability in various industries like galvanoplasty industries, foundry, and mining industries. This attractive commercial waste-cleaning method of fungal dead biomass biosorption can produce metallic nanoparticles in an economical and environmentally friendly way, opening up new markets for unique herbal nanotechnologies. Promising dazzling advancements have been shown alongside the effective application of nano remediation in the feature of the current review.

REFERENCES

Agnihotri M., Joshi S., Kumar A. R., Zinjarde S. and Kulkarni S. (2009). Biosynthesis of gold nanoparticles by the tropical marine yeast *Yarrowia lipolytica* NCIM 3589. *Materials Letters*, **63**, 1231–1234, https://doi.org/10.1016/j.matlet.2009.02.042

Ahmad A., Senapati S., Khan M. I., Kumar R. and Sastry M. (2005). Extra−/intracellular biosynthesis of gold nanoparticles by an alkalotolerant fungus, Trichothecium sp. *J Biomed Nanotechnol*, **1**, 47–53.

Ahmad A., Senapati S., Khan M. I., Kumar R., Ramani R., Srinivas V. and Sastry M. (2003). Intracellular synthesis of gold nanoparticles by a novel alkalotolerant actinomycete *Rhodococcus species*. *Nanotechnology*, **14**, 824–828, https://doi.org/10.1088/0957-4484/14/7/323

Anand P., Isar J., Saran S. and Saxena R. K. (2006). Bioaccumulation of copper by Trichoderma viride. *Bioresour Technol*, **97**, 1018–1025.

Armendariz V., Herrera I. and Peralta-Videa J. R. (2004). Size controlled gold nanoparticle formation by Avena sativa biomass: use of plants in nanobiotechnology. *J Nanopart Res*, **6**(4), 377–382.

Aziz N., Pandey R., Barman I. and Prasad R. (2016). Leveraging the attributes of *Mucor hiemalis*-derived silver nanoparticles for a synergistic broad-spectrum antimicrobial platform. *Frontiers in Microbiology*, **7**, 1984, https://doi.org/10.3389/fmicb.2016.01984

Baer D. R. (2011). Surface characterization of nanoparticles: critical needs and significant challenges. *J Surf Anal*, **17**(3), 163–169.

Balaji D. S., Basavaraja S., Deshpande R., Mahesh D. B., Prabhakar B. K. and Venkataraman A. (2009). Extracellular biosynthesis of functionalized silver nanoparticles by strains of Cladosporium cladosporioides fungus. *Colloid Surf B*, **68**, 88–92.

Bansal V., Rautaray D., Ahmad A. and Sastry M. (2004). Biosynthesis of zirconia nanoparticles using the fungus Fusarium oxysporum. *J Mater Chem*, **14**, 3303–3305.

Bansal V., Rautaray D., Bharde A., Ahire K., Sanyal A., Ahmad A. and Sastry M. (2005). Fungus-mediated biosynthesis of silica and titania particles. *J Mater Chem*, **15**, 2583–2589.

Bao H., Hao N., Yang Y. and Zhao D. (2010). Biosynthesis of biocompatible cadmium telluride quantum dots using yeast cells. *Nano Res*, **3**, 481–489.

Basavaraja S., Balaji S. D., Lagashetty A., Rajasab A. H. and Venkataraman A. (2008). Extracellular biosynthesis of silver nanoparticles using the fungus Fusarium semitectum. *Mat Res Bullet*, **43**, 1164–1170.

Baskar G., Chandhuru J., Fahad K. S. and Praveen A. S. (2013). Mycological synthesis, characterization and antifungal activity of zinc oxide nanoparticles. *Asian Journal of Pharmacy*, **3**(4), 142–146.

Bezbaruah A. N., Krajangpan S., Chisholm B. J., Khan E. and Bermudez J. J. E. (2009). Entrapment of iron nanoparticles in calcium alginate beads for groundwater remediation applications. *Journal of Hazardous Materials*, **166**, 1339–1343, https://doi.org/10.1016/j.jhazmat.2008.12.054

Bedekar P. A., Saratale R., Saratale G. D. and Govindwar S. (2014). Oxidative stress response in dye degrading bacterium Lysinibacillus sp. RGS exposed to Reactive Orange 16, degradation of RO16 and evaluation of toxicity. *Environ Sci Pollut Res*, **21**(18), 11075–11085.

Bhainsa K. C. and D'souza S. F. (2006). Extracellular biosynthesis of silver nanoparticles using the fungus Aspergillus fumigatus. *Colloid Surf B*, **47**, 160–164.

Bharde A., Rautaray D., Bansal V., Ahmad A., Sarkar I., Yusuf S. M. and Sastry M. (2006). Extracellular biosynthesis of magnetite using fungi. *Small*, **2**, 135–141.

Birla S. S., Tiwari V. V., Gade A. K., Ingle A. P., Yadav A. P. and Rai M. K. (2009). Fabrication of silver nanoparticles by Phoma glomerata and its combined effect against Escherichia coli, Pseudomonas aeruginosa and Staphylococcus aureus. *Lett Appl Microbiol*, **48**, 173–179.

Bishnoi N. R. G. (2005). Fungus – an alternative for bioremediation of heavy metals containing wastewater: a review. *Journal of Scientific & Industrial Research*, **64**, 93–100.

Bootharaju M.S. and Pradeep T. (2010). Uptake of Toxic Metal Ions from Water by Naked and Monolayer Protected Silver Nanoparticles: An X-Ray Photoelectron Spectroscopic Investigation. *The Journal of Physical Chemistry C*, **114**, 8328–8336. https://doi.org/10.1021/jp101988h

Castro-Longoria E., Vilchis-Nestor A. R. and Avalos-Borja M. (2011). Biosynthesis of silver, gold and bimetallic nanoparticles using the filamentous fungus Neurospora crassa. *Colloid Surf B*, **83**, 42–48.

Cheng W., Ding C., Sun Y. and Wang X. (2015). Fabrication of fungus/attapulgite composites and their removal of U (VI) from aqueous solution. *Chem Eng J*, **269**, 1–8.

Chen J. C., Lin Z. H. and Ma X. X. (2003). Evidence of the production of silver nanoparticles via pretreatment of Phoma sp. 3.2883 with silver nitrate. *Lett Appl Microbiol*, **37**, 105–108.

Chozhavendhan S., Rajamehala M., Karthigadevi G., Kumar R. P., Bharathiraja B. and Jayakumar M. (2021). Application of nanotechnology for the sustainable development of algal biofuel industries. In Nanomaterials, (pp. 401–410). Academic Press.

Dameron C. T., Reese R. N., Mehra R. K., Kortan A. R., Carroll P. J., Steigerwald M. L. *et al.* (1989). Biosynthesis of cadmium sulfide quantum semiconductor crystallites. *Nature*, **338**, 596–597, https://doi.org/10.1038/338596a0

Darroudi M., Ahmad M. B., Zamiri R., Zak A. K., Abdullah A. H. and Ibrahim N. A. (2011). Time-dependent effect in green synthesis of silver nanoparticles. *Int J Nanomedicine*, **6**(1), 677–681.

Das S. K., Das A. R. and Guha A. K. (2009). Gold nanoparticles: microbial synthesis and application in water hygiene management. *Langmuir*, **25**, 8192–8199.

Dhankhar R. and Guriyan R. B. (2011). Bacterial biosorbents for detoxification of rapid extra-/intracellular biosynthesis of gold nanoparticles by the fungus *Penicillium* sp. *Journal of Nanoparticle Research*, **13**(3), 921–930, https://doi.org/10.1007/s11051-010-0165-2

Dhillon G. S., Brar S. K., Kaur S. and Verma M. (2012). Green approach for nanoparticle biosynthesis by fungi: current trends and applications. *Crit Rev Biotechnol*, **32**, 49–73.

Ding C., Cheng W., Sun Y. and Wang X. (2015). Novel fungus-Fe_3O_4 bio-nanocomposites as high performance adsorbents for the removal of radionuclides. *J Hazard Mater*, **295**, 127–137.

Du L., Xian L. and Feng J.-X. (2011). Rapid extra-/intracellular biosynthesis of gold nanoparticles by the fungus Penicillium sp. *J Nanopart Res*, **13**(3), 921–930.

Fatima F., Verma S. R., Pathak N. and Bajpai P. (2016). Extracellular mycosynthesis of silver nanoparticles and their microbicidal activity. *Journal of Global Antimicrobial Resistance*, **7**, 88–92, https://doi.org/10.1016/j.jgar.2016.07.013

Fayaz M., Tiwary C. S., Kalaichelvan P. T. and Venkatesan R. (2010). Blue orange light emission from biogenic synthesized silver nanoparticles using Trichoderma viride. *Colloid Surf B*, **75**, 175–178.

Feng Y., Gong J. L., Zeng G. M., Niu Q. Y., Zhang H. Y., Niu C. G., Deng J. H. and Ming Y. (2010). Adsorption of Cd (II) and Zn (II) from aqueous solutions using magnetic hydroxyapatite nanoparticles as adsorbents. *Chem Eng J*, **162**(2), 487–494.

Gad G. M. (2010). Metals, minerals, and microbes: geomicrobiology and bioremediation. *Microbiology*, **156**, 609–643, https://doi.org/10.1099/mic.0.037143-0

Gade A. K., Bonde P., Ingle A. P., Marcato P. D., Duran N. and Rai M. K. (2008). Exploitation of Aspergillus niger for synthesis of silver nanoparticles. *J Biobase Mat Bioenerg*, **2**, 243–247.

Gardea-Torresdey J. L., Tiemann K. J., Gamez G., Dokken K. and Pingitore N. E. (1999). Recovery of gold (III) by alfalfa biomass and binding characterization using X-ray microfluorescence. *Adv Environ*, **3**(1), 83–93.

Gupta V. K., Rastogi A. and Nayak A. (2010). Adsorption studies on the removal of hexavalent chromium from aqueous solution using a low cost fertilizer industry waste material. *J Colloid Interface Sci*, **342**(1), 135–141.

Hanif M. A., Nadeem R., Bhatti H. N., Ahmad M. R. and Ansari T. M. (2007). Nickel biosorption by *Cassia fistula* (golden shower) biomass. *Journal of Hazardous Materials*, **139**, 345–355, https://doi.org/10.1016/j.jhazmat.2006.06.040

Han R., Zhang J., Zou W., Xiao H., Shi J. and Liu H. (2006). Biosorption of copper (II) and lead (II) from aqueous solution by chaff in a fixed-bed column. *J Hazard Mater*, **133**, 262–268.

Hodson M. E. (2010). The need for sustainable soil remediation. *Elements*, **6**, 363–336, https://doi.org/10.2113/gselements.6.6.363

Honary S., Barabadi H., Gharaei-Fathabad E. and Naghibi F. (2013). Green synthesis of silver nanoparticles induced by the fungus Penicillium citrinum. *Trop J Pharm Res*, **12**, 7–11.

Honary S., Barabadi H. and Gharaei-Fathabad N. F. (2012). Green synthesis of copper oxide nanoparticles using Penicillium aurantiogriseum, Penicillium citrinum and Penicillium waksmanii. *Dig J Nanomater Bios*, **7**, 999–1005.

Huang D. L., Wang C., Xu P., Zeng G. M., Lu B. A., Li N. J., Huang C., Lai C., Zhao M. H., Xu J. J. and Luo X. Y. (2015). A coupled photocatalytic-biological process for phenol degradation in the Phanerochaete chrysosporium-oxalate-Fe_3O_4 system. *Int Biodeterior Biodegrad*, **97**, 115–123.

Ingle A., Rai M., Gade A. and Bawaskar M. (2009). *Fusarium solani*: a novel biological agent for the extracellular synthesis of silver nanoparticles. *Journal of Nanoparticle Research*, **11**, 2079–2085, https://doi.org/10.1007/s11051-008-9573-y

Ingle A., Gade A., Pierrat S., Sonnichsen C. and Rai M. (2008). Mycosynthesis of silver nanoparticles using the fungus Fusarium acuminatum and its activity against some human pathogenic bacteria. *Curr Nanosci*, **4**, 141–144.

Kar P. K., Murmu S., Saha S., Tandon V. and Acharya K. (2014). Anthelmintic Efficacy of Gold Nanoparticles Derived from a Phytopathogenic Fungus, *Nigrospora oryzae*. *PLoS ONE*, **9**, e84693, https://doi.org/10.1371/journal.pone.0084693

Kashyap P. L., Kumar S., Srivastava A. K. and Sharma A. K. (2013). Myconanotechnology in agriculture: a perspective. *World Journal of Microbiology and Biotechnology*, **29**(2), 191–207, https://doi.org/10.1007/s11274-012-1171-6

Kapoor A. and Viraraghavan T. (1996). Biosorption of heavy metals on fungal biomass. Proceeding, 4th CSCE Environmental Engineering Specialty Conference (Edmonton, Alberta), pp. 333–343.

Kitching M., Ramani M. and Marsili E. (2015). Fungal biosynthesis of gold nanoparticles: mechanism and scale up. *Microbial Biotechnology*, **8**, 904–917, https://doi.org/10.1111/1751-7915.12151

Kogej A. and Pavko A. (2001). Laboratory experiments of lead biosorption by self-immobilized *Rhizopus nigricans* pellets in the batch stirred tank reactor and the packed dead column. *Chemical and Biochemical Engineering*, **15**, 75–79.

Kowshik M., Ashtaputre S., Kulkani S. K. and Parknikar K. M. M. (2003). Extracellular synthesis of silver nanoparticles by a silver-tolerant yeast strain MKY3. *Nanotechnology*, **14**, 95–100.

Kuber C., Bhainsa S. F. and Souza D. (2006). Extracellular biosynthesis of silver nanoparticles using the fungus *Aspergillus fumigatus*. *Colloids and Surfaces B: Biointerfaces*, **47**, 160–164, https://doi.org/10.1016/j.colsurfb.2005.11.026

Kango S., Kalia S., Celli A., Njuguna J., Habibi Y. and Kumar R. (2013). Surface modification of inorganic nanoparticles for development of organic–inorganic nanocomposites-a review. *Prog Polym Sci*, **38**(8), 1232–126.

Karthigadevi G., Manikandan S., Karmegam N., Subbaiya R., Chozhavendhan S., Ravindran B. and Awasthi M. K. (2021). Chemico-nanotreatment methods for the removal of persistent organic pollutants and xenobiotics in water–A review. *Bioresource Technology*, **324**, 124678.

Kathiresan K., Alikunhi N. M., Pathmanaban S., Nabikhan A. and Kandasamy S. (2010). Analysis of antimicrobial silver nanoparticles synthesized by coastal strains of Escherichia coli and Aspergillus niger. *Can J Microbiol*, **56**, 1050–1059.

Kathiresan K., Manivannan S., Nabeel M. A. and Dhivya B. (2009). Studies on silver nanoparticles synthesized by a marine fungus, Penicillium fellutanum isolated from coastal mangrove sediment. *Colloid Surf B*, **71**, 133–137.

Khan A., Rashid R., Murtaza G. and Zahra A. (2014). Gold nanoparticles: synthesis and applications in drug delivery. *Trop J Pharm Res*, **13**(7), 1169–1177.

Kumar R. R., Priyadharsani K. P. and Thamaraiselvi K. (2012). Mycogenic synthesis of silver nanoparticles by the Japanese environmental isolate Aspergillus tamarii. *J Nanopart Res*, **14**, 1–7.

Kurwadkar S., Hoang T. V., Malwade K., Kanel S. R., Harper W. F. and Struckhoff G. (2019). Application of carbon nanotubes for removal of emerging contaminants of concern in engineered water and wastewater treatment systems. *Nanotechnology for Environmental Engineering*, **4**, 1–16.

Lengke M. F., Sanpawanitchakit C. and Southam G. (2011). Biosynthesis of gold nanoparticles: a review. In: Rai M. K., Duran N. (eds) Metal nanoparticles in microbiology. Springer, New York, pp. 37–74.

Lim H. A., Mishra A. and Yun S. I. (2011). Effect of pH on the extra cellular synthesis of gold and silver nanoparticles by Saccharomyces cerevisae. *J Nanosci Nanotechnol*, **11**, 518–522.

Li G., He D., Qian Y., Guan B., Gao S., Cui Y. and Wang L. (2011). Fungus-mediated green synthesis of silver nanoparticles using Aspergillus terreus. *Int J Mol Sci*, **13**, 466–476.

Li L., Xu M., Chubik M., Chubik M., Gromov A., Wei G. and Han W. (2015). Entrapment of radioactive uranium from wastewater by using fungus-Fe_3O_4 bio-nanocomposites. *RSC Adv*, **5**(52), 41611–41616.

Lodhia J., Mandarano G., Ferris N. J., Eu P. and Cowell S. F. (2010). Development and use of iron oxide nanoparticles (part 1): synthesis of iron oxide nanoparticles for MRI. *Biomed Imaging Inter J*, **6**(2), e12.

Malik A. (2004). Metal bioremediation through growing cells. *Environ Int*, **30**, 261–278.

Maliszewska I., Juraszek A. and Bielska K. (2014). Green synthesis and characterization of silver nanoparticles using ascomycota fungi Penicillium nalgiovense AJ12. *J Clust Sci*, **25**, 989–1004.

Martínkova L., Uhnakova B. and Patek M. (2009). Biodegradation potential of the genus Rhodococcus. *Environ Int*, **35**(1), 162–177.

Masciangioli T. and Zhang W. X. (2003). Peer reviewed: environmental technologies at the nanoscale. *Environ. Sci. Technol.*, 2003, **37**(5), 102A–108A.

Merrin J. S., Sheela R., Saswathi N., Prakasham R. S. and Ramakrishna S. V. (1998). Biosorption of chromium (VI) using *Rhizopus arrhizus*. *Indian Journal of Experimental Biology*, **36**, 1052–1055.

Mollea C., Bosco F. and Ruggieri B. (2005). Fungal biodegradation of naphthalene: microcosmos studies. *Chemosphere*, **60**(5), 636–643.

Moghaddam A. B., Namvar F., Moniri M., Tahir P. M., Azizi S. and Mohamad R. (2015). Nanoparticles biosynthesized by fungi and yeast: a review of their preparation, properties, and medical applications. *Molecules*, **20**, 16540–16565, https://doi.org/10.3390/molecules200916540

Moghaddam K. M. (2010). An introduction to microbial metal nanoparticle preparation methods. *Young Investig*, **19**, 1–7.

Mukherjee P., Senapati S., Mandal D., Ahmad A., Khan M. I., Kumar R. and Sastry M. (2002). Extracellular synthesis of gold nanoparticles by the fungus Fusarium oxysporum. *Chem Biol Chem*, **3**, 461–463.

Mukherjee R., Paul T., Soren J. P., Halder S. K., Mondal K. C., Pati B. R. *et al.* (2019). Acidophilic α-amylase production from Aspergillus Niger RBP7 using potato peel as substrate: a waste to value added approach. *Waste Biomass Valori*, **10**(4), 851–863.

Mukherjee P., Ahmad A., Mandal D., Senapati S., Sainkar S. R., Khan M. I., Parishcha R., Ajaykumar P. V., Alam M., Kumar R. and Sastry M. (2001). Fungus-mediated synthesis of silver nanoparticles and their immobilization in the mycelial matrix: a novel biological approach to nanoparticle synthesis. *Nano Letters*, **1**, 515–519, https://doi.org/10.1021/nl0155274

Mukherjee, Roy M., Mandal B. P., Dey G. K., Mukherjee P. K., Ghatak J. and Kale S. P. (2008). Green synthesis of highly stabilized nanocrystalline silver particles by a non-pathogenic and agriculturally important fungus T. asperellum. *Nanotechnology*, **19**(7), 075103.

Mishra A., Tripathy S. K. and Yun S. I. (2011). Biosynthesis of gold and silver nanoparticles from *Candida guilliermondii* and their antimicrobial effect against pathogenic bacteria. *Journal of Nanoscience and Nanotechnology*, **11**, 243–248, https://doi.org/10.1166/jnn.2011.3265

Nair C. I., Jayachandran K. and Shashidhar S. (2008). Biodegradation of phenol. *Afr J Biotechnol*, **7**(25), 4951–4958.

Nanda A. and Majeed S. (2014). Enhanced antibacterial efficacy of biosynthesized AgNPs from Penicillium glabrum (MTCC1985) pooled with different drugs. *Int J Pharm Tech Res*, **6**, 217–223.

Nayak R. R., Pradhan N., Behera D., Pradhan K. M., Mishra S., Sukla L. B. and Mishra B. K. (2011). Green synthesis of silver nanoparticle by Penicillium purpurogenum NPMF: the process and optimization. *J Nanopart Res*, **13**, 3129–3137.

Nithya R. and Ragunathan R. (2009). Synthesis of silver nanoparticle using Pleurotus sajor caju and its antimicrobial study. *Digest J Nanomater Biostructur*, **4**, 623–629.

Ono A. and Togashi H. (2004). Highly selective oligonucleotide-based sensor for mercury(II) in aqueous solutions. *Angew Chem Int*, Ed **43**, 4300–4302.

Ozer A. and Ozer D. (2003). Comparative study of the biosorption of Pb (II), Ni (II) and Cr (VI) ions onto *S. cerevisiae*: determination of biosorption hats. *Journal of Hazardous Materials*, **100**, 219–229, https://doi.org/10.1016/S0304-3894(03)00109-2

Pavani K. V., Kumar N. S. and Sangameswaran B. B. (2012). Synthesis of lead nanoparticles by Aspergillus species. *Pol J Microbiol*, **61**, 61–63.

Prasad K. and Jha A. K. (2010). Biosynthesis of CdS nanoparticles: an improved green and rapid procedure. *Journal of Colloid and Interface Science*, **342**(1), 68–72, https://doi.org/10.1016/j.jcis.2009.10.003

Prasad R. (2016). Advances and Applications Through Fungal Nanobiotechnology. Springer International Publishing, Springer Cham (ISBN: 978-3-319-42989-2), https://link.springer.com/book/10.1007/978-3-319-68424-6

Prasad R. (2017). Fungal nanotechnology: applications in agriculture, industry, and medicine. (978-3-319-68957-9)

Priyadarshini E., Pradhan N., Sukla L. B. and Panda P. K. (2014). Controlled synthesis of gold nanoparticles using Aspergillus terreus IF0 and its antibacterial potential against Gram negative pathogenic bacteria. *J Nanotechnol*, **2014**, 1–9. http://dx.doi.org/10.1155/2014/653198

Saharan P., Chaudhary G. R., Mehta S. K. and Umar A. (2014). Removal of water contaminants by iron oxide nanomaterials. *J Nanosci Nanotechnol*, **14**(1), 627–643.

Salvadori M. R., Lepre L. F., Ando R. A., Nascimento C. A. O. and Corrêa B. (2013). Biosynthesis and uptake of copper nanoparticles by dead biomass of *Hypocrea lixii* isolated from the metal mine in the Brazilian Amazon region. *PLoS One*, **8**, e80519, https://doi.org/10.1371/journal.pone.0080519

Salvadori M. R., Ando R. A., Nascimento C. A. O. and Corrêa B. (2015). Extra and intracellular synthesis of nickel oxide nanoparticles mediated by dead fungal biomass. *PLoS One*, **5**, 1–15.

Sarkar J., Chattopadhyay D., Patra S., Deo S. S., Sinha S., Ghosh M. and Acharya K. (2011). Alternaria alternata mediated synthesis of protein capped silver nanoparticles and their genotoxic activity. *Dig J Nanomat Biostruct*, **6**, 563–573.

Sastry M., Ahmad A., Khan M. I. and Kumar R. (2003). Biosynthesis of metal nanoparticles using fungi and actinomycete. *Curr Sci*, **85**(2), 162–170.

Southichak B., Nakano K., Noura M., Chiba N. and Nushimura O. (2006). Phragmites austral: a novel biosorbent for the removal of heavy metals from aqueous solution. *Water Res*, **40**(12), 2295–2302.

Salvadori M. R., Ando R. A., do Nascimento C. A. O. and Corrêa B. (2014a). Bioremediation from wastewater and extracellular synthesis of copper nanoparticles by the fungus *Trichoderma koningiopsis*. *Journal of Environmental Science and Health, Part A*, **49**, 1286–1295, https://doi.org/10.1080/10934529.2014.910067

Salvadori M. R., Ando R. A., do Nascimento C. A. O. and Corrêa B. (2014b). Intracellular biosynthesis and removal of copper nanoparticles by dead biomass of yeast isolated from the wastewater of a mine in the Brazilian Amazonia. *PLoS One*, **9**, e87968, https://doi.org/10.1371/journal.pone.0087968

Salvadori M. R., Ando R. A., Muraca D., Knobel M., Nascimento C. A. O. and Corrêa B. (2016). Magnetic nanoparticles of Ni/NiO nanostructured in film form synthesized by dead organic matrix of yeast. *RSC Advances*, **6**, 60683–60692, https://doi.org/10.1039/C6RA07274G

Sari A., Tuzen M., Uluuozlu O. D. and Soylak M. (2007). Biosorption of Pb (II) and Ni (II) from aqueous solution by lichen (*Cladonia furcata*) biomass. *Biochemical Engineering Journal*, **37**, 151–158, https://doi.org/10.1016/j.bej.2007.04.007

Shelar G. B. and Chavan A. M. (2014). Extracellular biological synthesis, characterization and stability of gold nanoparticles using the fungus *Helminthosporium tetramera*. *International Journal of Pure & Applied Bioscience*, **2**(3), 281–285.

Srivastava S. K., Yamada R., Ogino C. and Condo A. (2013). Biogenic synthesis and characterization of gold nanoparticles by *Escherichia coli* K12 and its heterogeneous catalysis in degradation of 4-nitrophenol. *Nanoscale Research Letters*, **8**, 70, https://doi.org/10.1186/1556-276X-8-70

Syed A. and Ahmad A. (2012). Extracellular biosynthesis of platinum nanoparticles using the fungus *F. oxysporum*. *Colloids and Surfaces B: Biointerfaces*, **97**, 27–31, https://doi.org/10.1016/j.colsurfb.2012.03.026

Sanghi R. and Verma P. A. (2009). A facile green extracellular biosynthesis of CdS nanoparticles by immobilized fungus. *Chemical Engineering Journal*, **155**, 886–891, https://doi.org/10.1016/j.cej.2009.08.006

Rai M., Ingle A. P., Gade A. and Duran N. (2014). Synthesis of silver nanoparticles by Phoma gardenia and in vitro evaluation of their efficacy against human disease-causing bacteria and fungi. *IET Nanobiotechnol*, **9**, 71–75.

Ranjbar-Navazi Z., Pazouki M. and Halek F. S. (2010). Investigation of culture conditions for biosynthesis of silver nanoparticles using Aspergillus fumigatus. *Iran J Biotechnol*, **8**, 56–61.

Riddin T. L., Gericke M. and Whiteley C. G. (2006). Analysis of the inter-and extracellular formation of platinum nanoparticles by Fusarium oxysporum f. sp. lycopersici using response surface methodology. *Nanotechnology*, **17**, 3482

Theng B. K. and Yuan G. (2008). Nanoparticles in the soil environment. *Elements*, **4**, 395–399, https://doi.org/10.2113/gselements.4.6.395

Tikariha S., Singh S. S., Banerjee S. and Vidyarthi A. S. (2012). Biosynthesis of gold nanoparticles, scope and application: a review. *Int J Pharm Sci Res*, **3**(6), 1603–1615.

Vala A. K., Shah S. and Patel R. (2014). Biogenesis of silver nanoparticles by marine derived fungus Aspergillus flavus from Bhavnagar coast, Gulf of Khambhat, India. *J Mar Biol Oceanogr*, **3**, 1–3.

Varshney R., Mishra A. N., Bhadauria S. and Gaur M. S. (2009). A novel microbial route to synthesize silver nanoparticles using fungus Hormoconis resinae. *Digest J Nanomater Biostruct*, **4**, 349–355.

Vaseashta A., Vaclavikova M., Vaseashta S., Gallios G., Roy P. and Pummakarnchana O. (2007). Nanostructures in environmental pollution detection, monitoring, and remediation. *Science and Technology of Advanced Materials*, **8**, 47–59, https://doi.org/10.1016/j.stam.2006.11.003

Verma V. C., Kharwar R. N. and Gange A. C. (2010). Biosynthesis of antimicrobial silver nanoparticles by the endophytic fungus Aspergillus clavatus. *Nanomedicine*, **5**, 33–40.

Veglio F. and Beolchini F. (1997). Removal of metals by biosorption: a review. *Hydrometallurgy*, **44**, 301–316, https://doi.org/10.1016/S0304-386X(96)00059-X

Vigneshwaran N., Kathe A. A., Varadarajan P. V., Nachane R. P. and Balasubramanya R. H. (2006). Biomimetics of silver nanoparticles by white rot fungus, Phaenerochaete chrysosporium. *Colloid Surf B*, **53**, 55–59.

Volesky B. (2022). Detoxification of metal-bearing effluents: biosorption for the next century. *Hydrometallurgy*, **59**, 203–216, https://doi.org/10.1016/S0304-386X(00)00160-2

Xu P., Zeng G. M., Huang D. L., Lai C., Zhao M. H., Wei Z., Li N. J., Huang C. and Xie G. X. (2012). Adsorption of Pb (II) by iron oxide nanoparticles immobilized Phanerochaete chrysosporium: equilibrium, kinetic, thermodynamic and mechanisms analysis. *Chem Eng J*, **203**, 423–431.

doi: 10.2166/9781789063783_0211

Chapter 11

Rhizoremediation of organic emerging soil contaminants: green technology

Prasann Kumar[1], Duppala Hema Latha[1] and Joginder Singh[2]*

[1]Department of Agronomy, School of Agriculture, Lovely Professional University, Phagwara 144411, Punjab, India
[2]Department of Microbiology, School of Bioengineering and Biosciences, Lovely Professional University, Phagwara 144 411, Punjab, India
**Corresponding author: prasann0659@gmail.com*

ABSTRACT

For a very long time there has been interest in creating in situ techniques for remediation of natural contaminants. Rhizoremediation, a specific type of phytoremediation that includes the two plants and their related rhizosphere microorganisms, can take place naturally or can be incited by intentionally presenting specific microorganisms. These microorganisms degrade contaminants and can also promote plant development under stress conditions. Currently, improvement is related to the rapid pace of urbanization and steady increase in population, which must be considered. These cycles lead to soil contamination, which is a result of dumping hazardous wastes into the environment. Any type of hazardous component such as radiation and sound or different resources like heavy metals or particulate substance can be termed a contaminant. Among several contaminants, natural contaminants attract more attention as they are highly water repellent and are highly toxic. Owing to these properties, the contaminants are capable of being embedded in the established environment for longer periods of time, which further increases their risk of causing damage to our environment. This study is focussed on the rhizoremediation process that can be used to clean up natural toxins from the soil.

Keywords: heavy metals, rhizosphere, plants, microorganism, sustainability, no poverty, zero hunger

11.1 INTRODUCTION

It is important to note that natural contaminants have a tremendously large range, and include: hydrocarbons, various solvents, oil-based products, pesticides, and herbicides (Oleszczuk, 2006; MacKinnon & Duncan, 2013; Moore *et al.*, 2006). Meager amounts of these contaminants are able to change the properties of soil by altering the microbial population as well as the metabolic changes that are executed with it. In spite of the fact that there are conventional methods to degrade these pollutants, the advancements are exorbitant, which further deters their emanating release (Alkorta & Garbisu, 2001). In order to eliminate natural toxins from contaminated soil, bioremediation not only provides an excellent solution but also is eco-friendly and practical; however, it has certain limitations which must be taken into account (Glazer & Nikaido, 1995). As an alternative to bioremediation, a part of it is called rhizoremediation, which is used to handle the limitations of bioremediation. In rhizoremediation, plants remove contaminants from soil with the assistance of organisms that live in the rhizospheres in order to remediate soil contamination. (Oberai & Khanna, 2018). By providing a carbon source through their root exudates, plants serve as an ecosystem similar to a carbon source for all organisms (Cannon & Bowles, 1962). Microorganisms are responsible for the debasement process, and the buildups are taken up by plants via water as media. Seeds and microorganisms exist in the soil, and also environmental toxins are observed in soil. In 2001, United Nations Environment Program (UNEP) made a distinction among modern synthetics, pesticides, and side effects of the soil, which are classified into three categories that include numerous natural toxins in the soil. Polychlorinated biphenyls (PCBs) have been included in the class of modern synthetic substances, polychlorinated dibenzo-*p*-dioxins and polychlorinated dibenzofurans have been placed under the result class. Natural contaminants are found in the soil in a variety of forms, but polycyclic aromatic hydrocarbons (PAHs) and PCBs attract a great deal of attention due to the unpredictable nature of their presence and the fact that they can be found in soil in high concentrations. There are two types of sand: those that repel water and those that accumulate water at an exceptionally slow pace (Fabietti *et al.*, 2010; Gupta *et al.*, 2018; Ma *et al.*, 2009; Zhang *et al.*, 2007). In the environment, PAHs can be introduced by burning natural materials, which results in a build-up of PAHs. The PAHs that are released due to ignition can either be caused by natural situations or by manmade intervention, whereas PCBs, on the other hand, can only be caused by human activity and are exclusively anthropogenic in nature. (Cachada *et al.*, 2012). As a result of an increase in population, the use of fertilizers and pesticides has also increased causing contamination of soil. Initially, the use of composts and pesticides proved beneficial but later showed adverse effects on the soil. (Merrington *et al.*, 2002). Globally, more than two million tons of chemical pesticides are being used on an annual basis. Atapattu and Kodituwakku (2009) report that South Asia uses a large amount of pesticides, with 14% of the country's land located in rural areas and the use of synthetically synthesized pesticides at a high rate. Prevalence of weeds in the countryside can cause a loss of anywhere between 37% and 79% in rural areas; in order to counteract this colossal loss, ranchers apply herbicides and weedicides in order to eliminate weeds (Behera & Singh, 1999). A majority of farmers rely on insecticides to avoid the adverse effects of weeds, bugs, and also ants in fields (Ross, 2005). Several studies have shown that these blended synthetics are less used in developed nations, however, they are still used at high rates in developing nations such as India (Eqani *et al.*, 2011; Syed & Malik, 2011; Tariq *et al.*, 2007). Pesticides and insecticides are extensively used in India for controlling vermin and insects that are harmful to the environment. As part of its efforts to achieve this goal, India began developing Benzene hexachloride (BHC) and dichloro diphenyl trichloroethane (DDT) in 1952, and the development of these synthetic compounds has continued uninterrupted since that time. According to Gupta (2004), it has been noted that India alone delivered over 5000 metric tons of pesticides in 1958, including BHC and DDT, which were two of the most prominent pesticides that were shipped out of the country. There were an estimated 145 pesticides added to the market during mid-1990s, which resulted in almost 85 000 metric tons of waste. As individuals become increasingly aware of what they are using and consuming, they are causing a decrease in the use of artificially orchestrated pesticides; ranchers are now moving toward greener alternatives like plant-derived products and other options (such as biopesticides), which are causing a decline in the use of artificially orchestrated pesticides. Currently, India produces colossal amounts of pesticides that places it in the highest position in Asia and in the twelfth position globally when it comes to pesticide production. It has been estimated that India produces 90 000 tons of pesticides annually, which is a staggering amount when looking at the global scenario. Pesticides are produced in India as well as other products that are exported to different countries (Pozo *et al.*, 2011). As it relates to toxins, there are different classifications that include synthetic compounds that are more likely to remain in the environment after they have been applied as well as those that tend to bioaccumulate and spread for an extremely long distance. The potential of bioaccumulation and spreading of industrial and natural contaminants is also known as persistent organic pollutants (POPs) (Buccini, 2003). According to some scientists, POPs have been found elsewhere in other places as well, where pesticides have not been used. It might be due to the fact that they have a dispersing property, which they share with two natural pollutants found in soil and their sources, rainfall and wind (Barra *et al.*, 2005; Zhang *et al.*, 2008). Vos *et al.* (2000) reported that a significant portion of the mixtures within the POP category are capable of altering the ability of chemicals by upsetting the subatomic structure of the compounds in order to alter their properties. The purpose of conducting such an examination is to make individuals more aware of the severity of the effects that these synthetic compounds have on the environment as a whole as a result of the investigation. Approximately 17 synthetic substances that fall under the POP classification have been identified by the UNEP as pesticides, modern synthetic chemicals, and hazardous substances that may cause adverse health effects to humans. In the UNEP's fourth meeting, which was held in 2009, a few more synthetic substances were added to the POP list in every classification. In its subsequent meeting in 2011, endosulfan and its connected isomer were also added to the list of POP chemicals. Despite the fact that the use of these synthetic compounds has been banned some time ago due to their harmful effects on the environment and on human health, they are still widely used in some countries throughout the world, namely in countries in South Asia, despite these restrictions. The use of DDT is routinely used in both India and Pakistan to control mosquitoes, as evidenced by several studies (Alamdar *et al.*, 2014; Chakraborty *et al.*, 2010; Eqani *et al.*, 2011; Syed & Malik, 2011; Syed *et al.*, 2013, 2014; Zhang *et al.*, 2008). Since POPs are capable of irreversible dissipation, they might have the capability of polluting immaculate regions of the earth, such as the South Pole and the North Pole, if they are not controlled. South Asian nations, according to Minh *et al.* (2006), lack adequate facilities to deal with metropolitan waste; in an effort to dispose of waste, these nations rely on open waste disposal sites that are inadequately equipped to handle the waste. As a result of this study, Syed and Malik (2011) have focused on a few areas of Pakistan where processing plants and storage facilities that contained pesticides like DDT have been closed. As a result of this study, it was demonstrated that spill over from these places can lead to contamination of surrounding soil and water bodies. It has been found that similar types of studies have been conducted by various researchers instructing individuals to consider the possible health risks for people living in impacted regions (Agusa *et al.*, 2003).

11.2 DESTINY OF NATURAL CONTAMINATIONS IN SOIL

In case natural contamination enters the biological system and becomes part of it, the development of this contamination within the biological system is determined by the synthetic nature of the contamination; on the whole, if the substance is biodegradable, it will eventually become mineralized with the passage of time; however, if it is not biodegradable, then it will likely persist in the environment. When everything works out over time, it is possible to calculate the convergence of contaminations stored in the system by estimating how similar the groups of dregs are. Bhattacharya *et al.* (2003) as well as Guzzella *et al.* (2005) have demonstrated that these compartments play important roles in the accumulation of these toxins. It is important to note that there are different types of dregs (natural, unnatural, and inorganic) that come from a variety of different sources and are composed of different elements. The mechanisms that are responsible for incorporating bioaccumulative and polycyclic aromatic hydrocarbons into the residual as well as vaporization, surface overflow, and drainage are those that are responsible for incorporating them into the residual (Sarkar *et al.*, 2008). As a result of the fact that POPs are capable of degrading slowly, it is easy to detect POPs by focusing on silt since they have such a low degrading potential in comparison with other pollutants (El Nemr *et al.*, 2013).

It has been reported that out of the various natural synthetic substances that are present in the soil in the Indian scenario, pesticides are the most widely known. They are used in rural areas to protect crops from pests, as well as in civil areas to prevent diseases by controlling bugs that may lead to disease. As a result, it should be possible to observe high levels of contaminants in a few places within a few radii of home-grown and modern trash (Pozo *et al.*, 2011) that are located on the edges of those release points. Ali *et al.* (2014) studied residues found in river Gomti, India, where the author observed that hexachlorocyclohexane (HCH) and DDT concentrations were very high (BDL-81.2 mg/g). The east and west shorelines of India were also examined by the researcher, where concentrations of HCHs and DDTs were detected as well (BDL 109 mg/g). The same can be said about Pandey *et al.* (2011), who focused on the residues from river Ganga and parts of different waterways associated with it in order to observe the progression of the centralization of POPs through different seasons, as well as in different waterways. There was a study in which OCP centralization was recorded during a storm, before a rainstorm (158e308 ng/g), and after a rainstorm (307e844 ng/g) seasons. It was concluded that the significant cause of DDT contamination in the silt was the endured and utilized soils from farming fields. The study by Malik *et al.* (2009) explains that the ill-conceived use of DDT may also have contributed to its extreme taint in riverine regions. Bhattacharya *et al.* (2003) observed elevated levels of DDTs, HCHs, and endosulfan sulfate in the air before the storm season. As a result, the centralization of the above was tracked and arranged as follows: 1. HCH, 2. N-endosulfan sulfate, 3. DDTs, 4. a-endosulfan. This shows that DDT is degraded through a dehydrochlorination process, which may be due to decay of nonliving or living organisms. Elevated degree of pesticides in certain regions is due to the after effect from horticultural lands, domestic wastes, and hazardous disposables. Khwaja *et al.* (2006) studied samples of soil, additional ingredients, mortar tests of plant walls, and harmed sacks due to ban of DDT production in Pakistan. Disposed of packs were found to have the most elevated convergence of DDT (2822e2841 mg/g). According to Khwaja *et al.*, 91% of the soil samples collected after one year had been profoundly contaminated with DDT, with the majority of the soil samples having DDT deposits over the permissible limit of 0.05 mg/g in soil, which is the limit at which DDT can be used in soil. The study by Jan *et al.* (2009) conducted in the same region found a similar type of contamination around the plant. The study found that 500 m of the area surrounding the plant were still contaminated with DDT deposits. Based on their research, the authors found that the distance and grouping of DDT metabolites, such as p', p'-DDT, from the fixed assembling unit had diminished with the depth and distance from the fixed assembly unit of the DDT metabolites. Yadav *et al.* (1981) discovered that the soil around a DDT-combining industry in Delhi had a high concentration of DDT, and that night crawlers were contaminated with DDT. The results of this study provided evidence that a region that is close to DDT-combining factories has an elevated concentration of DDT in its soil.

11.3 RHIZOREMEDIATION: A TRADITIONAL METHODOLOGY

Rhizoremediation is a method of removing soil impurities by utilizing microorganisms from the rhizosphere to degrade them and therefore degrade soil impurities. A rhizodegradation process would also be called an organism-assisted phytoremediation innovation, or an organism-assisted bioremediation process. It is widely accepted that the rhizosphere is one of the most dynamic areas around roots (i.e., length of 2 mm) impacted due to plant action (Brink, 2016; Dzantor, 2007; Liu *et al.*, 2014). In a majority of cases, it is the rhizospheric microorganisms that are the principal supporters, while the plants provide a carbon source that is essential for the rhizospheric microorganisms to degrade contaminants. Hence, microorganisms can provide plants with supplements that will help to improve their development, protect the plants from microbes, degrade impurities, as well as diminish the degree of stress chemicals produced by the plants (Macek *et al.*, 2000; Qixing *et al.*, 2011). It is important to note that study on rhizospheric microbial contamination has been focused on ensuring the safety of flora from agrochemicals (Hoagland *et al.*, 1994; Jacobsen, 1997; Zablotowicz *et al.*, 1994). According to the available evidence, there are only a few cases that can be attributed to degrading natural mixtures, like PAHs (Radwan *et al.*, 1995), trichloroethylene (Walton & Anderson, 1990), and PCBs (Brazil *et al.*, 1995), to the degradation of natural contaminants. In order to reach an accurate diagnosis, it was important to dissect the nitty gritty of these examinations and the arrangement of the microorganisms. In addition, there are no data available regarding the expansion, endurance, and movement of the microbial population within the rhizosphere. As a result of different investigations, it was found that leguminous plants are best for rhizoremediation (Kuiper *et al.*, 2001; Qiu *et al.*, 1994; Shann & Boyle, 1994). The variables that impact the outcome of rhizoremediation are the natural connections with different living beings, the essential and optional digestion, endurance, and the foundation of the system. It has been shown that the development of bacteria in the rhizosphere is largely influenced by root exudation, and this is considered to be the most important impetus for the development of rhizoremediation (Martin *et al.*, 2014; Phillips *et al.*, 2012). A significant part of plant root exudates are mixtures of sugars, amino acids, and natural acids that are present in the oozes (Vancura & Hovadik, 1965). Despite these mixtures as an add up to the root cells, adhesive discharge through cells of roots, the rot of entire roots, or the hunger of cells of roots, additional supplements can also be derived (Lugtenberg & de Weger, 1992; Lynch & Whipps, 1990; Rovira, 1956). In order to isolate root exudates according to the process of their creation, they can be divided into four distinct groups, auxiliary floral metabolites, lysates (from senescent tissues and roots), detached exudations, and adhesives (tip of root and epidermis). A study has demonstrated that carbohydrate components with a low subatomic weight (LMW), auxiliary metabolites, AAs, and natural anions which are soluble in water like components of oozes of roots, while natural polymers, on the other, have been considered insoluble components (Martin *et al.*, 2014; Walker *et al.*, 2003a, 2003b). At the time of their metabolic cycle, soil organisms use these mixtures of carbon and energy as a carbon and energy source to fuel their metabolic activity (Chaudhry *et al.*, 2005). It is believed that root exudates should be responsible for causing a change in the microbial local environment of tainted soils in two unique ways: first, by altering the way organisms articulate the catabolic quality of their cells, and second, by choosing the particular strains of these organisms (Baudoin *et al.*, 2003; Benizri *et al.*, 2002; Butler *et al.*, 2003; Hartmann *et al.*, 2009; Siciliano *et al.*, 2003). Biological processes such as the biodegradation of toxins like TPH are achieved by the connections between organisms and soil air circulation that are provided by plant pulls (Martin *et al.*, 2014) (Table 11.1).

Despite the controversies surrounding the role of root exudates in determining the composition of the microbiome of the rhizosphere in soils polluted with nitro-fragrance mixtures with petroleum (Siciliano *et al.*, 2003; Thijs *et al.*, 2016), there is no doubt about their significance. Miya and Firestone (2001) discovered after collecting the root exudates of the slim oat that this root exudate was able to cope with the most extreme populations of PHC-corrupting organisms, as well as help to improve the degradation

Table 11.1 In the process of rhizoremediation of pollutants, plant interactions with microbes are important.

Particulates from Organic Sources	Plants	Microbes	References
Polychlorinated biphenyl	Beetroot	*P. fluorescenes*	Brazil *et al.* (1995)
Polyacrylic aromatic hydrocarbons	Alfa	NA	Nichols *et al.* (1997)
2,4-D	Wheat	*P. putida*	Kingsley *et al.* (1994)
Total petroleum hydrocarbons	Rye	*Rhodococcus*	Ingrid *et al.* (2016)
Pyrene	*Salix*	*P. putida*	Khan *et al.* (2014)
Fixed nitrogen	Reed	*Nitrospira* sp.	El Haleem *et al.* (2000)
3-Methylbenzoate	*Zea mays*	*P. putida*	Ronchel and Ramos (2001)
2,4-D	Barley	*B. cepacian*	Jacobsen (1997)
Trichloroethylene	Wheat	*P. fluorescens*	Yee *et al.* (1998)
Polyacrylic aromatic hydrocarbons	Wheat	*Rhizophagus*	Ingrid *et al.* (2016)
Naphthalene	Grass	*P. putida*	Kuiper *et al.* (2001)
Polycyclic aromatic hydrocarbons	*Poplus*	*Micrococcus* sp.	Bisht *et al.* (2010, 2014)
Mefenoxam	*Z. augustifolia*	*P. fluorescenes*	Pai *et al.* (2001)
1,4-dioxane	*Poplar*	*Actinomyces*	Kelley *et al.* (2001)

of phenanthrene. As a result, it exposed that combination of engineered oozes from roots and mineral supplements improved the bio-deprivation fastness of 3–5 ring polycyclic aromatic hydrocarbons (Joner *et al.*, 2002). A study by Yoshitomi and Shann (2001) affirmed that the microbial mineralization of pyrene in corn roots (*Zea mays* L.) was caused by sterile exudates produced by the roots. Regardless, for the purpose of determining whether any specific mixture of carboxylates, sugars, auxiliary metabolites, and amino acids plays a role in the exudate of the root, no reasonable investigations were conducted in order to determine the role of any particular mixture in the exudate of the root. It is also important to note that from time to time, it has also been discovered that root exudates also have a suppressive effect on the debasement of PCCs to some extent. As outlined by Phillips *et al.* (2012), it has been found that root exudates have an adverse effect on the mineralization fastness of n-hexadecane, naphthalene, and phenanthrene in horse feed (*Medicago sativa* L.) and wild rye (*Elymus angustus* Trin.) when exposed to the root exudates. Root exudate intensities of individual plants can be exceptionally diverse, depending on factors such as the growth phase of the plants, the availability of supplements, the amount of sunlight, the pH level of the soil, and the degree of climate (Gransee & Wittenmayer, 2000; Hodge *et al.*, 1997; Hütsch *et al.*, 2002; Leigh *et al.*, 2002; Neumann, 2007; Rovira, 1959).

Moreover, it has also been observed that organic boundaries, such as mycorrhizal parasites, can additionally affect the rate at which the roots exude; in particular, there are several examples of plants that show a slowdown in root exudation as a result of these parasites (Marschner *et al.*, 1997; Ryan *et al.*, 2012). It is observed that there is a fluctuation in the convergence of exudates at different parts of the root. As it has been reported, there are many exudates that are delivered to the end of the roots, as well as to the horizontal stretch of the roots at the tip of the roots (Marschner *et al.*, 2011; McDougall & Rovira, 1970; Neumann, 2007). Ryegrass (*Lolium perenne* L.) is one of the plants that contain a significant amount of PHC degrading microorganisms that are present close to the root surface (within 3 mm of the root surface), which results in the most intense deprivation of PHC at that site (Corgie *et al.*, 2004).

11.4 A COMPONENT OF THE RHIZOREMEDIATION PROCESS FOR THE REMOVAL OF NATURAL CONTAMINANTS

Debasements of bacteria for natural impurities at the rhizospheric area happen by two vigorous and anaerobic cycles in regular soils (Boopathy, 2004). Root oozes increment corruption of natural toxins, for example TPH and PHC, through four different routes, causing direct degradation of the rhizosphere by chemicals obtained from the plant, thereby increasing the bioavailability of impurities, metabolic interaction, and feelings as a consequence of the energy/supplement stream (Martin *et al.*, 2014). According to the exploration of the rhizoremediation of natural foreign substances, the accompanying significant systems are assumed to be engaged with this cycle. There is no doubt that compounds of microbial origin are fundamentally facilitating natural pollution corruption pathways; nevertheless, plant exudates contain various extracellular proteins that are capable of accelerating the degradation of a number of polyaromatic hydrocarbons within a short period of time, including peroxidases and laccases (Martin *et al.*, 2014). It has also been suggested that mycorrhizal parasites may play a role in the treatment of PHC particles by delivering extracellular compounds to the cells (Gramss *et al.*, 1999). It is generally accepted that the number of xenobiotics absorbed by a plant from a given environmental condition is directly dependent on the logarithm of the compound's octanol–water segment coefficient (K_{ow}) for that particular plant. Briggs *et al.* (1982) utilized the primary movement relationship as the happening stream focus factor (TSCF) as a mechanism for examining the relationship between the uptake of an active compound and its log K_{ow} esteem in a soilless framework, as a connection between the uptake and its log K_{ow} esteem. At a log K_{ow} value of 2.1, the TSCF for natural mixtures reaches a limit of 0.8 at a log K_{ow} value of 0.8. Compositions with a log K_{ow} value greater than 2.1 are more hydrophobic and are tightly bound to the lipid film of the root, while mixtures with a log K_{ow} value lower than 2.1 are not capable of traveling through the lipid film of the root epidermis. Most natural contaminations are hydrophobic, which means that they are insoluble in soil, and therefore cannot be degraded effectively by microorganisms because they are not readily available to them. By increasing the solubility of toxins, the root exudates were able to enhance their bioavailability, which in turn allowed microorganisms to find them more readily available to attack.

11.5 BIOSURFACTANTS DEVELOPMENT

As far as soil bioremediation is concerned, the bioavailability of the contaminant presents a significant challenge. The hydrophobic way of natural toxins, those are ineffectively disintegrated in the polar atoms like H_2O, and large numbers of these foreign substances turn into big soil particles. These shortages of bioavailability normally diminish evacuation productivity (Johnsen *et al.*, 2005). With the assistance of various techniques, for example, the discharge of biosurfactants, development of bio films, and creation of extracellular polymeric substances (EPS), microorganisms increment the availability of water-repellent mixtures (i.e., PAHs). Biosurfactants are hydrophilic and hydrophobic particles in nature that structure micelles with natural pollutants; accordingly, bioavailability increments and upgrades the rhizoremediation cycle. It is due to the hydrophobic nature of the central area of micelles that the water-repellent pollutants in the micelles are diluted, resulting in an expansion in the movement of the compound from the solid stage to the water stage, which makes them more accessible to bacterial degradation, as they become more water repellent. As biosurfactants, glycolipids are important biomolecules that are produced by microorganisms. Rhamnolipids, which are the main constituents of glycolipids, have been shown to increase the speed at which toxins are absorbed (Cui *et al.*, 2008; Mulligan, 2005; Providenti *et al.*, 1995; Shreve *et al.*, 1995; Zhang & Miller, 1994). Kuiper *et al.* (2004) observed that two lipopeptide biosurfactants secreted by *Pseudomonas putida* at the PAH contaminated areas expanded the emulsion development with toluene due to the disengagement of these putisolvins from plant roots. In terms of bioremediation, the pursuit of rhizobacteria is of great importance as it assists in improving the bioavailability of pollutants, and as such is of great importance. There are a large number of microorganisms that follow a positive chemotaxis pattern toward toxins, making them a good candidate for further research (Parales & Haddock, 2004). Therefore, due to the common activity of chemotaxis and biosurfactants in contaminated soils, they help to promote the development of microbial populations and to promote bacteria proliferation, thereby encouraging more zones to be cleared.

11.6 NATURAL CORROSIVE CREATION

Different rhizospheric microorganisms secrete various natural acids, for example, gluconic corrosive, phytic corrosive, and so on, which bring down the pH of the soil, in this manner upgrading the dissolvability of impurities and expanding their debasement (Oberai & Khanna, 2018). In recent times, it has been demonstrated that certain LMW aliphatic carboxylates, for example, oxalate and citrate, increase the PHCs in soil by causing a deviation from the soil grid (An *et al.*, 2010, 2011; Gao *et al.*, 2010). The system of desorption, on the other hand, is quite insignificant, and it is speculated that delivered soil natural components might cause the liberation of sorbed natural impurities or because of a competition for adsorption spots among carboxylates and certain PHCs, the adsorption of toxins could be diminished (An *et al.*, 2010; Reilley *et al.*, 1996).

11.7 UNDERLYING RELATIONSHIP AND CO-METABOLISM

Different auxiliary metabolites discharged from the roots of plants, for example, flavonoids, showed a comparable primary likeness to natural contaminations, specifically to sweet-smelling hydrocarbons (Bais *et al.*, 2008; Singer, 2006). According to this underlying relationship, the hazard of contaminations is improved by impacting appropriate enzymatic pathways, notwithstanding expansion in metabolic processes (Fletcher & Hegde, 1995). As per Kanaly and Bartha (1999), the mineralization of benzo-a-pyrene took place in the presence of other essential substrata by metabolic methods.

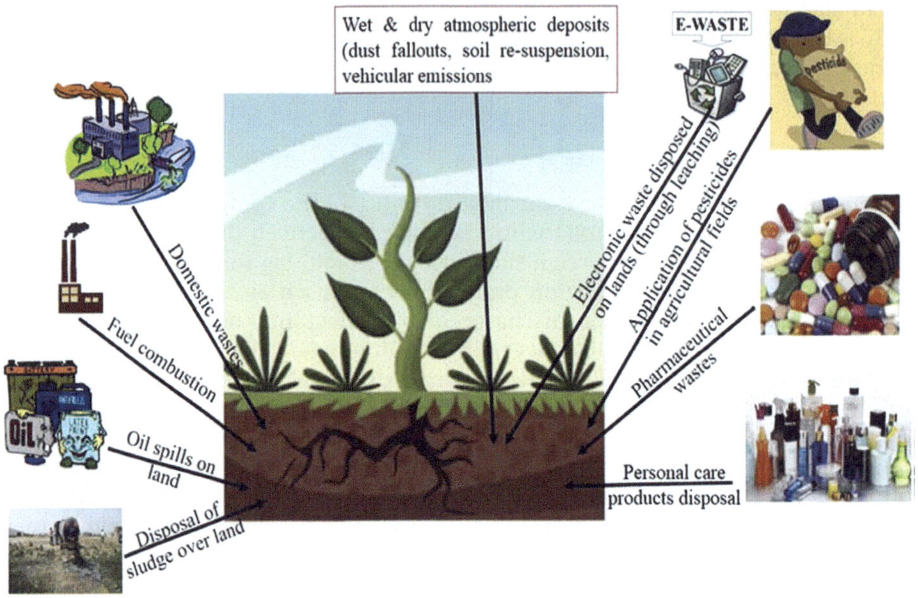

Figure 11.1 Wet and dry atmospheric depositions.

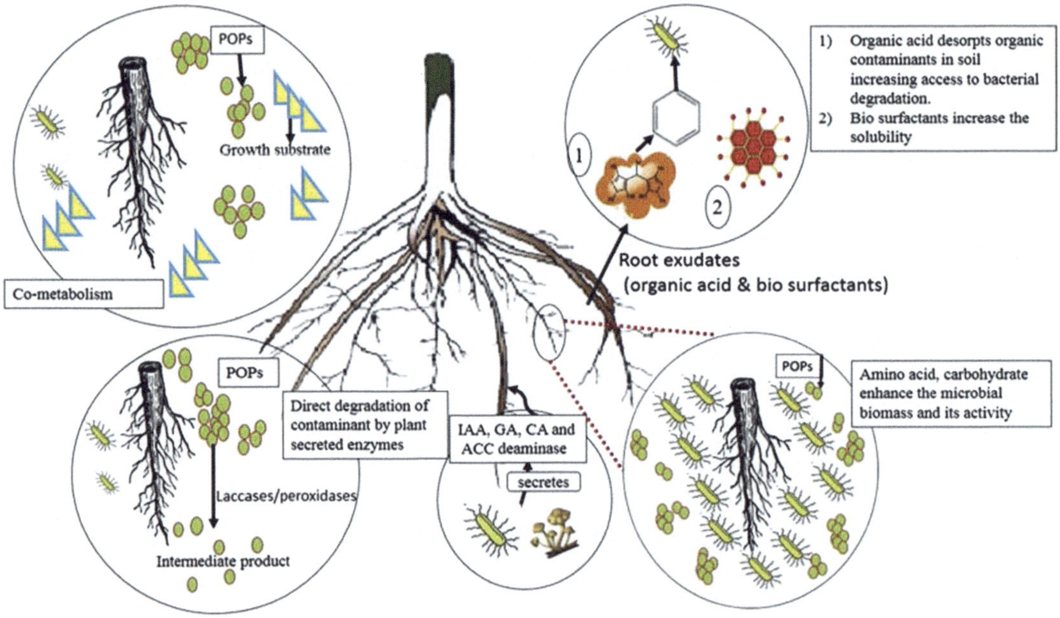

Figure 11.2 Interaction among plants and microbes at the time of biodegradation of organic contaminants.

11.8 LIVELINESS AND SUPPLEMENT STREAM

Root oozes of plants convey a consistent measure of supplements and energy to the microbes to upgrade the corruption of natural contamination (Anderson *et al.*, 1993; Kuiper *et al.*, 2004). Up to 10–40% of photosynthetic results of flora might return to the soil, which relates to approximately 15–60 kg nitrogen/ha/yr and 800–4500 kg carbon/ha/yr (Gerhardt *et al.*, 2009; Grayston *et al.*, 1997; Hütsch *et al.*, 2002; Lynch & Whipps, 1990; Singh *et al.*, 2004). Nitrogen and carbon are conveyed by oozes of roots to rhizospheric microorganisms like hydrophobic, high atomic weight (HMW) natural polymers or, as a solvent, LMW natural particles. The HMW natural polymers partake as a line of adhesive, shedding cells, and different types of root trash (Bais *et al.*, 2008). Since HMW natural polymers are generally nonsoluble and complexes naturally, it consequently has biodegradation rates to some degree (Kalbitz *et al.*, 2003; Kuzyakov, 2002). LMW exudates have higher solvency in water, and as a result of this explanation, microorganisms rapidly take up these mixtures. The principal parts of LMW root oozes are AAs, natural corrosive anions, and carbs. These parts are considered the fundamental wellspring of supplements and energy to the rhizosphere on the grounds that these are effectively decomposable (Van Hees *et al.*, 2005). Quick reaction is displayed regarding their breadth (inside of) by microbe networks when LMW root oozes are included in soil; notwithstanding, perception for quality articulation deviations might be required a few times (Darrah, 1991; Jones & Murphy, 2007). Da Silva *et al.* (2006) found the oozes delivered by roots of *Morus* (*Morus rubra* L.) that affected the development of PAH debasers in an equal way to that of all-out microbes. Nevertheless, there is a possibility of some catabolic restraint because of the presence of some additional more labile carbon bases in PHC-contaminated soils, as well as root oozes (Cases & de Lorenzo, 2005; Singer, 2006). Yuste *et al.* (1998) uncovered that *Pseudomonas* species contained high levels of carbon bases. As further concentration was initiated, it was found that when root concentrates and oozes were added to polluted soils, *P. putida* showed a decline in its ability to degrade phenanthrene (Rentz *et al.*, 2004).

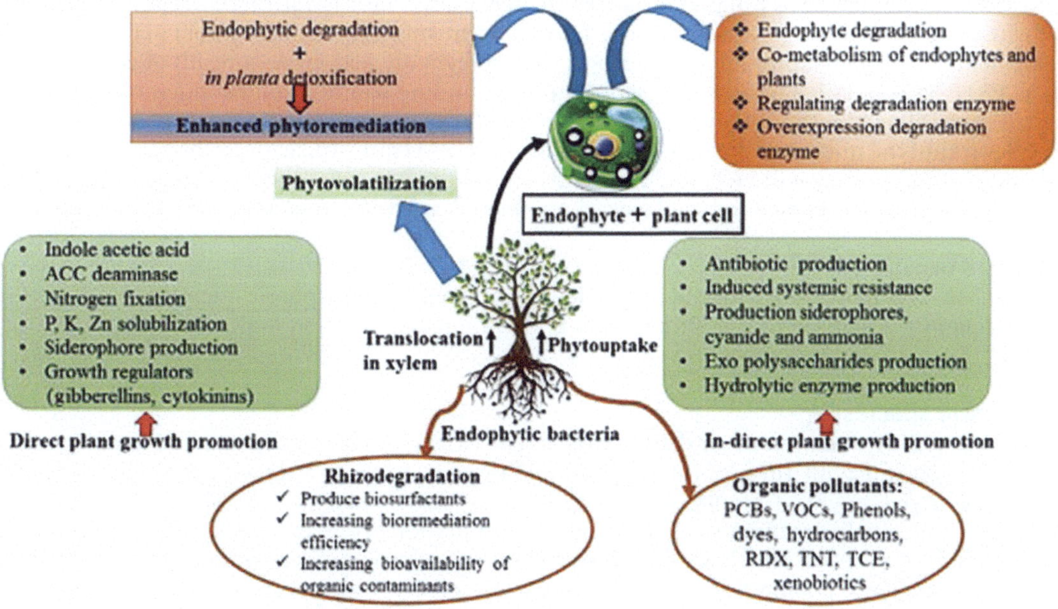

Figure 11.3 Rhizoremediation – a green technology.

11.9 RHIZOREMEDIATION: FACTORS INFLUENCING ITS EFFECTIVENESS

A number of factors influences the course of rhizoremediation, for example, natural variables, such as the soil temperature, pH, and soil organic matter, which are factors related to rhizoremediation. Other than this, different factors, for example, plant species and microbial populace associated with the strategy, additionally influence the cycle; bioavailability of the contaminations is another significant angle in the technique. These components eventually impact the contamination's appearance toward microbial variety.

11.10 ECOLOGICAL VARIABLES AND ITS POSSIBLE IMPACT

11.10.1 Temperature

Bandowe *et al.* (2014) expressed that it fills in as an urgent job in the redress of PAH by increasing its bioavailability as its dissolvability will, in general, ascent along the increment of temperature. The enzymatic exercises interceded due to the bacterial populace get accelerated along the climb of temperature to the ideal level, that is, Liang *et al.* (2003) led a trial in which they used to gage how much oxygen is required for the action of organisms during the most common way of treating the soil. At the time of this cycle, the convergence expanding oxygen seems a lot higher at 43°C rather than 22°C, 29°C, or 36°C. Similarly, Chung *et al.* (2007) found that temperature also affects the ability of bacterial cells and soil elements to adsorb and desorb toxins by influencing their efficiency of adsorption and desorption.

11.10.2 pH

In general, the majority of microorganism proteins released for the biodegradation of mixtures are reliant on pH levels that range from 6.5 to 7.5, which is about the same as the pH inside the cell. As described by Siddique *et al.* (2002), *Pandoraea* sp., which was isolated from an enhanced culture by Okeke *et al.* (2002), degrades HCH isomers at a pH range of 4–9 (Siddique *et al.*, 2002). The biodegradation of bio pesticides in the soil has been found to be at a low pH as opposed to nonpartisan and basic soils, as indicated by Singh *et al.* (2006) in their study. There is a strong correlation between pH and the remedy of PAH, since various kinds of microorganisms have different ideal pH levels for their development, thereby enabling them to perform their enzymatic action. As a result of contamination, such as PAHs, it is possible to alter the pH of the surrounding soil of the local microbiological zone, which in turn influences the R-O response and the solubility of toxins in the rhizosphere (Brito *et al.*, 2015). Several of the microbes show potential to degrade PAHs in situ, as described by Bamforth and Singleton (2005), yet they will be eminently obstructed if there is a significant change in pH, which is why change in pH is recommended in places where there is a significant amount of soil contamination.

11.10.3 Soil organic matter

Natural substances in the soil bring about supplement hotspot of rhizospheric microbial bio-organisms, and subsequently assist them in multiplication and enhance adsorption and desorption components of toxins.

11.10.4 Floral species

Flora has the ability to eliminate or debase toxins. For example, organics in soil and microorganisms exist in relationship with these flora add additional work to their assets. A few microorganisms advance the development of plants by their PGP movement; consequently, plants discharge a few metabolites, which help organisms engendering and evacuation proficiency of toxins. Other than this, flora discharge root oozes, which act as a C source to the microorganisms populace and assist them in accomplishing more populace in the rhizo zone; what's more, root oozes might have biosurfactants that upgrade the bio availability of the pollutants in rhizo zoned soil (Vergani *et al.*, 2017). Flora can influence natural toxins and particle matter levels in the air; McLachlan (1999) referred to it as 'backwoods channel impact'; yet it can bind to soil particles firmly with low solvency in water, which turn the accessibility difficult for degradation (Terzaghi *et al.*, 2018).

11.10.5 Movement of microorganisms

The movement of organisms in the rhizosphere can be impacted due to variables, sorting conditions, and species utilized, yet the existence of impurities can change the capability of microorganisms which normally takes place in the rhizosphere. There is a pressure state created by these toxins on the microbial local area, which drives the microbial local area to embrace and foster new characteristics against the pollutants in order to cope with these toxins. There is no doubt that species that are not a part of these polluted locales are capable of handling pollutants to a greater degree, and so these species can be used for bioremediation of impurities at sites where there is high contamination. However, they could replace the existing organisms at that site, and it would be unwise to ignore their opposition (Momose *et al.*, 2008). Aside from that, the microorganisms also possess resistance against toxins such as PAHs, or they can also mirror the capacity for degrading, so we can additionally be able to create the valuable opportunity for separating those qualities and using them in alternate ways of bioremediation (Mahmoudi *et al.*, 2011). Li *et al.* (2015) proved the beneficial properties of debasing arbitrates of PAHs from *Arthrobacter* SA02.

11.11 FACTORS INFLUENCING RHIZOREMEDIATION: ACCESSIBILITY OF TOXINS

11.11.1 Accessibility of toxins

The bio-accessibility of the impurities that are impacted by their dissolvability, focus on soil contamination, adsorption, and construction. Surfactants that are generally used improve the accessibility of contamination. For instance, sodium dodecyl benzenesulfonate was additional, D9 unsaturated fat desaturase focus was supported, which prompts enhanced ease of cell film by raising the unsaturated fat level, consequently strengthening the vehicle of impurity atom across the layer (Li *et al.*, 2015). Surfactants increment the corruption of PAHs in microbes by regulating certain qualities, that is, 1H2Nase and RHDase and guidelines of the cell film (Kuiper *et al.*, 2004; Ronchel & Ramos 2001).

11.12 CONCLUSION AND FUTURE POINTS OF VIEW

Natural substances that are contaminated pose a danger to the health of humans, as well as negatively affect our current circumstances, so it is urgent that a proficient arrangement be put in place to address this issue. In past examinations, a few plants like *M. sativa*, *Hordeum vulgare*, *Triticum aestivum*, and *Phragmites australis*, accounted for the effective expulsion of natural toxins from the soil (Kuiper *et al.*, 2004). For the rhizoremediation process, a rhizosphere microbial strain ought to be immunized with the plant seeds by covering them with microorganisms, so organisms can develop in relationship with the underlying foundations of that plant effectively and speed up remediation of contamination. This sort of vaccination of microorganisms will assist the organisms to spread along the underground part and increment the possibility of spreading in the soil. Kuiper *et al.* (2001) have screened a plant–organism connection, which showed the capability to debase naphthalene and the ability to give soil seeds some resistance to the presence of high concentrations of this toxin when sown. At the point when the seeds of grasses were covered with this bacterial strain and afterwards developed, it was recorded that the local bacterial area had the option to develop and spread further in soil with the germination and growth of the plant. Currently, *Pseudomonas* with recombinant DNA innovation has the ability to endure toxins in the rhizosphere, which was impractical for the parent bacterial strain. Despite the fact that there are many studies indicating rhizoremediation as a feasible method for determining removal of natural contaminants from soil, further research studies, newer plants, and microbial strains are required for the implementation of this method on a larger scale. As a general rule, we can conclude that rhizoremediation has demonstrated to be a successful approach for removing natural toxins from soil. In spite of this, most of the ongoing research is being carried out on a laboratory scale, and more efforts will be made in the near future to enhance this method to a field setting. Recent decades have seen a tremendous emphasis on rhizoremediation technologies, which are cost-effective and eco-friendly, as well as the utilization of plant-related microbes to decontaminate and eradicate engineered natural mixtures from ecological soils. There is a possibility that this innovation might provide an effective, financial, and reasonable green remediation innovation for the reduction of natural toxins from the soil in the event that is viewed in a serious way.

ACKNOWLEDGEMENT

The authors wish to express their sincere appreciation to Lovely Professional University's Department of Agronomy for the constant support and encouragement provided throughout the research process.

AUTHOR CONTRIBUTIONS

As well as contributing to the outline, the author is responsible for leading the writing process and editing the manuscript after it has been drafted. Because of extensive literature searches, D.H.L., P.K., and J.S. contributed to the drafting of the sections as well as the construction of figures and tables. In addition to providing professional advice, P.K. and J.S. helped revise the final version of the document. The final version of the chapter was read and approved by all authors before it was submitted for publication.

CONFLICTS OF INTEREST

None.

REFERENCES

Agusa T., Kunito T., Nakashima E., Minh T. B., Tanabe S., Subramanian A. and Viet P. H. (2003). Preliminary studies on trace element contamination in dumping sites of municipal wastes in India and Vietnam. In: Journal de Physique IV (Proceedings). EDP Sciences, France, Vol. **107**, pp. 21–24, https://doi.org/10.1051/jp4:20030233

Alamdar A., Syed J. H., Malik R. N., Katsoyiannis A., Liu J., Li J., Zhang G. and Jones K. C. (2014). Organochlorine pesticides in surface soils from obsolete pesticide dumping ground in Hyderabad City, Pakistan: contamination levels and their potential for air-soil exchange. *The Science of the Total Environment*, **470**, 733–741, https://doi.org/10.1016/j.scitotenv.2013.09.053

Ali U., Syed J. H., Malik R. N., Katsoyiannis A., Li J., Zhang G. and Jones K. C. (2014). Organochlorine pesticides (OCPs) in South Asian region: a review. *The Science of the Total Environment*, **476**, 705–717, https://doi.org/10.1016/j.scitotenv.2013.12.107

Alkorta I. and Garbisu C. (2001). Phytoremediation of organic contaminants in soils. *Bioresource Technology*, **79**(3), 273–276, https://doi.org/10.1016/S0960-8524(01)00016-5

An C., Huang G., Yu H., Wei J., Chen W. and Li G. (2010). Effect of short-chain organic acids and pH on the behaviors of pyrene in soilewater system. *Chemosphere*, **81**(11), 1423–1429, https://doi.org/10.1016/j.chemosphere.2010.09.012

An C. J., Huang G. H., Wei J. and Yu H. (2011). Effect of short-chain organic acids on the enhanced desorption of phenanthrene by rhamnolipid biosurfactant in soil water environment. *Water Research*, **45**(17), 5501–5510, https://doi.org/10.1016/j.watres.2011.08.011

Anderson T. A., Guthrie E. A. and Walton B. T. (1993). Bioremediation in the rhizosphere. *Environment Science Technology*, **27**(13), 2630–2636, https://doi.org/10.1021/es00049a001

Atapattu S. S. and Kodituwakku D. C. (2009). Agriculture in South Asia and its implications on downstream health and sustainability: a review. *Agricultural Water Management*, **96**(3), 361–373, https://doi.org/10.1016/j.agwat.2008.09.028

Bais H. P., Broeckling C. D. and Vivanco J. M. (2008). Root exudates modulate plant and microbe interactions in the rhizosphere. In: Secondary Metabolites in Soil Ecology, P. Karlovsky (ed.), Springer, Berlin, Heidelberg, pp. 241–252.

Bamforth S. M. and Singleton I. (2005). Bioremediation of polycyclic aromatic hydrocarbons: current knowledge and future directions. *Journal of Chemical Technology and Biotechnology*, **80**(7), 723–736, https://doi.org/10.1002/jctb.1276

Bandowe B. A. M., Bigalke M., Boamah L., Nyarko E., Saalia F. K. and Wilcke W. (2014). Polycyclic aromatic compounds (PAHs and oxygenated PAHs) and trace metals in fish species from Ghana (West Africa): bioaccumulation and health risk assessment. *Environment International*, **65**, 35–146.

Barra R., Popp P., Quiroz R., Bauer C., Cid H. and von Tümpling W. (2005). Persistent toxic substances in soils and waters along an altitudinal gradient in the Laja River Basin, Central Southern Chile. *Chemosphere*, **58**(7), 905–915, https://doi.org/10.1016/j.chemosphere.2004.09.050

Baudoin E., Benizri E. and Guckert A. (2003). Impact of artificial root exudates on the bacterial community structure in bulk soil and maize rhizosphere. *Soil Biology and Biochemistry*, **35**, 1183–1192, https://doi.org/10.1016/S0038-0717(03)00179-2

Behera B. and Singh G. S. (1999). Studies on weed management in monsoon season crop of tomato. *Indian Journal of Weed Science*, **31**(1 and 2), 67–70.

Benizri E., Dedourge O., Dibattista-Leboeuf C., Piutti S., Nguyen C. and Guckert A. (2002). Effect of maize rhizodeposits on soil microbial community structure. *Applied Soil Ecology*, **21**, 261–265, https://doi.org/10.1016/S0929-1393(02)00094-X

Bhattacharya B., Sarkar S. K. and Mukherjee N. (2003). Organochlorine pesticide residues in sediments of a tropical mangrove estuary, India: implications for monitoring. *Environment International*, **29**(5), 587–592, https://doi.org/10.1016/S0160-4120(03)00016-3

Bisht S., Pandey P., Sood A., Sharma S. and Bisht N. S. (2010). Biodegradation of naphthalene and anthracene by chemotactically active rhizobacteria of *Populus deltoides*. *Brazilian Journal of Microbiology*, **41**(4), 922–930, https://doi.org/10.1590/S1517-83822010000400011

Bisht S., Pandey P., Kaur G., Aggarwal H., Sood A., Sharma S., Kumar V. and Bisht N. S. (2014). Utilization of endophytic strain *Bacillus* sp. SBER3 for biodegradation of polyaromatic hydrocarbons (PAH) in soil model system. *European Journal of Soil Biology*, **60**, 67–76, https://doi.org/10.1016/j.ejsobi.2013.10.009

Boopathy R. (2004). Anaerobic biodegradation of no. 2 diesel fuel in soil: a soil column study. *Bioresource Technology*, **94**(2), 143–151, https://doi.org/10.1016/j.biortech.2003.12.006

Brazil G. M., Kenefick L., Callanan M., Haro A., De Lorenzo V., Dowling D. N. and O'gara F. (1995). Construction of a rhizosphere pseudomonad with potential to degrade polychlorinated biphenyls and detection of bph gene expression in the rhizosphere. *Applied and Environmental Microbiology*, **61**(5), 1946–1952, https://doi.org/10.1128/aem.61.5.1946-1952.1995

Briggs G. G., Bromilow R. H. and Evans A. A. (1982). Relationships between lipophilicity and root uptake and translocation of non-ionised chemicals by barley. *Pesticide Science*, **13**(5), 495–504, https://doi.org/10.1002/ps.2780130506

Brink S. C. (2016). Unlocking the secrets of the rhizosphere. *Trends in Plant Science*, **21**(3), 169–170, https://doi.org/10.1016/j.tplants.2016.01.020

Brito E. M., De la Cruz Barrón M., Caretta C. A., Goñi-Urriza M., Andrade L. H., Cuevas-Rodríguez G., Malm O., Torres J. P., Simon M. and Guyoneaud R. (2015). Impact of hydrocarbons, PCBs and heavy metals on bacterial communities in Lerma River, Salamanca, Mexico: investigation of hydrocarbon degradation potential. *The Science of the Total Environment*, **521–522**, 1–10, https://doi.org/10.1016/j.scitotenv.2015.02.098

Buccini J. (2003). The development of a global treaty on persistent organic pollutants (POPs). In: Persistent Org Pollut, H. Fiedler (ed.), Springer, Berlin, Heidelberg, pp. 13–30.

Butler J. L., Williams M. A., Bottomley P. J. and Myrold D. D. (2003). Microbial community dynamics associated with rhizosphere carbon flow. *Applied and Environmental Microbiology*, **69**(11), 6793–6800, https://doi.org/10.1128/AEM.69.11.6793-6800.2003

Cachada A., Pato P., Rocha-Santos T., da Silva E. F. and Duarte A. C. (2012). Levels, sources and potential human health risks of organic pollutants in urban soils. *The Science of the Total Environment*, **430**, 184–192, https://doi.org/10.1016/j.scitotenv.2012.04.075

Cannon H. L. and Bowles J. M. (1962). Contamination of vegetation by tetraethyl lead. *Science*, **137**(3532), 765–766, https://doi.org/10.1126/science.137.3532.765

Cases I. and de Lorenzo V. (2005). Genetically modified organisms for the environment: stories of success and failure and what we have learned from them. *International Microbiology*, **8**(3), 213–222.

Chakraborty P., Zhang G., Li J., Xu Y., Liu X., Tanabe S. and Jones K. C. (2010). Selected organochlorine pesticides in the atmosphere of major Indian cities: levels, regional versus local variations, and sources. *Environmental Science and Technology*, **44**(21), 8038–8043, https://doi.org/10.1021/es102029t

Chaudhry Q., Blom-Zandstra M., Gupta S. K. and Joner E. (2005). Utilising the synergy between plants and rhizosphere microorganisms to enhance breakdown of organic pollutants in the environment. *Environmental Science and Pollution Research International*, **12**, 34–48, https://doi.org/10.1065/espr2004.08.213

Chung M. K., Tsui M. T., Cheung K. C., Tam N. F. and Wong M. H. (2007). Removal of aqueous phenanthrene by brown seaweed *Sargassum hemiphyllum*: sorption-kinetic and equilibrium studies. *Separation and Purification Technology*, **54**(3), 355–362, https://doi.org/10.1016/j.seppur.2006.10.008

Corgie S. C., Beguiristain T. and Leyval C. (2004). Spatial distribution of bacterial communities and phenanthrene degradation in the rhizosphere of *Lolium perenne* L. *Applied and Environmental Microbiology*, **70**(6), 3552–3557, https://doi.org/10.1128/AEM.70.6.3552-3557.2004

Cui C. Z., Zeng C., Wan X., Chen D., Zhang J. Y. and Shen P. (2008). Effect of rhamnolipids on degradation of anthracene by two newly isolated strains, *Sphingomonas* sp. 12A and *Pseudomonas* sp. 12B. *Journal of Microbiology and Biotechnology*, **18**(1), 63–66.

Da Silva M. L., Kamath R. and Alvarez P. J. (2006). Effect of simulated rhizodeposition on the relative abundance of polynuclear aromatic hydrocarbon catabolic genes in a contaminated soil. *Environmental Toxicology and Chemistry*, **25**(2), 386–391, https://doi.org/10.1897/05-321R.1

Darrah P. R. (1991). Models of the rhizosphere. *Plant and Soil*, **133**(2), 187–199, https://doi.org/10.1007/BF00009191

Dzantor E. K. (2007). Phytoremediation: the state of rhizosphere 'engineering' for accelerated rhizodegradation of xenobiotic contaminants. *Journal of Chemical Technology and Biotechnology*, **82**(3), 228–232, https://doi.org/10.1002/jctb.1662

El Haleem D. A., Von Wintzingerode F., Moter A., Moawad H. and Göbel U. B. (2000). Phylogenetic analysis of rhizosphere-associated b-subclass proteobacterial ammonia oxidizers in a municipal wastewater treatment plant based on rhizoremediation technology. *Letters in Applied Microbiology*, **31**(1), 34–38, https://doi.org/10.1046/j.1472-765x.2000.00760.x

El Nemr A., Moneer A. A., Khaled A. and El-Sikaily A. (2013). Levels, distribution, and risk assessment of organochlorines in surficial sediments of the Red Sea coast, Egypt. *Environmental Monitoring and Assessment*, **185**(6), 4835–4853, https://doi.org/10.1007/s10661-012-2907-3

Eqani S. A. M. A. S., Malik R. N. and Mohammad A. (2011). The level and distribution of selected organochlorine pesticides in sediments from river Chenab, Pakistan. *Environmental Geochemistry and Health*, **33**(1), 33–47, https://doi.org/10.1007/s10653-010-9312-z

Fabietti G., Biasioli M., Barberis R. and Ajmone-Marsan F. (2010). Soil contamination by organic and inorganic pollutants at the regional scale: the case of Piedmont, Italy. *Journal of Soils and Sediments*, **10**(2), 290–300, https://doi.org/10.1007/s11368-009-0114-9

Fletcher J. S. and Hegde R. S. (1995). Release of phenols by perennial plant roots and their potential importance in bioremediation. *Chemosphere*, **31**(4), 3009–3016, https://doi.org/10.1016/0045-6535(95)00161-Z

Gao Y., Ren L., Ling W., Kang F., Zhu X. and Sun B. (2010). Effects of low-molecular-weight organic acids on sorption and desorption of phenanthrene in soils. *Soil Science Society of America Journal*, **74**(1), 51–59, https://doi.org/10.2136/sssaj2009.0105

Gerhardt K. E., Huang X. D., Glick B. R. and Greenberg B. M. (2009). Phytoremediation and rhizoremediation of organic soil contaminants: potential and challenges. *Plant Science*, **176**(1), 20–30, https://doi.org/10.1016/j.plantsci.2008.09.014

Glazer A. N. and Nikaido H. (1995). Biomass. In: Microbial Biotechnology: Fundamentals of Applied Microbiology, A. N. Glazer (ed.), Freeman and Co, New York, p. 662.

Gramss G., Kirsche B., Voigt K., Gunther T. and Fritsche W. (1999). Conversion rates of five polycyclic aromatic hydrocarbons in liquid cultures of fifty-eight fungi and the concomitant production of oxidative enzymes. *Mycological Research*, **103**, 1009–1018, https://doi.org/10.1017/S0953756298008144

Gransee A. and Wittenmayer L. (2000). Qualitative and quantitative analysis of water-soluble root exudates in relation to plant species and development. *Journal of Plant Nutrition and Soil Science*, **163**, 381–385, https://doi.org/10.1002/1522-2624(200008)163:4<381::AID-JPLN381>3.0.CO;2-7

Grayston S. J., Vaughan D. and Jones D. (1997). Rhizosphere carbon flow in trees, in comparison with annual plants: the importance of root exudation and its impact on microbial activity and nutrient availability. *Applied Soil Ecology*, **5**(1), 29–56, https://doi.org/10.1016/S0929-1393(96)00126-6

Gupta P. K. (2004). Pesticide exposure – Indian scene. *Toxicology*, **198**(1–3), 83–90, https://doi.org/10.1016/j.tox.2004.01.021

Gupta A., Singh M., Pandey K. D. and Kumar A. (2018). Biodegradation of xenobiotics using cyanobacteria. *Journal of Scientific Research ISSN*, **62**, 35–40.

Guzzella L., Roscioli C., Vigano L., Saha M., Sarkar S. K. and Bhattacharya A. (2005). Evaluation of the concentration of HCH, DDT, HCB, PCB and PAH in the sediments along the lower stretch of Hugli estuary, West Bengal, northeast India. *Environment International*, **31**(4), 523–534, https://doi.org/10.1016/j.envint.2004.10.014

Hartmann A., Schmid M., Tuinen D. and Berg G. (2009). Plant-driven selection of microbes. *Plant and Soil*, **321**, 235–257, https://doi.org/10.1007/s11104-008-9814-y

Hoagland R. E., Zablotowicz R. M. and Locke M. A. (1994). Propanil metabolism by rhizosphere microflora. In: Bioremediation Through Rhizosphere Technology, T. A. Anderson and J. R. Coats (eds.), American Chemical Society, Washington, DC, pp. 160–183.

Hodge A., Paterson E., Thornton B., Millard P. and Killham K. (1997). Effects of photon flux density on carbon partitioning and rhizosphere carbon flow of *Lolium perenne*. *Journal of Experimental Botany*, **48**, 1797–1805, https://doi.org/10.1093/jxb/48.10.1797

Hütsch B. W., Augustin J. and Merbach W. (2002). Plant rhizodeposition: an important source for carbon turnover in soils. *Journal of Soil Science and Plant Nutrition*, **165**(4), 397–407, https://doi.org/10.1002/1522-2624(200208)165:4<397::AID-JPLN397>3.0.CO;2-C

Ingrid L., Sahraoui A. L. H., Frédéric L., Yolande D. and Joël F. (2016). Arbuscular mycorrhizal wheat inoculation promotes alkane and polycyclic aromatic hydrocarbon biodegradation: microcosm experiment on aged-contaminated soil. *Environmental Pollution*, **213**, 549–560, https://doi.org/10.1016/j.envpol.2016.02.056

Jacobsen C. S. (1997). Plant protection and rhizosphere colonization of barley by seed inoculated herbicide degrading *Burkholderia* (*Pseudomonas*) *cepacia* DBO1 (pRO101) in 2, 4-D contaminated soil. *Plant and Soil*, **189**(1), 139–144, https://doi.org/10.1023/A:1004296615446

Jan M. R., Shah J., Khawaja M. A. and Gul K. (2009). DDT residue in soil and water in and around abandoned DDT manufacturing factory. *Environmental Monitoring and Assessment*, **155**(1–4), 31–38, https://doi.org/10.1007/s10661-008-0415-2

Johnsen A. R., Wick L. Y. and Harms H. (2005). Principles of microbial PAH-degradation in soil. *Environmental Pollution*, **133**(1), 71–84, https://doi.org/10.1016/j.envpol.2004.04.015

Joner E. J., Corgié S. C., Amellal N. and Leyval C. (2002). Nutritional constraints to degradation of polycyclic aromatic hydrocarbons in a simulated rhizosphere. *Soil Biology and Biochemistry*, **34**, 859–864, https://doi.org/10.1016/S0038-0717(02)00018-4

Jones D. L. and Murphy D. V. (2007). Microbial response time to sugar and amino acid additions to soil. *Soil Biology and Biochemistry*, **39**(8), 2178–2182, https://doi.org/10.1016/j.soilbio.2007.03.017

Kalbitz K., Schmerwitz J., Schwesig D. and Matzner E. (2003). Biodegradation of soil-derived dissolved organic matter as related to its properties. *Geoderma*, **113**(3–4), 273–291, https://doi.org/10.1016/S0016-7061(02)00365-8

Kanaly R. A. and Bartha R. (1999). Cometabolic mineralization of benzo [a] pyrene caused by hydrocarbon additions to soil. *Environmental Toxicology and Chemistry*, **18**(10), 2186–2190.

Kelley S. L., Aitchison E. W., Deshpande M., Schnoor J. L. and Alvarez P. J. (2001). Biodegradation of 1, 4-dioxane in planted and unplanted soil: effect of bioaugmentation with *Amycolata* sp. CB1190. *Water Research*, **35**(16), 3791–3800, https://doi.org/10.1016/S0043-1354(01)00129-4

Khan Z., Roman D., Kintz T., delas Alas M., Yap R. and Doty S. (2014). Degradation, phytoprotection and phytoremediation of phenanthrene by endophyte *Pseudomonas putida*, PD1. *Environmental Science and Technology*, **48**(20), 12221–12228, https://doi.org/10.1021/es503880t

Kingsley M. T., Fredrickson J. K., Metting F. B. and Seidler R. J. (1994). Environmental Restoration Using Plant-Microbe Bioaugmentation. Bioremediation of Chlorinated and Polyaromatic Hydrocarbon Compounds. Lewis Publishers, Boca Raton, Florida, pp. 287–292.

Kuiper I., Bloemberg G. V. and Lugtenberg B. J. (2001). Selection of a plant-bacterium pair as a novel tool for rhizostimulation of polycyclic aromatic hydrocarbon-degrading bacteria. *Molecular Plant-Microbe Interactions*, **14**(10), 1197–1205, https://doi.org/10.1094/MPMI.2001.14.10.1197

Kuiper I., Lagendijk E. L., Bloemberg G. V. and Lugtenberg B. J. (2004). Rhizoremediation: a beneficial plant-microbe interaction. *Molecular Plant-Microbe Interactions*, **17**(1), 6–15, https://doi.org/10.1094/MPMI.2004.17.1.6

Kuzyakov Y. (2002). Factors affecting rhizosphere priming effects. *Journal of Plant Nutrition and Soil Science*, **165**(4), 382–396, https://doi.org/10.1002/1522-2624(200208)165:4<382::AID-JPLN382>3.0.CO;2-#

Leigh M. B., Fletcher J. S., Fu X. and Schmitz F. J. (2002). Root turnover: an important source of microbial substrates in rhizosphere remediation of recalcitrant contaminants. *Environmental Science and Technology*, **36**, 1579–1583, https://doi.org/10.1021/es015702i

Li F., Zhu L., Wang L. and Zhan Y. (2015). Gene expression of an *Arthrobacter* in surfactant-enhanced biodegradation of a hydrophobic organic compound. *Environmental Science and Technology*, **49**(6), 3698–3704, https://doi.org/10.1021/es504673j

Liang C., Das K. C. and McClendon R. W. (2003). The influence of temperature and moisture contents regimes on the aerobic microbial activity of a biosolids composting blend. *Bioresource Technology*, **86**(2), 131–137, https://doi.org/10.1016/S0960-8524(02)00153-0

Liu R., Xiao N., Wei S., Zhao L. and An J. (2014). Rhizosphere effects of PAH-contaminated soil phytoremediation using a special plant named Fire Phoenix. *The Science of the Total Environment*, **473**, 350–358, https://doi.org/10.1016/j.scitotenv.2013.12.027

Löffler F. E., Ritalahti K. M. and Zinder S. H. (2013). Dehalococcoides and reductive dechlorination of chlorinated solvents. In: Bioaugmentation for Groundwater Remediation, H. F. Stroo, A. Leeson, and C. H. Ward (eds.), Springer, New York, NY, pp. 39–88.

Lugtenberg B. J. J. and de Weger L. A. (1992). Plant root colonization by *Pseudomonas* spp. In: Pseudomonas: Molecular Biology and Biotechnology, E. Galli, S. Silver and B. Witholt (eds.), American Society for Microbiology, Washington, DC, pp. 13–19.

Lynch J. M. and Whipps J. M. (1990). Substrate flow in the rhizosphere. *Plant and Soil*, **129**(1), 1–10, https://doi.org/10.1007/BF00011685

Ma W. L., Li Y. F., Sun D. Z. and Qi H. (2009). Polycyclic aromatic hydrocarbons and polychlorinated biphenyls in topsoils of Harbin, China. *Archives of Environmental Contamination and Toxicology*, **57**(4), 670, https://doi.org/10.1007/s00244-009-9314-y

Macek T., Mackova M. and Kás J. (2000). Exploitation of plants for the removal of organics in environmental remediation. *Biotechnology Advances*, **18**(1), 23–34, https://doi.org/10.1016/S0734-9750(99)00034-8

MacKinnon G. and Duncan H. J. (2013). Phytotoxicity of branched cyclohexanes found in the volatile fraction of diesel fuel on germination of selected grass species. *Chemosphere*, **90**(3), 952–957, https://doi.org/10.1016/j.chemosphere.2012.06.038

Mahmoudi N., Slater G. F. and Fulthorpe R. R. (2011). Comparison of commercial DNA extraction kits for isolation and purification of bacterial and eukaryotic DNA from PAH-contaminated soils. *Canadian Journal of Microbiology*, **57**(8), 623–628, https://doi.org/10.1139/w11-049

Malik A., Ojha P. and Singh K. P. (2009). Levels and distribution of persistent organochlorine pesticide residues in water and sediments of Gomti River (India) tributary of the Ganges River. *Environmental Monitoring and Assessment*, **148**(1–4), 421–435, https://doi.org/10.1007/s10661-008-0172-2

Marschner P., Crowley D. and Higashi R. (1997). Root exudation and physiological status of a root-colonizing fluorescent pseudomonad in mycorrhizal and non-mycorrhizal pepper (*Capsicum annuum* L.). *Plant and Soil*, **189**, 11–20, https://doi.org/10.1023/A:1004266907442

Marschner P., Crowley D. and Rengel Z. (2011). Rhizosphere interactions between microorganisms and plants govern iron and phosphorus acquisition along the root axis-model and research methods. *Soil Biology and Biochemistry*, **43**(5), 883–894, https://doi.org/10.1016/j.soilbio.2011.01.005

Martin B. C., George S. J., Price C. A., Ryan M. H. and Tibbett M. (2014). The role of root exuded low molecular weight organic anions in facilitating petroleum hydrocarbon degradation: current knowledge and future directions. *The Science of the Total Environment*, **472**, 642–653, https://doi.org/10.1016/j.scitotenv.2013.11.050

McDougall B. M. and Rovira A. D. (1970). Sites of exudation of ^{14}C-labelled compounds from wheat roots. *New Phytologist*, **69**, 999–1003, https://doi.org/10.1111/j.1469-8137.1970.tb02479.x

McLachlan M. S. (1999). Framework for the interpretation of measurements of SOCs in plants. *Environmental Science and Technology*, **33**(11), 1799–1804, https://doi.org/10.1021/es980831t

Merrington G., Nfa L. W., Parkinson R., Redman M. and Winder L. (2002). Agricultural Pollution: Environmental Problems and Practical Solutions. CRC Press, London.

Minh N. H., Minh T. B., Kajiwara N., Kunisue T., Subramanian A., Iwata H., Tana T. S., Baburajendran R., Karuppiah S., Viet P. H. and Tuyen B. C. (2006). Contamination by persistent organic pollutants in dumping sites of Asian developing countries: implication of emerging pollution sources. *Archives of Environmental Contamination and Toxicology*, **50**(4), 474–481, https://doi.org/10.1007/s00244-005-1087-3

Miya R. K. and Firestone M. K. (2001). Enhanced phenanthrene biodegradation in soil by slender oat root exudates and root debris. *Journal of Environmental Quality*, **30**(6), 1911–1918, https://doi.org/10.2134/jeq2001.1911

Momose Y., Hirayama K. and Itoh K. (2008). Competition for proline between indigenous *Escherichia coli* and *E. coli* O157: H7 in gnotobiotic mice associated with infant intestinal microbiota and its contribution to the colonization resistance against E. coli O157: H7. *Antonie van Leeuwenhoek*, **94**(2), 165–171, https://doi.org/10.1007/s10482-008-9222-6

Moore F. P., Barac T., Borremans B., Oeyen L., Vangronsveld J., Van Der Lelie D., Campbell C. D. and Moore E. R. (2006). Endophytic bacterial diversity in poplar trees growing on a BTEX-contaminated site: the characterisation of isolates with potential to enhance phytoremediation. *Systematic and Applied Microbiology*, **29**(7), 539–556, https://doi.org/10.1016/j.syapm.2005.11.012

Mulligan C. N. (2005). Environmental applications for biosurfactants. *Environmental Pollution*, **133**(2), 183–198, https://doi.org/10.1016/j.envpol.2004.06.009

Neumann G. (2007). Root exudates and nutrient cycling. In: Nutrient Cycling in Terrestrial Ecosystems, P. Marschner and Z. Rengel (eds.), Springer, Berlin, pp. 123–157.

Nichols T. D., Wolf D. C., Rogers H. B., Beyrouty C. A. and Reynolds C. M. (1997). Rhizosphere microbial populations in contaminated soils. *Water, Air, and Soil Pollution*, **95**(1–4), 165–178.

Oberai M. and Khanna V. (2018). Rhizoremediation plant microbe interactions in the removal of pollutants. *International Journal of Current Microbiology and Applied Science*, **7**(1), 2280–2287, https://doi.org/10.20546/ijcmas.2018.701.276

Okeke B. C., Siddique T., Arbestain M. C. and Frankenberger W. T. (2002). Biodegradation of g-hexachlorocyclohexane and a-hexachlorocyclohexane in water and soil slurry by *Pandoraea* sp. *Journal of Agricultural and Food Chemistry*, **50**, 2548–2555, https://doi.org/10.1021/jf011422a

Oleszczuk P. (2006). Persistence of polycyclic aromatic hydrocarbons (PAHs) in sewage sludge-amended soil. *Chemosphere*, **65**(9), 1616–1626, https://doi.org/10.1016/j.chemosphere.2006.03.007

Pai S. G., Riley M. B. and Camper N. D. (2001). Microbial degradation of mefenoxam in rhizosphere of *Zinnia angustifolia*. *Chemosphere*, **44**(4), 577–582, https://doi.org/10.1016/S0045-6535(00)00368-4

Pandey P., Khillare P. S. and Kumar K. (2011). Assessment of organochlorine pesticide residues in the surface sediments of river Yamuna in Delhi, India. *Journal of Environmental Protection*, **2**(05), 511, https://doi.org/10.4236/jep.2011.25059

Parales R. E. and Haddock J. D. (2004). Biocatalytic degradation of pollutants. *Current Opinion in Biotechnology*, **15**(4), 374–379, https://doi.org/10.1016/j.copbio.2004.06.003

Phillips L. A., Greer C. W., Farrell R. E. and Germida J. J. (2012). Plant root exudates impact the hydrocarbon degradation potential of a weathered-hydrocarbon contaminated soil. *Applied Soil Ecology*, **52**, 56–64, https://doi.org/10.1016/j.apsoil.2011.10.009

Pozo K., Harner T., Lee S. C., Sinha R. K., Sengupta B., Loewen M., Geethalakshmi V., Kannan K. and Volpi V. (2011). Assessing seasonal and spatial trends of persistent organic pollutants (POPs) in Indian agricultural regions using PUF disk passive air samplers. *Environmental Pollution*, **159**(2), 646–653, https://doi.org/10.1016/j.envpol.2010.09.025

Providenti M. A., Flemming C. A., Lee H. and Trevors J. T. (1995). Effect of addition of rhamnolipid biosurfactants or rhamnolipid-producing *Pseudomonas aeruginosa* on phenanthrene mineralization in soil slurries. *FEMS Microbiology Ecology*, **17**(1), 15–26, https://doi.org/10.1111/j.1574-6941.1995.tb00123.x

Qiu X., Shah S. I., Kendall E. W., Sorensen D. L., Sims R. C. and Engelke M. C. (1994). Grass-enhanced bioremediation for clay soils contaminated with polynuclear aromatic hydrocarbons. In: Bioremediation Through Rhizosphere Technology, T. A. Anderson and J. R. Coats (eds.), American Chemical Society, Washington, DC, pp. 142–157.

Qixing Z., Zhang C., Zhineng Z. and Weitao L. (2011). Ecological remediation of hydrocarbon contaminated soils with weed plant. *Molecular Ecology Resources*, **2**(2), 97–105.

Radwan S., Sorkhoh N. and Ei-Nemr I. (1995). Oil biodegradation around roots. *Nature*, **376**, 302, https://doi.org/10.1038/376302a0

Reilley K. A., Banks M. K. and Schwab A. P. (1996). Dissipation of polycyclic aromatic hydrocarbons in the rhizosphere. *Journal of Environmental Quality*, **25**(2), 212–219, https://doi.org/10.2134/jeq1996.00472425002500020002x

Rentz J. A., Alvarez P. J. and Schnoor J. L. (2004). Repression of *Pseudomonas putida* phenanthrene-degrading activity by plant root extracts and exudates. *Environmental Microbiology*, **6**(6), 574–583, https://doi.org/10.1111/j.1462-2920.2004.00589.x

Ronchel M. C. and Ramos J. L. (2001). Dual system to reinforce biological containment of recombinant bacteria designed for rhizoremediation. *Applied and Environmental Microbiology*, **67**(6), 2649–2656, https://doi.org/10.1128/AEM.67.6.2649-2656.2001

Ross G. (2005). Risks and benefits of DDT. *The Lancet*, **366**(9499), 1771–1772, https://doi.org/10.1016/S0140-6736(05)67722-7

Rovira A. D. (1956). Plant root excretions in relation to the rhizosphere effect. *Plant and Soil*, **7**, 209–217, https://doi.org/10.1007/BF01666119

Rovira A. D. (1959). Root excretions in relation to the rhizosphere effect. *Plant and Soil*, **11**, 53–64, https://doi.org/10.1007/BF01394753

Ryan M. H., Tibbett M., Edmonds-Tibbett T., Suriyagoda L. D., Lambers H., Cawthray G. R. and Pang J. (2012). Carbon trading for phosphorus gain: the balance between rhizosphere carboxylates and arbuscular mycorrhizal symbiosis in plant phosphorus acquisition. *Plant, Cell and Environment*, **35**, 2170–2180, https://doi.org/10.1111/j.1365-3040.2012.02547.x

Sarkar S. K., Bhattacharya B. D., Bhattacharya A., Chatterjee M., Alam A., Satpathy K. K. and Jonathan M. P. (2008). Occurrence, distribution and possible sources of organochlorine pesticide residues in tropical coastal environment of India: an overview. *Environment International*, **34**(7), 1062–1071, https://doi.org/10.1016/j.envint.2008.02.010

Shann J. R. and Boyle J. J. (1994). Influence of plant species on in situ rhizosphere degradation. In: Bioremediation Through Rhizosphere Technology, T. A. Anderson and J. R. Coats (eds.), American Chemical Society, Washington, DC, pp. 70–81.

Shreve G. S., Inguva S. and Gunnam S. (1995). Rhamnolipid biosurfactant enhancement of hexadecane biodegradation by *Pseudomonas aeruginosa*. *Molecular Marine Biology and Biotechnology*, **4**(4), 331–337.

Siciliano S. D., Germida J. J., Banks K. and Greer C. W. (2003). Changes in microbial community composition and function during a polyaromatic hydrocarbon phytoremediation field trial. *Applied and Environmental Microbiology*, **69**(1), 483–489, https://doi.org/10.1128/AEM.69.1.483-489.2003

Siddique T., Okeke B. C., Arshad M. and Frankenberger W. T. Jr. 2002. Temperature and pH effects on biodegradation of hexachlorocyclohexane isomers in water and soil slurry. *Journal of Agricultural and Food Chemistry*, **50**, 5070–5076, https://doi.org/10.1021/jf0204304

Singer A. C. (2006). The chemical ecology of pollutant biodegradation: bioremediation and phytoremediation from mechanistic and ecological perspectives. In: Phytoremediation Rhizoremediation, M. Mackova, D. Dowling, T. Macek (eds.), Springer, Dordrecht, pp. 5–21.

Singh B. K., Millard P., Whiteley A. S. and Murrell J. C. (2004). Unravelling rhizosphere microbial interactions: opportunities and limitations. *Trends in Microbiol*, **12**(8), 386–393, https://doi.org/10.1016/j.tim.2004.06.008

Singh B. K., Walker A. and Wright D. J. (2006). Bioremedial potential of fenamiphos and chlorpyrifos degrading isolates: influence of different environmental conditions. *Soil Biology and Biochemistry*, **38**, 2682–2693, https://doi.org/10.1016/j.soilbio.2006.04.019

Syed J. H. and Malik R. N. (2011). Occurrence and source identification of organochlorine pesticides in the surrounding surface soils of the Ittehad Chemical Industries Kalashah Kaku, Pakistan. *Environmental Earth Sciences*, **62**(6), 1311–1321, https://doi.org/10.1007/s12665-010-0618-z

Syed J. H., Malik R. N., Liu D., Xu Y., Wang Y., Li J., Zhang G. and Jones K. C. (2013). Organochlorine pesticides in air and soil and estimated air and soil exchange in Punjab, Pakistan. *The Science of the Total Environment*, **444**, 491–497, https://doi.org/10.1016/j.scitotenv.2012.12.018

Syed J. H., Malik R. N., Li J., Chaemfa C., Zhang G. and Jones K. C. (2014). Status, distribution and ecological risk of organochlorines (OCs) in the surface sediments from the Ravi River, Pakistan. *The Science of the Total Environment*, **472**, 204–211, https://doi.org/10.1016/j.scitotenv.2013.10.109

Tariq M. I., Afzal S., Hussain I. and Sultana N. (2007). Pesticides exposure in Pakistan: a review. *Environment International*, **33**(8), 1107–1122, https://doi.org/10.1016/j.envint.2007.07.012

Terzaghi E., Zanardini E., Morosini C., Raspa G., Borin S., Mapelli F., Vergani L. and Di Guardo A. (2018). Rhizoremediation half-lives of PCBs: role of congener composition, organic carbon forms, bioavailability, microbial activity, plant species and soil conditions, on the prediction of fate and persistence in soil. *The Science of the Total Environment*, **612**, 544–560, https://doi.org/10.1016/j.scitotenv.2017.08.189

Thijs S., Sillen W., Rineau F., Weyens N. and Vangronsveld J. (2016). Towards an enhanced understanding of plant microbiome interactions to improve phytoremediation: engineering the metaorganism. *Frontiers in Microbiology*, **7**, 341, https://doi.org/10.3389/fmicb.2016.00341

Van Hees P. A., Jones D. L., Finlay R., Godbold D. L. and Lundström U. S. (2005). The carbon we do not seed the impact of low molecular weight compounds on carbon dynamics and respiration in forest soils: a review. *Soil Biology and Biochemistry*, **37**(1), 1–13, https://doi.org/10.1016/j.soilbio.2004.06.010

Vancura V. and Hovadik A. (1965). Root exudates of plants: II. Composition of root exudates of some vegetables. *Plant and Soil*, **22**, 21–32, https://doi.org/10.1007/BF01377686

Vergani L., Mapelli F., Zanardini E., Terzaghi E., Di Guardo A., Morosini C., Raspa G. and Borin S. (2017). Phytorhizoremediation of polychlorinated biphenyl contaminated soils: an outlook on plant-microbe beneficial interactions. *The Science of the Total Environment*, **575**, 1395–1406, https://doi.org/10.1016/j.scitotenv.2016.09.218

Vos J. G., Dybing E., Greim H. A., Ladefoged O., Lambré C., Tarazona J. V., Brandt I. and Vethaak A. D. (2000). Health effects of endocrine-disrupting chemicals on wildlife, with special reference to the European situation. *Critical Reviews in Toxicology*, **30**(1), 71–133, https://doi.org/10.1080/10408440091159176

Walker T. S., Bais H. P., Grotewold E. and Vivanco J. M. 2003a. Root exudation and rhizosphere biology. *Plant Physiology*, **132**(1), 44–51, https://doi.org/10.1104/pp.102.019661

Walker T. S., Bais H. P., Halligan K. M., Stermitz F. R. and Vivanco J. M. 2003b. Metabolic profiling of root exudates of *Arabidopsis thaliana*. *Journal of Agricultural and Food Chemistry*, **51**, 2548–2554, https://doi.org/10.1021/jf021166h

Walton B. T. and Anderson T. A. (1990). Microbial degradation of trichloroethylene in the rhizosphere: potential application to biological remediation of waste sites. *Applied and Environmental Microbiology*, **56**(4), 1012–1016, https://doi.org/10.1128/aem.56.4.1012-1016.1990

Yadav D. V., Mittal P. K., Agarwal H. C. and Pillai M. K. (1981). Organochlorine insecticide residues in soil and earthworms in the Delhi area, India, AugusteOctober, 1974. *Pesticides Monitoring Journal*, **15**(2), 80–85.

Yee D. C., Maynard J. A. and Wood T. K. (1998). Rhizoremediation of trichloroethylene by a recombinant, root-colonizing *Pseudomonas fluorescens* strain expressing toluene ortho-monooxygenase constitutively. *Applied and Environmental Microbiology*, **64**(1), 112–118, https://doi.org/10.1128/AEM.64.1.112-118.1998

Yoshitomi K. J. and Shann J. R. (2001). Corn (*Zea mays* L.) root exudates and their impact on 14C-pyrene mineralization. *Soil Biology and Biochemistry*, **33**, 1769–1776, https://doi.org/10.1016/S0038-0717(01)00102-X

Yuste L., Canosa I. and Rojo F. (1998). Carbon-source-dependent expression of the PalkB promoter from the *Pseudomonas oleovorans* alkane degradation pathway. *Journal of Bacteriology*, **180**(19), 5218–5226, https://doi.org/10.1128/JB.180.19.5218-5226.1998

Zablotowicz R. M., Hoagland R. E. and Locke M. A. (1994). Glutathione S-transferase activity in rhizosphere bacteria and the potential for herbicide detoxification. In: Bioremediation Through Rhizosphere Technology, T. A. Anderson and J. R. Coats (eds.), American Chemical Society, Washington, DC, pp. 184–198.

Zhang G., Chakraborty P., Li J., Sampathkumar P., Balasubramanian T., Kathiresan K., Takahashi S., Subramanian A., Tanabe S. and Jones K. C. (2008). Passive atmospheric sampling of organochlorine pesticides, polychlorinated biphenyls, and polybrominated diphenyl ethers in urban, rural, and wetland sites along the coastal length of India. *Environmental Science and Technology*, **42**(22), 8218–8223, https://doi.org/10.1021/es8016667

Zhang Y. and Miller R. M. (1994). Effect of a *Pseudomonas* rhamnolipid biosurfactant on cell hydrophobicity and biodegradation of octadecane. *Applied and Environmental Microbiology*, **60**(6), 2101–2106, https://doi.org/10.1128/aem.60.6.2101-2106.1994

Zhang H. B., Luo Y. M., Wong M. H., Zhao Q. G. and Zhang G. L. (2007). Concentrations and possible sources of polychlorinated biphenyls in the soils of Hong Kong. *Geoderma*, **138**(3–4), 244–251, https://doi.org/10.1016/j.geoderma.2006.11.016

doi: 10.2166/9781789063783_0231

Chapter 12

Role of microorganisms in reclaiming 1,4-dioxane-contaminated sites: perspective analysis

Anil Kumar Singh[1] *and Ria Rautela Rana*[2]

[1]*Department of Agriculture, Sant Baba Bhag Singh University, Khiala, Punjab 144030, India*
[2]*Gujarat Ecological Education and Research (GEER) Foundation, Gandhinagar, Gujarat 382007, India*
**Corresponding author: anilimtech@gmail.com*

ABSTRACT

1,4-Dioxane is an emerging organic pollutant with suspected carcinogenic effect. It is produced as a by-product in several petrochemical-based industrial processes. It has also been utilized as a stabilizer in several chlorinated solvents. Physical and chemical properties of 1,4-dioxane has facilitated widespread contamination of land, surface, and groundwater resources. The extensive occurrence of 1,4-dioxane has highlighted the need for developing treatment methods that remediate contaminated water sources and prevent human exposure to 1,4-dioxane through drinking water. Removing 1,4-dioxane from contaminated sites and refurbishing it to uncontaminated condition is a challenging task. Remediation of 1,4-dioxane-contaminated sites can be achieved by physical, chemical, and biological processes. However, large-scale *in situ* remediation is not feasible by physical and chemical processes. Exploiting metabolic capabilities of 1,4-dioxane utilizing microorganisms for restoration and reclamation of contaminated sites is often seen as a most realistic option. Microorganisms can metabolize or co-metabolize 1,4-dioxane and bring about bioremediation of contaminated sites. The present chapter provides an overview and future perspective of microorganism-assisted reclamation of 1,4-dioxane-polluted sites. This chapter also discusses chemistry of dioxane, toxicity, and fate of dioxane in environment, reclamation strategies, different factors influencing reclamation, dioxane-degrading monooxygenase enzymes, and microbial degradation pathways.

Keywords: carcinogenic, biostimulation, bioattenuation, bioaugmentation, monooxygenases.

12.1 INTRODUCTION

Environment pollution is a serious problem faced by mankind. In simple words, environment pollution can be explained as the introduction of any contaminant that has the ability to adversely affect the biological, physical, and chemical properties of natural surroundings. Broadly speaking, pollution can have natural or anthropogenic origin. Natural pollution is caused by natural sources such as volcanic eruption, release of methane from paddy fields, and cattle. Generally, pollution caused by natural sources does not execute a severe hazardous effect on environment. In comparison to natural pollutants, man-made pollutants cause more serious problems as they have persistent nature and ability to get transported long distances. Owing to their persistent nature, man-made pollutants stay longer in the natural environment and enter the food chain and thus, affect the health of animals and humans. Man-made pollutants can also alter microbe-assisted natural processes by adversely affecting microbial population and diversity.

Man-made pollution can be inorganic or organic in nature. Inorganic pollutants cannot be degraded but transformed from toxic to non-toxic forms. Contrarily, organic pollutants can be degraded to simpler and non-toxic forms by natural processes. In spite of degradability, pollution due to man-made organic pollutants has become a common feature of present day human society. Unplanned and excessive expansion of industries has added an implausible amount of organic pollutants to air, soil, and water. Modern agriculture practice uses several agrochemicals that can adversely alter environmental condition by adding organic pollutants to the surroundings. In developing countries, aggressive agriculture practices, overpopulation, facile rule enforcement, and unplanned urbanization have augmented the problem of environment pollution.

Generally under natural conditions, organic pollutants are mitigated from environment by physical, chemical, and biological degradation processes. Degradation of organic pollutants under natural conditions may take a few hours to several years. Pollutants with short half-life do not cause any serious problem to environment. Some of the pollutants with short half-life may produce more toxic and persistent degradation intermediates resulting in a serious secondary pollution (Gheorghe & Ion, 2011). Contrarily, organic pollutants with longer half-life get accumulated in environment causing serious problems to all living organisms. Persistent man-made pollutants are long lasting, penetrable, and unbreakable making them more vulnerable to humans as well as all other life forms on earth. Their ability to get transported longer distances has made it a global problem. Moreover, their presence in the food chain has brought a lot of worries to mankind. Organic pollutants have infiltrated through the lowest member of the food chain making their way to the highest member. Gradual accumulation of organic pollutants in environment alters population dynamics and disturbs interaction of organisms at various tropic levels. This further aggravates to disturb the natural community structure of the polluted ecosystem. Man-made organic pollutants may have immunotoxicity, neurotoxicity, carcinogenicity, mutagenicity, and endocrine disruption potentials toward animals (Ossai *et al.*, 2020). Plants are also adversely affected by man-made organic pollutants. In recent past, organic pollutants have drawn scientific and political apprehension due to their wide distribution, bioaccumulation behavior, and potential harmful effects.

Conventionally, remediation of organic pollutant-contaminated sites can be achieved by physical, chemical, and biological processes (Ossai *et al.*, 2020). Very often single process fails to achieve a desirable level of toxic neutralization due to recalcitrant nature of pollutant and generation of toxic degradation intermediates. In such scenario, combination of different processes helps in achieving a desirable level of toxic neutralization.

Use of biological systems such as microorganisms, plants, and their products for reclamation is a sustainable and cost-effective method (Ossai *et al.*, 2020). Reclamation of organic pollutant-contaminated sites by biodegradation and bioremediation techniques uses metabolic capabilities of living organisms for rendering hazardous pollutants to non-hazardous forms (Ossai *et al.*, 2020). Despite decades of research, successful reclamation of organic pollutant-contaminated sites by biological means remains a challenging task. 1,4-Dioxane is one such organic pollutant that has contaminated soil and water resources at several places throughout the world. It is often suspected as a carcinogenic pollutant. The Agency for Toxic Substances and Disease Registry (ATSDR) included 1,4-dioxane as one of the priority substances for removal from drinking water (ATSDR, 2017).

This chapter focuses on microorganism-assisted reclamation of 1,4-dioxane-polluted sites. It also discusses the chemistry of 1,4-dioxane, distribution in environment, toxicity, reclamation strategies, microbial degradation pathways, 1,4-dioxane-degrading monooxygenase, and factors influencing reclamation.

12.2 CHEMICAL NATURE OF DIOXANE

1,4-Dioxane (CAS no. 123-91-1) is a synthetic heterocyclic organic compound with faint sweet odor. It has a molar mass of 88.106 g/mol and its chemical formula is $C_4H_8O_2$. 1,4-Dioxacyclohexane is a systematic International Union of Pure and Applied Chemistry (IUPAC) name of 1,4-dioxane. In 1,4-dioxane, oxygen atom is directly positioned opposite to each other and this results in the formation of stable symmetric ether linkage (Surprenant, 2000). It is simply known as dioxane because the other dioxane isomers (1,2- and 1,3-) are rarely encountered. 1,4-Dioxane structure is very stable to reaction with acids, oxides, and oxidizing agents (Surprenant, 2000). Unlike other organic pollutants, 1,4-dioxane is highly miscible in water.

Historically, 1,4-dioxane was used as a stabilizer in chlorinated industrial solvents. It is extensively used as a solvent during the production of a variety of organic chemicals present in deodorants, dyes, paints, antifreeze, shampoos, and cosmetics. 1,4-Dioxane is also part of pesticide formulations, food additives, and food-packaging adhesives (Surprenant, 2000).

12.3 DISTRIBUTION OF 1,4-DIOXANE IN THE ENVIRONMENT

Extensive use and unregulated disposal has made 1,4-dioxane contamination a global problem. 1,4-Dioxane can contaminate air, water, and soil (Godri Pollitt *et al.*, 2019). It is non-volatile, hence chances of air pollution by 1,4-dioxane is very rare. Moreover, in the gaseous phase, 1,4-dioxane has half-life of 35 h due to photo-oxidation. It has high water solubility of nearly 100 mg/mL, hence chances of water getting polluted is very high. Federal agencies in the United States have recognized 1,4-dioxane as emerging drinking water contaminant at several sites (Godri Pollitt *et al.*, 2019). Anderson *et al.* (2012) reported that 1,4-dioxane is present in 17.4% of the groundwater samples examined by their group. 1,4-Dioxane has been reported as a second most prevalent prior contaminant in public water supplies of United the States (Godri Pollitt *et al.*, 2019). The presence of 1,4-dioxane in the natural water system has also been reported from several other countries such as Japan, Korea, Canada, United Kingdom, and Germany. The occurrence of 1,4-dioxane in the groundwater has been often associated with other chlorinated solvents such as 1,1,1-trichloroethane (1,1,1-TCA), 1,1-dichloroethane (1,1-DCA), and trichloroethylene (Adamson *et al.*, 2015).

Consumer products can also become a source of 1,4-dioxane exposure and pollution. Personal care products containing ethoxylated surfactants may have 1,4-dioxane as contaminate. Black *et al.* (2001) reported the presence of 1,4-dioxane in various cosmetic products (279 mg/kg) and children's shampoo (85 mg/kg) used in the United States. It is estimated that women have greater chances of 1,4-dioxane exposure than men or children (Godri Pollitt *et al.*, 2019). The high exposure is mainly attributed to inhalation of the volatilized 1,4-dioxane from personal care products. Rarely, 1,4-dioxane get absorbed in the body during topical application of 1,4-dioxane-contaminated personal care product.

1,4-Dioxane adsorb weakly into mineral phases and organic matter (Jackson & Dwarakanath, 1999). This causes 1,4-dioxane to easily leach into water systems and subsurface environments. 1,4-Dioxane has half-life of 2–5 years and 56 days in groundwater and surface water, respectively (Adamson *et al.*, 2015). Shorter half-life of 1,4-dioxane in surface water as compared to subsurface water may be linked to microbial degradation in the former environment (Godri Pollitt *et al.*, 2019). Microbial population and diversity is relatively more in surface water as compared to subsurface water. Thus, there are more chances of microbial degradation of 1,4-dioxane in surface water bodies.

12.4 TOXIC EFFECT OF DIOXANE

1,4-Dioxane is documented hazardous substance for humans and animals (Göen *et al.*, 2016). The United States Environment Protection Agency has classified dioxane as a probable human carcinogen. International Agency for Research on Cancer has classified dioxane as a group 2B agent. This means it is possibly carcinogenic to humans due to the inadequate evidence in humans but sufficient research evidence in experimental animals for carcinogenicity. The average half-life of 1,4-dioxane in human body is approximately 1 h (Göen *et al.*, 2016). An 8 h of inhalation can result in the detection of 1,4-dioxane and 2-hydroxyethoxyacetic acid (HEAA) within first few hours of exposure. Dioxane and HEAA can also be detected in urine after 4 and 24 h after exposure, respectively (Göen *et al.*, 2016).

Studies have shown that dioxane primarily affect liver and kidney (Godri Pollitt *et al.*, 2019). Oral administration of dioxane-contaminated water to rats and mice induced dose-dependent liver toxicity, degeneration of hepatic cells, and development of pre-neoplastic lesions (Kano *et al.*, 2009; Stott *et al.*, 1981). Chronic exposure of dioxane-contaminated drinking water has also been reported to adversely affect kidney (EPA 2013; Woo *et al.*, 1977). As evident from Figure 12.1, 1,4-dioxane metabolism results in the formation of HEAA and diglycolic acids. HEAA formation takes place through multiple pathways occurring in liver, kidney, and nasal mucosa (Woo *et al.*, 1977). Cytochrome P450 plays a key role in converting 1,4-dioxane to HEAA and diglycolic acids (Nannelli *et al.*, 2005). Another view is that initially dioxane is metabolized to diethylene glycol and then converted to 2-hydroxyethoxyacetaldehyde which is rapidly metabolized to HEAA (CDCP, 2017; Schier *et al.*, 2013).

Toxicological studies in rats and mice demonstrated that 1,4-dioxane increased the occurrence of tumors in these animals (Kano *et al.*, 2009). Chronic exposure through inhalation or ingestion induces tumor formation in liver and nasal cavity (Dourson *et al.*, 2017; Kano *et al.*, 2009). However, *in vitro* studies did not exhibit any genotoxic effects even after prolong exposure to dioxane (EPA, 2013). Undoubtedly, dioxane can induce cancer but the mechanisms responsible for carcinogenic effect are not very clear (EPA, 2013). HEAA is a primary metabolite detected in urine after exposure to dioxane. HEAA is not known to have high toxic or carcinogenic effect on humans. Other metabolites such as diethylene glycol and diglycolic acid are known to have toxic effect on liver and kidney (Landry *et al.*, 2011, 2015; Schier *et al.*, 2013).

Oral administration of dioxane has also been reported to result in the formation of cytochrome P450 isoforms. Cytochrome P450 isoforms along with toxic metabolites such as diethylene glycol and diglycolic acid can cause hepatic tumor formation (Nannelli *et al.*, 2005). Co-contaminates of dioxane such as 1,1,1-TCA and 1,1-DCA are known carcinogens in animals. Metabolism of co-contaminates

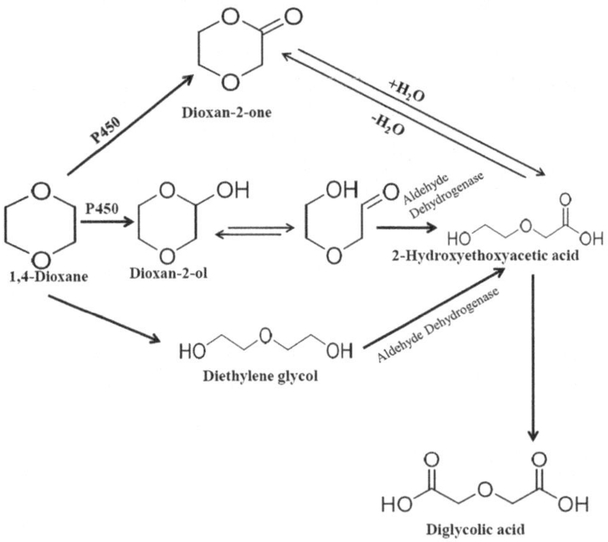

Figure 12.1 Proposed metabolic degradation pathways of 1,4-dioxane in humans (Woo *et al.*, 1977).

produces metabolites that can induce cancer in liver and kidney. Some researchers are of the opinion that dioxane acts as a tumor promoter rather than tumor inducer (Inoue *et al.*, 1989).

12.5 STRATEGIES FOR RECLAIMING DIOXANE-CONTAMINATED SITES

The distinctive physical and chemical characteristics of 1,4-dioxane confer challenges in the removal and reclamation of contaminated sites (Stepien *et al.*, 2014). Effective remediation of contaminated sites must convert toxic pollutants to non-toxic or less toxic form, approximately to pre-polluted conditions (Kumar *et al.*, 2015). The methods selected for remediation play a crucial role in complete removal, cleaning, reclamation, and restoration of 1,4-dioxane-contaminated sites. Unlike most of the organic pollutants, dioxane exhibits good water solubility as a result it favors partitioning into the aqueous phase making common physical separation methods such as air stripping or adsorption inefficient (Godri Pollitt *et al.*, 2019). Dioxane removal by membrane filtration is also not very effective. Linares *et al.* (2011) reported that reverse osmosis is not efficient in removing dioxane from contaminated water. Oxidation is considered as a most effective method for removing dioxane from contaminated sites (Godri Pollitt *et al.*, 2019). Dioxane oxidation can be achieved by both abiotic and biotic processes.

Conventional oxidation methods are not effective due to the formation of more toxic degradation intermediates. An advanced oxidation process that uses reactive radicals such as hydroxyl ($^{\cdot}OH$) and sulfate radicals ($SO_4^{\cdot-}$) has been demonstrated to mineralize 1,4-dioxane (Lee *et al.*, 2015). As evident from Figure 12.2, degradation of dioxane by advanced oxidation processes results in the formation of low molecular weight fatty acids such as glycolic, formic acid, glyoxal, acetic, and oxalic acids

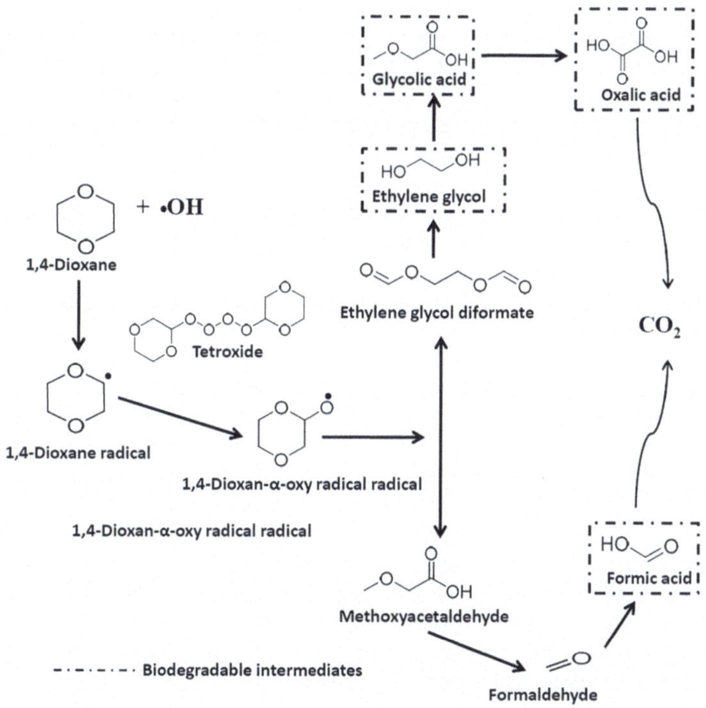

Figure 12.2 1,4-Dioxane degradation by advanced oxidation processes. Molecules shown in the dotted box can be assimilated by microorganisms under available natural conditions.

(Merayo *et al.*, 2014). Low molecular weight fatty acids can be removed by conventional biological treatments (Merayo *et al.*, 2014). However, feasibility of applying advanced oxidation processes at the contaminated sites is a challenging task (Lee *et al.*, 2015). Oxidation of dioxane can also be carried out in electrochemical cells. However, electrochemical oxidation is expensive and lacks selectivity. Boron-doped diamond electrodes used in electrochemical cells are expensive (Jasmann *et al.*, 2016). Photocatalytic advanced oxidation processes coupled with membrane filtration has also been demonstrated to efficiently remove dioxane from contaminated water. Extensive application of the abovementioned method is restricted due to high-energy requirement for generating ultraviolet and pressure (Hwangbo *et al.*, 2019).

Oxidation by biological means is a most suitable method for *in situ* removal of dioxane from contaminated sites due to environment friendly and inexpensive nature. Biological oxidation utilizes metabolic capabilities of microorganisms to transform toxic contaminates to benign form. Biodegradation, bioattenuation, biostimulation, and bioaugmentation have been demonstrated as effective strategies for *in situ* removal of 1,4- dioxane from contaminated sites (Godri Pollitt *et al.*, 2019). Unlike aforementioned abiotic strategies, biological methods of 1,4-dioxane removal do not require high-energy input or require expensive set ups. Moreover, biological degradation methods do not generate any toxic degradation intermediates (Godri Pollitt *et al.*, 2019).

12.6 DIOXANE-DEGRADING MICROORGANISMS AND DEGRADATION PATHWAYS

Dioxane is a man-made compound, hence microorganisms dwelling in soil and water have not encounter it in their pristine natural surroundings. Thus, chances of finding 1,4-dioxane degrader in natural environment are grim. Contrarily, in the past few decades, several researchers have reported isolation of 1,4-dioxane-degrading microorganisms from environmental samples (Kim *et al.*, 2009; Parales *et al.*, 1994). Probably, microorganisms have developed 1,4-dioxane utilization ability due to exposure to compounds with similar structure. Dioxane-degrading bacteria have been reported to degrade other cyclic ethers such as tetrahydrofuran (THF) and some naturally occurring ether compounds (Godri Pollitt *et al.*, 2019). For instance, cyclic ethers containing 5- and 6-membered rings are extensively present in natural environment. These cyclic ethers are utilized by microorganisms for biosynthesizing certain antibiotics (Martın *et al.*, 2014). Microorganisms having capacity to degrade these antibiotics have survival advantage over other non-degrading microbes (Topp *et al.*, 2013). Several dioxane- and THF-degrading microorganisms produce lactones as degradation intermediates (Sales *et al.*, 2013). Under natural conditions microorganisms encounter lactones as chemical signal molecules for accomplishing quorum sensing (Rehman & Leiknes, 2018). Many bacteria produce lactones degrading enzymes to disrupt signals and subsequently interfere with quorum sensing. Disrupting quorum sensing enables microorganisms to compete against the bacteria that produce them (Rehman & Leiknes, 2018). Aforementioned finding suggests that biodegradation of simple cyclic ethers such as THF and dioxane by microorganisms may be a survival outcome of backward evolution of catabolic pathways. Lactones assimilating microorganisms growing under depleted lactone conditions are likely to experience selective pressure for transforming other compounds such as dioxane into lactones as their survival strategy (He *et al.*, 2017).

Microorganisms can degrade 1,4-dioxane by metabolizing it as a source of carbon and energy (Godri Pollitt *et al.*, 2019). There are also reports that some microorganisms co-metabolize 1,4-dioxane while growing on other carbon sources such as propane, THF, toluene, and methane (He *et al.*, 2017). 1,4-Dioxane is assimilated as a primary substrate and co-substrate in metabolizing and co-metabolizing microorganisms, respectively (Figure 12.3). As 1,4-dioxane co-metabolizing microorganisms do not use co-substrates as major carbon or energy sources, they can often degrade co-substrates to concentrations well below those that can be achieved by organisms that metabolize and grow on these co-substrates.

Pseudonocardia dioxanivorans CB1190 is well-studied 1,4-dioxane-degrading bacteria. Initially, it was isolated as *Actinomycete* CB1190 from industrial-activated sludge (Parales *et al.*, 1994). Strain

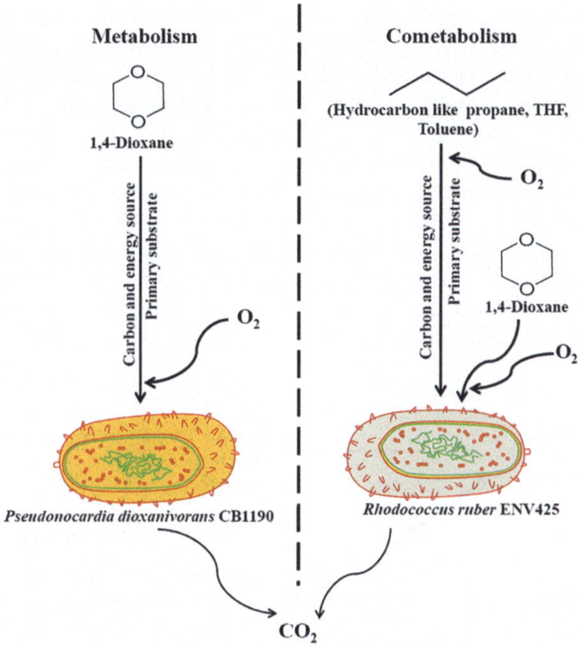

Figure 12.3 Simple illustration of microbial metabolism and co-metabolism of 1,4-dioxane.

CB1190 can use 1,4-dioxane as a sole carbon and energy source. Strain CB1190 has capability to metabolize 1,4-dioxane at the rate of 0.33 mg dioxane/h/mg protein (Parales *et al.*, 1994).

1,4-Dioxane metabolizing *Mycobacterium* sp. PH-06 was isolated from contaminated river sediments (Kim *et al.*, 2009). Strain PH-06 has ability to metabolize 1,4-dioxane by the formation of 1,4-dioxane-2-ol and ethylene glycol as major metabolites. Sei *et al.* (2013) reported isolation of 1,4-dioxane metabolizing *Afipia* sp. D1 from drainage of a chemical factory. Strain D1 was able to tolerate co-contaminant ethylene glycol when grown on 1,4-dioxane (Sei *et al.*, 2013). Vainberg *et al.* (2006) reported isolation of *Pseudonocardia* sp. strain ENV478 by enrichment culturing of liquor collected from a membrane bioreactor system that was treating wastes of fine chemical manufacturer. Strain ENV478 degraded 1,4-dioxane to HEAA at a rate of 21 mg/h/g total suspended solids. 1,4-Dioxane degradation was initiated after growth on THF, sucrose, lactate, yeast extract, 2-propanol, and propane (Vainberg *et al.*, 2006).

Aoyagi *et al.* (2018) used stable isotope probing (SIP) of rRNA for identifying active and diverse 1,4-dioxane-degrading microorganisms in activated sludge. There group reported a total of 9 OTUs that increased the expression of 16S rRNAs after SIP incubation, thus indicating their involvement in ^{13}C-labeled 1,4-dioxane degradation. During the study, Aoyagi *et al.* (2018) observed *P. dioxanivorans*, an ammonia-oxidizing bacterium and phylogenetically novel bacteria as dioxane degraders.

Generally, contaminated environment is believed to be a good source for isolating pollutant-degrading microorganisms. Contrarily, He *et al.* (2017) reported isolation of two consortia from a pristine environment. Consortia were capable of removing dioxane even at a low concentration of 300 μg/L within 3 days. Consortium A was primarily constituted by *Aminobacter* sp., *Ralstonia* spp, *Chitinophaga* spp, *Cupriavidus* spp, *Afipia* spp, *Ensifer* spp, and *Sediminibacterium* spp. Consortium B was mainly composed of *Chitinophaga* spp, *Aminobacter* spp, *Terrimonas* spp, *Acidovorax* spp, *Variovorax* spp, *Brevundimonas* spp, and *Sediminibacterium* spp (He *et al.*, 2017).

Most of the 1,4-dioxane degradation studies have been reported under aerobic conditions. There are very few studies showing anaerobic 1,4-dioxane degradation (Godri Pollitt *et al.*, 2019). Shen *et al.* (2008) reported anaerobic growth of iron-reducing bacterium in 1,4-dioxane-containing medium. In one of the recent microcosm studies, Ramalingam and Cupples (2020) reported anaerobic biodegradation of 1,4-dioxane in sediments collected from contaminated sites. Research was also carried out in uncontaminated agricultural soils and river sediments. Most efficient 1,4-dioxane biodegradation was reported in the nitrate-amended treatments. The soil microcosms were primarily dominated by *Rhodanobacter* with less abundance of *Pseudomonas*, *Mycobacterium*, and *Acinetobacter* (Ramalingam & Cupples, 2020). Study by Ramalingam and Cupples (2020) also highlighted the fact that natural attenuation is suitable for *in situ* cleanup technology when time is not constrained.

Literature suggests that bacteria are more dominant 1,4-dioxane degrader compared to fungi (Godri Pollitt *et al.*, 2019). Nakamiya *et al.* (2005) reported isolation of 1,4-dioxane-metabolizing

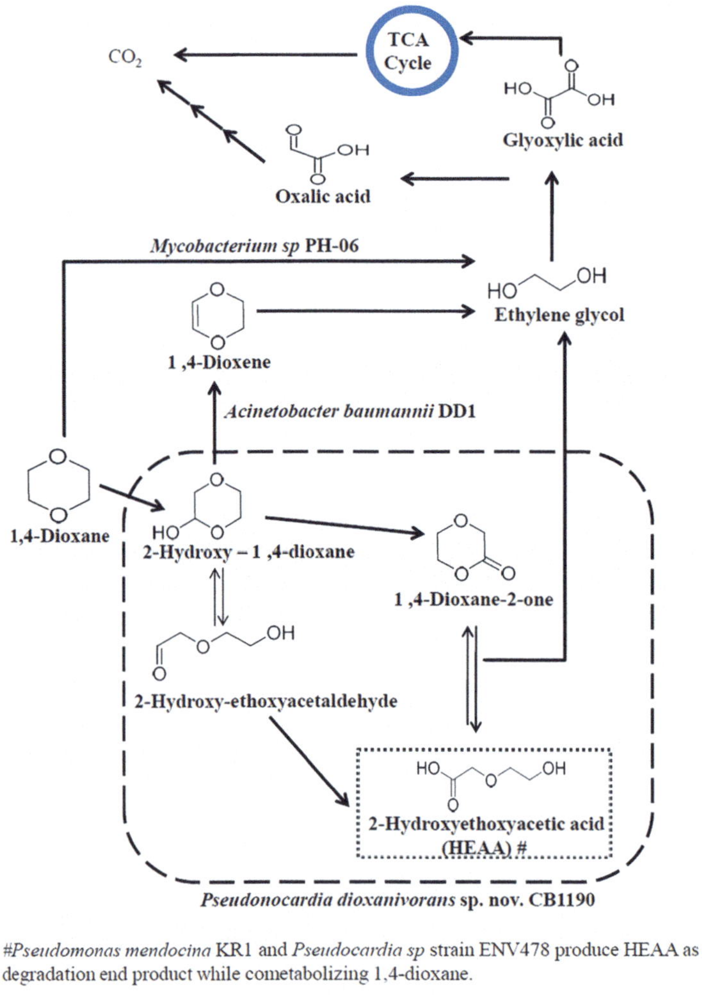

Figure 12.4 Simple illustration of biological degradation pathways of 1,4-dioxane. The degradation intermediates shown in the pathways are key recognized products observed during microbial degradation of 1,4-dioxane.

fungus *Cordyceps sinensis* strain A from garden soil of the National Institute for Environmental Studies. Strain A exhibited optimum growth on 0.034 M 1,4-dioxane. Fungal metabolism of 1,4-dioxane resulted in the formation of ethylene glycol, glycolic acid, and oxalic acid as degradation intermediates. Strain A was also able to utilize several other cyclic ethers such as THF, tetrahydropyran, 1,3-dioxane, and *s*-trioxane (Nakamiya *et al.*, 2005). Fungi have also been known to exhibit co-metabolic degradation of 1,4-dioxane (Skinner *et al.*, 2009). Growth of *Aureobasidium pullulans* and *Graphium* sp. on THF has been reported to induce 1,4-dioxane degradation ability in both fungi (Patt & Abebe, 1995; Skinner *et al.*, 2009). Skinner *et al.* (2009) reported that propane and THF induced monooxygenase activity in fungus *Graphium* sp. and degraded 1,4-dioxane. Propane-grown mycelia of fungus *Graphium* sp. could consume 50% of 1 mM 1,4-dioxane within 12 h of exposure (Skinner *et al.*, 2009).

Microbial metabolism as well as co-metabolism of 1,4-dioxane result in the formation of degradation intermediates such as 2-HEAA, 1,4-dioxane-2-ol, ethylene glycol, glycolic acid, glyoxylic acid, and oxalic acid (Figure 12.4). Glyoxylic acid or oxalic acid is further consumed in the tricarboxylic acid cycle to complete biodegradation of 1,4-dioxane (Nakamiya *et al.*, 2005). Overall, the 1,4-dioxane biodegradation pathway suggests that microbial metabolites generated during degradation are unlikely to have persistent or toxic nature (McElroy *et al.*, 2019).

12.7 RECLAMATION OF CONTAMINATED SITES BY *IN SITU* BIOREMEDIATION

Exploiting *in situ* bioremediation for reclaiming 1,4-dioxane-contaminated sites is simplest, cost-efficient, environmentally safe, and sustainable method. Godri Pollitt *et al.* (2019) highlighted *in situ* biodegradation of 1,4-dioxane as the best method of removing contamination from large area and diluted plume. *Ex situ* technologies are expensive and inefficient under environmental conditions. However, successful field-scale accomplishment using bioremediation to remove 1,4-dioxane are still scarce. 1,4-Dioxane metabolizing and 1,4-dioxane co-metabolizing microorganisms can be exploited for reclaiming contaminated sites (Godri Pollitt *et al.*, 2019).

Microorganisms metabolizing 1,4-dioxane do not require additional nutrient amendments, oxygen requirement is also low, and chances of aquifer getting clogged due to excessive growth is very less. However, plumes with low 1,4-dioxane concentration may not support growth of microorganisms, and thus restricting the effectiveness of bioremediation with the aid of 1,4-dioxane-metabolizing microbes. Contrarily, 1,4-dioxane co-metabolizing microorganisms can carry out bioremediation even at low concentrations of 1,4-dioxane. Co-metabolizing microorganisms degrade 1,4-dioxane as a co-substrate by utilizing primary substrates for growth and reproduction (Godri Pollitt *et al.*, 2019). 1,4-Dioxane co-metabolizing microorganisms can be employed in biostimulation and bioaugmentation for removing 1,4-dioxane from contaminated sites (Zenker *et al.*, 2004). There are several microorganisms that can co-metabolize and degrade dioxane (Godri Pollitt *et al.*, 2019; Vainberg *et al.*, 2006). Mahendra and Alvarez-Cohen (2006) reported that *Burkholderia cepacia* G4 degraded 1,4-dioxane at a rate of 0.1 ± 0.006 mg/h/mg protein by toluene-induced monooxygenase. *Pseudonocardia* sp. strain ENV478 degraded 1,4-dioxane by utilizing THF as the primary growth substrate (Vainberg *et al.*, 2006).

According to most of the guidelines safe groundwater must have dioxane concentrations below 1 g/L. However, achieving such low concentrations by exploiting microbial metabolic capabilities is a challenging task. Microbial growth rate at environmentally relevant dioxane concentrations is very slow. Dioxane co-metabolizing microorganisms help in overcoming this hurdle as they utilize primary substrates for growth and reproduction. Rolston *et al.* (2019) reported that *Rhodococcus rhodochrous* strain ATCC 21198 could transform low and environmentally relevant concentrations of dioxane while growing on isobutene. Microcosms constructed with aquifer solids from Fort Carson, Colorado were biostimulated with isobutane and inorganic nutrients. Biostimulation of native microorganisms help in transforming dioxane to acceptable concentration. Bioaugmentation with strain 21198 also helped to achieve acceptable 1,4-dioxane concentration in the microcosms (Rolston *et al.*, 2019).

Biostimulation enhances microbial degradation activity by addition of microbial growth supporting material and optimizing of various parameters that limits remediation at contaminated sites (Wu *et al.*, 2016). Li *et al.* (2010) reported 1,4-dioxane (50 mg/L) degradation in microcosm consisting of contaminated groundwater and soil from Alaska. Abiotic processes such as evaporation, adsorption, and diffusion into fine grain caused $15.0 \pm 8.7\%$ removal of 1,4-dioxane over a period of 6 months. Biosimulation of indigenous bacteria by adding 1-butanol and inorganic nutrients significantly enhanced the removal of 1,4-dioxane. This observation by Li *et al.* (2010) suggests that Arctic groundwater samples contain 1,4-dioxane degraders.

Bioattenuation exploits a variety of chemical, physical, and biological processes to remove, neutralize, transform, and reduce toxicity or concentration of dioxane at contaminated sites (Ossai *et al.*, 2020). Indigenous microbial population degrades or transforms dioxane to less toxic form depending upon their metabolic capabilities. Bioattenuation is useful at contaminated sites having low concentration of dioxane, where adopting other remediation methods is not feasible and time is not a limiting factor (Vásquez-Murrieta *et al.*, 2016). Li *et al.* (2015) demonstrated natural attenuation of dioxane in groundwater microcosms. A site impacted by 1,1,1-TCA and 1,1-dichloroethene (1,1-DCE) exhibited nearly 100% removal of dioxane in 20 weeks (Li *et al.*, 2015). Natural attenuation has been demonstrated to play a key role in cleaning dioxane-contaminated sites located in California (Adamson *et al.*, 2015).

Bioaugmentation involves the addition of exogenous microbial cultures, microbial consortia, or genetically engineered microorganisms for removing dioxane from contaminated sites. Microorganisms selected for bioaugmentation must be adapted to the environmental conditions prevalent at contaminated sites and have proven record to degrade dioxane. Kelley *et al.* (2001) reported that poplar root extract (40 mg/L) stimulated dioxane degradation in soil bioaugmented with *Amycolata* sp. CB1190. Within 45 days of incubation nearly 100 mg/L dioxane was removed from contaminated soil. Bioaugmentation of microcosms with *P. dioxanivorans* CB1190 exhibited rapid removal of dioxane at a rate of 0.16 ± 0.04 mg dioxane/day/mg protein (Li *et al.*, 2010). In another experiment, Li *et al.* (2014) carried out a bench-scale microcosm study in 1,4-dioxane-contaminated groundwater samples collected from California. The highest rate of 1,4-dioxane removal in the microcosm study was reported to be 3449 ± 459 μg/L/week (Li *et al.*, 2014).

In one of the microcosm studies, Lippincott *et al.* (2015) applied propane biosparging and *Rhodococcus ruber* ENV425 bioaugmentation to enhance *in situ* biodegradation of dioxane at Vandenberg Air Force Base site. Biosparging and bioaugmentation demonstrated dioxane decay rate in the range of 0.021–0.036/day. Dioxane concentration in wells reduced from 1090 to <2 μg/L after the reclamation process (Lippincott *et al.*, 2015).

Microbial consortia have also been reported to remove 1,4 dioxane from contaminated samples (Saul, 2012; Xiong *et al.*, 2020). Microbial consortia-based bioremediation of 1,4-dioxane is often considered better than pure culture-based remediation. Biological reactors used for water treatments comprise microbial consortia as maintaining pure culture free from contamination is very difficult. Also some bacteria are unable to completely mineralize 1,4-dioxane and produce degradation intermediates. These intermediates can be further mineralized by microbial members of consortia (Xiong *et al.*, 2020). Saul (2012) patented a method for removing 1,4-dioxane from contaminated groundwater with the help of microbial consortia. Bacterial consortium consisting of *Pseudomonas putida* strain B, *P. putida* stain E, and *Pseudomonas fluorescens* strain G was able to co-metabolize 1,4-dioxane and chlorinated solvents (Saul, 2012).

12.8 FACTORS INFLUENCING RECLAMATION OF PETROLEUM HYDROCARBON-CONTAMINATED SITES

Biological cleaning of 1,4-dioxane-contaminated sites is influenced by several biotic and abiotic factors. Microorganisms employed for cleaning of contaminated sites is the most important biotic factor influencing reclamation processes. In natural ecosystems, microbial community grows together in a synergistic relationship and produce bioactive metabolites (including oxidative and hydrolytic enzymes) to implicate the mineralization of dioxane (Adamson *et al.*, 2015; Li *et al.*, 2014). Selection of microorganisms for *in situ* remediation is decided keeping in view the environmental conditions and locally available resources. Several microcosm studies have been carried out to understand the influence of environmental conditions on reclamation processes (Li *et al.*, 2010; Lippincott *et al.*, 2015). Apart from nutrients, pH, temperature, oxygen, heavy metals, and co-contaminates are important abiotic factors influencing reclamation processes.

Oxygen is prerequisite for aerobic biodegradation of 1,4-dioxane. There are very few reports suggesting anaerobic biodegradation of 1,4-dioxane (Ramalingam & Cupples, 2020; Shen *et al.*, 2008). Generally, a low concentration of dissolved oxygen particularly in groundwater (below 3 mg/L) significantly reduces the efficiency of bioremediation (Lee *et al.*, 2014). Da Silva *et al.* (2020) observed that bioaugmentation efficiency increased by aeration of flow-through aquifer columns. To overcome problem arising due to anoxia, oxygen-releasing compounds (such as calcium dioxide, magnesium peroxide) are added to enhance *in situ* biodegradation (Kao *et al.*, 2003). However, such additions are likely to alter pH and adversely affect degradation processes. Lee *et al.* (2014) designed oxygen-releasing alginate beads (ORABs) to overcome oxygen deficiency during 1,4-dioxane degradation. In the presence of ORABs, 99.0% and 68.1% degradation of 1,4-dioxane was achieved under batch and column conditions, respectively. Batch and column without ORABs showed 34.3% and 18% degradation of 1,4-dioxane, respectively (Lee *et al.*, 2014).

The pH and temperature of growth medium or surrounding medium affect 1,4-dioxane-cleaning processes by influencing microbial growth, oxygen availability, and enzyme activity (Lee *et al.*, 2014). He *et al.* (2017) reported maximum 1,4-dioxane degradation by microbial consortium at pH 7 and 30°C. Generally, temperature supporting optimum microbial growth is conducive for pollutant degradation. Most of the studies have reported that 30°C as suitable temperature for 1,4-dioxane degradation (He *et al.*, 2017; Inoue *et al.*, 2018). Dioxane degradation at low temperatures (14 and 4°C) has been reported by Li *et al.* (2010). At low temperature (4°C), microcosms demonstrated a lag period of 3 months before a notable degradation of 1,4-dioxane (Li *et al.*, 2010). Inoue *et al.* (2018) reported that *Rhodococcus aetherivorans* JCM 14343 can degrade 1,4-dioxane at a wide range of temperatures (5–40°C) and pH (4–9). Strain JCM 14343 exhibited maximum degradation rate at pH 7 and 28°C (Inoue *et al.*, 2018). *Pseudonocardia* sp. N23 exhibited 1,4-dioxane degradation under a wide pH (3.8–8.2) and temperature (20–35°C) range. Maximum 1,4-dioxane degradation was observed at pH 7 and 30°C (Inoue *et al.*, 2018).

Microbial encounter with metals cannot be ruled out in natural environment. Usually, metal toxicity hinders microbial growth and reduces or even prevents biodegradation of pollutants (Inoue *et al.*, 2020). High metal concentrations in growth medium execute cellular toxicity and suppress enzymatic activity. Transition metals namely Co(II), Cu(II), Fe(II), Mo(VI), Ni(II), and Zn(II) reduce rate of dioxane degradation (Inoue *et al.*, 2020). Johnson *et al.* (2020) reported that *Mycobacterium austroafricanum* JOB5 can metabolize dioxane in planktonic pure cultures and biofilms even in the presence of 10 mg/L of Cr(VI).

Metal ions may not always have inhibitory effect on dioxane degradation. Inoue *et al.* (2020) reported that Fe(III) enhanced 1,4-dioxane degradation by *Mycobacterium* sp. D6 and *R. aetherivorans* JCM 14343. The presence of Mn(II) in growth medium enhanced 1,4-dioxane degradation by Pseudonocardia sp. D17 and Pseudonocardia sp. N23. Most of the 1,4-dioxane degraders have soluble di-iron monooxygenases (SDIMOs) to catalyze initial steps of biodegradation. Transition metals such as Mn and Fe act as cofactors in the catalytic site of SDIMOs (Inoue *et al.*, 2020).

1,4-Dioxane degradation is significantly affected by co-contaminates such as 1,1-DCE, *cis*-1,2-DCE, trichloroethene, and Cu(II) ions (Adamson *et al.*, 2015; Li *et al.*, 2015; Zhao *et al.*, 2018). The co-occurring compounds can influence reclamation processes by altering the activity of 1,4-dioxane-degrading enzymes. Zhao *et al.* (2018) observed that co-contaminants, 1,1-DCE and Cu(II) ions inhibited 1,4-dioxane degradation in soil columns. Cyclic ethers analogues of 1,4-dioxane are known to induce competitive inhibition in degradative enzymes (Mahendra *et al.*, 2013). Co-contaminates can inhibit dioxane biodegradation via complex mechanisms. It is believed that competition, damage of dioxane-degrading enzymes (i.e., SDIMOs), suppression of dioxane-degrading gene expression, and toxic effects on dioxane-degrading bacteria can inhibit 1,4-dioxane degradation (Mahendra *et al.*, 2013).

Some compounds (such as butanol and propane) can induce dioxane degradation (Hand *et al.*, 2015). Hand *et al.* (2015) reported 1-butanol as a better inducer compared to propane for *in situ* bioaugmentation of 1,4-dioxane-contaminated groundwater. *Mycobacterium vaccae* JOB5 and *Rhodococcus jostii* RHA1 were grown in Reasoner's 2A broth medium until an optical density of 0.8–1.0 was achieved. Bacterial cells collected from RHA1 medium were induced for expressing either butane- or propane-monooxygenases by exposing them to 1-butanol or propane, respectively. Bacterial cells induced by butane performed better than propane-induced cells (Hand *et al.*, 2015).

12.9 SOLUBLE DI-IRON MONOOXYGENASES

The microbial degradation pathways indicate that the rate-limiting step in 1,4-dioxane biodegradation is the initial oxidation of the carbon atom adjacent to the oxygen atom. Oxidation of carbon results in the cleavage of high-energy C–O bond of cyclic ether. Initial oxidation of 1,4-dioxane is catalyzed by monooxygenase enzymes known as SDIMOs. Monooxygenases are defined as a group of enzymes that catalyze the incorporation of one atom of molecular oxygen into a substrate, with the other being reduced to water. Dioxane biodegradation in metabolizing and co-metabolizing microorganisms is initiated by SDIMOs. 1,4-Dioxane-oxidizing SDIMOs are divided into six groups based on the arrangement of components, sequence similarity, and substrate specificity. However, monooxygenase enzymes responsible for metabolizing 1,4-dioxane are limited to group-5 (dioxane/THF monooxygenase) and group-6 SDIMOs (propane monooxygenase) (He *et al.*, 2017; Ramos-Garcia *et al.*, 2018). Mahendra and Alvarez-Cohen (2006) reported the presence of other two types of SDIMOs namely soluble methane monooxygenases and toluene monooxygenases in some dioxane-co-metabolizing microbial strains. SDIMOs are family of multi-component enzymes consisting of three or four components namely an oxygenase, a reductase, a coupling protein (in most cases), and additional ferredoxin (in several cases). An active site of SDIMOs contains a di-iron center (Li *et al.*, 2013).

In renowned dioxane degrader *P. dioxanivorans* CB1190, monooxygenase enzyme (THF monooxygenase or dioxane monooxygenase) expressing gene cluster *thmADBC* is located on plasmid pPSED02 (Sales *et al.*, 2013). Mahendra and Alvarez-Cohen (2006) reported *Escherichia coli* TG1/pBS(Kan) containing recombinant plasmids derived from toluene monooxygenases of *B. cepacia* G4, *Pseudomonas mendocina* KR1, and *Ralstonia pickettii* PKO1 was able to co-metabolize dioxane by utilizing toluene as a primary substrate. The recombinant strains transformed 0.17–0.60 mg of 1,4-dioxane/h/mg protein (Mahendra & Alvarez-Cohen, 2006).

Genes encoding monooxygenase enzymes have been used as biomarkers for detecting 1,4-dioxane degradation (Li *et al.*, 2013). This approach is effective for monitoring the known microorganisms employing monooxygenase-based 1,4-dioxane degradation. However, this approach cannot report new and unknown 1,4-dioxane degraders that lack monooxygenase-encoding genes or possess some novel degradation pathway involving new types of the genes (Aoyagi *et al.*, 2018). He *et al.* (2018) did not observe THF or dioxane monooxygenase gene in bacterial consortia that has ability to utilize dioxane as a carbon and energy source.

12.10 CONCLUSIONS

1,4-Dioxane has not only polluted land but also groundwater and surface water at several places throughout the world. Effective measures must be taken to manage and reclaim dioxane-contaminated sites. Among limited treatment options available, use of biological processes for remediating dioxane-contaminated sites is the most effective and environment-friendly option. Most of the biodegradation studies reported to date have focused on laboratory-cultured microorganisms. Dioxane-contaminated environment may contain many unknown and unidentified microorganisms. Theses dioxane degraders can play a critical role in the remediation of 1,4-dioxane and associated co-contaminants. Phylogenetic diversity of 1,4-dioxane degraders makes it difficult to identify functional microorganisms with a single biomarker. Use of modern approaches such as metagenomics and high-throughput sequencing can aid in discovery of novel degrading microorganisms, enzymes, and biotransformation pathways.

REFERENCES

Adamson D. T., Anderson R. H., Mahendra S. and Newell C. J. (2015). Evidence of 1,4-dioxane attenuation at groundwater sites contaminated with chlorinated solvents and 1,4-dioxane. *Environmental Science & Technology*, **49**, 6510–6518, https://doi.org/10.1021/acs.est.5b00964

Anderson R. H., Anderson J. K. and Bower P. A. (2012). Co-occurrence of 1,4-dioxane with trichloroethylene in chlorinated solvent groundwater plumes at US Air Force installations: fact or fiction. *Integrated Environmental Assessment and Management*, **8**, 731–737, https://doi.org/10.1002/ieam.1306

Aoyagi T., Morishita F., Sugiyama Y., Ichikawa D., Mayumi D., Kikuchi Y., Ogata A., Muraoka K., Habe H. and Hori T. (2018). Identification of active and taxonomically diverse 1,4-dioxane degraders in a full-scale activated sludge system by high-sensitivity stable isotope probing. *The ISME Journal*, **12**, 2376–2388, https://doi.org/10.1038/s41396-018-0201-2

ATSDR (2017). Support document to the 2017 substance priority list (candidates for toxicological profiles). Division of Toxicology and Human Health Sciences, Atlanta, GA.

Black R. E., Hurley F. J. and Havery D. C. (2001). Occurrence of 1,4-dioxane in cosmetic raw materials and finished cosmetic products. *Journal of AOAC International*, **84**, 660–670, https://doi.org/10.1093/jaoac/84.3.666

Centers for Disease Control and Prevention (2017). National health and nutrition examination survey: volatile organic compounds (VOCs) and trihalomethanes/MTBE-blood-special sample (VOCWBSH). 2013–2014 Data Documentation, Codebook, and Frequencies.

Da Silva M. L. B., He Y., Mathieu J. and Alvarez P. J. J. (2020). Enhanced long-term attenuation of 1,4-dioxane in bioaugmented flow-through aquifer columns. *Biodegradation*, **31**, 201–211, https://doi.org/10.1007/s10532-020-09903-0

Dourson M. L., Higginbotham J., Crum J., Burleigh-Flayer H., Nance P., Forsberg N. D., Lafranconi M. and Reichard J. (2017). Update: mode of action (MOA) for liver tumors induced by oral exposure to 1,4-dioxane. *Regulatory Toxicology and Pharmacology*, **88**, 45–55, https://doi.org/10.1016/j.yrtph.2017.02.025

EPA US (2013). Toxicological Review of 1,4-Dioxane (with Inhalation Update). United States Environmental Protection Agency, Washington, DC.

Gheorghe I. F. and Ion B. (2011). The effects of air pollutants on vegetation and the role of vegetation in reducing atmospheric pollution. In: The Impact of Air Pollution on Health, Economy, Environment and Agricultural Sources, M. K. Khallaf (ed.), Intech Croatia, 241–280.

Godri Pollitt K. J., Kim J. H., Peccia J., Elimelech M., Zhang Y., Charkoftaki G., Hodges B., Zucker I., Huang H., Deziel N. C., Murphy. K., Ishii M., Johnson C. H., Boissevain A., O'Keefe E., Anastas P. T., Orlicky D., Thompson D. C. and Vasiliou V. (2019). 1,4-Dioxane as an emerging water contaminant: state of the science and evaluation of research needs. *Science of the Total Environment*, **690**, 853–866, https://doi.org/10.1016/j.scitotenv.2019.06.443

Göen T., von Helden F., Eckert E., Knecht U., Drexler H. and Walter D. (2016). Metabolism and toxicokinetics of 1,4-dioxane in humans after inhalational exposure at rest and under physical stress. *Archives of Toxicology*, **90**, 1315–1324, https://doi.org/10.1007/s00204-015-1567-9

Hand S., Wang B. and Chu K.-H. (2015). Biodegradation of 1,4-dioxane: effects of enzyme inducers and trichloroethylene. *Science of the Total Environment*, **520**, 154–159, https://doi.org/10.1016/j.scitotenv.2015.03.031

He Y., Mathieu J., da Silva M. L. B., Li M. and Alvarez P. J. J. (2018). 1,4-Dioxane-degrading consortia can be enriched from uncontaminated soils: prevalence of *Mycobacterium* and soluble di-iron monooxygenase genes. *Microbial Biotechnology*, **11**(1), 189–198, https://doi.org/10.1111/1751-7915.12850

Hwangbo M., Claycomb E. C., Liu Y., Alivio T. E. G., Banerjee S. and Chu K. H. (2019). Effectiveness of zinc oxide-assisted photocatalysis for concerned constituents in reclaimed wastewater: 1,4-dioxane, trihalomethanes, antibiotics, antibiotic resistant bacteria (ARB), and antibiotic resistance genes (ARGs). *Science of the Total Environment*, **649**, 1189–1197, https://doi.org/10.1016/j.scitotenv.2018.08.360

Inoue O., Seiji K., Kawai T., Jin C., Liu Y. T., Chen Z., Cai S. X., Yin S. N., Li G. L. and Nakatsuka H. (1989). Relationship between vapor exposure and urinary metabolite excretion among workers exposed to trichloroethylene. *American Journal of Industrial Medicine*, **15**, 103–110, https://doi.org/10.1002/ajim.4700150111

Inoue D., Tsunoda T., Yamamoto N., Ike M. and Sei K. (2018). 1,4-Dioxane degradation characteristics of *Rhodococcus aetherivorans* JCM 14343. *Biodegradation*, **29**(3), 301–310, https://doi.org/10.1007/s10532-018-9832-2

Inoue D., Tsunoda T., Sawada K., Yamamoto N., Sei K. and Ike M. (2020). Stimulatory and inhibitory effects of metals on 1,4-dioxane degradation by four different 1,4-dioxane-degrading bacteria. *Chemosphere*, **238**, 124606, https://doi.org/10.1016/j.chemosphere.2019.124606

Jackson R. E. and Dwarakanath V. (1999). Chlorinated decreasing solvents: physical-chemical properties affecting aquifer contamination and remediation. *Groundwater Monitoring & Remediation*, **19**, 102–110, https://doi.org/10.1111/j.1745-6592.1999.tb00246.x

Jasmann J. R., Borch T., Sale T. C. and Blotevogel J. (2016). Advanced electrochemical oxidation of 1,4-dioxane via dark catalysis by novel titanium dioxide (TiO_2) pellets. *Environmental Science & Technology*, **50**, 8817–8826, https://doi.org/10.1021/acs.est.6b02183

Johnson N. W., Gedalanga P. B., Zhao L., Gu B. and Mahendra S. (2020). Cometabolic biotransformation of 1,4-dioxane in mixtures with hexavalent chromium using attached and planktonic bacteria. *Science of the Total Environment*, **706**, 135734, https://doi.org/10.1016/j.scitotenv.2019.135734

Kano H., Umeda Y., Kasai T., Sasaki T., Matsumoto M., Yamazaki K., Nagano K., Arito H. and Fukushima S. (2009). Carcinogenicity studies of 1,4-dioxane administered in drinking-water to rats and mice for 2 years. *Food and Chemical Toxicology*, **47**, 2776–2784, https://doi.org/10.1016/j.fct.2009.08.012

Kao C. M., Chen S. C., Wang J. Y., Chen Y. L. and Lee S. Z. (2003). Remediation of 20 PCE-contaminated aquifer by an in situ two-layer biobarrier: laboratory batch 21 and column studies. *Water Research*, **37**, 27–38, https://doi.org/10.1016/S0043-1354(02)00254-3

Kelley S. L., Aitchison E. W., Deshpande M., Schnoor J. L. and Alvarez P. J. J. (2001). Biodegradation of 1,4-dioxane in planted and unplanted soil: effect of bioaugmentation with *Amycolata* sp CB1190. *Water Research*, **35**, 3791–3800, https://doi.org/10.1016/S0043-1354(01)00129-4

Kim Y. M., Jeon J. R., Murugesan K., Kim E. J. and Chang Y. S. (2009). Biodegradation of 1,4-dioxane and transformation of related cyclic compounds by a newly isolated *Mycobacterium* sp. PH-06. *Biodegradation*, **20**, 511–519, https://doi.org/10.1007/s10532-008-9240-0

Kumar R., Das A. J. and Juwarkar A. A. (2015). Reclamation of petrol oil contaminated soil by rhamnolipids producing PGPR strains for growing *Withania somnifera* a medicinal shrub. *World Journal of Microbiology and Biotechnology*, **31**, 307–313, https://doi.org/10.1007/s11274-014-1782-1

Landry G. M., Martin S. and McMartin K. E. (2011). Diglycolic acid is the nephrotoxic metabolite in diethylene glycol poisoning inducing necrosis in human proximal tubule cells in vitro. *Toxicological Sciences*, **124**, 35–44, https://doi.org/10.1093/toxsci/kfr204

Landry G. M., Dunning C. L., Abreo F., Latimer B., Orchard E. and McMartin K. E. (2015). Diethylene glycol-induced toxicities show marked threshold dose response in rats. *Toxicology and Applied Pharmacology*, **282**, 244–251, https://doi.org/10.1016/j.taap.2014.12.010

Lee C. S., Le Thanh T., Kim E. J., Gong J., Chang Y. Y. and Chang Y. S. (2014). Fabrication of novel oxygen-releasing alginate beads as an efficient oxygen carrier for the enhancement of aerobic bioremediation of 1,4-dioxane contaminated groundwater. *Bioresource Technology*, **171**, 59–65, https://doi.org/10.1016/j.biortech.2014.08.039

Lee K. C., Beak H. J. and Choo K. H. (2015). Membrane photoreactor treatment of 1,4-dioxane containing textile wastewater effluent: performance, modeling, and fouling control. *Water Research*, **86**, 58–65, https://doi.org/10.1016/j.watres.2015.05.017

Li M. Y., Fiorenza S., Chatham J. R., Mahendra S. and Alvarez P. J. J. (2010). 1,4-Dioxane biodegradation at low temperatures in Arctic groundwater samples. *Water Research*, **44**, 2894–2900, https://doi.org/10.1016/j.watres.2010.02.007

Li M., Mathieu J., Yang Y., Fiorenza S., Deng Y., He Z., Zhou J. and Alvarez P. J. (2013). Widespread distribution of soluble di-iron monooxygenase (SDIMO) genes in Arctic groundwater impacted by 1,4-dioxane. *Environmental Science & Technology*, **47**, 9950–9958, https://doi:10.1021/es402228x

Li M., Van Orden E. T., DeVries D. J., Xiong Z., Hinchee R. and Alvarez P. J. (2015). Bench-scale biodegradation tests to assess natural attenuation potential of 1,4-dioxane at three sites in California. *Biodegradation*, **26**, 39–50, https://doi.org/10.1007/s10532-014-9714-1

Linares R. V., Yangali-Quintanilla V., Li Z. and Amy G. (2011). Rejection of micropollutants by clean and fouled forward osmosis membrane. *Water Research*, **45**, 6737–6744, https://doi.org/10.1016/j.watres.2011.10.037

Lippincott D., Streger S. H., Schaefer C. E., Hinkle J., Stormo J. and Steffan R. J. (2015). Bioaugmentation and propane biosparging for in situ biodegradation of 1,4-dioxane. *Groundwater Monitoring & Remediation*, **35**, 81–92, https://doi.org/10.1111/gwmr.12093

Mahendra S. and Alvarez-Cohen L. (2006). Kinetics of 1,4-dioxane biodegradation by monooxygenase-expressing bacteria. *Environment Sciences Technology*, **40**, 5435–5442, https://doi:10.1021/es060714v

Mahendra S., Grostern A. and Alvarez-Cohen L. (2013). The impact of chlorinated solvent co-contaminants on the biodegradation kinetics of 1,4-dioxane. *Chemosphere*, **91**, 88–92, https://doi.org/10.1016/j.chemosphere.2012.10.104

Martın T., Padron J. I. and Martın V. S. (2014). Strategies for the synthesis of cyclic ethers of marine natural products. *Synlett*, **25**, 12–32.

McElroy A. C., Hyman M. R. and Knappe D. R. U. (2019). 1,4-Dioxane in drinking water: emerging for 40 years and still unregulated. *Current Opinion in Environmental Science & Health*, **7**, 117–125, https://doi.org/10.1016/j.coesh.2019.01.003

Merayo N., Hermosilla D., Cortijo L. and Blanco A. (2014). Optimization of the Fenton treatment of 1,4-dioxane and on-line FTIR monitoring of the reaction. *Journal of Hazardous Materials*, **268**, 102–109, https://doi.org/10.1016/j.jhazmat.2014.01.008

Nakamiya K., Hashimoto S., Ito H., Edmonds J. S. and Morita M. (2005). Degradation of degradation of 1,4-dioxane and cyclic ethers by an isolated fungus. *Applied and Environmental Microbiology*, **71**, 1254–1258, https://doi.org/10.1128/AEM.71.3.1254-1258.2005

Nannelli A., De Rubertis A., Longo V. and Gervasi P. G. (2005). Effects of dioxane on cytochrome P450 enzymes in liver, kidney, lung and nasal mucosa of rat. *Archives of Toxicology*, **79**, 74–82, https://doi.org/10.1007/s00204-004-0590-z

Ossai I. C., Ahmed A., Hassan A. and Hamid F. S. (2020). Remediation of soil and water contaminated with petroleum hydrocarbon: a review. *Environmental Technology & Innovation*, **17**, 100526, https://doi.org/10.1016/j.eti.2019.100526

Parales R. E., Adamus J. E., White N. and May H. D. (1994). Degradation of 1,4-dioxane by an *Actinomycete* in pure culture. *Applied and Environmental Microbiology*, **60**, 4527–4530, https://doi.org/10.1128/aem.60.12.4527-4530.1994

Patt T. E. and Abebe H. M. (1995). Microbial degradation of chemical pollutants; Patent Number: 5399495.

Ramalingam V. and Cupples A. M. (2020). Anaerobic 1,4-dioxane biodegradation and microbial community analysis in microcosms inoculated with soils or sediments and different electron acceptors. *Applied Microbiology and Biotechnology*, **104**, 4155–4170, https://doi.org/10.1007/s00253-020-10512-3

Ramos-Garcia A. A., Shankar V., Saski C. A., Hsiang T. and Freedman D. L. (2018). Draft genome sequence of the 1,4-dioxane-degrading bacterium *Pseudonocardia dioxanivorans* BERK-1. *Genome Announcements*, **6**, e00207-18, https://doi.org/10.1128/genomeA.00207-18

Rehman Z. U. and Leiknes T. (2018). Quorum-quenching bacteria isolated from Red Sea sediments reduce biofilm formation by *Pseudomonas aeruginosa*. *Frontiers in Microbiology*, **9**, 1354, https://doi.org/10.3389/fmicb.2018.01354

Rolston H. M., Hyman M. R. and Semprini L. (2019). Aerobic cometabolism of 1,4-dioxane by isobutane-utilizing microorganisms including *Rhodococcus rhodochrous* strain 21198 in aquifer microcosms: experimental and modeling study. *Science of the Total Environment*, **694**, 133688, https://doi.org/10.1016/j.scitotenv.2019.133688

Sales C. M., Grostern A., Parales J. V., Parales R. E. and Alvarez-Cohen L. (2013). Oxidation of the cyclic ethers 1,4-dioxane and tetrahydrofuran by a monooxygenase in two *Pseudonocardia* species. *Applied and Environmental Microbiology*, **79**, 7702–7708, https://doi.org/10.1128/AEM.02418-13

Saul M. T. (2012). Bioaugmentation to remediate dioxane in groundwater. US20100140168 A1.

Schier J. G., Hunt D. R., Perala A., McMartin K. E., Bartels M. J., Lewis L. S., McGeehin M. A. and Flanders W. D. (2013). Characterizing concentrations of diethylene glycol and suspected metabolites in human serum, urine, and cerebrospinal fluid samples from the Panama DEG mass poisoning. *Clinical Toxicology*, **51**, 923–929, https://doi.org/10.3109/15563650.2013.850504

Sei K., Miyagaki K., Kakinoki T., Fukugasako K., Inoue D. and Ike M. (2013). Isolation and 816 characterization of bacterial strains that have high ability to degrade 1,4-dioxane as a sole carbon and energy source. *Biodegradation*, **24**, 665–674, https://doi.org/10.1007/s10532-012-9614-1

Shen W., Chen H. and Pan S. (2008). Anaerobic biodegradation of 1,4-dioxane by sludge enriched with iron-reducing microorganisms. *Bioresource Technology*, **99**, 2483–2487, https://doi.org/10.1016/j.biortech.2007.04.054

Skinner K., Cuiffetti L. and Hyman M. (2009). Metabolism and cometabolism of cyclic ethers by a filamentous fungus, a *Graphium* sp. *Applied and Environmental Microbiology*, **75**, 5514–5522, https://doi.org/10.1128/AEM.00078-09

Stepien D. K., Diehl P., Helm J., Thoms A. and Puttmann W. (2014). Fate of 1,4-dioxane in the aquatic environment: from sewage to drinking water. *Water Research*, **48**, 406–419, https://doi.org/10.1016/j.watres.2013.09.057

Stott W. T., Quast J. F. and Watanabe P. G. (1981). Differentiation of the mechanisms of oncogenicity of 1,4-dioxane and 1,3-hexachlorobutadiene in the rat. *Toxicology and Applied Pharmacology*, **60**, 287–300, https://doi.org/10.1016/0041-008X(91)90232-4

Surprenant K. S. (2000). Dioxane. In Ullmann's encyclopedia of industrial chemistry (ed.).

Topp E., Chapman R., Devers-Lamrani M., Hartmann A., Marti R., Martin-Laurent F., Sabourin L., Scott A. and Sumarah M. (2013). Accelerated biodegradation of veterinary antibiotics in agricultural soil following long-term exposure, and isolation of a sulfamethazine-degrading sp. *Journal of Environmental Quality*, **42**, 173–178, https://doi.org/10.2134/jeq2012.0162

Vainberg S., McClay K., Masuda H., Root D., Condee C., Zylstra G. J. and Steffan R. J. (2006). Biodegradation of ether pollutants by *Pseudonocardia* sp. strain ENV478. *Applied and Environmental Microbiology*, **72**, 5218–5224, https://doi.org/10.1128/AEM.00160-06

Vásquez-Murrieta M. S., Hernández-Hernández O. J., Cruz-Maya J. A., Cancino-Díaz J. C. and Jan-Roblero J. (2016). Approaches for removal of PAHs in soils: bioaugmentation, biostimulation and bioattenuation. In: Environmental Sciences. Soil Contamination – Current Consequences and Further Solutions, M. L. Larramendy and S. Soloneski (eds.), InTech, Croatia, 329–340.

Woo Y., Arcos J. C., Argus M. F., Griffin G. W. and Nishiyama K. (1977). Metabolism in vivo of dioxane: identification of *p*-dioxane-2-one as the major urinary metabolite. *Biochemical Pharmacology*, **26**, 1535–1538, https://doi.org/10.1016/0006-2952(77)90430-0

Wu M., Dick W. A., Li W., Wang X., Yang Q., Wang T., Xu L., Zhang M. and Chen L. (2016). Bioaugmentation and biostimulation of hydrocarbon degradation and the microbial community in a petroleum contaminated soil. *International Biodeterioration & Biodegradation*, **107**, 158–164, https://doi.org/10.1016/j.ibiod.2015.11.019

Xiong Y., Mason O. U., Lowe A., Zhang Z., Zhou C., Chen G., Villalonga M. J. and Tang Y. (2020). Investigating promising substrates for promoting 1,4-dioxane biodegradation: effects of ethane and tetrahydrofuran on microbial consortia. *Biodegradation*, **31**, 171–182, https://doi.org/10.1007/s10532-020-09901-2

Zenker M. J., Borden R. C. and Barlaz M. A. (2004). Biodegradation of 1,4-dioxane using trickling filter. *Journal of Environmental Engineering*, **130**, 926–931, https://doi.org/10.1061/(ASCE)0733-9372(2004)130:9(926)

Zhao L., Lu X., Polasko A., Johnson N. W., Miao Y., Yang Z., Mahendra S. and Gu B. (2018). Co-contaminant effects on 1,4-dioxane biodegradation in packed soil column flow-through systems. *Environmental Pollution*, **243**(Pt A), 573–581, https://doi.org/10.1016/j.envpol.2018.09.018

doi: 10.2166/9781789063783_0247

Chapter 13

Role of biomass-based biorefinery in mitigating environmental pollution

Pankaj Garkoti[1], Harshit Tiwari[1], Priyanka Sarkar[2]* and Biswanath Bhunia[3]

[1]Department of Hydro and Renewable Energy, Indian Institute of Technology, Roorkee 247667, India
[2]Department of Biotechnology, Techno India University, Salt Lake, Sector 5, Kolkata 700091, West Bengal, India
[3]Department of Bio Engineering, National Institute of Technology, Agartala 799046, India
*Corresponding author: Priyanka Sarkar. Email address: Priyanka.g@technoindiaeducation.com; priyankasarkar91@gmail.com

ABSTRACT

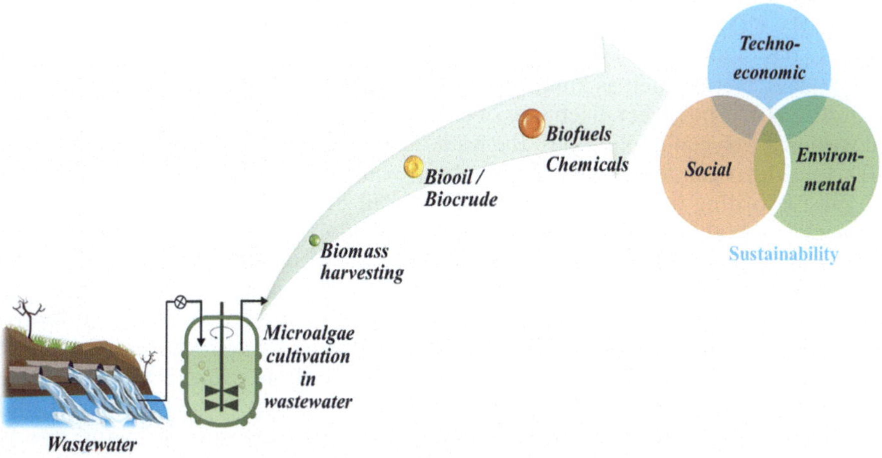

About 400 trillion liters of wastewater is generated yearly due to widespread and inefficient water use. The wastewater from different sources contains many nutrients, including nitrogen, phosphorous, and carbon. Therefore, these nutrients must be recovered using appropriate sustainable processes. The selection of the proper process depends upon the type of waste and its generation. Hence, there is a need to implement a biorefinery approach to utilize organic and biological pollutants to recover energy and bioproducts. For sustainable and inexpensive wastewater treatment, microalgae-based processes have garnered considerable interest. The biomass produced by microalgae can be used in bioenergy production and other value-added products. Thus, this chapter provides state-of-the art information regarding the potential of biomass-based integrated biorefineries for effluent treatment.

Keywords: microalgae, biorefinery, wastewater treatment, bioenergy, biomass, resource recovery

13.1 INTRODUCTION

Water is one of the essential resources on earth which plays a vital role in the survival and well-being of all living organisms (Wollmann *et al.*, 2019). In general, 97% of fresh water is found in the oceans, whereas 3% occurs in polar ice caps, glaciers, soil, and the atmosphere. However, 0.4–0.5% of freshwater is accessible for human consumption (Bureau of Reclamation, 2020). Population growth and industrialization consume vast amounts of freshwater, generating enormous volumes of effluent. Further, according to Corcoran (2010), rising population, urban expansion, and increasing agricultural and economic activities degrade natural water resources, especially in developing and developed nations. High-income nations treat an average of 70% of their produced wastewater, whereas middle and low-income countries treat 38% and 28%, respectively (United *et al.*, 2016). In addition, Qadir *et al.* (2020) reported that around 400 trillion liters of wastewater are produced annually across the world.

Wherein, Asia is the primary producer of wastewater (~160 trillion liters/year), followed by European and North American nations (~68 and 67 trillion liters/year, respectively). Moreover, Asia is a prominent producer of domestic effluent, which generates ~42% of the total effluent (Qadir *et al.*, 2020). This proportion is further projected to expand to 43.9% by the end of 2030. According to the Central Pollution Control Board, New Delhi, India, about 40 billion liters per day of wastewater was generated in 2011 (Singh *et al.*, 2016). Likewise, Bhuvaneshwari and Devika (2005) stated that the amount of effluent generation in India's metropolitan cities (~120 L/day) was twice as high as in other small cities (~60 L/day).

Wastewater can be classified according to its source, such as municipal wastewater, industrial wastewater, and agricultural runoff. However, their nutrient load varies with varying wastewater. In general, the BOD, COD, total nitrogen, and total phosphate are in the range of 27–96,000 mg/ L, 80–134,400 mg/L, 0.6–4200 mg/L, and 100–3030 mg/L, respectively. Moreover, wastewater contains various types of heavy metals such as zinc, copper, mercury, selenium, arsenic, and chromium, and so on) (Pancha *et al.*, 2019). The direct discharge of effluent into the environment endangered aquatic and terrestrial ecosystems (oxygen depletion in the aquatic environment). The presence of excessive levels of BOD, COD, total nitrogen, and total phosphate in water bodies leads to eutrophication (Ibrahim & Heikal, 2019). According to Qu and Fan (2010), about 80% of rivers in China are contaminated due to the presence of high concentrations of nitrate, phosphate, organic matter, heavy metals, and micronutrients. Eutrophication results in the proliferation of toxic microbes and damaging effects on aquatic creatures. Besides, industrial effluents such as polyaromatic hydrocarbons, petroleum chemicals, phenolic compounds, pesticides, polychlorinated biphenyls, and antibiotics, threaten terrestrial and aquatic ecosystems. According to the studies by the United Nations (WWAP, 2017), about 80% of effluent is directly discharged into open water bodies without prior treatment. Sato *et al.* (2013) reported that developed or highly developed nations treat more than 70% of industrial, and municipal wastewater, whereas developing and less-developed countries treat only 38% and 8%, respectively, due to technological availability and economic feasibility. Further, in 2012, ~13 million individuals died because of various forms of pollution, including water, air, soil, and chemical exposure (Neira & Prüss-Ustün, 2016). Therefore, each form of pollution must be treated before being released into the environment. Several previous treatments such as physical, chemical, and biological processes have been developed for the removal of toxic substances from wastewater; however, they have several disadvantages (Akubude *et al.*, 2019; Udaiyappan *et al.*, 2017). Moreover, conventional treatment methods are not equitable due to the following reasons (Abdelgadir *et al.*, 2014; Ezugbe & Rathilal, 2020; Gurung *et al.*, 2018; Li *et al.*, 2019; Pancha *et al.*, 2019; Qadir *et al.*, 2020; Sato *et al.*, 2013):

(i) It requires a high energy input (4000–3,91,000 Wh)
(ii) Time-consuming (120–1800 h)
(iii) Less efficient in nutrient removal (removal efficiency ~10–90%)
(iv) Produce an excessive amount of by-product 'sludge' that requires further treatment.

According to a report published by SAMCO (2016), the cost of wastewater treatment facilities varies from $200,000 to $700,000 at peak flow rates of 21 and 114 m^3/min, respectively. In Europe, wastewater treatment facilities consume roughly 1% of the overall electrical usage. The power used to remove 1 kg of COD ranged between 850 and 970 W-h/kg, with an average electricity cost rate of 140 $/W-h (Longo *et al.*, 2016; Panepinto *et al.*, 2016). However, advanced treatment methods require less energy (200 W-h/t) compared with conventional technologies (Li *et al.*, 2017). The demand for the development of effective and cost-effective technologies is critical in the middle- and low-income countries because of insufficient access to the above elements. Overall, conventional methods are neither environmentally viable nor capable of resolving the issue of water scarcity. Furthermore, recent research has progressively emphasized the recovery of several valuable resources from wastewater for bioenergy generation. Microalgae-mediated wastewater treatments and bioenergy generation are highly advantageous owing to its significant nutrient bioaccumulation efficiency, high biomass productivity, and biomolecule content (especially lipid and protein) that can be utilized for bioenergy production (Rawat *et al.*, 2013). Microalgae are aquatic microorganisms that utilize nutrients from multiple wastes to generate new biomass that can be used as feedstock for the bioenergy production and value-added bioproducts (Yang *et al.*, 2015). This is one of the viable techniques for sustainable socio-economic development based on the concept of circular bioeconomy.

13.2 BIOREFINERY

The increasing demand for energy and chemicals has led the researcher's attention toward renewable energy. Renewable resource utilization is the key to moving toward a sustainable economy from today's fossil fuel-based economy. It is strongly advised that biofuels (ethanol, biodiesel, butanol, Fischer–Tropsch diesel) be utilized specifically in heavy-duty vehicles and aviation. However, biofuel production is not profitable compared to current crude oil prices. The promising approach for reducing the biofuel production cost is constructing and implementing a biorefinery model. The co-production of value-added products (chemicals, materials) can reduce the production cost of biofuels by 30% using biorefinery models. International Energy Agency defines Biorefining as 'the sustainable processing of biomass into bio-based products and bioenergy.' The biorefinery model includes different steps, such as the collection of biomass, its pretreatment, hydrolysis, transformation, and product recovery. The major goal of the biorefinery concept is to replace fossil fuels and reduce negative environmental emissions produced through the disposal of organic matter. The biorefinery concept can complete the constant demand for energy, building block chemicals, and employment locally. The critical factors of a successful biorefinery are the integration of the biorefinery model with existing industrial infrastructure. The types of biorefinery are classified based on the biomass type and its origin. The wastewater biorefinery relies on the wastewater or sludge produced by households or different industries. Similarly, the agricultural, industrial, and municipal biorefinery depends on agriculture, industrial waste, municipal waste, and algae biomass, respectively. Different types of biorefineries are as follows:

(i) Agriculture-based biorefinery
(ii) Forestry-based biorefinery
(iii) Industrial waste biorefinery
(iv) Food waste biorefinery
(v) Animal waste biorefinery
(vi) Wastewater biorefinery
(vii) Plastic waste biorefinery
(viii) Algae-based biorefinery

13.3 WASTEWATER BIOREFINERY

The wastewater contains different biological, chemical, and physical pollutants; removing these pollutants is crucial to minimize water pollution and environmental security. The wastewater biorefinery aims to close the material and energy flow to improve economic and environmental sustainability. The most traditional way to produce energy from wastewater is to utilize co-produced sludge. The sludge, after thermal hydrolysis, can be fed to the anaerobic digester to produce biogas. The methane-rich biogas may be used directly in combined heat and power (CHP) units or purified to further use for transportation. The anaerobic treatment processes are the core of the wastewater biorefinery concept. The nutrients from the wastewater can be used as biofertilizers in agriculture. The nitrogen and phosphorous recovery from the wastewater are essential as it promotes eutrophication in the aquatic environment. Other than these nutrients, there is cellulosic sludge separated by upstream dynamic sieving, and polyhydroxyalkanoate (PHA) can be recovered from primary or secondary sludge. The cellulosic sludge is a feed for an anaerobic digester, and PHA can be used for the synthesis of various chemicals. Table 13.1 presents the different anaerobic bioreactor processes for treating wastewater and recovering energy as biogas. It can be noticed from the table that the anaerobic bioreactor used to treat wastewater are upflow anaerobic sludge blanket reactor (UASB), fixed-bed reactor (FBR), anaerobic baffled reactor (ABR), anaerobic moving biofilm reactor (AMBR), and continuous stirred tank reactor (CSTR) which have COD removal efficiency within range of 80%–95%. The biogas production from the UASB is observed to be more than other types of reactors using municipal or domestic wastewater (Akyol *et al.*, 2020; Hu *et al.*, 2018).

The other way to generate energy from wastewater is by producing biohydrogen using a dark fermentation process (Arooj *et al.*, 2007; Kim *et al.*, 2008; Mu *et al.*, 2006). Different studies have been performed to utilize wastewater for biohydrogen production. The high organic-rich cassava wastewater for biohydrogen production study shows that the increase in organic load reduces biohydrogen production. The highest hydrogen production achieved was 2.41 mol H_2/mol glucose (Mari *et al.*, 2020). A similar study for dairy wastewater showed that rate of the substrate to biohydrogen production becomes slow when protein degradation occurs. The study observed that dairy industry wastewater has low potential to generate biohydrogen (Xing *et al.*, 2010). However, adding nanoparticles (hematite or nickel oxide) to the dairy wastewater showed an improvement in biohydrogen yield by 43%. It is concluded from these studies that biohydrogen production from industrial wastewater is good from an economic point of view.

The wastewater biorefinery utilizes biological agents to remove these pollutants and utilize them to produce value-added products (ethanol, biodiesel, polylactic acid). The cultivation of microorganisms or algae in wastewater allows the removal of pollutants (a rich source of nitrogen, phosphorous, and ammonium) cost-effectively and biomass as a co-product which can be utilized for the production of bioproducts (carbohydrates, lipids, and proteins). The cultivation of *Saccharomyces cerevisiae* for olive mill wastewater treatment showed decolorization and phenol removal efficiency of about 63% and 34%, respectively. The microbial culture after treatment showed the potential for lipid and bioethanol production. The co-culturing of microalgae and bacteria for wastewater treatment has been studied in a biorefinery approach. The author concluded that the COD removal efficiency was about 82.7% and the biomass productivity was 343.3 mg/L/d.

The available methods for resource recovery from wastewater require chemicals and catalysts. Therefore, more attention is now given to the biological approach to recover resources to make the process more sustainable. The biological approach includes the utilization of different microorganisms such as bacteria, fungi, algae, and plants. Among all these, microalgae-based wastewater treatment seems to be more promising due to algae biosorption/bioaccumulation efficiency, high biomass productivity, and nutrients (nitrogen and phosphorous) utilization. The algae-based wastewater treatment has been studied by various researchers and is discussed in the next section.

Table 13.1 Anaerobic bioreactor for wastewater treatment and biogas production.

Country	Wastewater	Reactor Type	Temperature (°C)	HRT (h)	COD Removal Efficiency (%)	Methane (m^3 CH_4/kg COD Removed)	References
Spain	Municipal wastewater	UASB	18 ± 2	7	80	0.24	Gouveia *et al.* (2015)
Australia	Synthetic domestic wastewater	UAGB	20	/	91	0.16	Chen *et al.* (2017)
China	Domestic wastewater	UASB	22–25	8	92	0.45	Hu *et al.* (2018)
Poland	Synthetic dairy wastewater	AMBR	20	132	86.6	0.002	Zielińska *et al.* (2018)
Germany	Raw dairy effluent	CSTR, ABR	38	7.6	91	0.264	Jürgensen *et al.* (2018)
Spain	Real municipal wastewater	–	10–25	10	94	0.14	Peña *et al.* (2019)
United States	Municipal wastewater	–	12.7–31.5	6–20	88	0.12	Lim *et al.* (2019)
United States	Domestic wastewater	FBR	13–32	3.9 ± 1	90	0.17	Evans *et al.* (2019)
Taiwan	Domestic wastewater	FBR	20–35	12	95	0.37	Chen *et al.* (2019)
Singapore	Domestic wastewater	CSTR	30 ± 3	17	91	0.07	Xu *et al.* (2020)
Italy	Municipal wastewater	UASB	30	5–6	83	0.56	Akyol *et al.* (2020)

13.4 ALGAE-BASED WASTEWATER BIOREFINERY

The algae growth requires nutrients and water, which can be completed with wastewater. The microalgae accumulate phosphorous and nitrogen from the wastewater and utilize CO_2 for their growth using a process known as bio-fixation (Nizami *et al.*, 2017). A pilot project on palm oil mill effluent using a microalgae system with biogas production in Indonesia is designed to generate 1 MW of electricity (Ichsan *et al.*, 2014). Biohydrogen using dark fermentation is another way of generating bioenergy. Literature suggests that the wastewater treatment with microalgae reduced the COD by 70–90%. The energy consumption analysis concluded that the microalgae-based wastewater treatment consumes 10 times (0.2 kW/h/m^3) less energy compared to conventional wastewater treatment (2 kW/h/m^3) (Sivaramakrishnan *et al.*, 2022). The microalgae can grow in both open ponds and in the photobioreactor. The photobioreactor has a low contamination risk, easy to operate, and has high biomass productivity. Microalgae cultivation in the open pond is preferable from an economic point of view (Wijffels *et al.*, 2010). Table 13.2 presents the economic feasibility of microalgae-based wastewater treatment. The comparative economic study of microalgae cultivation in an open pond and photobioreactor shows that algae oil production in the open pond is 2.12 times cheaper than in a photobioreactor (Davis *et al.*, 2011). A biorefinery approach for producing glycerol, biogas, and biodiesel from microalgae showed that the system's payback period is around 3.3 years (Wijffels *et al.*, 2010). The production of biofuels and other value-added from microalgae depends on the composition of the strain. Therefore, selecting microalgal strains and coupling different bioprocesses to produce different bio-products becomes much more critical for the profitability of microalgal biorefinery. Freshwater-based microalgae cultivation is expensive and threatens water security. The wastewater for microalgae cultivation not only provides nutrients but also reduces the demand of freshwater, making the process cost-effective. Microalgae cultivation in wastewater and flue gas increases biomass productivity and bioenergy and bioactive compounds. It is reported that a microalgae–bacteria consortium is a novel approach to treating wastewater. The result showed that the consortia helped in reducing COD and NO_3 with greater efficiency than the monoculture. The immobilization of microalgae is another approach to treating wastewater efficiently (Goswami *et al.*, 2021). The immobilization of microalgae on sodium alginate showed 23% higher COD reduction efficiency than non-immobilized algal cells (Chu *et al.*, 2020). Table 13.2 presents different case studies and their economic feasibility. The sustainability of microalgae biorefinery can be increased if it produces multiple products such as fertilizer, fuels, food, pharmaceuticals, and cosmetics. It can be seen from the table that there are different routes to utilize the produced biomass, such as liquification, transesterification, anaerobic digestion, and pyrolysis. The utilization of biomass will produce bioproducts such as bio-oil, biochar, syngas, and biogas (Figure 13.1).

13.5 POTENTIAL PRODUCTS FROM WASTEWATER BIOREFINERY

Sewage sludge management is a cost-intensive part of the wastewater treatment process. In commercial-scale wastewater plants, the sewage sludge is usually utilized in the anaerobic digestion plants to produce biogas and utilized to produce heat and electricity, which completes the industrial plant's heat and electricity demand. However, the recovery of nutrients from the sludge in the biorefinery approach can help increase the wastewater plants' profitability.

13.5.1 Volatile fatty acid

The acidogenic fermentation of the biomass yields volatile fatty acids. The sludge from the wastewater is gaining attention of researchers as a potential feedstock for VFA production (Peces *et al.*, 2016). The VFA is a precursor for different bioproducts such as PHA, esters, ketones, alkanes, and biofuels (Lin *et al.*, 2018). Thus, makes it a high economic value-added product. The important properties of VFA are its high flexibility, low stiffness, and high tensile strength. The utilization of carbon from sludge to VFA is an energy and cost-effective strategy. However, VFA production from the sludge could reduce the energy recovery from the sludge in the form of biogas (Peces *et al.*, 2016). The optimization of both

Table 13.2 Economic feasibility of microalgae-based biorefineries.

Strain	Feedstock	Cultivation / Biomass Utilized	Value-added Products	Capital Cost (million $)	Operating Cost (million $)	NPV (million $)	Reference
Dunaliella salina	FW, CO_2, salts	OP	Fertilizer, carotene	11.87	14.20	9.22	Thomassen *et al.* (2016)
Chlorella vulgaris	Municipal WW, CO_2	PBR/PL	Bio-oil, syngas, biochar	5.35	1.24	–	Xin *et al.* (2016)
-	FW, CO_2, salts	OP/TE	Diesel	381.2	292.4	6.27/gallon	Hoffman *et al.* (2017)
Chlorella vulgaris	WW, CO_2	PBR/PL	Bio-oil Syngas Biochar	47.07	8.94	$1.85/gallon bio-oil	Xin *et al.* (2018)
Arthrospira	FW, CO_2, salts	–	Phycocyanin Polysaccharide	1273	–	Negative	Chaiklahan *et al.* (2018)
Chlorella vulgaris	WW, CO_2	OP/ LF	Bio-oil	1.56	4.19	173.82	Musa *et al.* (2019)
Synechocystis GMO	FW, CO_2, salts	PBR	Bioethanol Fertilizer Biogas	38.37	–	Negative	Lopes *et al.* (2019)
Chlorella vulgaris	WW, CO_2	OP/LF	Bio-oil	1.34	22.52	74.82	Musa *et al.* (2019)
Chlorella vulgaris	Paper WW, flue gas	OP/AD	Lipids Biogas	22.47	4.40	16.29	Silva *et al.* (2021)

FW = freshwater; WW = wastewater; OP = open pond; PL = pyrolysis; PBR = photobioreactor; TE = transesterification; LF = liquefaction; AD = anaerobic digestion.

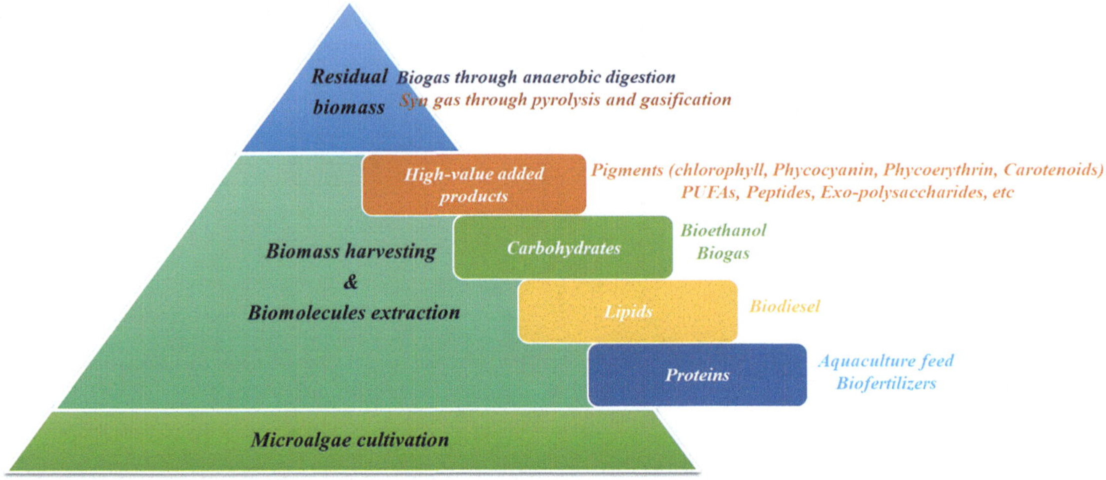

Figure 13.1 Microalgae and their by-products.

processes has been studied, and it concluded that the optimization should focus on the cost of VFA production and the impact on methane generation (Akyol *et al.*, 2020).

13.5.2 Nutrients

The energy-efficient nutrient recovery by using the biorefinery concept in wastewater can decrease the cost and increase the sustainability of wastewater treatment plants. Nitrogen and phosphorous as nutrients, PHA as a biopolymer, and cellulose are currently verified biorefinery products from wastewater treatment plants (Raheem *et al.*, 2018). The process of crystallization of phosphorous and magnesium salt usually accomplishes phosphorus recovery. The process is also known as struvite recovery. The phosphorus recovery reduces environmental disposal issues and increases the sustainability of the plant as its market value is 0.5 times of imported phosphorous (Ghosh *et al.*, 2019). The struvite as fertilizer is preferred in agricultural applications as it releases Mg, N, and P simultaneously and with slow rate than other fertilizers (Puchongkawarin *et al.*, 2015). The struvite recovery from the sludge also decreases the volume of sludge produced, which saves the sludge disposal cost of the plant. The phosphorous recovery from wastewater has been carried out commercially by various technologies such as CalPrex™ reactor, AirPrex™, WASSTRIP™, and Ostara Pearl™ (Akyol *et al.*, 2020). Other than phosphorous, nitrogen recovery is necessary as it also causes eutrophication. The conventional method to remove nitrogen from the wastewater is by nitrification, denitrification, and anammox process. However, the traditional process mostly focuses on the removal of nutrients rather than their recovery (Rodríguez Arredondo *et al.*, 2019). The recovered nitrogen in the form of ammonia has fertilizer application. Therefore, its recovery makes the wastewater plant a sustainable process. Various studies have used microbial fuel cells (MFC) for nitrogen recovery (Yadav *et al.*, 2022). The nitrogen recovery from the anode and cathode is achieved 47% and 83% when an output voltage of 600–700 mV and external load of 500 Ω is supplied in three-chamber MFC (Xiao *et al.*, 2016). In another study, the single-chamber MFC recovered 85 mg N/m^2 (Littfinski *et al.*, 2022). The data show that only 20% of total nitrogen is recovered through recovery (Yadav *et al.*, 2022). Various other processes are in the development phase, such as ion exchange, ammonia stripping, and nanofiltration. Microalgal-based wastewater treatment with nitrogen recovery has been studied, and algal biomass is used as animal feed (Sharma *et al.*, 2022). The recovery of these nutrients using microalgae is reported to be cost-effective. However, energy and sustainability analysis is required before implementation on a commercial scale (Khan *et al.*, 2022).

13.5.3 Polyhydroxyalkanoate

Polyhydroxyalkaonate (PHA) is produced from a renewable source and has properties similar to petroleum-based plastics (Raheem *et al.*, 2018). The advantage of PHA is its biodegradable nature and non-toxicity (Mannina *et al.*, 2020). It has been reported that PHA production cost using pure cultures is more than mixed cultures (Puyol *et al.*, 2017). The microbial cultures that produce these polymers are well-grown in activated sludge and used as a carbon source. The PHA production in full-scale activated sludge at the pilot scale reported a production efficiency of about 0.47 kg PHA/kg VSS (Bengtsson *et al.*, 2017). Various studies have reported that PHA can be produced from waste in sufficient quantities. The economic and environmental aspects of the production of PHA should be evaluated to implement the technology at a commercial scale (Fernández-Dacosta *et al.*, 2015).

13.5.4 Cellulose

Domestic or municipal wastewater contains about 30–50% cellulose fiber (Behera *et al.*, 2018). Cellulose biodegradation is difficult, but if recovered, it has a great potential to increase the economy of the plant (Crutchik *et al.*, 2018). The application of recovered fiber is in the paper industry as a building material, biofuel, and bioplastics (Van Der Hoek *et al.*, 2016). The cellulose enters the wastewater majorly in the form of toilet paper. The cellulose fiber can be recovered from the wastewater by sieving. The recovered fiber is dewatered, which minimizes chemical consumption, reduces electricity consumption, and reduces the phosphate release and volume of the sludge. The technical feasibility and energy cost to recover cellulose should be evaluated before implementation as the recovery process might affect the efficiency of the wastewater plant (Lin *et al.*, 2015).

13.6 POTENTIAL PRODUCTS FROM ALGAL BIOMASS

The wastewater treatment through microalgae will also have algae biomass other than sludge. Microalgae biomass may be used to make a variety of bio-products such as biogas, bioethanol, bio-oil, charcoal, and biodiesel. The bioproducts made from microalgae have been discussed below.

13.6.1 Biogas

Biogas production from algal biomass is more technically feasible than other bioenergy products. However, for its commercial use, the microalgae feedstock price should be decreased by 75% of its current market value (Goswami *et al.*, 2021). The produced biomass can be utilized in the anaerobic digester for biogas production. Biogas is a mixture of methane, carbon dioxide, and trace amounts of hydrogen sulfide and other gases (Singh *et al.*, 2022). The biogas production is reported to be more for the algal biomass having lipid content of 40% lower. The biogas generated may be utilized in CHP plants to generate heat and power in wastewater treatment plants. Preheating the algae biomass has been reported to increase the biomethane content in the biogas. The preheating of microalgae biomass at 70°C–90°C enhanced the biomethane production by 40% (Wang *et al.*, 2016; Wollmann *et al.*, 2019).

13.6.2 Biodiesel

The lipid is converted into biodiesel by a process known as transesterification. The process of transesterification involves the transformation of lipids into low-molecular-weight fatty acid alkyl esters (Trivedi *et al.*, 2015). The catalyst used for the reaction could be an alkaline catalyst, enzyme catalyst, heterogeneous inorganic catalyst, or acid catalyst (Goswami *et al.*, 2021). The lipid concentration in microalgal cells varies from 10% to 50% and, therefore, can be used for biodiesel production. The lipid content in the microalgal cell varies among the different species. The *N. oculate* species reported to have 16% lipid (Converti *et al.*, 2009), *C. protothecoides* have 55% lipid (Xu *et al.*, 2006), *Picochlorum* sp. contains 25% lipid (Pereira *et al.*, 2013), and *Chlorella* sp. BTA contains about 25% lipid (Mondal *et al.*, 2016). The environmental stress can increase the lipid concentration in the microalgal biomass by 38–40%. During biodiesel production, glycerol is a by-product that can be further utilized by yeast or fungi to convert it into organic acids, mannitol, or single-cell oil.

13.6.3 Bioethanol

The microalgae cells are rich in carbohydrate molecules, and fermentation results in bioethanol. The carbohydrate concentration varies from 15% to 50% in microalgae, depending on the species (Markou & Nerantzis, 2013). Carbohydrates may be found in many forms, including cellulose, mannitol, and agar, and can be extracted via physical, chemical, or biological processes. The major species used for bioethanol production are *Cholorococcum*, *Chlorella*, and *Chlamydomonas* (Trivedi *et al.*, 2015). Several studies have reported that brown algae might be used for making bioethanol due to their high carbohydrate content. Despite these benefits, bioethanol production from microalgae is not yet commercialized due to the high downstream process cost (Goswami *et al.*, 2021).

13.6.4 Syngas

The syngas can be produced through the thermochemical combustion of algae biomass. Different thermochemical processes are gasification, pyrolysis, and direct combustion (Milano *et al.*, 2016; Razzak *et al.*, 2013). Syngas is a mixture of methane, carbon dioxide, hydrogen, and carbon monoxide gas. The species that are reported to have the highest syngas production rate (50–90%) are *Chlorella* sp., and *Scenedesmus* sp. (Raheem *et al.*, 2018). The syngas can be directly burnt or used to produce electricity. The experimental study on the gasification of microalgal biomass in the presence of CaO and Fe_2O_3 (oxygen carrier) resulted in a higher production rate of syngas (Liu *et al.*, 2018).

13.6.5 Other applications

Various microalgal species have been reported to have pharmaceutical, nutraceutical, antimicrobial, anti-inflammatory, and antiproliferative properties (Goswami *et al.*, 2021). However, few of these species can grow on wastewater as well as produce high-quality biomass. The microalgae *Haematococcus pluvialis* cultivation on piggery wastewater showed an astaxanthin concentration in the cell of about 83.9 ppm. The author reported that biomass could not be further used for producing biomolecules or biomass due to the presence of toxic compounds (Kang *et al.*, 2006). The removal of these toxic compounds requires a high purification level, therefore, high operating cost. On the contrary, several studies reported that microalgae cultivated in wastewater could be used as a biofertilizer. The study on the cultivation of *C. minutissima* in brewery wastewater reported that the algal biomass could be used as a biofertilizer to promote organic farming (Amenorfenyo *et al.*, 2019).

13.7 CHALLENGES AND FUTURE PERSPECTIVE

Conventional wastewater treatment requires much energy and cost to remove contaminants and their treatment. Microalgae-based biorefinery is cost-effective as it utilizes atmospheric CO_2 and solar energy for its growth. Microalgae are reported to be 10–15 times faster than plants to fix carbon dioxide (Onyeaka *et al.*, 2021). The biorefinery presently focuses on producing a single type of value-added product, such as biofuel. The merits of the single-step biorefinery are high-quality products, easy recovery, and well-established (Gifuni *et al.*, 2019). However, high cost is the major challenge for its commercialization. The feasibility analysis of microalgae cultivation for biofuel production reported that the biodiesel cost for an open pond system was $9.84/gallon and for a photobioreactor was $20.53/gallon (Davis *et al.*, 2011). The production cost was reported to be much higher than the present fossil fuel prices. The reason for the high price (20–30% of the capital cost) was accounted by the cost of nutrients and freshwater for algae cultivation and the high downstream processing cost (50–60% of capital cost (Javed *et al.*, 2019; Lam *et al.*, 2018). Other than the economic aspects, the presence of bacteria, fungi, and protozoa also affects microalgae growth. The contamination can be removed with the pretreatment step but is cost and energy intensive. Different wastewater contains different types of contaminants. Therefore, there is a need for the selection of microalgal species based on the type of wastewater. The species should be selected based on nutrient removal capacity, high lipid accumulation, and biomass productivity. The low C:N ratio of wastewater is preferred for microalgae biomass productivity. However, a high C:N ratio is preferred for lipid accumulation. Another challenge for wastewater treatment using microalgae is internal shading. The light could not penetrate inside the wastewater due to the presence of suspended solids. It reduces the growth of microalgae inside the water. Therefore, highly efficient reactors are required to be designed. Process parameter such as temperature also affects microalgal growth. Hence, for cold regions, the heating requirement is one of the major challenges. A heating source like solar energy can be utilized to overcome the heating issue, but it requires a large amount of capital cost. The high hydraulic retention time (HRT) of microalgae is another major challenge. The major commercial technologies for wastewater treatment have an HRT of less than 1 day (Acién Fernández *et al.*, 2018; Matamoros *et al.*, 2015). The extraction of the biomolecule is still complex and expensive. The challenges are required to be overcome for the commercialization of microalgal-based biorefinery.

REFERENCES

Abdelgadir A., Chen X., Liu J., Xie X., Zhang J., Zhang K., Wang H. and Liu N. (2014). Characteristics, process parameters, and inner components of anaerobic bioreactors. *BioMed Research International*, **2014**, 1–10, https://doi.org/10.1155/2014/841573

Acién Fernández F. G., Gómez-Serrano C. and Fernández-Sevilla J. M. (2018). Recovery of nutrients from wastewaters using microalgae. *Frontiers in Sustainable Food Systems*, **2**(September), 1–13, https://doi.org/10.3389/fsufs.2018.00059

Akubude V. C., Nwaigwe K. N. and Dintwa E. (2019). Production of biodiesel from microalgae via nanocatalyzed transesterification process: a review. *Materials Science for Energy Technologies*, **2**(2), 216–225, https://doi.org/10.1016/j.mset.2018.12.006

Akyol Ç, Foglia A., Ozbayram E. G., Frison N., Katsou E., Eusebi A. L. and Fatone F. (2020). Validated innovative approaches for energy-efficient resource recovery and re-use from municipal wastewater: from anaerobic treatment systems to a biorefinery concept. *Critical Reviews in Environmental Science and Technology*, **50**(9), 869–902, https://doi.org/10.1080/10643389.2019.1634456

Amenorfenyo D. K., Huang X., Zhang Y., Zeng Q., Zhang N., Ren J. and Huang Q. (2019). Microalgae brewery wastewater treatment: potentials, benefits and the challenges. *International Journal of Environmental Research and Public Health*, **16**(11), 1910, https://doi.org/10.3390/ijerph16111910

Arooj M. F., Han S. K., Kim S. H., Kim D. H. and Shin H. S. (2007). Sludge characteristics in anaerobic SBR system producing hydrogen gas. *Water Research*, **41**(6), 1177–1184, https://doi.org/10.1016/j.watres.2006.11.052

Behera C. R., Santoro D., Gernaey K. V. and Sin G. (2018). Organic carbon recovery modeling for a rotating belt filter and its impact assessment on a plant-wide scale. *Chemical Engineering Journal*, **334**(November 2017), 1965–1976, https://doi.org/10.1016/j.cej.2017.11.091

Bengtsson S., Werker A., Visser C. and Korving L. (2017). PHARIO: Stepping Stone to a Sustainable Value Chain for PHA Bioplastic Using Municipal Activated Sludge. Stichting Toegepast Onderzoek Waterbeheer Amersfoort, The Netherlands.

Bhuvaneshwari B. and Devika R. (2005). Studies on the physico-chemical and biological characteristics of Coovum river. *Asian Journal of Microbiology, Biotechnology and Environmental Sciences*, **7**(3), 449–451.

Bureau of Reclamation (2020). Water Facts – Worldwide Water Supply. https://www.usbr.gov/mp/arwec/water-facts-ww-water-sup.html

Chaiklahan R., Chirasuwan N., Loha V., Tia S. and Bunnag B. (2018). Stepwise extraction of high-value chemicals from *Arthrospira* (*Spirulina*) and an economic feasibility study. *Biotechnology Reports*, **20**, e00280, https://doi.org/10.1016/j.btre.2018.e00280

Chen, Guo W., Ngo H. H., Chang S. W., Duc Nguyen D., Dan Nguyen P., Bui X. T. and Wu Y. (2017). Impact of reactor configurations on the performance of a granular anaerobic membrane bioreactor for municipal wastewater treatment. *International Biodeterioration and Biodegradation*, **121**, 131–138, https://doi.org/10.1016/j.ibiod.2017.03.021

Chen W.-H., Tsai C.-Y., Chen S.-Y., Sung S. and Lin J.-G. (2019). Treatment of campus domestic wastewater using ambient-temperature anaerobic fluidized membrane bioreactors with zeolites as carriers. *International Biodeterioration & Biodegradation*, **136**(October 2018), 49–54, https://doi.org/10.1016/j.ibiod.2018.10.010

Chu F.-J., Wan T.-J., Chen H., Wu C.-H. and Kao P.-M. (2020). Magnetophoretic harvesting of *Nannochloropsis oculata* using iron oxide immobilized beads. *Water*, **12**(1), 236, https://doi.org/10.3390/w12010236

Converti A., Casazza A. A., Ortiz E. Y., Perego P. and Del Borghi M. (2009). Effect of temperature and nitrogen concentration on the growth and lipid content of *Nannochloropsis oculata* and *Chlorella vulgaris* for biodiesel production. *Chemical Engineering and Processing: Process Intensification*, **48**(6), 1146–1151, https://doi.org/10.1016/j.cep.2009.03.006

Corcoran E., Nellemann C., Bos R., Osborn D. and Savelli H. (2010). Sick water? The central role of wastewater management in sustainable development: a rapid response assessment. In UNEP, 23–73, www.grida.no.

Crutchik D., Frison N., Eusebi A. L. and Fatone F. (2018). Biorefinery of cellulosic primary sludge towards targeted short chain fatty acids, phosphorus and methane recovery. *Water Research*, **136**, 112–119, https://doi.org/10.1016/j.watres.2018.02.047

Davis R., Aden A. and Pienkos P. T. (2011). Techno-economic analysis of autotrophic microalgae for fuel production. *Applied Energy*, **88**(10), 3524–3531, https://doi.org/10.1016/j.apenergy.2011.04.018

Evans P. J., Parameswaran P., Lim K., Bae J., Shin C., Ho J. and McCarty P. L. (2019). A comparative pilot-scale evaluation of gas-sparged and granular activated carbon-fluidized anaerobic membrane bioreactors for domestic wastewater treatment. *Bioresource Technology*, **288**(November 2018), 120949, https://doi.org/10.1016/j.biortech.2019.01.072

Ezugbe E. O. and Rathilal S. (2020). Membrane technologies in wastewater treatment: a review. *Membranes*, **10**(5), 89, https://doi.org/10.3390/membranes10050089

Fernández-Dacosta C., Posada J. A., Kleerebezem R., Cuellar M. C. and Ramirez A. (2015). Microbial community-based polyhydroxyalkanoates (PHAs) production from wastewater: techno-economic analysis and ex-ante environmental assessment. *Bioresource Technology*, **185**, 368–377, https://doi.org/10.1016/j.biortech.2015.03.025

Ghosh S., Lobanov S. and Lo V. K. (2019). An overview of technologies to recover phosphorus as struvite from wastewater: advantages and shortcomings. *Environmental Science and Pollution Research*, **26**(19), 19063–19077, https://doi.org/10.1007/s11356-019-05378-6

Gifuni I., Pollio A., Safi C., Marzocchella A. and Olivieri G. (2019). Current bottlenecks and challenges of the microalgal biorefinery. *Trends in Biotechnology*, **37**(3), 242–252, https://doi.org/10.1016/j.tibtech.2018.09.006

Goswami R. K., Mehariya S., Verma P., Lavecchia R. and Zuorro A. (2021). Microalgae-based biorefineries for sustainable resource recovery from wastewater. *Journal of Water Process Engineering*, **40**(July 2020), 101747, https://doi.org/10.1016/j.jwpe.2020.101747

Gouveia J., Plaza F., Garralon G., Fdz-Polanco F. and Peña M. (2015). Long-term operation of a pilot scale anaerobic membrane bioreactor (AnMBR) for the treatment of municipal wastewater under psychrophilic conditions. *Bioresource Technology*, **185**, 225–233, https://doi.org/10.1016/j.biortech.2015.03.002

Gurung K., Tang W. Z. and Sillanpää M. (2018). Correction to: unit energy consumption as benchmark to select energy positive retrofitting strategies for Finnish wastewater treatment plants (WWTPs): a case study of Mikkeli WWTP. *Environmental Processes*, **5**(3), 667–681. In Environmental Processes. https://doi.org/10.1007/s40710-018-0340-5, https://doi.org/10.1007/s40710-018-0310-y

Hoffman J., Pate R. C., Drennen T. and Quinn J. C. (2017). Techno-economic assessment of open microalgae production systems. *Algal Research*, **23**, 51–57, https://doi.org/10.1016/j.algal.2017.01.005

Hu Y., Yang Y., Yu S., Wang X. C. and Tang J. (2018). Psychrophilic anaerobic dynamic membrane bioreactor for domestic wastewater treatment: effects of organic loading and sludge recycling. *Bioresource Technology*, **270**(June), 62–69, https://doi.org/10.1016/j.biortech.2018.08.128

Ibrahim and Heikal S. M. (2019). Bioremediation of wastewater from an agricultural drain by the use of microalgae. *SVU–International Journal of Basic Sciences*, **1**(1), 6–15.

Ichsan, Hadiyanto H. and Hendroko R. (2014). Integrated biogas-microalgae from waste waters as the potential biorefinery sources in Indonesia. *Energy Procedia*, **47**, 143–148, https://doi.org/10.1016/j.egypro.2014.01.207

Javed F., Aslam M., Rashid N., Shamair Z., Khan A. L., Yasin M., Fazal T., Hafeez A., Rehman F., Rehman M. S. U., Khan Z., Iqbal J. and Bazmi A. A. (2019). Microalgae-based biofuels, resource recovery and wastewater treatment: a pathway towards sustainable biorefinery. *Fuel*, **255**(February), 115826, https://doi.org/10.1016/j.fuel.2019.115826

Jürgensen L., Ehimen E. A., Born J. and Holm-Nielsen J. B. (2018). A combination anaerobic digestion scheme for biogas production from dairy effluent – CSTR and ABR, and biogas upgrading. *Biomass and Bioenergy*, **111**, 241–247, https://doi.org/10.1016/j.biombioe.2017.04.007

Kang C. D., An J. Y., Park T. H. and Sim S. J. (2006). Astaxanthin biosynthesis from simultaneous N and P uptake by the green alga *Haematococcus pluvialis* in primary-treated wastewater. *Biochemical Engineering Journal*, **31**(3), 234–238, https://doi.org/10.1016/j.bej.2006.08.002

Khan M. J., Das S., Vinayak V., Pant D. and Ghangrekar M. M. (2022). Live diatoms as potential biocatalyst in a microbial fuel cell for harvesting continuous diafuel, carotenoids and bioelectricity. *Chemosphere*, **291**(P1), 132841, https://doi.org/10.1016/j.chemosphere.2021.132841

Kim D. H., Kim S. H., Ko I. B., Lee C. Y. and Shin H. S. (2008). Start-up strategy for continuous fermentative hydrogen production: early switchover from batch to continuous operation. *International Journal of Hydrogen Energy*, **33**(5), 1532–1541, https://doi.org/10.1016/j.ijhydene.2008.01.012

Lam G. P., Vermuë M. H., Eppink M. H. M., Wijffels R. H. and van den Berg C. (2018). Multi-product microalgae biorefineries: from concept towards reality. *Trends in Biotechnology*, **36**(2), 216–227, https://doi.org/10.1016/j.tibtech.2017.10.011

Li W., Li L. and Qiu G. (2017). Energy consumption and economic cost of typical wastewater treatment systems in Shenzhen, China. *Journal of Cleaner Production*, **163**, S374–S378, https://doi.org/10.1016/j.jclepro.2015.12.109

Li, Liu Q., Fang F., Luo R., Lu Q., Zhou W., Huo S., Cheng P., Liu J., Addy M., Chen P., Chen D. and Ruan R. (2019). Microalgae-based wastewater treatment for nutrients recovery: a review. *Bioresource Technology*, **291**, 121934, https://doi.org/10.1016/j.biortech.2019.121934

Lim K., Evans P. J. and Parameswaran P. (2019). Long-term performance of a pilot-scale gas-sparged anaerobic membrane bioreactor under ambient temperatures for holistic wastewater treatment. *Environmental Science and Technology*, **53**(13), 7347–7354, https://doi.org/10.1021/acs.est.8b06198

Lin, Nierop K. G. J., Girbal-Neuhauser E., Adriaanse M. and van Loosdrecht M. C. M. (2015). Sustainable polysaccharide-based biomaterial recovered from waste aerobic granular sludge as a surface coating material. *Sustainable Materials and Technologies*, **4**, 24–29, https://doi.org/10.1016/j.susmat.2015.06.002

Lin L., Li R. H. and Li X. Y. (2018). Recovery of organic resources from sewage sludge of Al-enhanced primary sedimentation by alkali pretreatment and acidogenic fermentation. *Journal of Cleaner Production*, **172**, 3334–3341, https://doi.org/10.1016/j.jclepro.2017.11.199

Littfinski T., Beckmann J., Gehring T., Stricker M., Nettmann E., Krimmler S., Murnleitner E., Lübken M., Pant D. and Wichern M. (2022). Model-based identification of biological and pH gradient driven removal pathways of total ammonia nitrogen in single-chamber microbial fuel cells. *Chemical Engineering Journal*, **431**(P1), 133987, https://doi.org/10.1016/j.cej.2021.133987

Liu G., Liao Y., Wu Y. and Ma X. (2018). Synthesis gas production from microalgae gasification in the presence of Fe_2O_3 oxygen carrier and CaO additive. *Applied Energy*, **212**(January), 955–965, https://doi.org/10.1016/j.apenergy.2017.12.110

Longo S., d'Antoni B. M., Bongards M., Chaparro A., Cronrath A., Fatone F., Lema J. M., Mauricio-Iglesias M., Soares A. and Hospido A. (2016). Monitoring and diagnosis of energy consumption in wastewater treatment plants. A state of the art and proposals for improvement. *Applied Energy*, **179**, 1251–1268, https://doi.org/10.1016/j.apenergy.2016.07.043

Lopes T. F., Cabanas C., Silva A., Fonseca D., Santos E., Guerra L. T., Sheahan C., Reis A. and Gírio F. (2019). Process simulation and techno-economic assessment for direct production of advanced bioethanol using a genetically modified *Synechocystis* sp. *Bioresource Technology Reports*, **6**(February), 113–122, https://doi.org/10.1016/j.biteb.2019.02.010

Mannina G., Presti D., Montiel-Jarillo G., Carrera J. and Suárez-Ojeda M. E. (2020). Recovery of polyhydroxyalkanoates (PHAs) from wastewater: a review. *Bioresource Technology*, **297**(November 2019), 122478, https://doi.org/10.1016/j.biortech.2019.122478

Mari A. G., Andreani C. L., Tonello T. U., Leite L. C. C., Fernandes J. R., Lopes D. D., Rodrigues J. A. D. and Gomes S. D. (2020). Biohydrogen and biomethane production from cassava wastewater in a two-stage anaerobic sequencing batch biofilm reactor. *International Journal of Hydrogen Energy*, **45**(8), 5165–5174, https://doi.org/10.1016/j.ijhydene.2019.07.054

Markou G. and Nerantzis E. (2013). Microalgae for high-value compounds and biofuels production: a review with focus on cultivation under stress conditions. *Biotechnology Advances*, **31**(8), 1532–1542, https://doi.org/10.1016/j.biotechadv.2013.07.011

Matamoros V., Gutiérrez R., Ferrer I., García J. and Bayona J. M. (2015). Capability of microalgae-based wastewater treatment systems to remove emerging organic contaminants: a pilot-scale study. *Journal of Hazardous Materials*, **288**, 34–42, https://doi.org/10.1016/j.jhazmat.2015.02.002

Milano J., Ong H. C., Masjuki H. H., Chong W. T., Lam M. K., Loh P. K. and Vellayan V. (2016). Microalgae biofuels as an alternative to fossil fuel for power generation. *Renewable and Sustainable Energy Reviews*, **58**, 180–197, https://doi.org/10.1016/j.rser.2015.12.150

Mondal M., Ghosh A., Sharma A. S., Tiwari O. N., Gayen K., Mandal M. K. and Halder G. N. (2016). Mixotrophic cultivation of *Chlorella* sp. BTA 9031 and *Chlamydomonas* sp. BTA 9032 isolated from coal field using various carbon sources for biodiesel production. *Energy Conversion and Management*, **124**, 297–304, https://doi.org/10.1016/j.enconman.2016.07.033

Mu Y., Yu H. Q. and Wang Y. (2006). The role of pH in the fermentative H_2 production from an acidogenic granule-based reactor. *Chemosphere*, **64**(3), 350–358, https://doi.org/10.1016/j.chemosphere.2005.12.048

Musa M., Doshi A., Brown R. and Rainey T. J. (2019). Microalgae dewatering for biofuels: a comparative techno-economic assessment using single and two-stage technologies. *Journal of Cleaner Production*, **229**, 325–336, https://doi.org/10.1016/j.jclepro.2019.05.039

Neira M. and Prüss-Ustün A. (2016). Preventing disease through healthy environments: a global assessment of the environmental burden of disease. *Toxicology Letters*, **259**, S1–S3, https://doi.org/10.1016/j.toxlet.2016.07.028

Nizami A. S., Rehan M., Waqas M., Naqvi M., Ouda O. K., Shahzad K., Miandad R., Khan M. Z., Syamsiro M., Ismail I. M. I. and Pant D. (2017). Waste biorefineries: enabling circular economies in developing countries. *Bioresource Technology*, **241**, 1101–1117, https://doi.org/10.1016/j.biortech.2017.05.097

Onyeaka H., Miri T., Obileke K., Hart A., Anumudu C. and Al-Sharify Z. T. (2021). Minimizing carbon footprint via microalgae as a biological capture. *Carbon Capture Science & Technology*, **1**(October), 100007, https://doi.org/10.1016/j.ccst.2021.100007

Pancha I., Chokshi K. and Mishra S. (2019). Industrial wastewater-based microalgal biorefinery: a dual strategy to remediate waste and produce microalgal bioproducts. *In Application of Microalgae in Wastewater Treatment*, **2**, 173–193, https://doi.org/10.1007/978-3-030-13909-4_8

Panepinto D., Fiore S., Zappone M., Genon G. and Meucci L. (2016). Evaluation of the energy efficiency of a large wastewater treatment plant in Italy. *Applied Energy*, **161**, 404–411, https://doi.org/10.1016/j.apenergy.2015.10.027

Peces M., Astals S., Clarke W. P. and Jensen P. D. (2016). Semi-aerobic fermentation as a novel pre-treatment to obtain VFA and increase methane yield from primary sludge. *Bioresource Technology*, **200**, 631–638, https://doi.org/10.1016/j.biortech.2015.10.085

Peña M., do Nascimento T., Gouveia J., Escudero J., Gómez A., Letona A., Arrieta J. and Fdz-Polanco F. (2019). Anaerobic submerged membrane bioreactor (AnSMBR) treating municipal wastewater at ambient temperature: operation and potential use for agricultural irrigation. *Bioresource Technology*, **282**(March), 285–293, https://doi.org/10.1016/j.biortech.2019.03.019

Pereira H., Barreira L., Custódio L., Alrokayan S., Mouffouk F., Varela J., Abu-Salah K. M. and Ben-Hamadou R. (2013). Isolation and fatty acid profile of selected microalgae strains from the red sea for biofuel production. *Energies*, **6**(6), 2773–2783, https://doi.org/10.3390/en6062773

Puchongkawarin C., Gomez-Mont C., Stuckey D. C. and Chachuat B. (2015). Optimization-based methodology for the development of wastewater facilities for energy and nutrient recovery. *Chemosphere*, **140**, 150–158, https://doi.org/10.1016/j.chemosphere.2014.08.061

Puyol D., Batstone D. J., Hülsen T., Astals S., Peces M. and Krömer J. O. (2017). Resource recovery from wastewater by biological technologies: opportunities, challenges, and prospects. *Frontiers in Microbiology*, **7**(JAN), 1–23, https://doi.org/10.3389/fmicb.2016.02106

Qadir M., Drechsel P., Jiménez Cisneros B., Kim Y., Pramanik A., Mehta P. and Olaniyan O. (2020). Global and regional potential of wastewater as a water, nutrient and energy source. *Natural Resources Forum*, **44**(1), 40–51, https://doi.org/10.1111/1477-8947.12187

Qu J. and Fan M. (2010). The current state of water quality and technology development for water pollution control in China. *Critical Reviews in Environmental Science and Technology*, **40**(6), 519–560, https://doi.org/10.1080/10643380802451953

Raheem A., Sikarwar V. S., He J., Dastyar W., Dionysiou D. D., Wang W. and Zhao M. (2018). Opportunities and challenges in sustainable treatment and resource reuse of sewage sludge: a review. *Chemical Engineering Journal*, **337**(October 2017), 616–641, https://doi.org/10.1016/j.cej.2017.12.149

Rawat I., Ranjith Kumar R., Mutanda T. and Bux F. (2013). Biodiesel from microalgae: a critical evaluation from laboratory to large scale production. *Applied Energy*, **103**, 444–467, https://doi.org/10.1016/j.apenergy.2012.10.004

Razzak S. A., Hossain M. M., Lucky R. A., Bassi A. S. and De Lasa H. (2013). Integrated CO_2 capture, wastewater treatment and biofuel production by microalgae culturing – a review. *Renewable and Sustainable Energy Reviews*, **27**, 622–653, https://doi.org/10.1016/j.rser.2013.05.063

Rodríguez Arredondo M., Kuntke P., ter Heijne A. and Buisman C. J. N. (2019). The concept of load ratio applied to bioelectrochemical systems for ammonia recovery. *Journal of Chemical Technology and Biotechnology*, **94**(6), 2055–2061, https://doi.org/10.1002/jctb.5992

SAMCO (2016). How Much Does a Wastewater Treatment System Cost? (Pricing, Factors, Etc.). https://samcotech.com/cost-wastewater-treatment-system/

Sato T., Qadir M., Yamamoto S., Endo T. and Zahoor A. (2013). Global, regional, and country level need for data on wastewater generation, treatment, and use. *In Agricultural Water Management*, **130**(1), 1–13, https://doi.org/10.1016/j.agwat.2013.08.007

Sharma R., Mishra A., Pant D. and Malaviya P. (2022). Recent advances in microalgae-based remediation of industrial and non-industrial wastewaters with simultaneous recovery of value-added products. *Bioresource Technology*, **344**(PB), 126129, https://doi.org/10.1016/j.biortech.2021.126129

Silva M. I., Gonçalves A. L., Vilar V. J. P. and Pires J. C. M. (2021). Experimental and techno-economic study on the use of microalgae for paper industry effluents remediation. *Sustainability*, **13**(3), 1314, https://doi.org/10.3390/su13031314

Singh Tiwari A. and Das M. (2016). Phyco-remediation of industrial waste-water and flue gases with algal-diesel engenderment from micro-algae: a review. *In Fuel*, **173**, 90–97, https://doi.org/10.1016/j.fuel.2016.01.031

Singh R., Paritosh K., Pareek N. and Vivekanand V. (2022). Integrated system of anaerobic digestion and pyrolysis for valorization of agricultural and food waste towards circular bioeconomy: review. *Bioresource Technology*, **360**(July), 127596, https://doi.org/10.1016/j.biortech.2022.127596

Sivaramakrishnan R., Suresh S., Kanwal S., Ramadoss G., Ramprakash B. and Incharoensakdi A. (2022). Microalgal biorefinery concepts' developments for biofuel and bioproducts: current perspective and bottlenecks. *International Journal of Molecular Sciences*, **23**(5), 2623, https://doi.org/10.3390/ijms23052623

Thomassen G., Egiguren Vila U., Van Dael M., Lemmens B. and Van Passel S. (2016). A techno-economic assessment of an algal-based biorefinery. *Clean Technologies and Environmental Policy*, **18**(6), 1849–1862, https://doi.org/10.1007/s10098-016-1159-2

Trivedi J., Aila M., Bangwal D. P., Kaul S. and Garg M. O. (2015). Algae based biorefinery – how to make sense? *Renewable and Sustainable Energy Reviews*, **47**, 295–307, https://doi.org/10.1016/j.rser.2015.03.052

Udaiyappan A. F., Abu Hasan H., Takriff M. S. and Sheikh Abdullah S. R. (2017). A review of the potentials, challenges and current status of microalgae biomass applications in industrial wastewater treatment. *Journal of Water Process Engineering*, **20**, 8–21, https://doi.org/10.1016/j.jwpe.2017.09.006

United T., World N., Development W., Tran M., Koncagul E. and Connor R. (2016). The United Nations World Water Development Report 2016: WATER AND JOBS Facts and Figures. Report: UN WWDR4-2016.

Van Der Hoek J. P., De Fooij H. and Struker A. (2016). Wastewater as a resource: strategies to recover resources from Amsterdam's wastewater. *Resources, Conservation and Recycling*, **113**, 53–64, https://doi.org/10.1016/j.resconrec.2016.05.012

Wang Y., Ho S. H., Cheng C. L., Guo W. Q., Nagarajan D., Ren N. Q., Lee D. J. and Chang J. S. (2016). Perspectives on the feasibility of using microalgae for industrial wastewater treatment. *Bioresource Technology*, **222**, 485–497, https://doi.org/10.1016/j.biortech.2016.09.106

Wijffels R. H., Barbosa M. J. and Eppink M. H. M. (2010). Microalgae for the production of bulk chemicals and biofuels. *Biofuels, Bioproducts and Biorefining*, **4**(3), 287–295, https://doi.org/10.1002/bbb.215

Wollmann F., Dietze S., Ackermann J. U., Bley T., Walther T., Steingroewer J. and Krujatz F. (2019). Microalgae wastewater treatment: biological and technological approaches. *Engineering in Life Sciences*, **19**(12), 860–871, https://doi.org/10.1002/elsc.201900071

WWAP (2017). Wastewater: The Untapped Resource. In The United Nations World Water Development Report. https://wedocs.unep.org/20.500.11822/20448

Xiao Y., Zheng Y., Wu S., Yang Z. H. and Zhao F. (2016). Nitrogen recovery from wastewater using microbial fuel cells. *Frontiers of Environmental Science and Engineering*, **10**(1), 185–191, https://doi.org/10.1007/s11783-014-0730-5

Xin C., Addy M. M., Zhao J., Cheng Y., Cheng S., Mu D., Liu Y., Ding R., Chen P. and Ruan R. (2016). Comprehensive techno-economic analysis of wastewater-based algal biofuel production: a case study. *Bioresource Technology*, **211**, 584–593, https://doi.org/10.1016/j.biortech.2016.03.102

Xin C., Addy M. M., Zhao J., Cheng Y., Ma Y., Liu S., Mu D., Liu Y., Chen P. and Ruan R. (2018). Waste-to-biofuel integrated system and its comprehensive techno-economic assessment in wastewater treatment plants. *Bioresource Technology*, **250**(August 2017), 523–531, https://doi.org/10.1016/j.biortech.2017.11.040

Xing Y., Li Z., Fan Y. and Hou H. (2010). Biohydrogen production from dairy manures with acidification pretreatment by anaerobic fermentation. *Environmental Science and Pollution Research*, **17**(2), 392–399, https://doi.org/10.1007/s11356-009-0187-4

Xu H., Miao X. and Wu Q. (2006). High quality biodiesel production from a microalga *Chlorella protothecoides* by heterotrophic growth in fermenters. *Journal of Biotechnology*, **126**(4), 499–507, https://doi.org/10.1016/j.jbiotec.2006.05.002

Xu Ng T. C. A., Huang S. and Ng H. Y. (2020). Effect of quorum quenching on EPS and size-fractioned particles and organics in anaerobic membrane bioreactor for domestic wastewater treatment. *Water Research*, **179**, 115850, https://doi.org/10.1016/j.watres.2020.115850

Yadav A., Rene E. R., Sharma M., Jatain I., Mandal M. K. and Dubey K. K. (2022). Valorization of wastewater to recover value-added products: a comprehensive insight and perspective on different technologies. *Environmental Research*, **214**(P2), 113957, https://doi.org/10.1016/j.envres.2022.113957

Yang J. S., Cao J., Xing G. L. and Yuan H. L. (2015). Lipid production combined with biosorption and bioaccumulation of cadmium, copper, manganese and zinc by oleaginous microalgae *Chlorella minutissima* UTEX2341. *Bioresource Technology*, **175**, 537–544, https://doi.org/10.1016/j.biortech.2014.10.124

Zielińska M., Zieliński M. and Dębowski M. (2018). Organic compounds and phosphorus removal from dairy wastewater by biofilm on iron-containing supports. *Journal of Environmental Engineering*, **144**(1), 1, https://doi.org/10.1061/(ASCE)EE.1943-7870.0001309

doi: 10.2166/9781789063783_0263

Chapter 14

Role of nanotechnology in environmental cleaning

Jayato Nayak[1] and Sankha Chakrabortty[2]*

[1]Center for Life Sciences, Mahindra University, Hyderabad, Telangana 500043, India
[2]Department of Chemical Technology, KIIT University, Bhubaneswar, Odisha 751024, India
**Corresponding author: nayak.jayato@gmail.com*

ABSTRACT

Air, water, and soil pollution pose serious threats to ecosystem health and should be prioritized as a global issue of paramount importance. Emerging technologies such as 'nanoremediation,' which embraces the concept of nanotechnology through the development of nanoparticles (NPs), join established procedures such as thermal treatment, pump-and-treat, and chemical oxidation in the arsenal of remediation approaches. Nanoremediation is the process of using designed NPs to clean up polluted medium; it is more efficient and cost-effective than conventional approaches. Nanomaterials' unique properties such as their nanostructure and cost-effectiveness are the main factors of their growing popularity for use in environmental clean-up. NPs have excellent electrical characteristics, catalytic behavior, and sensitivity because of their large surface area to mass ratio. Remediation by NPs can primarily occur through catalysis and chemical reduction. Because of their high surface area, random distribution of active sites, and versatility in coating changes, NPs have also been used in adsorption-based removal processes. The ability of NPs to spread in tight places also increases their usefulness for cleaning up polluted water and soil. Water nanofiltration utilizes nanomaterial-based membranes because the membrane pores can trap substantial components in water waste streams. In addition, the membrane contact selectively separates the less abundant molecules. Metal oxides, carbon nanotubes, quantum dots, and biopolymers are just a few of the nanomaterials used in water, soil, and air purification. Environmentally, NPs are useful for cleaning soil and water. Organic solvents, pesticides, metals, and fertilizers such as nitrates can all be dechlorinated, detoxified, and transformed by injecting it into the affected area. Besides carbon nanotubes and bimetallic NPs, there are other NPs that can be used to address these problems. Food processing, disinfectants, water purification, and even common household appliances all make use of NPs. The use of nanotechnology in pollution prevention, monitoring, and remediation has been gaining popularity.

Keywords: nanotechnology, wastewater treatment, adsorption, catalysis, filtration, nanoremediation

14.1 INTRODUCTION

Currently, nanomaterial-based emerging technologies have turned quintessential to guarantee no waste generation and environmental friendliness in the face of the rising tide of anthropogenic discarded volumes of wastewater from industrial houses that link to various health issues of living beings (Lin *et al.*, 2020). A number of infectious disorders, including parasitic attacks, diarrhea, and dermatitis, are the typical issues rising from such water contaminations (Adegoke *et al.*, 2018). There are four main categories of treatment: physical, chemical, biological, and sludge. Adsorption, ion exchange, and reverse osmosis are just few of the conventional technologies that aren't very efficient or reusable. But, in-fact, separation of exhausted adsorbents is reckoned as a crucial problem till date. Emerging technologies include fuel cell development using microbes, treatments using algae and membrane separations, and so on. But, the slowness of the processes which include living species turns out to be the major drawback where membrane separation gets affected due to fouling issues. Under such circumstances, nanotechnology in wastewater decontamination offers a promising alternative solution that can be used to address the problems now facing society.

Nanotechnology's versatility, in terms of both surface area and reactivity, makes it the ideal treatment approach (Crane & Scott, 2012). Nanoparticles are highly efficient in removing pollutants such heavy metals, dyes, and organic and inorganic chemicals (Kumar *et al.*, 2014). A variety of nanostructures, including graphene, silver nanoparticles, nanomagnets, and nanotubes, have shown promise as wastewater treatment options. Nanoadsorbents such as silicas and chitosan-based particles (Vunain *et al.*, 2016), were employed in water treatment for the heavy metal removal. Titanium dioxide (TiO_2) photodegradation is garnering a lot of attention from scientists in recent years as a potential method for reducing the impact of these hazardous chemicals (Al-Mamun *et al.*, 2019). Under solar or ultraviolet irradiation, these nanocatalytic semiconductor materials show remarkable aptitude for photocatalytic remediation (Feizpoor & Habibi-Yangjeh, 2018). TiO_2 is the first transition metal oxide to undergo oxidative deterioration when exposed to sunlight because of its low-energy band gap and sufficiently positive valence band. The other such widely used nanocatalytic particles utilized in wastewater treatment are zinc oxide (ZnO) and magnesium oxide. One of the promising methods for the synthesis of nanomaterials is the use of ultrasound methodology (Low *et al.*, 2021a, 2021b, 2022). Heavy metals in wastewater are detected using nanomaterial-based biosensors (Hara & Singh, 2021). One of the capable materials for adsorptive decontamination of wastewater is biomass-derived renewable biochar, by the virtue of its microporous volume, active functional groups, and appreciable specific surface area which could be tailor-made using temperature and choice of feed stocks (Ahmad *et al.*, 2014; Mohanty *et al.*, 2013; Tan *et al.*, 2021). Biochar made through pyrolysis is useful for removing non-polar organic contaminants because of its high surface area (Mohanty *et al.*, 2013). In this chapter, a special focus on nanotechnology-based waste remediation therapies has been described.

14.2 BIOMASS-DERIVED ADSORPTIVE PROCESS FOR WASTE TREATMENT

Thermochemical degradation of diverse biomass feedstocks yields biochar, a porous carbonaceous substance. The processes of pyrolysis, gasification, and hydrothermal carbonization are the means by which biochar is manufactured (Park *et al.*, 2016). Adsorption onto biochar is an affordable and sustainable process (Zazycki *et al.*, 2018). Biochar is an appealing substrate for wastewater treatment because of its high surface area and abundance of surface functional groups. Feedstock choice, pretreatment method, and thermochemical decomposition method all play significant roles in determining biochar's adsorption capacity (Yu *et al.*, 2019). Biochar's porous characteristics and adsorptivity can be improved after being functionalized with acid or base. Impregnating certain materials with biochar and then allowing them to cure produces biochar composites with improved surface properties. Biochar's surface can also be altered with the use of steam or gases. Recently, surface modification techniques have included the use of ultrasound, plasma, and microwave energy (Karim *et al.*, 2017). Thus, biochar functions as an effective adsorbent, demonstrating a level of adsorption ability that is on par with, and in some cases even exceeds, that of commercial activated carbon. Bagasse biochar efficiently removes lead from wastewater created during battery production (Bharti & Kumar, 2018). In heavy metal removal, mixed biochar–chitosan substrates excel. Biochar made from digested sludge has been shown to be effective in the treatment of municipal wastewater for the removal of ammonia (Tang *et al.*, 2019). Biochars made from municipal solid wastes have been employed extensively as adsorbents to eliminate chromium, ammonia nitrogen (NH_4^+-N), phosphate (PO_4^{3-}-P), F^-, acetaminophen, and methylene blue from wastewater (Zhou *et al.*, 2019). Persulfate activation using biochar produced from municipal solid waste has been used to detoxify water of organic contaminants such as azo dyes and antibiotics (Meng *et al.*, 2020). The elimination of sulfamethazine by steam-activated biochar was reported by Rajapaksha *et al.* (2015). Agricultural wastewater has had Cu^{2+}, As^{5+}, Cd^{2+}, and Pb^{2+} successfully removed by biochar adsorption technology (Higashikawa *et al.*, 2016). The effectiveness of rice-husk biochar against malachite green was 99.98%. Biochar-based filters have become increasingly common for use in wastewater treatment systems in recent years (Kaetzl *et al.*, 2019; Perez-Mercado *et al.*, 2019). Biochar from cotton gin waste has been shown to remove halogens, dyes, and chlorinated hydrocarbons from aqueous solutions (Haque *et al.*, 2020). The physicochemical properties of biochar derived from cotton gin waste and walnut shells possess the desired characteristics needed to remove pharmaceuticals from aqueous solution (Nodun *et al.*, 2020). Biochar is a carbon sink, soil fertility booster, and a catalyst that has been used in wastewater treatment. Regeneration has a significant impact on the cost-effectiveness of biochar adsorption. The effectiveness of regeneration is affected by three factors: the adsorbent employed, the regeneration process, and the regeneration dose (Ambaye *et al.*, 2021). pH and temperature are adjusted, and chemicals such as aniline and dyes are added, all in the name of regeneration (Fagbohungbe *et al.*, 2017). Biochar adsorption is mostly determined by pH (Shang *et al.*, 2017). Degradation efficiency for organic pollutants is improved by biochar-supported photocatalysts such as Au–TiO_2 composite (Ambaye *et al.*, 2021). TiO_2-impregnated biochar with nanoparticles of softwood, bamboo, reed straw, or corn cob improved its sorption capacity by boosting its physicochemical qualities and active site.

14.3 SEPARATION OF CONTAMINANTS BY NANOFILTRATION

In recent years, thin film composite (TFC) polyamide (PA) membranes have been the standard commercial nanofiltration (NF) membrane. These membranes are created by interfacial polymerization (IP) of two monomers (such as piperazine and tricarbonyl chloride) over a porous substrate (Li *et al.*, 2021). As a consequence of the development of nanotechnology, the incorporation of functional nanoparticles into PA layers has become a common practice for tuning the performance of TFC membranes (Liao *et al.*, 2021). Adding nanomaterials to an aqueous or organic phase and subsequently performing IP are two methods that have been demonstrated to significantly improve the hydrophilicity, fouling resistance, and separation properties of TFC membranes (Zhao *et al.*, 2021). As per the reported research (Jeong *et al.*, 2007), the membrane's water permeability can be improved by incorporating nanoscale particles of the mineral clay zeolite into the PA matrix. The size of the nanoparticle is a significant component in influencing the quality of the thin film nanofilter membranes (Lind *et al.*, 2009). Linde Type A (LTA) zeolites and a PA matrix are utilized to generate nanofilters by utilizing IP. The manufacturing of rougher membranes was the consequence of utilizing larger nanoparticles, whilst the formation of thin film membranes with high permeability was the result of using smaller nanoparticles. It is feasible to manufacture a nanoparticle from a nanomaterial that has improved performance by making slight adjustments to the production technique. During the process of producing a modified membrane filter, an IP technique was utilized to produce a PA material, and polyether sulfone was used as a support for nanoparticles (Dong *et al.*, 2016). This was accomplished by fabricating the PA material from polyether sulfone. They discovered that by making this modification to the production process, it is possible to produce a nanocomposite that has nanoparticles more evenly distributed throughout it and that has more constant NF performance. As a result of their capacity to eradicate bacteria, silver nanoparticles have established themselves as a lucrative business. The membrane that is utilized in commercial nanofilters has been modified in order to accommodate the presence of silver nanoparticles. The *in situ* synthesis of silver nanoparticles on the surface of the membrane made this outcome attainable. Following the application of polyvinyl alcohol as a coating, silver salts were next added to the membrane. It was discovered that silver nanoparticles ranged in size from 10 to 20 nanometers. Utilizing this refined method for the fabrication of membranes, we were able to produce a membrane with anti-biofouling qualities and minimize the pace at which silver nanoparticles leached out. Adding graphene oxide (GO) particles to the membrane matrix improved the physicochemical properties of the polysulfone (PSF) substrate and the PA selective layer, capitalizing on the strong hydrophilicity and electronegativity of GO to boost the membrane's water flux and salt rejection performance (Lai *et al.*, 2016). However, the high price of GO makes industrial scale-up by heavy doping with the filler prohibitive. Emerging nanomaterial O-MoS2 has a structure comparable to GO but at a far more reasonable cost. In our prior work, we suggested using high-energy ultrasonication in conjunction with the Hummers' approach to fabricate O-MoS2. It has high electronegativity and is highly hydrophilic. It also has good non-swelling qualities and a stable chemical composition. Improved water flux and salt rejection were achieved in a TFC NF membrane produced by Yang *et al.* by adding hydrophilic O-MoS2 into the PA layer (Yang *et al.*, 2021). Hollow fiber membranes made from the more mechanically and chemically stable O-MoS2 have been shown to have improved performance and resistance to fouling (Jiang *et al.*, 2019). Embedding O-MoS2 into the substrate layer in order to make use of O-MoS2's characteristics holds great promise for enhancing the separation performance and stability of the TFC NF membrane. Development of such an NF membrane is shown in Figure 14.1.

In addition to nanoparticles, carbon nanotubes (CNTs) have been evaluated as potential components of a nanofilter composite. Poly (dopamine)-modified multi-walled CNTs are incorporated into a polyethylenimine/trimethyl carbonate-based polymeric system developed by Zhao *et al.* (2017). This increased water flux was evidence of the high-water permeability of the NF membrane. In order to increase the efficiency of nanofilter membranes, a lot of time and energy is spent researching material chemistry and creating new membrane synthesis procedures that can accept nanoparticles.

14.4 DECONTAMINATION USING CNTS AND CARBON NANOPARTICLES (CNPS)

Many studies are being conducted, each one concentrating on one of the various nanoadsorbents currently in use. Pores in CNTs serve as binding sites, making them more approachable than activated carbon for big compounds like tetracycline (Zhang *et al.*, 2019). CNTs can be used as adsorbents for both persistent pollutants and detecting poisons, in addition to their pre-concentration role. Ionic interaction and complex formation are the mechanisms by which CNTs adsorb metal ions. In addition, CNTs' antibacterial activities manifest themselves via producing oxidative stress in pathogens and causing cellular membrane breakdown. In contrast to conventional cleaning methods such as chlorination and ozonation, no toxic metabolic end-products would be created if chemical reactions took place (Liu *et al.*,

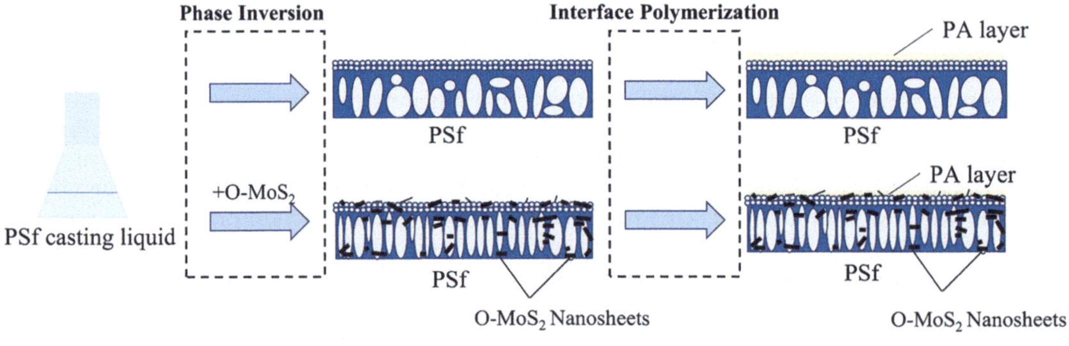

Figure 14.1 Development scheme of a PSF/O–MoS2 nanocomposite substrate-supported TFC NF membrane. Reproduced with copyright from Jiang *et al.* (2023).

2013). To improve their effectiveness, CNT standards are constantly upgraded with newer technologies. Examples of such advancements include the design of new types of multi-wall carbon nanotubes (MWCNTs) and the incorporation of magnetic characteristics, such as polyacrylamide-functionalized MWCNTs that selectively dissolved humic acid and lead (II) (Kausar, 2016). Functionalization of CNTs employing oxygen-containing agents (nitric acid, sulfuric acid, potassium permanganate) imposes surface charge negativity that is responsible in improvization of performance through electrostatic contact (Jeon *et al.*, 2011). Surface-modified CNT exhibited the following heavy metal ion removal pattern as: $Pb2+ > Ni2+ > Zn2+ > Cu2+ > Cd2+$ (Rao *et al.*, 2007). It has also been discovered that CNTs can be utilized to get rid of organic compounds and colors. The removal of the reactive red M-2BE textile dye from water using MWCNT was optimized. Rather than commercially available powdered activated carbon, the developed MWCNT showed better color removal (about 99%) (Machado *et al.*, 2011). Kuo *et al.* (2008) found that adsorption efficiency is proportional to target molecule concentration when studying the removal of C.I. Direct Yellow 86 (DY86) and C.I. Direct Red (DR224). They found that increasing concentrations of CNTs and ionic strengths resulted in greater dye removals. Adsorption, substance degradation, photodegradation, and compound sanitation are some of the methods used to treat or remediate water. Photocatalytic degradation mechanism using CNT has been represented in Figure 14.2. Nanomaterials made of metal and non-metal oxides are commonly used to remove harmful toxins from wastewater. Ferric oxide nanomolecules are useful for a variety of applications due to the abundance of iron and the simple amalgamation process required to create them (Joshi *et al.*, 2019). Innovations in nanotechnology point to fresh approaches to addressing these issues, such as increasing energy efficiency and lowering water desalination costs. Adsorption-based methods are crucial and simple to apply for water-filtering devices, but they have limited capacity to evacuate salts compared to conventional desalination techniques, which are energy-intensive and time-consuming. When comparing plasma-modified ultra-long CNTs to conventional carbon-based water treatment systems, the former's substantially greater explicit adsorption limit concerning salt (over 4-fold by weight) stands out.

Heavy metals, hazardous chemicals, and dyes all find their way into the wastewater produced by sectors as diverse as mining, chemistry, and textiles. One reliable technique to purify the contaminated discharge and turning it benign prior to its return to the natural water cycle is through adsorptive capturing of these pollutants utilizing CNP. Physisorption is the term used to describe physical interactions through weak molecular forces or electrostatic forces. Chemisorption is the process of binding to an adsorbent through chemical means. Both chemisorption and physisorption are temperature-dependent, with the former increasing at higher temperatures and the latter being

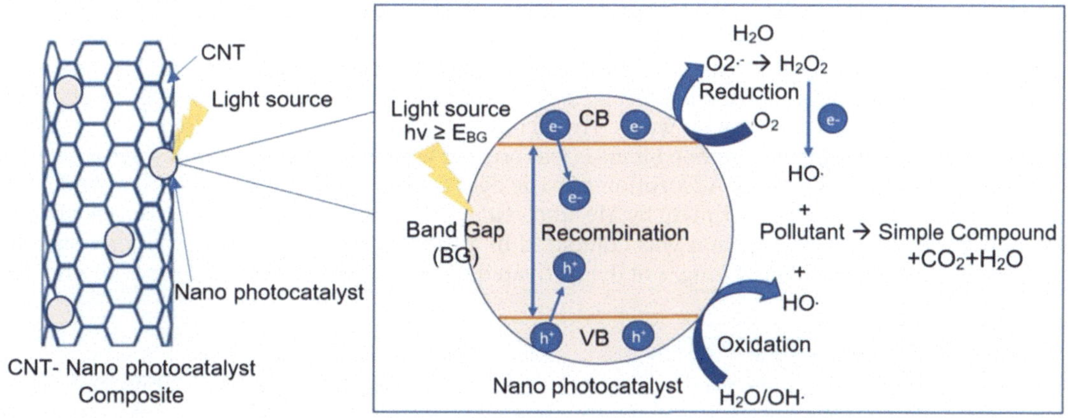

Figure 14.2 Photocatalytic degradation using CNT. Reproduced with copyright from Mishra and Sundaram (2023).

affected by them. In order to use adsorption in wastewater treatments, one must first determine the optimal parameters for doing so, such as the dosage ratio of pollutant to CNPs, the treatment period, the pH, and the temperature (Krishnamurthy & Agarwal, 2013). CNPs have a large surface area, making them ideal for removing pollutants during treatment. Also, in physisorption, the electrostatic connection between the nanoparticle and the pollutant is considerably greater than the weaker van der Waals forces because functionalization could be performed on the nanoparticle to match the molecule's charge, whether it be positive or negative. Researchers looked into the efficacy of CNPs with added functions (Mubaraka *et al.*, 2012). According to the results of their research, CNTs functionalized with sulfuric acid and nitric acid employed for the treatment of wastewater for 101 min at 90°C and at a pH of 5 were more effective in removing heavy metals, particularly copper. Utilization of CNPs as photocatalysts has been shown in Table 14.1. Exploitation of CNTs as an adsorbent has been exhibited in Table 14.2 and that for membrane filtration is displayed in Table 14.3.

14.5 REMEDIATION USING GRAPHENE AND NANOPOROUS-ACTIVATED CARBON

Graphene's enhanced mechanical and electrical stability, as well as their ability to absorb pollutants, can be attributed to their two-dimensional structure. Deng *et al.* (2010) produced functional graphene nanosheets in potassium hexafluorophosphate electrolyte solution and utilized them to remove heavy metals such as lead and cadmium from wastewater. Lead (406 mg/g) and cadmium (73.42 mg/g) were both removed using the GNSPF6 graphene nanosheet that was produced. Treatment of graphene sheets with heat improved their ability to adsorb heavy metals, as reported by Huang *et al.* (2011). Several studies have shown that graphene may be used to effectively remove colors from wastewater. Dye removal capacity between graphite and graphene oxide (GO) was evaluated by Bradder *et al.* (2011), and GO was found to have a 140% better efficiency than graphite. Due to the rising cost of carbon nanomaterials, GO nanocomposites are being used as a replacement for metallic and organic contaminants. Graphene-based materials are efficient adsorbers of a variety of anions, including fluoride and perchlorates. When it comes to eliminating cationic dyes, GO and its composites perform exceptionally well, while graphene and its composites perform best when dealing with anionic dyes. Metal-embedded nanomaterials (Me/MeO ENMs) are one kind of nanomaterials, while nanomaterials composed of two or more metals and oxides form a second class. There are several types of Me/MeO ENMs, but nanosized zerovalent iron, TiO_2 (Ren *et al.*, 2013), and ZnO are currently the most widely employed for treating water/natural remediation. By adding oxygen or nitrogen functionalizing groups to the activated carbon with acid or alkali, the carbon can be transformed into nanoporous-activated carbon. Such activated carbon will have pores measuring around 1–2 nm in size. Nanoporous-activated carbon has a stellar reputation for its ability to purge wastewaters of unsightly impurities. Textiles benefit greatly from the lightfastness of azo dyes. The dye's azo bonds are responsible for its stability and resistance to degradation under extreme conditions. Reactive dyes can be eliminated with the use of nanoporous-activated carbon in such situations. When trying to get rid of reactive black dye, Ahmad and Hameed (2010) produced granules of activated charcoal made from bamboo waste. An 80 mM tall adsorption column filled with granules of activated charcoal derived from bamboo biomass was created where the effective ratio of dye to activated carbon was found to be 39.9 mg/g. Adsorption of basic dyes onto granular-activated carbon and neutral zeolite was investigated in an experiment by Meshko *et al.* (2001). The adsorbent effectiveness of neutral zeolite and granular-activated carbon was compared in a batch study. The adsorptive performance of granular-activated charcoal was 77 mg/g of dye-activated carbon, demonstrating its superior efficiency.

14.6 RISKS IN NANOREMEDIATION

Tailor-made nanomaterials can be used in a variety of cutting-edge applications, but their use in treatment of water and industrial discharges can have unintended consequences for both the environment and human health (Yao *et al.*, 2013). According to Westerhoff *et al.* (2018), when nanotechnology is applied

Table 14.1 Effect of CNT as a photocatalyst in wastewater treatment.

Photocatalyst	Synthesis Method	Light Source	Target Pollutant	Operating Parameters	Removal Efficiency	Reference
MWCNT + TiO_2	Sol–gel	Visible	Methylene blue (MB)	MWCNT: TiO_2 molar ratio (by weight) = 0.05:1 Light intensity = 15 W Calcination temperature = 400°C Time = 180 min Pollutant concentration = 1×10^{-5} M	With TiO_2 only = 22% MWCNT + TiO_2 = 70%	Wongaree *et al.* (2015)
MWCNT + C_3N_4	Hydrothermal	Visible	Rhodamine B (RhB) MB Methylene orange (MO)	Light intensity = 300 W $\lambda \geq 400$ nm Catalyst dose = 5 mg/mL Irradiance time = 1.5 h for MB and 3 h for MO and RhB	For MB = 66.5% MO = 89.7% RhB = 85.4%	Xu *et al.* (2013)
CNT + P-TiO_2	Hydrothermal	UV and visible	MO	Mass ratio of CNT: P-TiO_2 = 5:100 Light intensity for UV = 250 W and for visible = 400 W Irradiation time = 80 min	~100%	Wang and Zhou (2011)
CNT + g-C_3N_4 + $BiVO_4$	Wet impregnation	Simulated sunlight	Phenolic compound	Irradiation time = 2 h Pseudo first-order kinetics Temkin adsorption model Phenol concentration = 10 mg/L Light intensity = 500 W	80.6%	Samsudin *et al.* (2019)
CNT + TiO_2 + Al_2O_3	Hydrothermal	UV	Metamifop herbicide	Catalyst dose = 10 mg Metamifop herbicide concentration = 5 ppm	84.54%	Yusoff *et al.* (2021)

(*Continued*)

Table 14.1 Effect of CNT as a photocatalyst in wastewater treatment. (*Continued*)

Photocatalyst	Synthesis Method	Light Source	Target Pollutant	Operating Parameters	Removal Efficiency	Reference
CNT + TiO_2 + PAN polymer	Electro-spinning	UV	Phenol	pH = 5 Irradiation time = 7 min Light intensity = 100 W Catalyst dose = 20 mg Phenol concentration = 10 ppm Recycle and reuse = 3 cycles	99.2%	Mohamed *et al.* (2020)
CuO + CNT	Chemical method	UV	Direct red (DR) 31 dye Reactive red (RR) 120 dye	Light intensity = 9 W First-order kinetics Reaction rate 0.0105 and 0.0137/min for RR120 and DR31 Catalyst dose = 5 mg	87% for RR 120 89% for DR 31	Mahmoodi *et al.* (2016)
Ti^{3+} + TiO_2 + SWCNT	Two-step chemical precipitation and calcination	Visible	MB	Irradiation time = 25 min Reaction rate = 0.0083/min	83%	Akter *et al.* (2021)
MWCNT + TiO_2 + SiO_2	Sol–gel	UV	Carbamazepine (CBZ) and Bisphenol A (BPA)	Pseudo first-order kinetics Rate constant = 0.0131–0.0743/min for CBZ and 0.0827–0.1751/min for BPA Irradiation time = 20 min Catalyst concentration = 500 mg/L	50%	Czech and Buda (2015)
MWCNT + Ag-ZnO	Precipitation method	Visible	Congo red (CR)	Catalyst concentration = 150 mg/L Light intensity = 40 W pH = 6 Congo red concentration = 15 mg/L First-order kinetics Rate constant = 0.0023/min	99%	Yibeltal *et al.* (2020)

*Source:*Reproduced with copyright from Mishra and Sundaram (2023).

Table 14.2 Effect of CNT as an adsorbent in wastewater treatment.

Adsorbent	Synthesis Method	Target Pollutant	Operating Parameters	Removal Efficiency/ Adsorption Capacity	Reference
SWCNT	Aryl diazonium salt chemistry and cross-linking processes	1-Pyrene butyric acid, diquat dibromide, and MB	High cross-linking degree with thickness = 50 μm Pollutant concentration = 15 μg/mL	50% 28% 83%	Barrejón *et al.* (2019)
MWCNT	Commercially available	CR, Ponceau 4R (PR) dyes	pH = 6 Temperature = 25°C Dye concentration = 200 mg/L Dose = 1 g/L	PR = 67.9 mg/g CR = 256 mg/g	Ferreira *et al.* (2017)
MWCNT	Commercially available	CR	Contact time = 1 h pH = 11 Temperature = Endothermic Initial concentration = 200 ppm Adsorbent dose = 50 mg	92%	Zare *et al.* (2015)
MWCNT	Catalytic chemical vapor deposition (CVD)	Yellow 81, red 159, and blue 116 dye	Initial concentration of dye = 100 mg/L CNT dose = 60 mg Kinetics = pseudo second order Isotherm model Blue 16 = Temkin model Red 159 = Langmuir and Freundlich Yellow 81 = Freundlich and Temkin	33 mg/g of CNT	Vuono *et al.* (2017)
MWCNT	Catalytic CVD	MB and MO	Contact time = 1 h Catalyst dose = 20 mg Initial concentration of dye = 10 mg/L Temperature = 298 K pH = 6 Kinetics = pseudo second order	MB = 6.96 mg/g MO = 5.464 mg/g	Robati *et al.* (2016)
Chitosjan + SiO_2 + MWCNT	Gelation method	Reactive blue 19 (RB19) and direct blue 71 (DB 71)	pH = 6.8 and 2 for DB71 and RB 19 respectively Isotherm model = Langmuir Kinetic = pseudo second order R^2 = 0.996 and 0.998 for DB71 and RB19	DB71 = 61.4 mg/g RB19 = 97.1 mg/g	Abbasi (2017)

(Continued)

Table 14.2 Effect of CNT as an adsorbent in wastewater treatment. (*Continued*)

Adsorbent	Synthesis Method	Target Pollutant	Operating Parameters	Removal Efficiency/ Adsorption Capacity	Reference
	–	Disperse red 1 (DR 1), acid blue 113 (AB 113), MB, and MO	Kinetics = pseudo second order Isotherm = Langmuir model Initial concentration of dye = 20 mg/L Temperature = 25°C Dose (mg/g) = 19.4 for DR1 and MB = 18.4 for AB113 = 19 for MO pH = 8, 7, 6, and 7 for MB, AB113, MO, and DR1, respectively Contact time = 15 min for MB, MO, AB113 and 10 min for DR1, respectively	DR1 = 500 mg/g MB = 91 mg/g MO = 96 mg/g AB113 = 172 mg/g	Mohammadi and Veisi (2018)
Fe_3O_4/acrylic acid/*N*-isopropyl acrylamide/ MWCNTs	–	MB, RhB, and crystal violet (CV)	Kinetics = pseudo second order Isotherm = Langmuir Initial concentration of dye = 50 mg/L Dose of adsorbent = 50 mg pH = 8 Temperature = 25°C	CV = 287 mg/g MB = 302 mg/g RhB = 231 mg/g	Hosseinzadeh *et al.* (2018)
Gelatin + iron oxide + MWCNT	Co-precipitation and emulsification	MB, direct red 80 (DR 80)	Isotherm = Freundlich model Kinetic = pseudo second order Temperature = 65°C Time = 6 h Initial concentration = 500 mg/g	DR81 = 96.1% MB = 76.3%	Saber-Samandari *et al.* (2017)
Microwave-induced MWCNT		Cu(II)	Contact time = 35 min Dose of adsorbent = 100 mg pH = 5.5 Agitation speed = 160 r/min Isotherm = Freundlich and Langmuir Kinetics = pseudo second order	99.9%	Mubarak *et al.* (2016)

MWCNT	Tubular microwave CVD	Pb(II)	pH = 5 Initial concentration of Pb(II) = 10 mg/L Agitation speed = 2.8 r/s Dose of adsorbent = 100 mg Contact time = 22.5 min Isotherm = Langmuir and Freundlich Kinetics = pseudo second order	99.9%	Mubarak *et al.* (2015)
Functionalized MWCNT	–	Cu(II)	Contact time = 1 h Dose of adsorbent = 10 mg/L Initial concentration = 20 mg pH = 3 Isotherm = Langmuir Kinetics = pseudo second order	93%	Gupta *et al.* (2017)
CNT + waste cooking palm oil + Fe	Thermal CVD	Zn (II), Cu (II), and Fe (II)	Dose of adsorbent = 1800 mg/L pH = 7 Initial concentration of heavy metal = 100 mg/L Isotherm model = Langmuir for Cu (II)	99.2–99.9%	Abu Bakar *et al.* (2021)
MoS_2 + thiol-functionalized MWCNT	Tubular microwave CVD	Pb(II)Cd (II)	Isotherm = Freundlich Kinetics = pseudo second order Dose of adsorbent = 2 mg/mL pH = 6 Contact time = 1 h	Cd(II) = 66.6 mg/gPb (II) = 90 mg/g	Gusain *et al.* (2019)
MWCNT-COOH and MWCNT-OH	Catalytic CVD	Cr(VI)	Initial concentration of Cr(VI) = 5–60 mg/L pH = 5 Dose of adsorbent = 25 mg Kinetic = intra-particle diffusion model and pseudo second order	8.09 mg/g by MWCNT-COOH 7.85 mg/g by MWCNT-OH	Zhang *et al.* (2018)

*Source:*Reproduced with copyright from Mishra and Sundaram (2023).

Table 14.3 CNT in membrane and filtration for wastewater treatment.

Membrane/Filter Type	Material	Synthesis Method	Membrane Performance	Reference
VACNT	CNT + polyethersulfone (PES)	CVD	Used for ultrafiltration (UF) Water transportation speed = 100 $L/m^2.h$ at 60 Psi 3 times faster than random oriented CNT + PES 10 times faster than PES membrane	Li *et al.* (2014)
VACNT	CNT + epoxy	Water-assisted thermal CVD	3 times higher water flux than UF2 log Less bacterial concentration in VCNT (4×10^5 CFU/mL) as compared to UF (8×10^7 CFU/mL) Permeate flux reduction after 600 min = 67% for VCNT and 55% for UF	Baek *et al.* (2014)
VACNT	CNT + stainless-steel mesh	Thermal CVD	Separate water layer and diesel and even surfactant-stabilized emulsions High hydrophobicity and oleophobicity	Lee and Baik (2010)
SWCNT	CNT + poly vinylidene fluoride	Vacuum filtration	Complete removal of viral pathogens and bacteria 5×10^5 *Escherichia coli* cells were removed Cell inactivation = 79% after contact time of 20 min Metabolically active *E. coli* = 6% as compared to over 70% in control filter Damage *E. coli* cell membrane	Brady-Estévez *et al.* (2008)
VACNT	CNT + poly tetrafluoro ethylene + Si substrate	Water-assisted CVD	Millimeter-thick UF membrane Water permeability = 30 000/Lm^2/h/bar Inhibit bacterial growth and resist biofilm formation	Lee *et al.* (2015)

VACNT	MWCNT + Fe + Al_2O_3 + Si	Thermal CVD	The proportion of irreversible fouling to total resistance for UF and VCNT membranes were 24% and 69%, respectively, while the proportion of reversible fouling resistance was 38% and 5%, respectively BSA (bovine serum albumin) removal improved from 71% to 90% after surface modification by graft-polymerization of methacrylic acid	Park *et al.* (2014)
Vertically aligned double-wall CNT	Double-wall CNT + Si wafer substrate	CVD	NaCl rejection = 41.4% Water flux = 1.31×10^{-3} to 62.7×10^{-3} L/cm^2/day/MPa	Matsumoto *et al.* (2017)
CNT bucky paper	CNT	CVD, UV/ozone treatment and alkoxysilanization	Salt rejection $\geq$95% Life span improved by 50%	Dumée *et al.* (2011)
VACNT	MCNT + polydimethylsiloxane + Si wafer	CVD	Salt rejection = 96.5% at operational pressure of 2 bar	Trivedi and Alameh (2016)
Graphene oxide coated VACNT (GO-VACNT) and polyamide coated VA-CNT (PC-VACNT)	CNT + epoxy	–	Salt rejection = 45% and 65% for GO-VCNT and PO-VCNT at operational pressure of 15.5 bar, respectively	Baek *et al.* (2016)
Polyamide/outer-wall VA-CNT	CNT + epoxy	Interfacial polymerization	Reverse osmosis membrane Flux = 128.6/Lm^2/h Salt rejection = 98.3% at operational pressure of 15.5 bar	Li *et al.* (2019)

Source: Reproduced with copyright from Mishra and Sundaram (2023).

to wastewater treatment, the concept of water is modified by modifying the organic oxygen demand and the material oxygen demand of the water. The potential risks to human health posed by using nanotechnology to purify water are one of the factors limiting the case's scope. As nanotechnology retains substantial water additions and raises the degree of poisonousness in the water, it should be used only for cleaning wastewater. The effect of CNTs on living species is an uncharted topic, despite the fact that several literatures declare and support their usage in wastewater treatment. Several studies have demonstrated CNTs' cytotoxicity toward mammalian cells. Mice were used to investigate the impact of inhaled MWCNTs on their biological systems. Mice were given MWCNT doses of 0.3 and 1 ppm where the reduction of immunity was observed at 1 ppm. CNTs are more hazardous than quartz and asbestos on mammalian cells. Several studies have demonstrated that agglomerated CNTs are more hazardous than their dispersed counterparts. Nevertheless, CNT agglomeration can be prevented by modifying the CNTs' surfaces. Based on their research comparing raw CNTs to surface-modified CNTs, it was concluded that the latter is having higher solubility. To fully understand the impacts and repercussions of utilizing CNP in wastewater treatment, additional, in-depth preclinical investigations are required. Thin film photocatalysts for the purification of wastewater have only been the subject of a small number of research projects. However, there is a paucity of data on the characteristics of photo-active nanomaterials regarding the toxicity levels of treated water sample containing spent catalysts, standard processes of toxicity tests, reusing and recirculation of spent catalysts, safety of treatment impacts, and economic assessment (Rueda-Marquez *et al.*, 2020). Due of nanoparticles' adaptability, interest in them has been generated from a wide range of sectors and individuals. Due to a lack of qualified individuals, Labfirm Fusion is unable to serve as a bridge across different industries and technologies. The water and sewage treatment industries have benefited greatly from this line of research. In addition to these sectors, others, such as those involved in healthcare, electrical instruments, building materials, and consumer goods, also reap substantial benefits. The novelty of nanotechnology means that information associated with the subject remains hidden to many water partners, making their work difficult. Manufactured nanoparticles can be used in the practical purification and disinfection of water and wastewater. Therefore, more effort must be devoted to enhancing this breakthrough. The government, along with private sectors, must coordinate, be pro-active, and include skills, resources, policies, and standard operating procedures to increase the safe application of nanoremediation. Nanotechnology needs to be widely adopted for economic, social, and strategic reasons, and there is a sizable market for this. Effective methods for removing nanoparticles from water are needed to mitigate or eliminate their harmful effects on humans and the environment.

14.7 CONCLUSIONS

Despite water's importance in every aspect of human life, it is becoming increasingly hard to come by in many places of the world. Challenges in providing safe drinkable water can be mitigated through the use of renewable resources to provide ultrafine freshwater, as well as through the conservation and protection of water bodies from contamination. Nanotechnology and its potential uses have been identified as promising avenues for addressing water scarcity and quality. Because of their huge surface areas and size- and shape-dependent catalytic capabilities, nanomaterials are the subject of a significant amount of research and investigation into their potential use in applications such as membrane separations, catalysis, and adsorption. Although it is expensive, membrane technology is increasingly being seen as the best option for purifying water. Nanoparticle-functionalized membranes have emerged as a promising new alternative to the conventional approaches. The versatility of today's technologies, such as membrane technology, nanotechnology, and biochar adsorption, also permits the creation of the most practical applications possible. While constructed biochar may have certain uses, the most crucial consideration when deciding on a wastewater treatment strategy is how to properly dispose of the sludge that is generated during the treatment process. Therefore, an effective and discharge-free solution is urgently needed to address the most challenging problem facing sustainable development. There are more upsides than downsides to employing nanostructures in wastewater treatment, making them a potentially beneficial alternative to the many existing treatment approaches. The water treatment procedure benefits greatly from the use of biochar made from Municipal Solid Waste (MSW). The low specific surface area of biochar made from mixed municipal solid waste could be raised through clay alteration. Biochar made from properly sorted MSW outperforms adsorption and advanced oxidation processes, respectively, in terms of efficiency and stability in the removal of pollutants. Biochar made from food waste in municipal solid waste is very good at cleaning the air through adsorption and oxidation processes. In order to get the most use out of the materials contained in municipal solid waste, it is possible to separate out things such as fruit peels, bones, and shells. Bone and shell biochar is extremely effective at removing organic pollutants thanks to its high adsorption capacity and catalytic activity. Biochar made from peels is very effective at activating persulfate. Biochar made from recycled paper is also effective for the adsorption removal of heavy metals. Biochar's adsorption capacity and catalytic activity inside oxidation systems were enhanced by modifying its surface structure and chemical composition utilizing various preparation processes. In order to enhance the adsorption system's ability to remove contaminants, an alkaline treatment was performed to the biochar that was produced from mixed municipal solid waste. Biochar produced by pyrolysis and acid treatment can be used to eliminate organic pollutants in oxidation systems. Bone and eggshells were used to create this biochar. To further facilitate the removal of pollutants via adsorption and oxidation processes, bone is frequently pre-carbonized prior to the formation of biochar via the process of pyrolysis. CNTs outperform traditional water and wastewater treatment methods in several key areas. Improved adsorption capacity, greater selectivity, and a rapid response time are all hallmarks of their performance as an efficient adsorbent. Filters made from CNTs are also effective in sterilizing the air. Water filtration and desalination are two more useful applications among them. Incorporating CNT into the membrane improves its water permeability, strength, and antifouling properties. Two major obstacles are lowering the cost of generating CNTs and increasing their efficiency. However, CNTs almost see only academic and laboratory applications. The feasibility of their usage in industrial and large-scale applications, as well as the development of low-cost, non-toxic methods of synthesis, warrants further investigation.

REFERENCES

Abbasi M. (2017). Synthesis and characterization of magnetic nanocomposite of chitosan/SiO_2/carbon nanotubes and its application for dyes removal. *Journal of Cleaner Production*, **145**, 105–113, https://doi.org/10.1016/j.jclepro.2017.01.046

Abu Bakar S., Jusoh N., Mohamed A., Muqoyyanah M., Othman M. H. D., Mamat M. H., Ahmad M. K., Mohamed M. A., Azlan M. N., Hashim N., Birowosuto M. D. and Soga T. (2021). Carbon nanotubes from waste cooking palm oil as adsorbent materials for the adsorption of heavy metal ions. *Environmental Science and Pollution Research*, **28**(46), 65171–65187.

Adegoke A. A., Amoah I. D., Stenström T. A., Verbyla M. E. and Mihelcic J. R. (2018). Epidemiological evidence and health risks associated with agricultural reuse of partially treated and untreated wastewater: a review. *Frontiers in Public Health*, **6**, 337, https://doi.org/10.3389/fpubh.2018.00337

Ahmad A. and Hameed B. (2010). Fixed-bed adsorption of reactive azo dye onto granular activated carbon prepared from waste. *Journal of Hazardous Materials*, **175**, 298–303.

Ahmad M., Rajapaksha A. U., Lim J. E., Zhang M., Bolan N., Mohan D., Vithanage M., Lee S. S. and Ok Y. S. (2014). Biochar as a sorbent for contaminant management in soil and water: a review. *Chemosphere*, **99**, 19–33.

Akter J., Hanif M., Islam M. A., Sapkota K. and Hahn J. (2021). Selective growth of Ti^{3+}/TiO_2/CNT and Ti^{3+}/TiO_2/C nanocomposite photocatalysts for enhanced visible-light utilization to degrade organic pollutants by lowering the bandgap of TiO2, https://doi.org/10.21203/rs.3.rs-199530/v1

Al-Mamun M. R., Kader S., Islam M. S., Khan M. Z. H. (2019). Photocatalytic activity improvement and application of UV-TiO_2 photocatalysis in textile wastewater treatment: A review. *Journal of Environmental Chemical Engineering*, **7**(1032), 1–17.

Ambaye T., Vaccari M., van Hullebusch E. D., Amrane A. and Rtimi S. (2021). Mechanisms and adsorption capacities of biochar for the removal of organic and inorganic pollutants from industrial wastewater. *International Journal of Environmental Science and Technology*, **18**, 3273–3294.

Baek Y., Kim C., Seo D. K., Kim T., Lee J. S., Kim Y. H., Ahn K. H., Bae S. S., Lee S. C., Lim J., Lee K. and Yoon J. (2014). High performance and antifouling vertically aligned carbon nanotube membrane for water purification. *Journal of Membrane Science*, **460**, 171–177.

Baek Y., Seo D. K., Choi J. H., Lee B., Kim Y. H., Park S. M., Jung J., Lee S. and Yoon J. (2016). Improvement of vertically aligned carbon nanotube membranes: desalination potential, flux enhancement, and scale-up. *Desalination and Water Treatment*, **57**(58), 28133–28140.

Barrejón M., Syrgiannis Z., Burian M., Bosi S., Montini T., Fornasiero P., Amenitsch H. and Prato M. (2019). Cross-linked carbon nanotube adsorbents for water treatment: tuning the sorption capacity through chemical functionalization. *ACS Applied Materials & Interfaces*, **11**(13), 12920–12930.

Bharti S. K. and Kumar N. (2018). Kinetic study of lead (Pb 2+) removal from battery manufacturing wastewater using bagasse biochar as biosorbent. *Applied Water Science*, **8**, 1–13.

Bradder P., Ling S. K., Wang S. and Liu S. (2011). Dye adsorption on layered graphite oxide. *Journal of Chemical & Engineering Data*, **56**, 138–141.

Brady-Estévez A. S., Kang S. and Elimelech M. (2008). A single-walled-carbon-nanotube filter for removal of viral and bacterial pathogens. *Small*, **4**(4), 481–484, https://doi.org/10.1002/smll.200700863

Crane R. A. and Scott T. B. (2012). Nanoscale zero-valent iron: future prospects for an emerging water treatment technology. *Journal of Hazardous Materials*, **211**, 112–125.

Czech B. and Buda W. (2015). Photocatalytic treatment of pharmaceutical wastewater using new multiwall-carbon nanotubes/TiO_2/SiO_2 nanocomposites. *Environmental Research*, **137**, 176–184, https://doi.org/10.1016/j.envres.2014.12.006

Deng X., Lü L., Li H. and Luo F. (2010). The adsorption properties of Pb (II) and Cd (II) on functionalized graphene prepared by electrolysis method. *Journal of Hazardous Materials*, **183**, 923–930.

Dong L.-X., Huang X.-C., Wang Z., Yang Z., Wang X.-M. and Tang C. Y. (2016). A thin-film nanocomposite nanofiltration membrane prepared on a support with in situ embedded zeolite nanoparticles. *Separation and Purification Technology*, **166**, 230–239.

Dumée L., Germain V., Sears K., Schütz J., Finn N., Duke M., Cerneaux S., Cornu D. and Gray S. (2011). Enhanced durability and hydrophobicity of carbon nanotube bucky paper membranes in membrane distillation. *Journal of Membrane Science*, **376**(1–2), 241–246.

Fagbohungbe M. O., Herbert B. M., Hurst L., Ibeto C. N., Li H., Usmani S. Q. and Semple K. T. (2017). The challenges of anaerobic digestion and the role of biochar in optimizing anaerobic digestion. *Waste Management*, **61**, 236–249.

Feizpoor S., Habibi-Yangjeh A. and Yubuta K. (2018). Integration of carbon dots and polyaniline with TiO_2 nanoparticles: Substantially enhanced photocatalytic activity to removal various pollutants under visible light. *Journal of Photochemistry and Photobiology A: Chemistry*, **367**, 94–104.

Ferreira G. M. D., Ferreira G. M. D., Hespanhol M. C., de Paula Rezende J., dos Santos Pires A. C., Gurgel L. V. A. and da Silva L. H. M. (2017). Adsorption of red azo dyes on multi-walled carbon nanotubes and activated carbon: A thermodynamic study. *Colloids and Surfaces A: Physicochemical and Engineering Aspects*, **529**, 531–540, https://doi.org/10.1016/j.colsurfa.2017.06.021

Gupta V. K., Agarwal S., Bharti A. K. and Sadegh H. (2017). Adsorption mechanism of functionalized multi-walled carbon nanotubes for advanced Cu (II) removal. *Journal of Molecular Liquids*, **230**, 667–673, https://doi.org/10.1016/j.molliq.2017.01.083

Gusain R., Kumar N., Fosso-Kankeu E. and Ray S. S. (2019). Efficient removal of Pb(II) and Cd(II) from industrial mine water by a hierarchical MoS2/SH-MWCNT nanocomposite. *ACS Omega*, **4**(9), 13922–13935.

Haque A. N. M. A., Remadevi R., Rojas O. J., Wang X. and Naebe M. (2020). Kinetics and equilibrium adsorption of methylene blue onto cotton gin trash bioadsorbents. *Cellulose*, **27**, 6485–6504.

Hara T. O. and Singh B. (2021). Electrochemical biosensors for detection of pesticides and heavy metal toxicants in water. *Recent Trends in Program*, **1**(3), 462–478, https://doi.org/10.1021/acsestwater.0c00125

Higashikawa F. S., Conz R. F., Colzato M., Cerri C. E. P. and Alleoni L. R. F. (2016). Effects of feedstock type and slow pyrolysis temperature in the production of biochars on the removal of cadmium and nickel from water. *Journal of Cleaner Production*, **137**, 965–972.

Hosseinzadeh S., Hosseinzadeh H., Pashaei S. and Khodaparast Z. (2018). Synthesis of magnetic functionalized MWCNT nanocomposite through surface RAFT co-polymerization of acrylic acid and N-isopropyl acrylamide for removal of cationic dyes from aqueous solutions. *Ecotoxicology and Environmental Safety*, **161**, 34–44, https://doi.org/10.1016/j.ecoenv.2018.05.063

Huang Z.-H., Zheng X., Lv W., Wang M., Yang Q.-H. and Kang F. (2011). Adsorption of lead (II) ions from aqueous solution on low-temperature exfoliated graphene nanosheets. *Langmuir*, **27**, 7558–7562.

Jeon I.-Y., Chang D. W., Kumar N. A. and Baek J.-B. (2011). Functionalization of carbon nanotubes. In: Carbon Nanotubes-Polymer Nanocomposites, InTech, United Kingdom, ISBN 978-953-307-498-6, pp. 91–110, https://doi.org/10.5772/18396

Jeong B.-H., Hoek E. M., Yan Y., Subramani A., Huang X., Hurwitz G., Ghosh A. K. and Jawor A. (2007). Interfacial polymerization of thin film nanocomposites: a new concept for reverse osmosis membranes. *Journal of Membrane Science*, **294**, 1–7.

Jiang Q. L., Tian H. L. and Zhang K. S. (2019). Enhanced performance of poly(m-phenylene isophthalamide) (PMIA) composite hollow fiber ultrafiltration membranes by O-MoS2 nanosheets modification. *Desalination and Water Treatment*, **166**, 245–258.

Jiang Q., Wang Y., Li Y., Luo J. and Xiong J. (2023). Nanocomposite substrate-supported nanofiltration membrane for efficient treatment of rare earth wastewater. *Results in Engineering*, **18**, 101040, https://doi.org/10.1016/j.rineng.2023.101040

Joshi S., Garg V. K. and Kataria N. (2019). Applications of Fe_3O_4 Ac nanoparticles for removal of dyes in simulated wastewater. *Chemosphere*, **236**, Article 124280, 1–11.

Kaetzl K., Lübken M., Uzun G., Gehring T., Nettmann E., Stenchly K. and Wichern M. (2019). On-farm wastewater treatment using biochar from local agroresidues reduces pathogens from irrigation water for safer food production in developing countries. *Science of the Total Environment*, **682**, 601–610.

Karim A. A., Kumar M., Singh S. K., Panda C. R. and Mishra B. K. (2017). Potassium enriched biochar production by thermal plasma processing of banana peduncle for soil application. *Journal of Analytical and Applied Pyrolysis*, **123**, 165–172.

Kausar A. (2016). Properties of polyacrylamide and functional multi-walled carbon nanotubes. *American Journal of Nanoscience Nanotechnology Research*, **4**(1), 1–8.

Krishnamurthy G. and Agarwal S. (2013). Optimization of reaction conditions for high yield synthesis of carbon nanotube bundles by low-temperature solvothermal process and study of their H2 storage capacity. *Bulletin of the Korean Chemical Society*, **34**, 3046–3054.

Kumar S., Ahlawat W., Bhanjana G., Heydarifard S., Nazhad M. M. and Dilbaghi N. (2014). Nanotechnology-based water treatment strategies. *Journal of Nanoscience and Nanotechnology*, **14**, 1838–1858.

Kuo C.-Y., Wu C.-H. and Wu J.-Y. (2008). Adsorption of direct dyes from aqueous solutions by carbon nanotubes: determination of equilibrium, kinetics and thermodynamics parameters. *Journal of Colloid and Interface Science*, **327**, 308–315.

Lai G. S., Lau W. J., Goh P. S., Ismail A. F., Yusof N. and Tan Y. H. (2016). Graphene oxide incorporated thin film nanocomposite nanofiltration membrane for enhanced salt removal performance. *Desalination*, **387**, 14–24.

Lee C. and Baik S. (2010). Vertically-aligned carbon nano-tube membrane filters with superhydrophobicity and superoleophilicity. *Carbon*, **48**(8), 2192–2197, https://doi.org/10.1016/j.carbon.2010.02.020

Lee B., Baek Y., Lee M., Jeong D. H., Lee H. H., Yoon J. and Kim Y. H. (2015). A carbon nanotube wall membrane for water treatment. *Nature Communications*, **6**(1), 1–7, https://doi.org/10.1038/ncomms8109

Li S., Liao G., Liu Z., Pan Y., Wu Q., Weng Y., Zhang X., Yang Z. and Tsui O. K. C. (2014). Enhanced water flux in vertically aligned carbon nanotube arrays and polyethersulfone composite membranes. *Journal of Materials Chemistry A*, **2**(31), 12171–12176.

Li K., Lee B. and Kim Y. (2019). High-performance reverse osmosis membrane with carbon nanotube support layer. *Journal of Membrane Science*, **592**, 117358.

Li X., Wang Z., Han X. L., Liu Y. Y., Wang C., Yan F. Z. and Wang J. X. (2021). Regulating the interfacial polymerization process toward high-performance polyamide thin-film composite reverse osmosis and nanofiltration membranes: a review. *Journal of Membrane Science*, **640**, Article 119765, 1–24.

Liao Z. P., Zhu J. Y., Li X. and Van der Bruggen B. (2021). Regulating composition and structure of nanofillers in thin film nanocomposite (TFN) membranes for enhanced separation performance: a critical review. *Separation and Purification Technology*, **266**, Article 118567, 1–21.

Lin Y. T., Kao F. Y., Chen S. H., Wey M. Y. and Tseng H. H. (2020). A facile approach from waste to resource: reclaimed rubber-derived membrane for dye removal. *Journal of the Taiwan Institute of Chemical Engineers*, **112**, 286–295.

Lind M. L., Jeong B.-H., Subramani A., Huang X. and Hoek E. M. (2009). Effect of mobile cation on zeolite-polyamide thin film nanocomposite membranes. *Journal of Materials Research*, **24**, 1624–1631.

Liu X., Wang M., Zhang S. and Pan B. (2013). Application potential of carbon nanotubes in water treatment: A review. *Journal of Environmental Sciences*, **25**(7), 1263–1280, https://doi.org/10.1016/S1001-0742(12)60161-2

Low S. S., Chen Z., Li Y., Lu Y. and Liu Q. (2021a). Design principle in biosensing: critical analysis based on graphitic carbon nitride (G-C3N4) photoelectrochemical biosensor. *TrAC Trends in Analytical Chemistry*, **145**, 116454. ISSN 0165-9936, https://doi.org/10.1016/j.trac.2021.116454

Low S. S., Lim C. N., Yew M., Chai W. S., Low L. E., Manickam S., Tey B. T. and Show P. L. (2021b). Recent ultrasound advancements for the manipulation of nanobiomaterials and nanoformulations for drug delivery. *Ultrasonics Sonochemistry*, **80**, 105805. ISSN 1350-4177, https://doi.org/10.1016/j.ultsonch.2021.105805

Low S. S., Yew M., Lim C. N., Chai W. S., Low L. E., Manickam S., Tey B. T. and Show P. L. (2022). Sonoproduction of nanobiomaterials – a critical review. *Ultrasonics Sonochemistry*, **82**, 105887. ISSN 1350-4177, https://doi.org/10.1016/j.ultsonch.2021.105887

Machado F. M., Bergmann C. P., Fernandes T. H. M., Lima E. C., Royer B., Calvete T. and Fagan S. B. (2011). Adsorption of reactive red M-2BE dye from water solutions by multi-walled carbon nanotubes and activated carbon. *Journal of Hazardous Materials*, **192**(3), 1122–1131.

Mahmoodi N. M., Rezaei P., Ghotbei C. and Kazemeini M. (2016). Copper oxide-carbon nanotube (CuO/CNT) nanocomposite: synthesis and photocatalytic dye degradation from colored textile wastewater. *Fibers and Polymers*, **17**(11), 1842–1848, https://doi.org/10.1007/s12221-016-6645-y

Matsumoto H., Tsuruoka S., Hayashi Y., Abe K., Hata K., Zhang S., Saito Y., Aiba M., Tokunaga T., Iijima T., Hayashi T., Inoue H. and Amaratunga G. A. J. (2017). Water transport phenomena through membranes consisting of vertically-aligned double-walled carbon nanotube array. *Carbon*, **120**, 358–365.

Meng H., Nie C., Li W., Duan X., Lai B., Ao Z., Wang S. and An T. (2020). Insight into the effect of lignocellulosic biomass source on the performance of biochar as persulfate activator for aqueous organic pollutants remediation: epicarp and mesocarp of citrus peels as examples. *Journal of Hazardous Materials*, **399**, 123043, https://doi.org/10.1016/j.jhazmat.2020.123043

Meshko V., Markovska L., Mincheva M. and Rodrigues A. (2001). Adsorption of basic dyes on granular acivated carbon and natural zeolite. *Water Research*, **35**, 3357–3366.

Mishra S. and Sundaram B. (2023). Efficacy and challenges of carbon nanotube in wastewater and water treatment. *Environmental Nanotechnology, Monitoring & Management*, **19**, 100764, https://doi.org/10.1016/j.enmm.2022.100764

Mohamed A., Yousef S., Nasser W. S., Osman T. A., Knebel A., Sánchez E. P. V. and Hashem T. (2020). Rapid photocatalytic degradation of phenol from water using composite nanofibers under UV. *Environmental Sciences, Europe*, **32**(1), 1–8, https://doi.org/10.1186/s12302-020-00436-0

Mohammadi A. and Veisi P. (2018). High adsorption performance of β-cyclodextrin-functionalized multi-walled carbon nanotubes for the removal of organic dyes from water and industrial wastewater. *Journal of Environmental Chemical Engineering*, **6**(4), 4634–4643, https://doi.org/10.1016/j.jece.2018.07.002

Mohanty P., Nanda S., Pant K. K., Naik S., Kozinski J. A. and Dalai A. K. (2013). Evaluation of the physiochemical development of biochars obtained from pyrolysis of wheat straw, Timothy grass and pinewood: effects of heating rate. *Journal of Analytical and Applied Pyrolysis*, **104**, 485–493.

Mubarak N. M., Sahu J. N., Abdullah E. C., Jayakumar N. S. and Ganesan P. (2015). Novel microwave-assisted multiwall carbon nanotubes enhancing Cu (II) adsorption capacity in water. *Journal of the Taiwan Institute of Chemical Engineers*, **53**, 140–152, https://doi.org/10.1016/j.jtice.2015.02.016

Mubarak N. M., Sahu J. N., Abdullah E. C. and Jayakumar N. S. (2016). Rapid adsorption of toxic Pb(II) ions from aqueous solution using multiwall carbon nanotubes synthesized by microwave chemical vapor deposition technique. *Journal of Environmental Sciences*, **45**, 143–155, https://doi.org/10.1016/j.jes.2015.12.025

Mubaraka N., Daniela S., Khalid M. and Tana J. (2012). Comparative study of functionalize and non-functionalized carbon nanotube for removal of copper from polluted water. *International Journal of Chemical and Environmental Engineering*, **3**, 314–317.

Ndoun M. C., Knopf A., Preisendanz H. E., Vozenilek N., Elliott H. A., Mashtare M. L., Velegol S., Veith T. L. and Williams C. F. (2023). Fixed bed column experiments using cotton gin waste and walnut shells-derived biochar as low-cost solutions to removing pharmaceuticals from aqueous solutions. *Chemosphere*, **330**, 138591, 1–9, https://doi.org/10.1016/j.chemosphere.2023.138591

Park S.-M., Jung J., Lee S., Baek Y., Yoon J., Seo D. K. and Kim Y. H. (2014). Fouling and rejection behavior of carbon nanotube membranes. *Desalination*, **343**, 180–186, https://doi.org/10.1016/j.desal.2013.10.005

Park J.-H., Ok Y. S., Kim S.-H., Cho J.-S., Heo J.-S., Delaune R. D. and Seo D.-C. (2016). Competitive adsorption of heavy metals onto sesame straw biochar in aqueous solutions. *Chemosphere*, **142**, 77–83.

Perez-Mercado L. F., Lalander C., Joel A., Ottoson J., Dalahmeh S. and Vinnerås B. (2019). Biochar filters as an on-farm treatment to reduce pathogens when irrigating with wastewater-polluted sources. *Journal of Environmental Management*, **248**, 109295.

Rajapaksha A. U., Vithanage M., Ahmad M., Seo D.-C., Cho J.-S., Lee S.-E., Lee S. S. and Ok Y. S. (2015). Enhanced sulfamethazine removal by steam-activated invasive plant-derived biochar. *Journal of Hazardous Materials*, **290**, 43–50.

Rao G. P., Lu C. and Su F. (2007). Sorption of divalent metal ions from aqueous solution by carbon nanotubes: a review. *Separation and Purification Technology*, **58**, 224–231.

Ren D., Colosi L. M. and Smith J. A. (2013). Evaluating the sustainability of ceramic filters for point-of-use drinking water treatment. *Environmental Science & Technology*, **47**(19), 11206–11213, https://doi.org/10.1021/es4026084

Robati D., Bagheriyan S., Rajabi M., Moradi O. and Peyghan A. A. (2016). Effect of electrostatic interaction on the methylene blue and methyl orange adsorption by the pristine and functionalized carbon nanotubes. *Physica E: Low-Dimensional Systems and Nanostructures*, **83**, 1–6, https://doi.org/10.1016/j.physe.2016.04.005

Rueda-Marquez J. J., Levchuk I., Fernández Ibañez P. and Sillanpää M. (2020). A critical review on application of photocatalysis for toxicity reduction of real wastewaters. *Journal of Cleaner Production*, **258**, 120694, https://doi.org/10.1016/j.jclepro.2020.120694

Saber-Samandari S., Saber-Samandari S., Joneidi-Yekta H. and Mohseni M. (2017). Adsorption of anionic and cationic dyes from aqueous solution using gelatin-based magnetic nanocomposite beads comprising carboxylic acid functionalized carbon nanotube. *Chemical Engineering Journal*, **308**, 1133–1144, https://doi.org/10.1016/j.cej.2016.10.017

Samsudin M. F. R., Bacho N., Sufian S. and Ng Y. H. (2019). Photocatalytic degradation of phenol wastewater over Z-scheme g-C3N4/CNT/BiVO4 heterostructure photocatalyst under solar light irradiation. *Journal of Molecular Liquids*, **277**, 977–988.

Shang J., Zong M., Yu Y., Kong X., Du Q. and Liao Q. (2017). Removal of chromium (VI) from water using nanoscale zerovalent iron particles supported on herb-residue biochar. *Journal of Environmental Management*, **197**, 331–337.

Tan X.-F., Zhu S.-S., Wang R.-p., Chen Y.-D., Show P.-L., Zhang F.-F. and Ho S.-H. (2021). Role of biochar surface characteristics in the adsorption of aromatic compounds: pore structure and functional groups. *Chinese Chemical Letters*, **32**(10), 2939–2946. ISSN 1001-8417, https://doi.org/10.1016/j.cclet.2021.04.059

Tang B., Shan L., Liang S. and Zhou J. (2019). Issues and opportunities facing aqueous zinc-ion batteries. *Energy & Environmental Science*, **12**, 3288–3304.

Trivedi S. and Alameh K. (2016). Effect of vertically aligned carbon nanotube density on the water flux and salt rejection in desalination membranes. *SpringerPlus*, **5**(1), 1–13, https://doi.org/10.1186/s40064-016-2783-3

Vunain E., Mishra A. and Mamba B. (2016). Dendrimers, mesoporous silicas and chitosan-based nanosorbents for the removal of heavy-metal ions: a review. *International Journal of Biological Macromolecules*, **86**, 570–586.

Vuono D., Catizzone E., Aloise A., Policicchio A., Agostino R., Migliori M. and Giordano G. (2017). Study of adsorption behavior of multi-walled carbon nanotubes towards dyes applied in textile applications. *Advanced Science Letters*, **23**, 5851–5854, https://doi.org/10.1166/asl.2017.9048

Wang S. and Zhou S. (2011). Photodegradation of methyl orange by photocatalyst of CNTs/P-TiO2 under UV and visible-light irradiation. *Journal of Hazardous Materials*, **185**(1), 77–85, https://doi.org/10.1016/j.jhazmat.2010.08.125

Westerhoff P., Hritvoski K. M., Lopes L. and Sergi Garcia S. (2018). Electrolytic reduction of nitrate fundamental to full scale water treatment. *Applied Catalysis B: Environment*, **236**, 546–568.

Wongaree M., Chiarakorn S. and Chuangchote S. (2015). Photocatalytic improvement under visible light in TiO_2 nanoparticles by carbon nanotube incorporation. *Journal of Nanomaterials*, **16**(1), 295, 1–10, https://doi.org/10.1155/2015/689306

Xu Y., Hui X., Wang L., Yan J., Li H., Song Y., Huang L. and Cai G. (2013). The CNT modified white C_3N_4 composite photocatalyst with enhanced visible-light response photoactivity. *Dalton Transactions (Cambridge, England: 2003)*, **42**, 7604, https://doi.org/10.1039/c3dt32871f

Yang S. S., Tian H. L., Hill M. R. and Zhang K. S. (2021). Effect and regulation mechanism of oxidation degrees on the O-MoS2 structure and separation performance of nanofiltration membrane. *Journal of Membrane Science*, **635**, Article 119468, 1–12.

Yao D., Chen Z., Zhao K., Yang Q. and Zhang W. (2013). Limitation and challenges faced to the researches on the environmental risk of nanotechnology. *Procedia Environmental Sciences*, **18**, 149–156, https://doi.org/10.1016/j.proenv.2013.04.020

Yibeltal A. W., Beyene B. B., Admassie S. and Taddesse A. M. (2020). MWCNTs/Ag-ZnO nanocomposite for efficient photocatalytic degradation of Congo red. *Bulletin of the Chemical Society of Ethiopia*, **34**(1), 55–66.

Yu H., Zou W., Chen J., Chen H., Yu Z., Huang J., Tang H., Wei X. and Gao B. (2019). Biochar amendment improves crop production in problem soils: a review. *Journal of Environmental Management*, **232**, 8–21.

Yusoff N., Lim P., Azmi N., Yusoff M. and Muhammad M. (2021). Effect of photocatalyst dosage and air loading in photocatalytic degradation of metamifop. *IOP Conference Series: Earth and Environmental Science*, **842**, Article 012061, 1–8, https://doi.org/10.1088/1755-1315/842/1/012061

Zare K., Sadegh H., Shahryari-hoshekandi R., Maazinejad B., Ali V., Tyagi I., Agarwal S. and Gupta V. K. (2015). Enhanced removal of toxic Congo red dye using multi-walled carbon nanotubes: kinetic, equilibrium studies and its comparison with other adsorbents. *Journal of Molecular Liquids*, **212**, 266–271.

Zazycki M. A., Godinho M., Perondi D., Foletto E. L., Collazzo G. C. and Dotto G. L. (2018). New biochar from pecan nutshells as an alternative adsorbent for removing reactive red 141 from aqueous solutions. *Journal of Cleaner Production*, **171**, 57–65.

Zhang Y., Yang J., Zhong L. and Liu L. (2018). Effect of multi-wall carbon nanotubes on Cr(VI) reduction by citric acid: implications for their use in soil remediation. *Environmental Science and Pollution Research*, **25**(24), 23791–23798, https://doi.org/10.1007/s11356-018-2438-8

Zhang M., Xu L., Qi C. and Zhang M. (2019). Highly effective removal of tetracycline from water by hierarchical porous carbon: batch and column adsorption. *Industrial & Engineering Chemistry Research*, **58**(43), 20036–20046.

Zhao N., Yang X., Zhang J., Zhu L. and Lv Y. (2017). Adsorption mechanisms of dodecylbenzene sulfonic acid by corn straw and poplar leaf biochars. *Materials*, **10**, 1119.

Zhao B., Guo Z. Q., Wang H. L., Wang L., Qian Y. R., Long X. L., Ma C., Zhang Z. H., Li J. J. and Zhang H. W. (2021). Enhanced water permeance of a polyamide thin-film composite nanofiltration membrane with a metal-organic framework interlayer. *Journal of Membrane Science*, **625**, Article 119154, 1–9.

Zhou J., Liu Y., Han Y., Jing F. and Chen J. (2019). Bone-derived biochar and magnetic biochar for effective removal of fluoride in groundwater: effects of synthesis method and coexisting chromium. *Water Environment Research*, **91**(7), 588–597, https://doi.org/10.1002/wer.1068

doi: 10.2166/9781789063783_0283

Chapter 15

Fungi: a veritable tool for refractory pollutant remediation

*Shivangi Mudaliar[1], Komal Agrawal[2] and Pradeep Verma[1]**

[1]*Bioprocess and Bioenergy Laboratory (BPEL), Department of Microbiology, Central University of Rajasthan, Bandarsindri, Kishangarh, Ajmer, Rajasthan 305817, India*
[2]*Department of Microbiology, School of Bio Engineering and Biosciences, Lovely Professional University, Phagwara, Punjab 144411, India*
**Corresponding author: pradeepverma@curaj.ac.in; vermaprad@yahoo.com*

ABSTRACT

The rapidly escalating population, urbanization, and industrial development are intensifying problems related to pollution. Soil and water pollution affect the health of organisms and the environment. Remarkable results in environmental cleaning have been obtained through the practice of bioremediation. Bioremediation is a sustainable and inexpensive method for wastewater treatment. Currently, the focus is on having clean technologies that generate less waste, with maximum output. Mycoremediation has several advantages over other bioremediation tools. Fungi are abundantly found in the environment and offer a non-toxic, zero maintenance, recyclable, less invasive, and secure approach for the treatment of wastewater. Fungus and their enzyme systems have been well explored for the degradation of toxic wastes and remediation of polluted sites. The extracellular enzyme system of fungus secretes various enzymes that break down the complex substances into simpler forms. Additionally, it is an in-situ remediation approach that exploits the capability of the fungus to reuse. This chapter deals with the idea of mycoremediation and its mechanism. The role of fungal cell walls in the biosorption, bioaccumulation, and biomineralization of heavy metals has also been discussed.

Keywords: bioremediation, mycoremediation, fungi, in situ, pollutants

15.1 INTRODUCTION

Anthropological and industrial actions, such as refinery, chemical industry, pharmaceutical industry, leather industry, heavy metals, dyes, mining, agrochemicals, detergents, food industry, and so on are progressively polluting the available source of potable water. According to the World Health Organization (WHO), over 30% of all human illnesses and 40% of all morbidity worldwide are due to contaminated water (Negi & Das, 2023). Wastewater must first go through the treatment process before it is disposed of into rivers (Akhtar & Mannan, 2020). Precipitation, electrocoagulation, adsorption, zerovalent iron, ion exchange, and membrane separation are some of the physical processes used in the treatment of wastewater

(GilPavas *et al.*, 2020). On the contrary, methods have numerous downsides, including high costs, high energy requirements, and the production of hazardous components as a result of the use of chemicals (Crini & Lichtfouse, 2019). Therefore, biological methods are favored over physicochemical procedures because biological methods are cost-effective and non-toxic to the ecosystem. Various microorganisms have been widely used in the industrial sector for wastewater treatment because they are inexpensive, use less energy, and yield useful products that can be utilized in the generation of energy as well as other processes (Grelska & Noszczyńska, 2020; Rajta *et al.*, 2020). Contrarily, fungi have attracted much less attention than bacteria for wastewater treatment. Fungi can release a variety of reductive enzymes that transform wastewater organics into valuable fungal proteins and biochemicals (Grelska & Noszczyńska, 2020). The biosorption technique, which can be used, depends on proteins found on the fungal surface and other compounds to filter out hazardous metallic elements and other inorganic materials from wastewater (Dang *et al.*, 2018). In this chapter, the focus is on the efficacy of using fungus in the treatment of wastewater. These organisms are capable of generating a huge number of reductive proteins and enzymes. The hydrolysis of a variety of environmental contaminants, including dyes, pharmaceuticals, heavy metals, trace organic contaminants, and endocrine disruptors, is carried out by a variety of fungal strains.

15.2 MYCOREMEDIATION: A SUSTAINABLE ALTERNATIVE

One of the aspects of bioremediation is mycoremediation where fungi are being used to remove pollutants sustainably. The first reported fungus that was able to degrade pollutants was *Phanerochaete chrysosporium* (Bumpus & Aust, 1984). Fungi can store the contaminants in their tissue. The intricate cell wall structure improves their potential of binding different metals through various functional groups and helps in metal sequestration. Ascomycota and Basidiomycota are the commonly found phyla of fungi in heavy metal-polluted sites (Hyde *et al.*, 2019). Moreover, they can target a broad range of organic and inorganic pollutants as they possess many different intracellular and extracellular enzymes with low substrate specificity (Bhadouria *et al.*, 2020). The properties of fungi such as their enzyme system, ability to overcome certain environmental conditions, and saprophytic nature help them in the utilization of waste compounds as carbon and energy sources. Diversified metabolic capabilities have made fungi an excellent selection for the bioremediation process (Deshmukh *et al.*, 2016; Noman *et al.*, 2019). Bioremediation of heavy metals and other toxic compounds using fungi generally occurs through biosorption, biotransformation, bioaccumulation, precipitation, and sequestration (Kumar *et al.*, 2021a). Besides, the enzymatic approaches are being used for several years and the commonly released enzyme by fungi is oxidoreductase. Furthermore, ligninolytic fungi, also offer the conversion of harmful chemicals without the need of transporting contaminants across cell membranes to meet cytosolic enzymes (Park & Choi, 2020). The mycelium of fungi has filtering abilities that target specific pollutant compounds. Mycelium stimulates microbial and enzymatic activity and thus, less toxins are produced. Not only in organic pollutants or heavy metal compounds but mycoremediation has also been applied for the removal of oil spills (Roshandel *et al.*, 2021) in soil or water. The intra- and extracellular enzymes and their metabolic mechanism have proven to be efficient in the removal of pollutants from water. This approach is worthwhile and an eco-friendly alternative for the exclusion of recalcitrant pollutants (Park & Choi, 2020).

15.3 MECHANISM FOR MYCOREMEDIATION OF REFRACTORY POLLUTANTS

A sustainable way to remediate pollutants is bioremediation and fungi seem to be a good agent for this, especially in treating heavy metal-polluted water. The high metal adsorption and accumulation ability of fungi could be used. The major pollutants in wastewater are heavy metals. Fungi can change or reduce the toxicity of metal through changes in pH, bioaccumulation, and biosorption. The most important biomechanisms used by fungi for heavy metal and other pollutant degradation are metabolically passive (bioadsorption) and metabolically active (bioaccumulation, biomineralization, biotransformation, and bioprecipitation) process (Dutta & Hyder, 2017; Kumar & Dwivedi, 2021b).

15.3.1 Biosorption

The presence of different functional groups on the wall of fungi makes it an important source of biosorbent. In a metabolically independent manner, metal ions bind to negatively charged free groups in different biopolymers. It is also known as mycoadsorption in the context of fungi and occurs on the surface of the biomass. The cell wall composition of fungi consists of 80–90% of polysaccharides, lipids, and proteins with various functional groups which help in the sorption of recalcitrant pollutants (Dhankhar & Hooda, 2011). The binding of metal occurs in two steps (1) interaction of metal and chemical groups of the cell wall, (2) mineral deposition of a large number of metals. The cell wall of fungus is the foremost thing that interacts with the metal ions (Negi & Das, 2023).

Biosorption can be categorized into various headings according to the dependence on the metabolism–metabolism-dependent and -independent process. Besides, the site from where the metal is removed, on that basis it can be classified as (a) cell surface sorption (occurs on the surface of fungi by physicochemical interaction, it is reversible) (b) intracellular accumulation (transportation of metals across the membrane occurs in viable cells only) (c) extracellular accumulation (uptake of metal takes place on the surface and the solution) (Dhankhar & Hooda, 2011) (Figure 15.1).

Biosorption is widely used for heavy metal removal due to its low cost, eco-friendly nature, easy handling, removing the smallest amount of toxic compounds, and high biomass production (Sharma *et al.*, 2018). The fungi used for biosorption are mostly non-pathogenic. The procedure is affected by several factors such as pH, inoculum amount, temperature (to a certain extent), and metal ion used. Amidst all other mechanisms for metal remediation, biosorption has come out as a promising tool as compared to the existing expensive methods. Mathur *et al.* (2018) evaluated the performance of *Aspergillus lentulus* and *Aspergillus fumigatus* for the decolorization of dye using bioaccumulation and biosorption and found that the biosorption followed by bioaccumulation was successfully removing reactive remazol red dye with 76% removal rate (Mathur *et al.*, 2018). Sawicka (2019), with the help of *Aspergillus* sp. and *Saccharomyces* sp. removed Pt (IV) and Pd (II). They found that the maximum competence of Pt (IV) on yeast and fungi was attained at acidic pH (pH 2.0) and pH 2.5–3.5 for Pd (II). *Aspergillus* sp. shows a higher adsorption capability of 5.49 mg/g for Pt (IV) and 4.28 mg/g for Pd (II). The study showed that fungi can eradicate the ions of metals from polluted water (Sawicka, 2019).

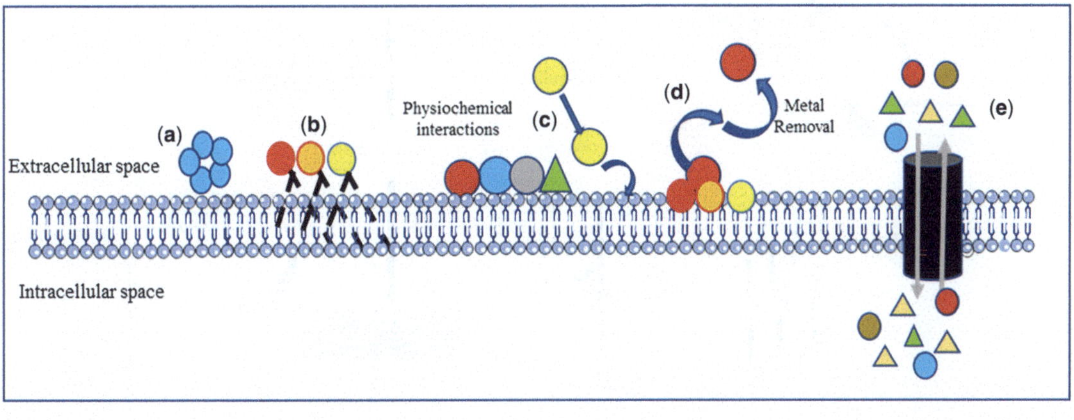

Figure 15.1 Pictorial representation of various categories of biosorption in the fungus cell wall: (a) mineral deposition, (b) metal and chemical interaction, (c) cell surface sorption, (d) extracellular accumulation, and (e) intracellular accumulation.

15.3.2 Bioaccumulation

Bioaccumulation is a metabolically energy-driven procedure where microbes with the help of a translocation system uptakes the heavy metal into their intracellular space. Once inside the intracellular space, the heavy metals can be sequestered by proteins and peptide ligands (Mishra & Malik, 2013). In the process of bioaccumulation, physical or chemical damage to the cell wall and lipid membrane is needed to acquire the heavy metals, thus the cells cannot be recycled. The process of bioaccumulation is followed by biosorption on the cell wall then the compounds are transported to intracellular space by transformation with the help of enzymes and ATP. With the use of energy, metals are transported through the cell membrane inside the cell. Bioaccumulation depends on the biological and chemical properties of the cell, internal structure, genetic adaptation of the organism, availability of metal, changes in the metal and toxicity (Vašinková *et al.*, 2021). Vašinková *et al.* (2021) compared the ability of *Aspergillus* genus from the affected site of Ostramo Lagoons (Ostrava, Czech Republic). The experimentation was done on six isolated fungal species. The media was supplemented with Cu (II), Ni (II), and Cr (III) and were separately added with spores of these fungi. It was found that the accumulation capacity of *A. ochraceus* exhibited the highest efficiency of 57.42, 56.88, and 37.73% for Cu (II), Zn (II), and Cr (III), respectively (Figure 15.2).

15.3.3 Biotransformation

In biotransformation, enzymes isolated from microorganisms are used for the conversion of toxic compounds. It is a harmless approach that produces nontoxic products, and leads to the almost complete removal of toxic compounds and is thus recognized as a better mechanism to degrade contaminants. Different microorganisms such as bacteria, plants, and fungi isolated have been used for biotransformation. The biotransformation converts toxic compounds into nontoxic by various conversion methods such as hydroxylation (addition of hydroxyl group to the molecule in the presence of hydroxylases which often increases the polarity and reduces toxicity), oxidation–reduction (oxidation of hydroxyl groups and reduction of carbonyl groups), de-epoxidation (helps in opening the epoxide ring which ultimately reduces

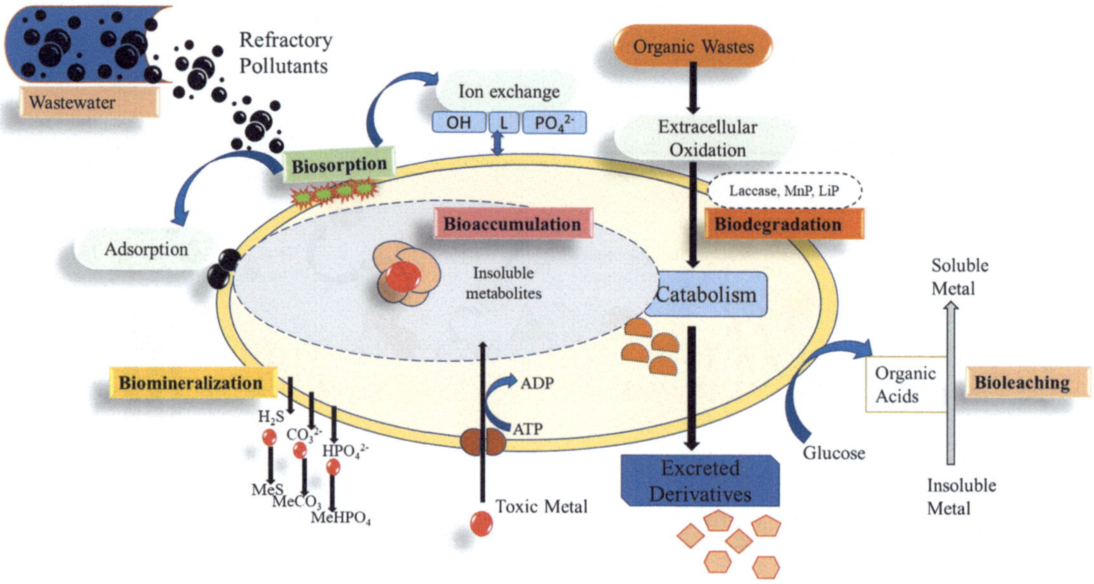

Figure 15.2 Various paths for the degradation of refractory pollutants through different biological mechanisms in fungi.

the toxicity of the compound), hydrogenation (reduction of C = C double bond by addition of hydrogen), glucuronidation/glycosylation (the process in which glucuronic acid or glucose is covalently attached to hydroxyl group), deamination (removal of amino group), demethylation (removal of methyl group), and sulfation (sulfotransferase-catalyzed conjugation of a sulfo group on a hydroxyl group) (Li *et al.*, 2020). About 14 major transformed products were identified in experiments from metoprolol biodegradation with the use of three fungi. Mainly hydroxylation, oxidation, and *o*-dealkylation mechanism were used and TP238, TP282A, TP284, TP300, TP316, and TP134 were the major compounds obtained. It was found that the maximum competence was attained through *Ganoderma lucidum* up to 51 and 77% for metoprolol and metoprolol acid, respectively (Jaén-Gil *et al.*, 2019). Also, in some reports biotransformation is done using biocatalysts. García-Morales *et al.* (2018), reported that laccase enzyme released by *Pycnoporus sanguineus* CS43 can efficiently transform pharmaceutically toxic pollutants acetaminophen (ACE) and diclofenac (DCF). ACE and DCF showed 90 and 50% of transformation, respectively, by free and immobilized laccase detoxification through biotransformation, which could be an important and sustainable approach for bioremediation (García-Morales *et al.*, 2018).

15.3.4 Biomineralization

Biomineralization is the process where microorganisms are used for the conversion of biomacromolecules into inorganic minerals. It differs from conventional mineralization using the participation of cells, metabolite, and organic matrix. Biomineralization has a significant role in the elimination of heavy metals from polluted location (Zhao *et al.*, 2020). The participation of fungi in this method is very well established. The process can be active or passive. Different types of biomineralization identified are: (a) biologically controlled mineralization (biologically controlled growth and morphology), (b) biologically induced mineralization (induces changes in the environment), (c) biologically influenced mineralization (biological matrix starts mineral nucleation) (Planý *et al.*, 2021).

15.4 ROLE OF FUNGAL ENZYMES IN MYCOREMEDIATION

The self-purification ability of ecosystems is greatly enhanced by the ability of the fungus to break down a wide variety of natural substances, and these abilities can be utilized to create ecological biotechnologies. Fungi are natural decomposers of the ecosystem as they release several enzymes naturally. These enzymes such as catalases, lipases, laccases, peroxidases, xylanases, and proteases are being used for bioremediation. Fungal enzymes are capable of degrading organic as well as inorganic compounds. Fungi produce ligninolytic enzymes that can hydrolyze lignin and are capable to break the toxic compounds. The enzyme can hydrolyze the chromophore structure of reactive dyes such as azo dyes (Sheam *et al.*, 2021), can transform heavy metals and metalloids (Kumar & Dwivedi, 2020), and can effectively degrade persistent organic pollutants (Šrédlová *et al.*, 2019) (Table 15.1).

15.4.1 Laccase

Laccases utilize molecular oxygen as a co-substrate in the reaction in addition to copper as a co-factor (Viswanath *et al.*, 2014). These oxidoreductases oxidize diphenols and other similar chemicals by accepting electrons from molecular oxygen. Most phenolic and non-phenolic compounds can be oxidized by laccases; the fungus *Trametes versicolor* has an activity that is 20 times greater than that of other species. The lignin is broken down by fungal laccases (Agrawal *et al.*, 2019; Agrawal & Verma, 2020b), which also remove any potentially dangerous phenols that are released. Additionally, dihydroxy naphthalene melanins, which are dark-colored polymers produced by species unaffected by ecological pressure, are thought to be released by fungal laccases. A varied range of applications for fungal laccases exist including pulp bleaching detoxification and effluent decolorization (Agrawal and Verma 2019a, 2019b, 2020a), phenolic removal in wines, organic synthesis biosensors, hindering dye transfer functions in washing powders and detergents, and the synthesis of composite medical compounds, a number of them are patented (Senthivelan *et al.*, 2016). Laccases have low substrate specificity that

Table 15.1 Fungal strains in the treatment of different wastewater contaminants using various approaches.

	Microbial Agents	Mechanism of Removal	Wastewater Chemical	Conditions optimization	pH	Temp. (°C)	Decolorization Efficiency (%)	References
Dyes	*Pleurotus ostreatus* HAUCC 162	Decolourization	Malachite Green	100 mg/L, 24 h	6	25	91.5	Zhuo *et al.* (2019)
	Pleurotus pulmonarius		Malachite green	100 mg/L, 36 h	–	–	68.6	Lallawmsanga Leo *et al.* (2019)
	Aspergillus iizukae 605EAN		Remazol Brilliant Blue R	100 mg/L, 5 days	7	–	90.43	Noman *et al.* (2020)
	Cyberlindnera fabianii		AR14	50 mg/L, 12 h	5	–	97	Danouche *et al.* (2021)
	Mucor hiemalis	Detoxification	Textile effluent	100 mg/L, 60 h	–	35	91.35	Molla and Khan (2018)
	Trametes versicolor	Biodegradation	Reactive blue 19	200 mg/L, 210 min	4	50	85	Dauda and Erkurt (2020)
	Aspergillus salinarus		Reactive Red HE7B	50 mg/L, 96 h	–	30	97.41	Sheam *et al.* (2021)
	Gloeophyllum trabeum		Methyl orange	75 mg/L, 28 days	–	30	46.67	Purnomo *et al.* (2020)
	Pichia kudriavzevii CR-Y103	Biodegradation and detoxification	Reactive orange 16	1 g/L,160 rpm, 24 h	6	30	100	Rosu *et al.* (2018)
Metals	*Paecilomyces* sp.	Biosorption	Co (II)	5 g, 100 rpm	5	28	100	Cárdenas González *et al.* (2019)
	Talaromyces sp. KM-31		Arsenic	1000 mg/L	5.3	28	>70	Nam *et al.* (2019)
	Penicillium polonicum		Lead (II)	4 mmol/L, 180 rpm	–	30	90.3	Xu *et al.* (2020b)
	Aspergillus flavus CR500	Bio reduction	Hexavalent chromium	50 mg/L	5–9	20–40	99	Kumar and Dwivedi (2019)
	Trichoderma lixii CR700	Bioaccumulation	Copper	10 mg/L, 120 h	5–8	–	84.6	Kumar and Dwivedi (2021a)
	Trichoderma fasciculatum and Trichoderma longibrachiatum	Bioadsorption	Cadmium	20 mg/L, 120 h	5	30	67.10 76.25	Kumar *et al.* (2020)
	Rhizopus oryzae		Lead (II) Cadmium (II)	200 rpm	6	–	94.25 93.38	Naeimi *et al.* (2018)

Agrochemicals	*Mucor racemosus* DDF	Biodegradation	Dieldrin	10 days	–	25	90	Kataoka (2018)
	Trametes versicolor		Diuron Bentazon	0.25 mL/g, 27 days	4.5	25	93 90	Beltrán-Flores *et al.* (2021)
	Rhizopus nodus *Aspergillus fumigatus* *Penicillium citreonigum*		Diazinon	35 mg/L, 21 days	–	28	91.1 76.4 72.2	Jahin *et al.* (2020)
	Trametes hirsuta MTCC-1171		Chlorpyrifos	150 mg/L, 150 rpm	6	30	>95	Patil *et al.* (2021)
	Serratia liquefaciens	Biotransformation	Kraft lignin	1000 mg/L, 120 rpm	–	30	65	Singh *et al.* (2019)
Organic compounds	*Phlebia acanthocystis* TMIC34875	Biodegradation	Pentachlorophenol	–	–	–	100	Xiao and Kondo (2020)
	Pleurotus sajor-caju LBM 105		Polychlorine biphenyl	217 mg/L		–	97.7	Sadañoski *et al.* (2020)
	Aspergillus ochraceus NRRL 3174		Petroleum	1 g/L, 7 days	5	–	80	Bilen Ozyurek *et al.* (2021)
	Pleurotus ostreatus *Pleurotus eryngii*		Trichloroethylene	140 mg/L	–	–	100	Mayans *et al.* (2021)
	Fusarium sp.		PAH	–	–	–	>80	Gupta *et al.* (2021)

lets them oxidize a variety of xenobiotic substances, such as chlorinated phenolics (Deng *et al.*, 2022), pesticides (Zeng *et al.*, 2017), and polycyclic aromatic hydrocarbons (Xu *et al.*, 2020a). Laccases have also been found to degrade polycyclic aromatic hydrocarbons, which are created by fossil fuels and natural oil sources. Laccase derived from Coriolopsis gallica UAMH8260 can oxidize carbazole, n-ethyl carbazole, fluorine, and dibenzothiophene (Bressler *et al.*, 2000). Laccases produced by fungi have various benefits that make them particularly interesting while using biotechnology to clean industrial wastewater. They can be used to decontaminate a number of agricultural goods, including coffee pulp, olive mill wastes, and bleaching kraft pulp, as per their adaptable substrate specificity. Both of these processes are possible. Laccase synthesis in fungi is nitrogen sensitive. Recombinants and nitrogen levels have a close relationship. Homologous or heterologous methods can be used to generate it (Viswanath *et al.*, 2014).

15.4.2 Catalase

It is a broadly dispersed catalyst that can split hydrogen peroxide into oxygen and water. It is a component of oxygen-dependent organisms and has a protective role against the adverse effects of reactive oxygen species (ROS), such as hydroxyl radicals and super-oxide anions (Vara & Karnena, 2020). Mono or bifunctional catalase enzymes are produced by fungi to combat the toxicity of ROS, which breaks down macromolecules in biological systems. ROS is broken down into harmless products by these enzymes. It has been demonstrated that the occurrence of heavy metals, including copper, cadmium, lead, and zinc causes bacteria to produce more ROS. In the consortium of *Aspergillus niger*, *Penicillium* sp., and *Rhizopus* sp. the catalase activity was found to be high when exposed to heavy metals including Pb and Cu at a concentration of 50 mg/L. It has been shown by Lin *et al.* (2009), that to precisely evaluate the remediation potential of fungi catalase activity may be utilized, and as the amount of oil increased in polluted environments this activity declines. The usage of catalase is an efficient technique for providing fungi with the tolerance abilities essential for the removal of contaminants from the ecosystem. As a result, fungal enzymes are essential elements in the process of bioremediation of areas polluted with metals (Negi & Das, 2023).

15.4.3 Peroxidase

Peroxidases are oxidoreductases that catalyze a variety of reactions such as the oxidation of numerous organic and inorganic molecules as well as the reduction of peroxides like hydrogen peroxide. These heme proteins have the protoporphyrin IX iron (III) prosthetic group. Three distinct categories of peroxidases can be identified based on their corresponding sources and activities: lignin peroxidase (LiP), manganese peroxidase (MnP), and versatile peroxidase (VP). LiP and MnP are two heme peroxidases that are dependent on the presence of hydrogen peroxidase and manganese to carry out their functions (Sellami *et al.*, 2022). This enzyme plays a role in the breakdown of potentially hazardous compounds (Zhou *et al.*, 2013). The VP enzymes have wide substrate specificity and can oxidize both phenolic and non-phenolic substances as they have three active oxidation sites. Initially, a diffusible mediator converts manganese (II) to manganese (III), followed by a heme-dependent binding pocket with a low redox potential, and finally, an electron transfer pathway connecting the surface-reactive tryptophan radical to the heme with a high redox potential (Barber-Zucker *et al.*, 2022). LiP and MnP have been the focus of significant research because of their ability to hydrolyze a wide range of industrial wastes, such as a complex and diverse mix of organic and hazardous chemicals.

15.5 DEGRADATION OF POLLUTANTS USING MYCOREMEDIATION

Fungi have shown some outbreaking outcomes in providing redressal for various refractory pollutants through biosorption and bioaccumulation procedures. The previous proofs or details of the degradation of refractory pollutants have been discussed in the below-mentioned sections (Table 15.2).

Table 15.2 Fungal strains and enzymes in bioremediation of refractory pollutants.

Fungal Strains	Type of Pollutants	Pollutants	Enzymes	Optimized Conditions	Degradation Efficiency (%)	References
Trametes versicolor	PAH	Chrysene Benzo pyrene	LAC	30 °C, 7 days	87.9 38.8	Vipotnik *et al.* (2022)
Phialophora fastigiata		Naphthalene	LAC	20 °C, pH 3	38	Simanjorang and Subowo (2018)
Lasiodiplodia theobromae		Benzo pyrene	LiP, LAC	–	–	Cao *et al.* (2020)
Coriolopsis byrsina		Phenanthrene	LiP, LAC, MnP	50 mg/L, 25°C, pH 6	99.90	Agrawal *et al.* (2018)
Byssochlamys spectabilis and *Aspergillus fumigates*		Chlorpyrifos	LAC	20 mg/L, 28°C, 150 rpm	98.4	Kumar *et al.* (2021b)
Trametes versicolor	Polar pesticide	Malathion	Cytochrome P450	25°C, 135 rpm, 7 days	100	Hu *et al.* (2022)
Coriolopsis gallica	Phenolics	Phenol *p*-Chlorophenol Bisphenol	LAC	60°C, pH 6–8	90.78 93.26 99.66	Cen *et al.* (2022)
Trichoderma citrinoviride AJAC3		17 β-estradiol	MnP, LiP	200 mg/L, 28°C, 120 rpm	99.6	Chatterjee and Abraham (2019)
Ganoderma applanatum and *Laetiporus sulphureus*	NSAID	Celecoxib Diclofenac ibuprofen	MnP, LiP, LAC	–	98 96 95	Bankole *et al.* (2020)
Pleurotus ostreatus	Triclosan	Triclosan	LAC MnP	–	90	Maadani Mallak *et al.* (2020)
Aspergillus arcoverdensis SSSIHL-01	Azo dye	Congo red	LAC LiP MnP	100 mg/L 30°C, 48 h	98.61	Skanda *et al.* (2023)

LAC, laccase, LiP, lignin peroxidase, MnP, manganese peroxidase.

15.5.1 Organic pollutants

The deliberate manufacturing of chemicals used in the industrial process, disease control, and agriculture is responsible for the release of environmentally degrading organic pollutants including dioxins, dichloro diphenyl trichloroethane (DDT), furan, phenol, pyridine, dimethyl phthalate (DMP), and trichloroethylene in the environment. The remaining of these pollutants in the environment is an alarming situation for human as well as wildlife health. Some commonly detected chemicals such as polycyclic aromatic hydrocarbon (PAH) and polychlorine biphenyl (PCB) are carcinogenic, persistent hydrocarbons and can cause harmful effects. Several procedures such as chlorination, ozonation, membrane process, flocculation, solvent extraction, and biological degradation have been developed to treat these harmful pollutants (Uthayakumar *et al.*, 2022). Howbeit, these approaches are quite expensive and can generate secondary pollutants that could be more toxic than the primary ones. And thus, the focus is on biological approaches as they are more sustainable and greener. Bioremediation is used for degradation purpose as it can restore the affected area as well as mineralizes the pollutants (Haripriyan *et al.*, 2022). Various organic pollutants and their degradation using fungi have been discussed in the below sections.

15.5.2 Polycyclic aromatic hydrocarbon

Polycyclic aromatic hydrocarbon is a diverse category of toxic organic compounds having two or more fused rings of benzene. Anthropogenic and natural activities such as pyrolysis of fossil fuels, oil, petroleum, gas, and so on are the chief source of their release into the ecosystem. PAHs are carcinogenic, teratogenic, and mutagenic, that need to be considered for appropriate treatment before release into the natural environment. One such approach is bioremediation, where microorganisms are used for the degradation or mineralization of pollutants. Much literature has been reported regarding the efficiency of bacteria and fungi in the remediation of PAHs. Several fungal species such as *Aspergillus*, *Cunninghamella*, and *Penicillium* are useful in transforming PAHs classified pollutants (Al-Hawash, 2018). Torres-Farradá *et al.* (2019) worked with seven *Ganoderma* strains to check their efficiency for the breakdown of PAHs – naphthalene, fluorene, and phenanthrene. They found that these strains were capable of degrading the PAHs compounds and amongst the different *Ganoderma* strains, UH-M strain exhibited the maximum degradation potential (~100%) for three PAHs. The degradation of PAHs from fungal strains showed enhanced degradation and thus are promising candidates for the bioremediation of refractory pollutants (Torres-Farradá *et al.*, 2019). Agrawal *et al.* (2018) testified that *Ganoderma lucidum* a white-rot fungi (WRF) has shown maximum biodegradation efficiency of phenanthrene (99.65%), and pyrene (99.58%) in mineral broth medium (Agrawal *et al.*, 2018). Hadibarata and Yuniarto (2020) screened 804 isolates on plates containing Poly R-478, RBBR, syringaldehyde and guaiacol. They identified a few strains with positive reactions and showed more than 90% results with indicators. The highest activity of laccase was found in *Coriolopsis caperata* BM-172 (880 U/L, 15–20 days, and 3500–3600 mg/L). The highest degradation rate for pyrene (89.5%), anthracene (85.2%), and phenanthrene (95.6%) was detected with *Pluteus chysophaeus* BM-792 after 25 days of incubation (Hadibarata & Yuniarto, 2020). More experimental studies will be required for the remediation of PAHs from wastewater and soil using fungi and other microorganisms.

15.5.3 Polychlorinated biphenyls

They are synthetically stable chlorinated organic pollutants and one of the most obstinate classes of xenobiotic contaminants. The physiochemical removal techniques are expensive and require a cost-effective approach for the degradation of pollutants. One such approach is bioremediation using different fungus is coming out as an effective tool for the degradation of PCBs. PCBs have low reactivity and can be stable in harsh environments and are not easily degradable. The metabolism by microbial enzymes could be an effective route for the biodegradation of PCBs. Mycoremediation using wood-rot fungi strain has been proven as an efficient approach for the bioremediation of PCBs. Sadanoski *et al.* (2019) showed the efficiency of *Pleurotus sajor-caju* (*P. sajor-caju*) LBM 105 in the remediation of PCBs. It was found that after 35 days of incubation, *P. sajor-caju* LBM 105 was able to eradicate 97.7 and 91.7% of the PCBs mixture and toxicity was reduced in complex and mineral media, respectively. They added PCBs to *P. sajor-caju* LBM 105 media and found that the laccase activity enhanced three-fold on the 28th day. By cultivating them in an N-limited medium they found that mycoremediation of PCBs is a cost-effective (0.69 $/g/d) method (Sadañoski *et al.*, 2019). Šrédlová *et al.* (2019) worked with the other species of *Pleurotus* and proved their efficiency by the degradation of PCBs. They took spent oyster mushroom substrate as the inoculum and the growth substrate as well. In continuous low flow setup, the highest biodegradation efficiency (87%) was recorded, and subsequently scaled up the volume by 500 L and treated 4000 L of real polluted groundwater containing 0.1–1 μg/L of PCBs. This setup was able to eliminate 82, 80, 65, and 30–50% of di, tri, tetra, and pentachlorinated PCBs congeners, respectively. *P. ostreatus* was able to hydrolyze PCBs in the presence of other fungal species and bacteria (Šrédlová *et al.*, 2019). Fungus ability to degrade aromatic compounds, and growth capability on PCBs was an indicator that they could degrade PCBs. Other than *Pleurotus* sp. there are several other fungi strain such as *Phanerochaete chrysosporium, Trametes versicolor, Coriolopsis polyzona*, and *Grifola frondosa* was found efficient in the degradation of PCBs (Elangovan *et al.*, 2019).

15.5.4 Phenolic compounds

Industries like pulp and paper, tanning, and olive oil mill processing are the major sources of phenolic contaminants in the environment. Phenolic compounds are recalcitrant by nature and need proper treatment for their degradation. Biocatalytic strategies for the biodegradation of phenolic pollutants have gained a lot of attention. Laccases and polyphenol oxidases are a group of enzymes having low substrate specificity, and efficient oxidation potential, and hence it is highly utilized in the biodegradation of phenolic pollutants. Utilization of laccases enzyme can be a better tool for the biodegradation of phenolic pollutants in wastewater. Moreover, they are available abundantly and also have a brilliant capability to oxidize a variety of organic compounds (Cen *et al.*, 2022). Cen *et al.* (2022) studied the efficiency of yellow laccase extracted from WRF *Coriolopsis gallica* (*C. gallica*) NCULAC F1 and checked its degradation efficiency against phenolic compounds. Initially, they observed an increase in enzyme activity to 10 690 U/L. Fe^{3+} and Mn^{2+} highly stimulated the *C. gallica* Lac enzymatic (laccase) activity by 162.56 and 226.05%, respectively. The activity of the enzyme was high when it was exposed to organic reagents and inhibitors. The pollutants selected for the experiment are phenol, *p*-chlorophenol and bisphenol. In the first 60 min, the elimination rates were more than 80% and in 120 min the elimination rates of chlorophenol, phenol, and bisphenol were 93.26, 90.78, 99.66%, respectively (Cen *et al.*, 2022). This study proves how efficiently the fungal enzymes could remove the phenolic compounds from wastewater and thus, could be used in environmental treatment.

15.5.5 1,1,1-Trichloro-2,2-bis(4-chlorophenyl) ethane

1,1,1-Trichloro-2,2-bis(4-chlorophenyl) ethane (DDT) is an organochlorine pesticide that is being used extensively to control insect-borne disease and agricultural pests. They can easily enter the environment through air and water and can enter into the food chain. They get accumulated through the food chain. Though DDT is banned due to its negative impacts on the environment, its residues are still detectable. They have been shown to remain in the ecosystem mainly in the form of DDT, 1,1-dichloro-2,2-bis(4-chlorophenyl) ethane, and 1,1-dichloro-2,2-bis-(4-chlorophenyl) ethylene (DDE). To eliminate DDT in a greener and harmless way, bioremediation is the best option. WRF have been reported for the biotransformation and biodegradation of DDT from water and soil. *Ganoderma lingzhi* is one of the most significant and broadly dispersed WRFs globally and is related to the degradation of a wide variety of woods. Grizca Boelan (2018) reported that *G. lingzhi* BMC9057 is degrading DDT in the PDB medium. They found that on the 7th day of incubation, 53% of DDT was eliminated. Their ability to produce ligninolytic enzymes helps them to degrade pollutants. Also, with fungus, they added 5 and 10 ml of *Bacillus subtilis* and found that the degradation was enhanced to 71.79 and 82.30%. Here the bacteria helped in increasing the solubility of *G. lingzhi* and thus enhanced the overall DDT degradation process. Thus, the use of fungus in the degradation of DDT or other pollutants has proven to be very efficient up to a large extent and with further improvements could be the future alternative for almost every remediation method (Grizca Boelan & Setyo Purnomo, 2018).

15.6 INORGANIC POLLUTANTS

The increasing population and human activities have increased the amount of waste and ultimately resulting in the accumulation of several recalcitrant pollutants. There are various methods to degrade these pollutants but still, the contamination is increasing. The major source of water pollutant is human and natural activities such as sewage, agriculture, chemical waste from industries, volcanic eruptions, and so on. These inorganic pollutants contain many organic metal complexes, different inorganic compounds, mineral acids, cyanides, metals, metals complexes, and so on. Aquatic life is affected by the accumulation of heavy metal ions and is the reason for diverse health effects on the kidney, liver, circulatory system, nervous system, blood, gastrointestinal system, skin, and bones (Gadd, 2010). Generally, inorganic pollutants comprise of cadmium, arsenic, fluorides, mercury, aluminum, chromium, nitrites, nitrates, and lead (Yadav *et al.*, 2021). Mostly they are persistent and resistant to degradation. Many physiochemical techniques have been reported for their degradation and removal but they have significant disadvantages and thus needed a more cost-effective and efficient alternative that could not harm to the environment as well (Dhankhar & Hooda, 2011). Out of all methods, bioremediation (mycoremediation) has shown considerable positive effects that fascinate researchers. As opposed to organic pollutants, which can be mineralized into carbon dioxide and water, heavy metals can only be changed into less harmful forms, making heavy metal bioremediation more challenging. The remediation of heavy metal polluted sites has been a matter of concern for decades. The need of the hour is to have a sustainable and eco-friendly process for the recovery and degradation of these pollutants. Mycoremediation can be used for the biodegradation and elimination of metal contaminants from the water or soil. These are cost-effective and also help in the restoration of the site, and seem as a better alternative compared to conventional remediation methods (Sharma *et al.*, 2018). Several fungi were reported for the bioremediation of metal. Cárdenas González *et al.* (2019) showed biosorption of Co (II) using fungal biomasses, out of which *Paecilomyces* showed the highest metal removal efficiency, 93% at 24 h of incubation, while *Penicillium* sp. and *Aspergillus niger* (*A. niger*) showed 77.5 and 70% metal removal efficiency at the same time of incubation, respectively. Also, the same fungi showed higher efficiency to remove metals when the concentration of bioadsorbents was increased by 100% for *Paecilomyces* sp. and *Penicillium* sp. and 96.4% for *A. niger* from contaminated water (Cárdenas González *et al.*, 2019). Dey *et al.* (2020) studied the potential of *Aspergillus fumigatus* (*A. fumigatus*) for the bioremediation of metal. They found the accumulation efficiency of the fungus with different metal ions and arranged them in an increasing order. They took lindane (pesticide) and a multi-metal mixture that included Cd, Cr, Cu, Ni, Pb, and Zn. The metal removal trend was Cd (100%) > Ni (99%) > Cu (98%) > Pb (91%) > Cr (81%) in the absence of lindane. The removal efficiency depends on the metal ion characteristics and the more electropositive metal ion will be more strongly attracted to the surface and will accumulate more. When they check accumulation in the presence of co-contaminant (multi-metal and lindane-amended medium), it was Zn (98%) > Pb (95%) > Cd (63%) > Ni (46%) > Cu (37%). This decrease could be because of the screening effect by pesticide lindane. This published literature has provided a comprehensive evidence on the strategy that *A. fumigatus* adapted to remove metals (Dey *et al.*, 2020). Similarly, in a study, *Trichoderma lixii* CR700 was investigated for its Cu^{2+} removal efficiency. In batch study, CR700 was able to remove 84.6% of Cu^{2+} at the concentration of 10 mg/L in 120 h. The CR700 uses simultaneous sorption and accumulation mechanism in the removal of Cu^{2+} (Kumar & Dwivedi, 2021a). One report showed the *in vitro* bioadsorption of Cd^{2+} ions using *Trichoderma fasciculatum* (*T. fasciculatum*) and *Trichoderma longibrachiatum* (*T. longibrachiatum*) fungal species. Under the optimized condition, *T. fasciculatum* showed a 67.10% removal rate, while *T. longibrachiatum* showed 76.25% removal of Cd^{2+} ions at pH 5, 120 h incubation, and 30°C. The morphological changes were detected by scanning electron microscope (SEM) and Fourier transform infrared spectroscopy (FTIR) and found that metal ions are binding efficiently on the surface of the fungus (Kumar *et al.*, 2020). This information helps in the optimization of the process and further developments to design remediation systems in the field safely and sustainably.

15.7 CHALLENGES AND PROSPECTS

Studies over the years have demonstrated that waste and harmful pollutants can be successfully remediated by either enhancing the local microbial population or by adding the necessary fungal strains to the soil. To have an accurate understanding of mycoremediation, it is essential to have a comprehensive knowledge of their degradation and detoxification capacity. For a larger-scale application, this technology still needs to be done with a lot of labor and effort. Subsequently, the process is successful, and thus it requires proper funding so that certain products can be developed and made accessible for licensing and commercialization. Scientists believe that the method could be more effective than other approaches because it is faster and inexpensive, although the full potential of fungi environmental cleanup is yet remaining underexploited. This presents a tremendous opportunity for the full use of nature's most potent decomposer in the future. The scientific community has to pay more attention to and put more effort into understanding the fundamental mechanism of fungi in the degradation of toxic compounds. To comprehend this mechanism, one can use contemporary methods like functional and full proteome analyses, which may reveal many genes and proteins that are essential to the mycoremediation process. With the use of this knowledge, it may be possible to model genetically enhanced fungi with a more effective and rapid hydrolyzing capacity. Additionally, there is a continual need to identify and characterize fungal strains with a better capacity to break down and detoxify different organic and inorganic pollutants (Pandey *et al.*, 2021).

15.8 CONCLUSION

The elimination and degradation of refractory contaminants through mycoremediation have shown promising activity. Mycoremediation is an eco-friendly and efficient way to remove pollutants from contaminated water. The mycelial structure of fungus has made it advantageous over bacteria and plants in the degradation process. Most of the studies focussed on the enzymatic system for the degradation of contaminants. But still, in the majority of cases, the basic machinery of mycoremediation needs further research and proteomic studies could help in this aspect. This information will help to develop genetically enhanced fungi that can efficiently degrade pollutants. Further studies are required to address the problems related to this method for the improvement of the remediation process. The evidences support the fact that mycoremediation can be a game changer in the removal of refractory pollutants in the coming future.

REFERENCES

Agrawal N., Verma P. and Shahi S. K. (2018). Degradation of polycyclic aromatic hydrocarbons (phenanthrene and pyrene) by the ligninolytic fungi *Ganoderma lucidum* isolated from the hardwood stump. *Bioresources and Bioprocessing*, **5**(1), 11–20, https://doi.org/10.1186/s40643-018-0197-5

Agrawal K. and Verma P. (2019a). Biodegradation of synthetic dye Alizarin Cyanine Green by yellow laccase producing strain Stropharia sp. ITCC-8422. *Biocatalysis and Agricultural Biotechnology*, **21**, 101291, https://doi.org/10.1016/j.bcab.2019.101291

Agrawal K. and Verma P. (2019b). Column bioreactor of immobilized Stropharia sp. ITCC 8422 on natural biomass support of L. cylindrica for biodegradation of anthraquinone violet R. *Bioresource Technology Reports*, **8**, p. 100345, https://doi.org/10.1016/j.biteb.2019.100345

Agrawal K., Bhardwaj N., Kumar B., Chaturvedi V. and Verma P. (2019). Process optimization, purification and characterization of alkaline stable white laccase from Myrothecium verrucaria ITCC-8447 and its application in delignification of agroresidues. *International Journal of Biological Macromolecules*, **125**, 1042–1055, https://doi.org/10.1016/j.ijbiomac.2018.12.108

Agrawal K. and Verma P. (2020a). Potential removal of hazardous wastes using white laccase purified by ATPS–PEG–salt system: an operational study. *Environmental Technology & Innovation*, **17**, p. 100556, https://doi.org/10.1016/j.eti.2019.100556

Agrawal K. and Verma P. (2020b). Production optimization of yellow laccase from Stropharia sp. ITCC 8422 and enzyme-mediated depolymerization and hydrolysis of lignocellulosic biomass for biorefinery application. *Biomass Conversion and Biorefinery*, pp. 1–20, https://doi.org/10.1007/s13399-020-00869-w

Akhtar N. and Mannan M. A. U. (2020). Mycoremediation: an unexplored gold mine. In: New and Future Developments in Microbial Biotechnology and Bioengineering: Recent Advances in Application of Fungi and Fungal Metabolites: Environmental and Industrial Aspects, J. Singh and P. Gehlot (eds). Elsevier BV, Amsterdam, Netherlands, https://doi.org/10.1016/B978-0-12-821007-9.00002-4

Al-Hawash A. B. (2018). Fungal degradation of polycyclic aromatic hydrocarbons. *International Journal of Pure & Applied Bioscience*, **6**(2), 8–24, https://doi.org/10.18782/2320-7051.6302

Bankole P. O., Adekunle A. A., Jeon B. H. and Govindwar S. P. (2020). Novel cobiomass degradation of NSAIDs by two wood rot fungi, *Ganoderma applanatum* and *Laetiporus sulphureus*: ligninolytic enzymes induction, isotherm and kinetic studies. *Ecotoxicology and Environmental Safety*, **203**, 110997, https://doi.org/10.1016/j.ecoenv.2020.110997

Barber-Zucker S., Mindel V., Garcia-Ruiz E., Weinstein J. J., Alcalde M. and Fleishman S. J. (2022). Stable and functionally diverse versatile peroxidases designed directly from sequences. *Journal of the American Chemical Society*, **144**(8), 3564–3571, https://doi.org/10.1021/jacs.1c12433

Beltrán-Flores E., Sarrà M. and Blánquez P. (2021). Pesticide bioremediation by *Trametes versicolor*: application in a fixed-bed reactor, sorption contribution and bioregeneration. *Science of the Total Environment*, **794**, 148386, https://doi.org/10.1016/j.scitotenv.2021.148386

Bhadouria R., Das S., Kumar A., Singh R. and Singh V. K. (2020). Mycoremediation of agrochemicals. In: Agrochemicals Detection, Treatment and Remediation: Pesticides and Chemical Fertilizers, M. N. Vara Prasad (ed.), Butterworth-Heinemann, United Kingdom. LTD, https://doi.org/10.1016/B978-0-08-103017-2.00022-2

Bilen Ozyurek S., Avcioglu N. H. and Seyis Bilkay I. (2021). Mycoremediation potential of *Aspergillus ochraceus* NRRL 3174. *Archives of Microbiology*, **203**(10), 5937–5950, https://doi.org/10.1007/s00203-021-02490-5

Bressler D. C., Fedorak P. M. and Pickard M. A. (2000). Oxidation of carbazole, N-ethylcarbazole, fluorene, and dibenzothiophene by the laccase of *Coriolopsis gallica*. *Biotechnology Letters*, **22**(14), 1119–1125, https://doi.org/10.1023/A:1005633212866

Bumpus J. A. and Aust S. D. (1984). Bumpus_et_al-1987-BioEssays. **6**(4), 166–170.

Cao H., Wang C., Liu H., Jia W. and Sun H. (2020). Enzyme activities during Benzo[a]pyrene degradation by the fungus *Lasiodiplodia theobromae* isolated from a polluted soil. *Scientific Reports*, **10**(1), 1–11, https://doi.org/10.1038/s41598-020-57692-6

Cárdenas González J. F., Rodríguez Pérez A. S., Vargas Morales J. M., Martínez Juárez V. M., Rodríguez I. A., Cuello C. M., Fonseca G. G., Escalera Chávez M. E. and Muñoz Morales A. (2019). Bioremoval of cobalt(ii) from aqueous solution by three different and resistant fungal biomasses. *Bioinorganic Chemistry and Applications*, **2019**, 1–8, https://doi.org/10.1155/2019/8757149

Cen Q., Wu X., Cao L., Lu Y., Lu X., Chen J., Fu G., Liu Y. and Ruan R. (2022). Green production of a yellow laccase by *Coriolopsis gallica* for phenolic pollutants removal. AMB Express, **12**, 96–111, https://doi.org/10.1186/s13568-022-01434-6

Chatterjee A. and Abraham J. (2019). Mycoremediation of 17 β-Estradiol using *Trichoderma citrinoviride* strain AJAC3 along with enzyme studies. *Environmental Progress and Sustainable Energy*, **38**(4), 1–9, https://doi.org/10.1002/ep.13142

Crini G. and Lichtfouse E. (2019). Advantages and disadvantages of techniques used for wastewater treatment. *Environmental Chemistry Letters*, **17**(1), 145–155, https://doi.org/10.1007/s10311-018-0785-9

Dang C., Yang Z., Liu W., Du P., Cui F. and He K. (2018). Role of extracellular polymeric substances in biosorption of Pb^{2+} by a high metal ion tolerant fungal strain *Aspergillus niger* PTN31. *Journal of Environmental Chemical Engineering*, **6**(2), 2733–2742. Elsevier BV, https://doi.org/10.1016/j.jece.2018.04.005

Danouche M., Ferioun M., Bahafid W. and El Ghachtouli N. (2021). Mycoremediation of azo dyes using *Cyberlindnera fabianii* yeast strain: application of designs of experiments for decolorization optimization. *Water Environment Research*, **93**(8), 1402–1416, https://doi.org/10.1002/wer.1499

Dauda M. Y. and Erkurt E. A. (2020). Investigation of reactive Blue 19 biodegradation and byproducts toxicity assessment using crude laccase extract from *Trametes versicolor*. *Journal of Hazardous Materials*, **393**, 121555, https://doi.org/10.1016/j.jhazmat.2019.121555

Deng W., Zhao W. and Yang Y. (2022). Degradation and detoxification of chlorophenols with different structure by LAC-4 laccase purified from white-rot fungus *Ganoderma lucidum*. *International Journal of Environmental Research and Public Health*, **19**(13), 8150–8191, https://doi.org/10.3390/ijerph19138150

Deshmukh R., Khardenavis A. A. and Purohit H. J. (2016). Diverse metabolic capacities of fungi for bioremediation. *Indian Journal of Microbiology*, **56**(3), 247–264, https://doi.org/10.1007/s12088-016-0584-6

Dey P., Malik A., Mishra A., Kumar D., Von Bergen M. and Jehmlich N. (2020). Mechanistic insight to mycoremediation potential of a metal resistant fungal strain for removal of hazardous metals from multimetal pesticide matrix. *Environmental Pollution*, **262**, 114255, https://doi.org/10.1016/j.envpol.2020.114255

Dhankhar R. and Hooda A. (2011). Fungal biosorption – an alternative to meet the challenges of heavy metal pollution in aqueous solutions. *Environmental Technology*, **32**(5), 467–491, https://doi.org/10.1080/09593330.2011.572922

Dutta S. D. and Hyder S. (2017). Mycoremediation – a potential tool for sustainable management. *Journal of Mycopathological Research*, **57**(1), 25–34.

Elangovan S., Pandian S. B. S., Geetha S. J. and Joshi S. J. (2019). Polychlorinated biphenyls (PCBs): environmental fate, challenges and bioremediation. In: Microbial Metabolism of Xenobiotic Compounds, P. K. Arora (ed.), Springer Nature, Singapore, pp. 165–188, https://doi.org/10.1007/978-981-13-7462-3_8

Gadd G. M. (2010). Metals, minerals and microbes: geomicrobiology and bioremediation. *Microbiology*, **156**(3), 609–643, https://doi.org/10.1099/mic.0.037143-0

García-Morales R., García-García A., Orona-Navar C., Osma J. F., Nigam K. D. P. and Ornelas-Soto N. (2018). Biotransformation of emerging pollutants in groundwater by laccase from *P. sanguineus* CS43 immobilized onto titania nanoparticles. *Journal of Environmental Chemical Engineering*, **6**(1), 710–717, https://doi.org/10.1016/j.jece.2017.12.006

GilPavas E., Dobrosz-Gómez I. and Gómez-García M. Á. (2020). Efficient treatment for textile wastewater through sequential electrocoagulation, electrochemical oxidation and adsorption processes: optimization and toxicity assessment. *Journal of Electroanalytical Chemistry*, **878**, 114578, https://doi.org/10.1016/j.jelechem.2020.114578

Godlewska-Żyłkiewicz B., Sawicka S. and Karpińska J. (2019). Removal of platinum and palladium from wastewater by means of biosorption on fungi Aspergillus sp. and yeast Saccharomyces sp. *Water (Switzerland)*, **11**(7). https://doi.org/10.3390/w11071522

Grelska A. and Noszczyńska M. (2020). White rot fungi can be a promising tool for removal of bisphenol A, bisphenol S, and nonylphenol from wastewater. *Environmental Science and Pollution Research*, **27**(32), 39958–39976, https://doi.org/10.1007/s11356-020-10382-2

Grizca Boelan E. and Setyo Purnomo A. (2018). Abilities of co-cultures of white-rot fungus *Ganoderma lingzhi* and bacteria *Bacillus subtilis* on biodegradation DDT. *Journal of Physics: Conference Series*, **1095**(1), 1–8, https://doi.org/10.1088/1742-6596/1095/1/012015

Gupta S., Fernandes S., Gupta H. and Rathod S. (2021). Assessment of mycoremediation potential of *Fusarium* spp. on polycyclic aromatic hydrocarbon in Western India. *International Journal of Scientific Research in Science and Technology, Mill*, **8**, 509–515, https://doi.org/10.32628/ijsrst218571

Hadibarata T. and Yuniarto A. (2020). Biodegradation of polycyclic aromatic hydrocarbons by high-laccase Basidiomycetes fungi isolated from tropical forest of Borneo. *Biocatalysis and Agricultural Biotechnology*, **28**, 101717, https://doi.org/10.1016/j.bcab.2020.101717

Haripriyan U., Gopinath K. P., Arun J. and Govarthanan M. (2022). Bioremediation of organic pollutants: a mini review on current and critical strategies for wastewater treatment. *Archives of Microbiology*, **204**(5), 1–9, https://doi.org/10.1007/s00203-022-02907-9

Hu K., Barbieri M. V., López-García E., Postigo C., López de Alda M., Caminal G. and Sarrà M. (2022). Fungal degradation of selected medium to highly polar pesticides by *Trametes versicolor*: kinetics, biodegradation pathways, and ecotoxicity of treated waters. *Analytical and Bioanalytical Chemistry*, **414**(1), 439–449, https://doi.org/10.1007/s00216-021-03267-x

Hyde K. D., Xu J., Rapior S., Jeewon R., Lumyong S., Niego A. G. T., Abeywickrama P. D., Aluthmuhandiram J. V. S., Brahamanage R. S., Brooks S., Chaiyasen A., Chethana K. W. T., Chomnunti P., Chepkirui C., Chuankid B., de Silva N. I., Doilom M., Faulds C., Gentekaki E., … Stadler M. (2019). The amazing potential of fungi: 50 ways we can exploit fungi industrially. *Fungal Diversity*, **97**(1), 1–136, https://doi.org/10.1007/s13225-019-00430-9

Jaén-Gil A., Castellet-Rovira F., Llorca M., Villagrasa M., Sarrà M., Rodríguez-Mozaz S. and Barceló D. (2019). Fungal treatment of metoprolol and its recalcitrant metabolite metoprolol acid in hospital wastewater: biotransformation, sorption and ecotoxicological impact. *Water Research*, **152**, 171–180, https://doi.org/10.1016/j.watres.2018.12.054

Jahin H. S., Gaber S. E. and Hussain M. T. (2020). Bioremediation of diazinon pesticide from aqueous solution by fungal-strains isolated from wastewater. *World Journal of Chemistry*, **15**(1), 15–23, https://doi.org/10.5829/idosi.wjc.2020.15.23

Kataoka R. (2018). Biodegradability and biodegradation pathways of chlorinated cyclodiene insecticides by soil fungi. *Journal of Pesticide Science*, **43**(4), 314–320, https://doi.org/10.1584/jpestics.J18-03

Kumar V. and Dwivedi S. K. (2019). Hexavalent chromium reduction ability and bioremediation potential of *Aspergillus flavus* CR500 isolated from electroplating wastewater. *Chemosphere*, **237**, 124567, https://doi.org/10.1016/j.chemosphere.2019.124567

Kumar V. and Dwivedi S. K. (2020). Multimetal tolerant fungus *Aspergillus flavus* CR500 with remarkable stress response, simultaneous multiple metal/loid removal ability and bioremediation potential of wastewater. *Environmental Technology and Innovation*, **20**, 101075, https://doi.org/10.1016/j.eti.2020.101075

Kumar V. and Dwivedi S. K. (2021a). Bioremediation mechanism and potential of copper by actively growing fungus *Trichoderma lixii* CR700 isolated from electroplating wastewater. *Journal of Environmental Management*, **277**, 111370, https://doi.org/10.1016/j.jenvman.2020.111370

Kumar V. and Dwivedi S. K. (2021b). Mycoremediation of heavy metals: processes, mechanisms, and affecting factors. *Environmental Science and Pollution Research*, **28**(9), 10375–10412, https://doi.org/10.1007/s11356-020-11491-8

Kumar R., Sharma P., Umar A., Kumar R., Singh N., Joshi P. K., AlAlharthi F. A., Alghamdi A. A. and Al-Zaqri N. (2020). In vitro bioadsorption of Cd^{2+} ions: adsorption isotherms, mechanism, and an insight to mycoremediation. *Processes*, **8**(9), 1085–1100, https://doi.org/10.3390/pr8091085

Kumar A., Sharma A., Chaudhary P. and Gangola S. (2021a). Chlorpyrifos degradation using binary fungal strains isolated from industrial waste soil. *Biologia*, **76**(10), 3071–3080, https://doi.org/10.1007/s11756-021-00816-8

Kumar A., Yadav A. N., Mondal R., Kour D., Subrahmanyam G., Shabnam A. A., Khan S. A., Yadav K. K., Sharma G. K., Cabral-Pinto M., Fagodiya R. K., Gupta D. K., Hota S. and Malyan S. K. (2021b). Myco-remediation: a mechanistic understanding of contaminants alleviation from natural environment and future prospect. *Chemosphere*, **284**, 131325, https://doi.org/10.1016/j.chemosphere.2021.131325

Lallawmsanga Leo V. V., Passari A. K., Muniraj I. K., Uthandi S., Hashem A., Abd_Allah E. F., Alqarawi A. A. and Singh B. P. (2019). Elevated levels of laccase synthesis by *Pleurotus pulmonarius* BPSM10 and its potential as a dye decolorizing agent. *Saudi Journal of Biological Sciences*, **26**(3), 464–468, https://doi.org/10.1016/j.sjbs.2018.10.006

Li P., Su R., Yin R., Lai D., Wang M., Liu Y. and Zhou L. (2020). Detoxification of mycotoxins through biotransformation. *Toxins*, **12**(2), 1–37, https://doi.org/10.3390/toxins12020121

Lin X., Li X., Sun T., Li P., Zhou Q., Sun L. and Hu, X. (2009). Changes in microbial populations and enzyme activities during the bioremediation of oil-contaminated soil. *Bulletin of Environmental Contamination and Toxicology*, **83**, 542–547, https://doi.org/10.1007/s00128-009-9838-x

Maadani Mallak A., Lakzian A., Khodaverdi E., Haghnia G. H. and Mahmoudi S. (2020). Effect of *Pleurotus ostreatus* and *Trametes versicolor* on triclosan biodegradation and activity of laccase and manganese peroxidase enzymes. *Microbial Pathogenesis*, **149**, 104473, https://doi.org/10.1016/j.micpath.2020.104473

Mathur M., Gola D., Panja R., Malik A. and Ahammad S. Z. (2018). Performance evaluation of two *Aspergillus* spp. for the decolourization of reactive dyes by bioaccumulation and biosorption. *Environmental Science and Pollution Research*, **25**(1), 345–352, https://doi.org/10.1007/s11356-017-0417-0

Mayans B., Camacho-Arévalo R., García-Delgado C., Alcántara C., Nägele N., Antón-Herrero R., Escolástico C. and Eymar E. (2021). Mycoremediation of soils polluted with trichloroethylene: first evidence of *Pleurotus* genus effectiveness. *Applied Sciences (Switzerland)*, **11**(4), 1–11, https://doi.org/10.3390/app11041354

Mishra A. and Malik A. (2013). Recent advances in microbial metal bioaccumulation. *Critical Reviews in Environmental Science and Technology*, **43**(11), 1162–1222, https://doi.org/10.1080/10934529.2011.627044

Molla A. H. and Khan H. I. (2018). Detoxification of textile effluent by fungal treatment and its performance in agronomic usages. *Environmental Science and Pollution Research*, **25**(11), 10820–10828, https://doi.org/10.1007/s11356-018-1361-3

Naeimi B., Foroutan R., Ahmadi B., Sadeghzadeh F. and Ramavandi B. (2018). Pb(II) and Cd(II) removal from aqueous solution, shipyard wastewater, and landfill leachate by modified *Rhizopus oryzae* biomass. *Materials Research Express*, **5**(4), 1–30, https://doi.org/10.1088/2053-1591/aab81b

Nam I. H., Murugesan K., Ryu J. and Kim J. H. (2019). Arsenic (As) removal using *Talaromyces* sp. KM-31 isolated from As-contaminated mine soil. *Minerals*, **9**(10), 1–11, https://doi.org/10.3390/min9100568

Negi B. B. and Das C. (2023). Mycoremediation of wastewater, challenges, and current status: a review. *Bioresource Technology Reports*, **22**, 101409, https://doi.org/10.1016/j.biteb.2023.101409

Noman E., Al-Gheethi A., Mohamed R. M. S. R. and Talip B. A. (2019). Myco-remediation of xenobiotic organic compounds for a sustainable environment: a critical review. *Topics in Current Chemistry*, **377**(17), 1–41, https://doi.org/10.1007/s41061-019-0241-8

Noman E., Al-Gheethi A. A., Talip B., Mohamed R. and Kassim A. H. (2020). Mycoremediation of Remazol Brilliant Blue R in greywater by a novel local strain of *Aspergillus iizukae* 605EAN: optimisation and mechanism study. *International Journal of Environmental Analytical Chemistry*, **100**(14), 1650–1668, https://doi.org/10.1080/03067319.2019.1657852

Pandey A. K., Gaur V. K., Udayan A., Varjani S., Kim S. H. and Wong J. W. C. (2021). Biocatalytic remediation of industrial pollutants for environmental sustainability: research needs and opportunities. *Chemosphere*, **272**, 129936, https://doi.org/10.1016/j.chemosphere.2021.129936

Park H. and Choi I. G. (2020). Genomic and transcriptomic perspectives on mycoremediation of polycyclic aromatic hydrocarbons. *Applied Microbiology and Biotechnology*, **104**(16), 6919–6928, https://doi.org/10.1007/s00253-020-10746-1

Patil P. D., Singh A. A. and Yadav D. G. (2021). Biodegradation of organophosphorus insecticide chlorpyrifos into a major fuel additive 2,4-bis(1,1 dimethylethyl) phenol using white-rot fungal strain *Trametes hirsuta* MTCC-1171. *Journal of the Indian Chemical Society*, **98**(9), 100120, https://doi.org/10.1016/j.jics.2021.100120

Planý M., Pinzari F., Šoltys K., Kraková L., Cornish L., Pangallo D., Jungblut A. D. and Little B. (2021). Fungal-induced atmospheric iron corrosion in an indoor environment. *International Biodeterioration and Biodegradation*, **159**, 105204, https://doi.org/10.1016/j.ibiod.2021.105204

Purnomo A. S., Andyani N. E. A., Nawfa R. and Putra S. R. (2020). Fenton reaction involvement on methyl orange biodegradation by brown-rot fungus *Gloeophyllum trabeum*. *AIP Conference Proceedings*, **2237**, 1–6, https://doi.org/10.1063/5.0005230

Rajta A., Bhatia R., Setia H. and Pathania P. (2020). Role of heterotrophic aerobic denitrifying bacteria in nitrate removal from wastewater. *Journal of Applied Microbiology*, **128**(5), 1261–1278, https://doi.org/10.1111/jam.14476

Roshandel F., Saadatmand S., Iranbakhsh A. and Ardebili Z. O. (2021). Mycoremediation of oil contaminant by *Pleurotus florida* (P. Kumm) in liquid culture. *Fungal Biology*, **125**(9), 667–678, https://doi.org/10.1016/j.funbio.2021.04.002

Rosu C. M., Avadanei M., Gherghel D., Mihasan M., Mihai C., Trifan A., Miron A. and Vochita G. (2018). Biodegradation and detoxification efficiency of azo-dye reactive orange 16 by *Pichia kudriavzevii* CR-Y103. *Water, Air, and Soil Pollution*, **229**(1), 1–18, https://doi.org/10.1007/s11270-017-3668-y

Sadañoski M. A., Benítez S. F., Fonseca M. I., Velázquez J. E., Zapata P. D., Levin L. N. and Villalba L. L. (2019). Mycoremediation of high concentrations of polychlorinated biphenyls with *Pleurotus sajor-caju* LBM 105 as an effective and cheap treatment. *Journal of Environmental Chemical Engineering*, **7**(6), 103453–103482, https://doi.org/10.1016/j.jece.2019.103453

Sadañoski M. A., Tatarin A. S., Barchuk M. L., Gonzalez M., Pegoraro C. N., Fonseca M. I., Levin L. N. and Villalba L. L. (2020). Evaluation of bioremediation strategies for treating recalcitrant halo-organic pollutants in soil environments. *Ecotoxicology and Environmental Safety*, **202**, 110929, https://doi.org/10.1016/j.ecoenv.2020.110929

Sellami K., Couvert A., Nasrallah N., Maachi R., Abouseoud M. and Amrane A. (2022). Peroxidase enzymes as green catalysts for bioremediation and biotechnological applications: a review. *Science of the Total Environment*, **806**, 150500, https://doi.org/10.1016/j.scitotenv.2021.150500

Senthivelan T., Kanagaraj J. and Panda R. C. (2016). Recent trends in fungal laccase for various industrial applications: an eco-friendly approach – a review. *Biotechnology and Bioprocess Engineering*, **21**(1), 19–38, https://doi.org/10.1007/s12257-015-0278-7

Sharma S., Tiwari S., Hasan A., Saxena V. and Pandey L. M. (2018). Recent advances in conventional and contemporary methods for remediation of heavy metal-contaminated soils. *3 Biotech*, **8**(4), 1–18, https://doi.org/10.1007/s13205-018-1237-8

Sheam M. M., Biswas S. K., Ahmed K. R., Syed S. B., Hossain M. S., Khan M. S. A., Hasan M. R., Zohra F. T. and Rahman M. M. (2021). Mycoremediation of reactive red HE7B dye by *Aspergillus salinarus* isolated from textile effluents. *Current Research in Microbial Sciences*, **2**, 100056, https://doi.org/10.1016/j.crmicr.2021.100056

Simanjorang S. W. and Subowo Y. B. (2018). The ability of soil-borne fungi to degrade polycyclic aromatic hydrocarbon (PAH). *AIP Conference Proceedings*, **2021**(2018), 030019, https://doi.org/10.1063/1.5062743

Singh A. K., Yadav P., Bharagava R. N., Saratale G. D. and Raj A. (2019). Biotransformation and cytotoxicity evaluation of kraft lignin degraded by ligninolytic *Serratia liquefaciens*. *Frontiers in Microbiology*, **10**, 1–11, https://doi.org/10.3389/fmicb.2019.02364

Skanda S., Bharadwaj P. S. J., Kar S. and Muthukumar V. S. (2023). Bioremoval capacity of recalcitrant azo dye Congo red by soil fungus *Aspergillus arcoverdensis* SSSIHL-01. *Bioremediation Journal*, **0**(0), 1–12, https://doi.org/10.1080/10889868.2021.1984198

Šrédlová K., Škrob Z., Filipová A., Mašín P. and Holecová J. (2019). Biodegradation of PCBs in contaminated water using spent oyster mushroom substrate and a trickle-bed bioreactor. Water Research, 115274, 150, https://doi.org/10.1016/j.watres.2019.115274

Torres-Farradá G., Manzano-León A. M., Rineau F., Ramos Leal M., Thijs S., Jambon I., Put J., Czech J., Guerra Rivera G., Carleer R. and Vangronsveld J. (2019). Biodegradation of polycyclic aromatic hydrocarbons by native *Ganoderma* sp. strains: identification of metabolites and proposed degradation pathways. *Applied Microbiology and Biotechnology*, **103**(17), 7203–7215, https://doi.org/10.1007/s00253-019-09968-9

Uthayakumar H., Radhakrishnan P., Shanmugam K. and Kushwaha O. S. (2022). Growth of MWCNTs from *Azadirachta indica* oil for optimization of chromium(VI) removal efficiency using machine learning approach. *Environmental Science and Pollution Research*, **29**(23), 34841–34860, https://doi.org/10.1007/s11356-021-17873-w

Vara S. and Karnena M. K. (2020). Fungal enzymatic degradation of industrial effluents – a review. *Current Research in Environmental and Applied Mycology*, **10**(1), 417–442, https://doi.org/10.5943/CREAM/10/1/33

Vašinková M., Dlabaja M. and Kučová K. (2021). Bioaccumulation of toxic metals by fungi of the genus *Aspergillus* isolated from the contaminated area of Ostramo Lagoons. *IOP Conference Series: Earth and Environmental Science*, **900**(1), 1–8, https://doi.org/10.1088/1755-1315/900/1/012048

Vipotnik Z., Michelin M. and Tavares T. (2022). Biodegradation of chrysene and benzo[a]pyrene and removal of metals from naturally contaminated soil by isolated *Trametes versicolor* strain and laccase produced thereof. *Environmental Technology and Innovation*, **28**, 102737, https://doi.org/10.1016/j.eti.2022.102737

Viswanath B., Rajesh B., Janardhan A., Kumar A. P. and Narasimha G. (2014). Fungal laccases and their applications in bioremediation. *Enzyme Research*, **2014**, 1–21, https://doi.org/10.1155/2014/163242

Xiao P. and Kondo R. (2020). Biodegradation and biotransformation of pentachlorophenol by wood-decaying white rot fungus Phlebia acanthocystis TMIC34875. *Journal of Wood Science*, **66**(1), 1–11, https://doi.org/10.1186/s10086-020-1849-6

Xu P., Du H., Peng X., Tang Y., Zhou Y., Chen X., Fei J., Meng Y. and Yuan L. (2020a). Degradation of several polycyclic aromatic hydrocarbons by laccase in reverse micelle system. *Science of the Total Environment*, **708**, 134970, https://doi.org/10.1016/j.scitotenv.2019.134970

Xu X., Hao R., Xu H. and Lu A. (2020b). Removal mechanism of Pb(II) by *Penicillium polonicum*: immobilization, adsorption, and bioaccumulation. *Scientific Reports*, **10**(1), 1–12, https://doi.org/10.1038/s41598-020-66025-6

Yadav P., Rai S. N., Mishra V. and Singh M. P. (2021). Mycoremediation of environmental pollutants: a review with special emphasis on mushrooms. *Environmental Sustainability*, **4**(4), 605–618, https://doi.org/10.1007/s42398-021-00197-3

Zeng S., Qin X. and Xia L. (2017). Degradation of the herbicide isoproturon by laccase-mediator systems. *Biochemical Engineering Journal*, **119**, 92–100, https://doi.org/10.1016/j.bej.2016.12.016

Zhao W., Zhu G., Daugulis A. J., Chen Q., Ma H., Zheng P., Liang J. and Ma X. (2020). Removal and biomineralization of Pb 2 þ in water by fungus *Phanerochaete chrysoporium*. *Journal of Cleaner Production*, **260**, 120980, https://doi.org/10.1016/j.jclepro.2020.120980

Zhou X. W., Cong W. R., Su K. Q. and Zhang Y. M. (2013). Ligninolytic enzymes from *Ganoderma* spp: current status and potential applications. *Critical Reviews in Microbiology*, **39**(4), 416–426, https://doi.org/10.3109/1040841X.2012.722606

Zhuo R., Zhang J., Yu H., Ma F. and Zhang X. (2019). The roles of *Pleurotus ostreatus* HAUCC 162 laccase isoenzymes in decolorization of synthetic dyes and the transformation pathways. *Chemosphere*, **234**, 733–745, https://doi.org/10.1016/j.chemosphere.2019.06.113

doi: 10.2166/9781789063783_0301

Chapter 16

Bioremediation: a green tool to remediate refractory pollutants

*Shivangi Mudaliar[1], Komal Agrawal[2] and Pradeep Verma[1]**

[1]Bioprocess and Bioenergy Laboratory (BPEL), Department of Microbiology, Central University of Rajasthan, Bandarsindri, Kishangarh, Ajmer, Rajasthan 305817, India

[2]Department of Microbiology, School of Bio Engineering and Biosciences, Lovely Professional University, Phagwara, Punjab 144411, India

**Corresponding author: pradeepverma@curaj.ac.in; vermaprad@yahoo.com*

ABSTRACT

The rapid escalation of pollution in the environment by anthropogenic and household activities is responsible for the contamination of various ecological niches. The result of which is an increased demand for the development of sustainable and innovative wastewater treatment technologies. Bioremediation has been regarded as the most efficient technique to remove heavy metals from polluted water and soil. A wide range of microorganisms including bacteria, fungi, and yeasts can actively modify the toxic contaminants and has a decisive part in the degradation, and mineralization of harmful substances into less lethal forms. Primarily bioremediation focuses on *ex-situ* as well as *in-situ* treatment and the ever-evolving technology that can be used to manage diverse pollutants. Thus, the chapter emphasizes on the sources of pollutants along with several bioremediation technologies for the treatment of wastewater. Additionally, new techniques such as nanotechnology, electro bioremediation, and constructed wetlands, and so forth that are being implemented to minimize the industry expenses for the treatment of wastewater have also been discussed in the chapter.

Keywords: bio-based remediation, refractory pollutants, sustainable, bioremediation

16.1 INTRODUCTION

Water is vital for the maintenance of life on earth and the accelerated expansion of industries and agriculture has polluted the ecosystem with varied contaminants, including toxic heavy metals (Kumar & Dwivedi, 2021), polychlorinated biphenyls (Sadañoski *et al.*, 2019), and various agrochemicals (Bhadouria *et al.*, 2020). Their existence in the ecosystem is of great concern, as they are toxic and non-biodegradable by nature (Kour *et al.*, 2021). Industrial wastes are directly discharged into the aquatic environment, disrupting the photosynthetic process and altering the dissolved oxygen and pH level of water. In addition, treating and managing waste has been a challenging task (Amulya *et al.*, 2016) and a major problem associated with the waste is its mineralization. Several conventional approaches have been applied to transform this waste; however, it is a tedious process, and an expensive task (De Donno Novelli *et al.*, 2021) and the production of intermediate by-products can be more damaging than the original compound resulting in more deleterious impacts. Thus, these contaminants and their by-products require proper eradication through sustainable methods along with simultaneous restoration of the polluted site. Bioremediation is cost-effective and green means for the removal of refractory pollutants by treating wastewater. The front runners of this process are the microbial community and microalgae that result in sludge and treated water or biomass formation. Particularly microbes are capable of degrading, cleansing, and even gathering toxic organic and inorganic substances. Heavy metals can enter the environment from a variety of sources, including natural, industrial solid waste, agricultural, atmospheric sources, inland effluent, and more. Most parts of the world have been contaminated by activities such as electroplating, mining, metallurgical smelting, and the use of agricultural pesticides and fertilizers. In the past two decades, bioremediation technology has developed with the ultimate aim of efficiently restoring the damaged sites in a greener way with low cost. There is not even one bioremediation method that can be used as a sure approach to restore the affected habitats, despite the fact that researchers have established and modelled various bioremediation strategies. The majority of the problems relating to the bioremediation and biodegradation of contaminating compounds can be resolved by using native microorganisms found in damaged environments, because the ecological conditions are favorable for their survival and metabolic growth (Medfu Tarekegn *et al.*, 2020). All pollutants found in the environment must be eliminated, biodegraded, or inactivated. Any environmentally friendly treatment methods that can be described and shown to be efficient, with low costs and elevated capacity to remove pollutants, will make a significant contribution to resolving the environmental crisis we are currently facing. Thus, the current chapter will confer about various sources of pollutants and their effects on the environment. This chapter gives us a glimpse on the various bioremediation technologies that could be implemented for the remediation of wastewater.

16.2 SOURCES OF REFRACTORY POLLUTANTS AND THEIR CONSEQUENCES

Refractory pollutants from textile, paper, and agricultural sectors are the major source of contributors that cause harmful effects on aquatic bodies and the environment (Wen *et al.*, 2017). In the below sections, different pollutants discharged from various sectors have been elaborated (Figure 16.1).

16.2.1 Textile and apparel industries

The textile sector is a major water-consuming industry, nearly 15% of used dyes are released along with the discharged water directly (Rajhans *et al.*, 2021). Textile wastewater is characterized by high

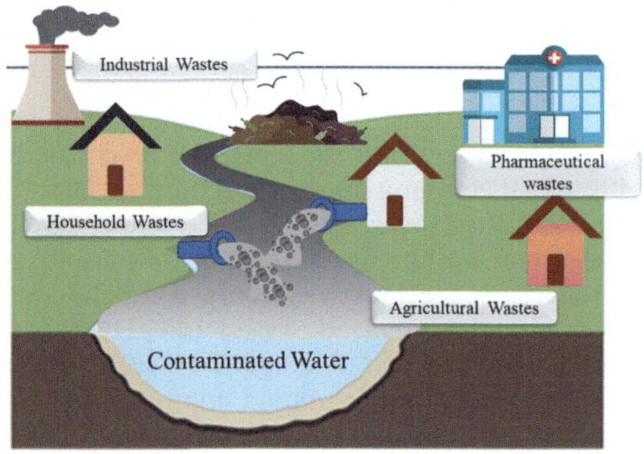

Figure 16.1 Major sources of refractory pollutants contaminating the soil and water.

pH, biochemical oxygen demand (BOD), chemical oxygen demand (COD), high dye content, total dissolved solids (TDS), total suspended particles (TSP), sulfates, and chlorides (Khan & Malik, 2018). Rajhans *et al.* (2021) showed for the first time the direct relationship between toxicity and COD, they have proved that the toxicity of wastewater was increasing with an increase in effluents COD in a very short period (Rajhans *et al.*, 2021). This contributes to the accumulation of toxic or xenobiotic pollutants in the environment that lasts for an extended duration and leads to severe health and environmental risks (Kishor *et al.*, 2021a). The most typical refractory pollutants from the textile industry are reactive azo dyes, aniline, and polyvinyl alcohol. Amidst pollutants aniline has been listed as a pollutant of concern by the United States Environmental Protection Agency (USEPA) as it is carcinogenic and mutagenic while polyvinyl alcohol being a water-soluble polymer causes the lack of dissolved oxygen. The removal of these pollutants through conventional methods is difficult due to their poor degradability (Yu *et al.*, 2022). The major contributor to textile wastewater is a reactive dye (Khan & Malik, 2018) coming out from the textile industries that contain different hazardous persistent coloring pollutants. Due to their several chromophore groups including azo (–N=N–), nitro (–N=O), carbonyl (–C=O), quinoid, and auxochrome groups like hydroxyl (–OH), carboxyl (–COOH), amine (–NH_2), and sulfonate (–SO_3H), textile dyes are aromatic, heterocyclic compounds that are persistent (Kishor *et al.*, 2021b). Compounds, such as Sudan I (azo dye) have toxic/carcinogenic character when transformed enzymatically to aromatic amines by intestinal bacteria (Piątkowska *et al.*, 2018). Synthetic dyes have been reported to cause cancer, skin diseases, and central nervous system disorders (Khan & Malik, 2018). The matrix of the textile industry's effluent is quite complex and contains a variety of aromatic compounds, color content and toxic metals, such as arsenic (As), lead (Pb), chromium (Cr), antimony (Sb), cadmium (Cd) and mercury (Hg) (Chandanshive *et al.*, 2020; Kishor *et al.*, 2021c). Khan and Malik (2018) showed that textile wastewater contains several heavy metals and organic compounds that inhibited the process of seed germination thus indicated the presence of toxic pollutants. Textile wastes contain a large number of contaminants that needs to be treated before their disposal into water. Many conventional and new treatments have been reported but their use is limited due to several drawbacks such as not being eco-friendly as well as causes harm to human health, so there is a need to opt for a biological approach. Biological tools have emerged as an effective way to degrade and detoxify textile industrial wastewater (Khandare & Govindwar, 2015; Kishor *et al.*, 2021a). Bioremediation technologies offer a sustainable and greener way for the removal of pollutants.

16.2.2 Paper and pulp industries

Paper and pulp industries release a huge amount of wastewater that remains toxic after secondary treatment and contain a large amount of lignin residues, and organic pollutants such as adsorbable organic halides (AOX) (Kumar *et al.*, 2020). AOX, are a group of a pollutant that includes chlorinated organic compounds, such as resin acids, phenols, chlorinated hydrocarbons, furans, and dioxin. Further, these organic compounds can be separated into two groups based on their molecular weight (MW), those having MW $>$ 1 kDa are high molecular weight (HMW) compounds, and those having MW $<$ 1 kDa are low molecular weight (LMW) compounds. Molecular weight is an important factor for evaluation of environmental behavior (Lindholm-Lehto *et al.*, 2015). These pollutants are discharged mainly by bleaching and pulping processes and affect both ecosystems as well as human health. Many severe respiratory disorders, cardiac problems, and eye and skin irritation have occurred because of the gaseous pollutants released by industries. Moreover, inorganic pollutants such as copper (Cu), zinc (Zn), nickel (Ni), and ferrous (Fe) have been found to cause neurotoxicity (Singh & Chandra, 2019). The physiochemical methods available are energy intensive, expensive, inefficient, and produce toxic sludge in bulk due to which the handling is difficult. To satisfy the discharge standards of environmental regulation, paper and pulp industries have started adopting biological methods for the remediation of wastewater (Kumar *et al.*, 2020; Dias *et al.*, 2020). Biological treatment using microorganisms, such as bacteria, fungi, algae, and yeast can help in reducing the amount of operation and easy handling. They are efficient in removing phenolics as well as color compounds unlike the conventional methods (Erkan & Engin, 2017; Qadir & Chhipa, 2015; Sharma *et al.*, 2020; Cen *et al.*, 2022). Howbeit, due to legislative pressures the water consumption has been reduced, also, by replacing the elemental chlorine with free chlorine the amount of AOX has been decreased. This has helped in reducing the number of pollutants generated by paper and pulp industries to a great extent. The technological advancements in the biodegradation approach could be an efficient method for the remediation of refractory pollutants. It could be an excellent, inexpensive tool for the removal of harmful organic wastes in a way (Singh *et al.*, 2018).

16.2.3 Agrochemical wastes

The rapidly growing population has increased the demand for food and food resources has made farmers utilize fertilizers and pesticides (Dhananjayan *et al.*, 2020; Nayak *et al.*, 2018) in an unregulated manner to meet product consumer needs globally. As a result, a large amount of contaminants gets washed off to nearby water bodies and affects the health of living bodies coming in contact (Akhtar & Mannan, 2020b). According to the American Heritage Science Dictionary, 'agrochemicals are substances such as hormones, fungicides, and insecticides that are frequently used to enhance the productivity of crops' (Bhadouria *et al.*, 2020). These chemical fertilizers consist of nitrogen, phosphorous, and potassium in their structure. Pesticides have subdivisions like insecticides (organochlorine, carbamate, pyrethrin, and rotenone), fungicides (sulfur metals, organomercurial, chlorophenols, phthalimides), herbicides (chlorates, arsenicals, chlorophenoxy), and so on (Gupta, 2019). It has been found that these chemicals affect the nervous system and also affect different parts of the animal and human body. Insecticides like organophosphate and carbamate are known acetylcholine inhibitors that primarily block the normal breakdown of acetylcholine. Besides, they show nicotinic and muscarinic effects such as manifested by salivation, lacrimation, urination, abdominal pain, vomiting, and diarrhea. While the nicotinic effect shows twitching of muscles, seizures, and tremors followed by convulsions (Gupta, 2019). Some chlorinated hydrocarbon compounds, heptachlor epoxide, lindane, and oxychlordane are also found in the fatty tissues after long exposure to these chemicals. Fungicides, such as chlorothalonil, captan, captafol, sulfur derivatives of dithiocarbamic acid, and dimethyldithiocarbamates are nontoxic and are being used widely by farmers. However, their hydrolyzed products such as ethylthiourea is carcinogenic (Gupta, 2019). Organic and inorganic pollutants of agriculture can cause more or less toxic effects to human as well as to the environment. Some compounds like anions (Cl^-, SO_4^{2-}) are recalcitrant in structure and requires further treatment before their final disposal into the environment. Microorganisms have the potential to degrade these agrochemicals even if they are present in low concentration.

16.2.4 Pharmaceutical and medical waste

Pharmaceutical compounds are biologically active, designed to prevent several different types of diseases in humans and animals. The usage of pharmaceutically active compounds (PhACs) has raised a lot of concern as they are vital for the maintenance of good health (Kumar & Bharadvaja, 2020). As a result, the discharge from pharmaceutical industries has also increased which ultimately pollutes the environment with incomplete transformation of the compounds. Luo *et al.* (2019) identified the composition of PhACs taken from different medical sites that showed the relative content of organic compounds. Wastewater from antibiotic pharmaceutical industries was reported to have 15.86% of 5-(phenylmethoxy)-1H-indole-3-carbaldehyde, 22.28% amyl acetate, and 14.27% N, N-dimethylformamide, including dodecyl mercaptan, decamethylcyclopentasiloxane with total of 53 organic compounds in the influent while only 30 in the effluent. By contrast, the effluent from traditional Chinese medicine pharmaceutical plant had 25 more organic compounds than the influent and this depicted that the incomplete degradation of organic matters in the treatment plants and increased the number of refractory pollutants in the effluent. Amongst all pollutants, the highest relative content in organic compound class was acids (63.65%), esters (44.96%), and heterocyclic compounds (30.24%) while in effluents it was esters (52.66%) and organosilicon compounds (42.46%) (Luo *et al.*, 2019).

Various techniques, such as membrane separation, advanced oxidation process, flocculation, sand filter process, and so on have been efficiently used for the elimination of contaminants. Likewise, bioremediation is also found to be effectively removing a small amount of recalcitrant compounds (Kumar & Bharadvaja, 2020) (Table 16.1).

16.3 APPROACHES FOR BIOREMEDIATION OF REFRACTORY POLLUTANTS

For the treatment of refractory pollutants, bioremediation is a promising technique and is a better alternative to conventional techniques. This process is extensively used as it is economical and eco-friendly technology. There are various approaches for bioremediation of pollutants. Some of the approaches have been discussed below.

16.3.1 Bioremediation

Removal of environmental pollutants by a biological process where microorganisms and their metabolic competence are being used for the breakdown of a broad range of chemicals is known as bioremediation. Microorganisms convert hazardous compounds into non-hazardous compounds. This approach is cost-effective and eco-friendly and it mineralizes the contaminants completely. Even the smallest number of toxic contaminants can be removed by bioremediation. Also, bioremediation helps in restoring the polluted site to its original condition without harming the environment (Jobby *et al.*, 2018). Microorganisms that live in groundwater can naturally degrade the harmful contaminants. The microbes convert these chemicals into water and harmless gases, for example, CO_2. The varying structures of the different components present in industrial wastes demand separate treatment methodologies. A natural system using microbes could be established to degrade these contaminants. The bioremediation could be classified into *ex situ* and *in situ* according to the treatment site, pollution range, pollutant category, and treatment cost (Haripriyan *et al.*, 2022) as depicted in the below-given picture. Fungi, bacteria, plants, actinomycetes, and algae are being used prominently for the elimination of contaminated soil and water (Zhang *et al.*, 2020). Microbes are capable of transforming or detoxifying the contaminants in the water in a sustainable way (Figure 16.2).

16.3.2 Fungi

In recent years, presence and persistence of organic and inorganic contaminants have augmented rapidly. It is necessary to devise a strategy that is capable of overcoming these challenges and offering *in-situ* treatment of the pollutants. The use of fungi or their derivatives to redress environmental contaminants can be a cost-effective, environment-friendly, and successful method in addressing the growing issue of soil and water pollution. Mycoremediation methods mostly rely on microorganisms (fungi) native to the contaminated sites (Akhtar & Mannan, 2020a). Mycoremediation is a significant procedure that uses fungus to eliminate petroleum hydrocarbons from the environment. Due to their complex enzymatic systems, quick ability to adapt to harmful organic contaminants, and harsh environmental circumstances, fungi have benefits. Bilen Ozyurek *et al.* (2021) investigated the potential of spore suspension, live biomass, and cell-free culture supernatant of *Aspergillus ochraceus* in the degradation of petroleum. Fungal discs showed 94% success rate than spore suspension which was 87% in petroleum degradation. The strain *A. ochraceus* has a high degradation capacity. Fungi have phenomenal competence to grow in the presence of heavy metals and pesticides. Dey *et al.* (2020) in their work proved the accumulation of metals by *Aspergillus fumigatus* in the order Zn (98%) > Pb (95%) > Cd (63%) > Cr (62%) > Ni (46%) > Cu (37%) from a cocktail of metals and lindane (30 mg/L). Furthermore, the results of transmission electron microscopy coupled with energy-dispersive x-ray analysis (TEM-EDAX) were used to determine the uptake of the metals Cd, Cu, and Pb in the membrane of the cytoplasm and the accumulation of the metals Cr, Ni, and Zn in the cytoplasm of the fungus. FTIR revealed the involvement of fungal cell walls in metal chelation (Dey *et al.*, 2020). In addition, the production of edible mushrooms from industrial and agricultural wastes may be a technique with

Table 16.1 Ecotoxicological effects of various pollutants from different industries.

Industries	Pollutants	Types of Pollutants	Ecotoxicological and Human Effects	References
Pharmaceuticals and medical	Endocrine disruptive compounds (EDC)	Chlororganics Imidazoles Triazoles Paraquat Bisphenol A Estrone Phthalates	• Thyroid dysfunction • Change of sex in aquatic animals • Antiandrogenic effects by mimicking estrogenic actions • Modifies the hormone levels in the body • Affects fertility by acting as a reproductive toxicant • Gene suppression, activation with alteration in the hormonal feedback system	Pironti *et al.* (2021)
		Polychlorinated biphenyls Dioxins Heavy metals Diclofenac Naproxen	• Causes acute renal failure • Inhibits the growth of cyanobacteria and algae • Neurotoxicity and behavioral changes are been observed for some drugs	Khan *et al.* (2020)
Paper and pulp	Adsorbable organic halogens (AOX)	Organochlorines Organophosphates	• Causes phytotoxicity and chromosomal aberrations • Various androgenic and carcinogenic components are being found in the effluent that results in the deteriorating health of organisms • AOX inhibits cytochrome oxidase and damages the electron transport system. • The bleaching process is responsible for respiratory, reproductive, genetic, chronic, and skin damage	Mandeep Gupta *et al.* (2019)
Textiles and apparel	Heavy metals	Chromium (Cr) Cadmium (Cd) Lead (Pb) Mercury (Hg) Nickel (Ni) Copper (Cu)	• Causes mutagenic, carcinogenic, or teratogenic effects on different species depending on the concentration • Exposure to Cd affects osmoregulation and Ca metabolism by affecting carbonic anhydrase (enzyme) in the tissues • Cd is highly toxic and thus decreases the number of freshwater mussels • Smoking and tobacco are the sources of Cd exposure in humans. • Exposure to Hg vapors in the process of amalgamation of gold • Welders exposed to welding fumes had increased levels of Cr Cd, Ni, and Pb in their blood which results in oxidative stress • Cd, Pb, and Hg are nephrotoxic, particularly in the renal cortex. The accumulation of metals in the body leads to several histopathological problems.	Ali *et al.* (2019)

(Continued)

Table 16.1 Ecotoxicological effects of various pollutants from different industries. (*Continued*)

Industries	Pollutants	Types of Pollutants	Ecotoxicological and Human Effects	References
	Dyes	Congo red Azure B Disperse orange Crystal violet	• Toxic to aquatic organisms • It stops the light from penetrating the photic zone and depletes oxygen levels thus hindering the biological function of aquatic flora and fauna • Potentially carcinogenic for humans • Can cause dermatitis and central nervous system problems • Long-term exposure can cause bladder cancer, colorectal cancer, and colon cancer	Al-Tohamy *et al.* (2022)
Agricultural	Pesticides	Insecticide	• Few are carcinogenic, they cause liver damage, muscle tremors, carcinogenicity, and convulsions • Endosulfan are persistent organic pollutants (POP) that cause endocrine disruption, sex hormone hindrance, and even cancer • Some enter the bloodstream and cause more severe effects on health such as vomiting, cramps, diarrhea, headache, and even death • Pesticides such as parathion disrupt the synthesis of macromolecules like DNA, RNA and proteins and can be neurotoxic	Pandey *et al.* (2018)
		Herbicide	• Affects the liver and kidney in animals when exposed to high levels of herbicides • Nonspecific alterations such as depression of the central nervous system, myocardium problems, gastrointestinal problems • Also, weakness, ataxia, fatigue, tachycardia, vomiting, sweating, abdominal pain • Application of herbicide affects the ecology, number and microorganism distribution in soil	Magnoli *et al.* (2020)
		Fungicide	• Affects liver cells • High reactive oxygen species production due to oxidative stress in humans • Can be a potential risk factor for the pathogenesis of non-alcoholic fatty liver disease in humans • Few are partly responsible for the development of Parkinson's disease	Kara *et al.* (2020); Kwon *et al.* (2021)

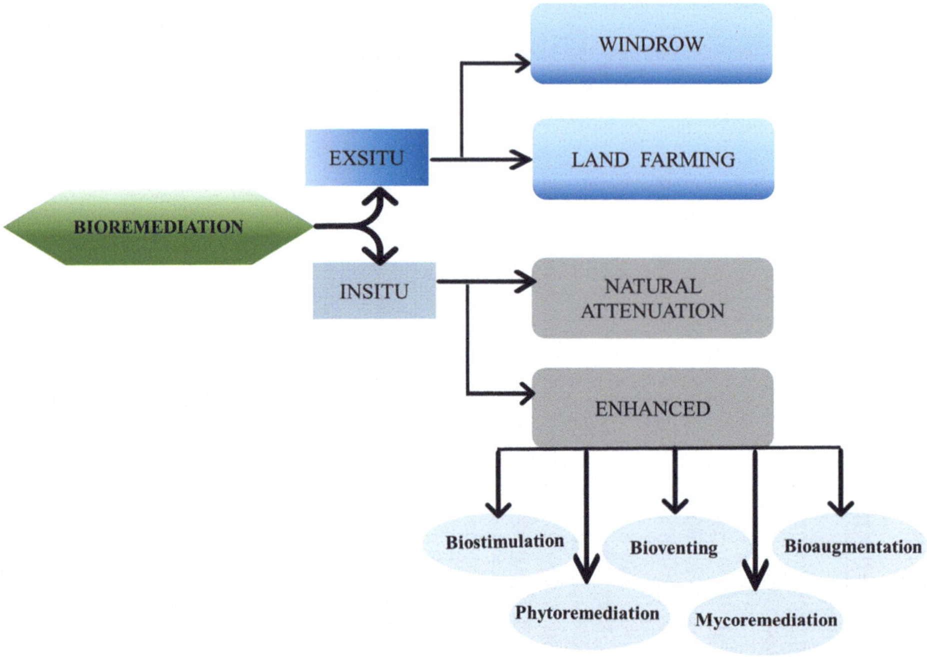

Figure 16.2 Pictorial illustration of the classification of bioremediation on the basis of the treatment site and their subclassifications.

added value that can turn what would otherwise be considered trash into diets and feedstuffs. In addition to producing healthy mushrooms, it reduces the genotoxicity and toxicity of mushroom species (Kulshreshtha *et al.*, 2014). Thus, more study is required to fully understand the capabilities of fungi as a remediating tool and their safety characteristics for consumption and the treatment of wastewater.

16.3.3 Bacteria

The emerging condition provided the stimulus to look for new sources for the treatment of textile dye. A better and more sustainable option is bacteria, which could be used for the remediation of pollutants from various sources contaminating soil and water (Gangola *et al.*, 2019). The textile industry without proper treatment contains a good amount of synthetic dye and their degradation in an eco-friendly manner is a global challenge. Afrin *et al.* (2021) showed the potential of a bacterial consortium of *Enterococcus faecium*, and *Pseudomonas aeruginosa* for the hydrolysis of Novacron dyes. The treated dyes were then used for agricultural purposes and found that it was non-toxic to agriculturally valuable bacteria (Afrin *et al.*, 2021). Other than synthetic dyes several hydrocarbons are also found in the wastewater and their degradation can be done biologically. Polycyclic aromatic hydrocarbon (PAH) adversely affects the health of humans. This study used nine natural bacterial strains isolated naturally for the degradation of four kinds of PAH. The study was conducted on several cultural factors and their effects on PAHs degradation. In the study, it was found that a bacterial mixture can increase the degradation rates of phenanthrene, and benzo pyrene and the rate was over 83%. They concluded that the efficient degradation was due to the cooperation of dioxygenase and dehydrogenase enzymes (Xu *et al.*, 2022). Industries are the dominant factors for the economic development of society and the need to have efficient biological remediation is persistent. To enhance the efficiency of degradation of bacteria, the study of genetic engineering has been introduced. Genetically engineered microorganisms (GEMs) facilitated methods for the elimination of metals/metalloids, which are thought to be both environmentally safe and economically practical strategies. At low concentrations, metals and metalloids are extremely poisonous and primarily contribute to the survival equilibrium of environmental activity. However, because they are not biodegradable, metals and metalloids remain in nature and bio-accumulate in the food chain (Sharma *et al.*, 2021). GEMs are projected to be an extremely effective process when compared to other procedures because they are highly capable of detoxifying (Table 16.2).

16.3.4 Plants

Plants can be found everywhere in different habitats and be a promising bioremediation tool for various waste effluents and meets all the requirements of the green purification process. This plant-based technology also helps in the restoration of habitats with reduced remedial costs. Zhang *et al.* (2021) while performing remediation with plants on slightly Cd-contaminated soil via three cropping systems found the highest Cd abstraction amount (148 h/ha) in monoculture. In addition, the net value of 1.88×10^4 US$ was achieved using the Monte-Carlo simulation. The metabolic and extractive potential of plants makes them capable of biodegradation. Recent developments in omics and nanotechnology are creating new avenues for the competent treatment of emerging contaminants effectively (Rai *et al.*, 2020). The uptake of organic/xenobiotic compounds in plants occurs through passive uptake. Several transporters like the phosphate transporter1 family (PHT1), a phosphate-H^+ symporter containing 520–550 amino acids with ~molecular mass of 58 kDa help transport arsenic (Rai *et al.*, 2020). The phytoremediation process takes place in wetlands also called natural marsh or root wetlands. Under reducing conditions like wetlands, one of the most poisonous and common inorganic forms of As is AsIII. It can attach to sulfhydryl groups in proteins and restrict general protein functioning (Abedin *et al.*, 2002). In comparison to other treatments, phytoremediation is easy to operate, can tolerate the lack of wastewater, as well as its maintenance cost is also low. Studies show that phytoremediation has almost 100% efficiency and is environmentally friendly in eliminating organic compounds from polluted water (Gatidou *et al.*, 2017). Singh *et al.* (2021) investigated the use of water macrophytes, that

Table 16.2 Microorganisms involved in the degradation of refractory pollutants.

Microorganism	Name	Pollutants	References
Bacteria	*Bacillus subtilis*	Endosulfan	Ahmad (2020)
	Bacillus velezensis	Slaughter wastewater	Li *et al.* (2018)
	Bacillus aryabhattai	Lignin and color	Zainith *et al.* (2019)
	Bacillus cohnni (RKS9)	Congo red	Kishor *et al.* (2021a)
	Bacillus albus MW407057	Methylene blue	Kishor *et al.* (2021b)
Fungi	*Aspergillus flavus* *Aspergillus niger* *Penicillium chrysogenum*	Endosulfan	Ahmad (2020)
	Aspergillus sp.	Platinum and palladium	Godlewska-Żyłkiewicz *et al.* (2019)
	Aspergillus sydowii *Fusarium lichenicola*	Polyaromatic hydrocarbon compounds (PAH)	Obire *et al.* (2020)
	Aspergillus sp.	Heavy metals	Vašinková *et al.* (2021)
	Ganoderma luicdum	Phenanthrene Chrysene	Agrawal *et al.* (2018)
	Ganoderma applanatum *Laetiporus sulphureus*	NSAIDs	Bankole *et al.* (2020)
Microalgae	*Desmodesmus* sp. *WR1*	17-α Estradiol	Wang *et al.* (2020)
	Chlorella vulgaris	Sulfamethoxazole	Xie *et al.* (2020)

is *Pistia stratiotes* aided in removing the pollutants of plant mill effluent (PME). They found that using 75% of PME treatment gave the best result in terms of removal with the highest biomass and chlorophyll enrichment. Their work was innovative in terms of phytoremediation of significant physicochemical and microbiological pollutants of paper mill effluent (PME) by *P. stratiotes*. Textile wastes are also prone to be toxic to the environment. Mahajan *et al.* (2019) showed the phytoremediation potential of *Chara vulgaris* for the treatment of textile wastes. *C. vulgaris* competently decreased BOD, COD, pH, EC, and TDS of the 10–5% of concentrated textile effluent within 120 h of treatment. It also can be used for the tertiary-level treatment of textile effluents. While performing the process one needs to consider higher efficiency such as the type of plant to be used, process time, number, also the type of soil and wetland. Also, the plant used should have a high rate of biomass growth, high content of salts, phosphate, and nitrogen, and be less sensitive to varying compositions of soil and wastewater (Justin *et al.*, 2010). The installation of this system occurs in two types: first is abiotic, based on hydrolysis, adsorption, and photodegradation and second is biological, through bioaccumulation and metabolism in tissue (Gatidou *et al.*, 2017). The different biological processes of phytoremediation are phytovolatilization, phytodegradation, phytoaccumulation, phytostabilization, and rhizofiltration (Polińska *et al.*, 2021). The drawback of this method is that the accessibility of plants all-round the year is difficult and consequently, this is considered a secondary or tertiary form of treatment. It is a slow process and that's why it is suggested that it should be combined with other agents to augment the process (Khandare & Govindwar, 2015). This approach has emerged as an eco-friendly, passive, cost-effective, and solar energy-driven method for cleaning the environment in comparison to other physio-chemical techniques. Amidst remediation approaches, phytoremediation is a good option for the future but still, the studies remain on the laboratory scale. There is a need to take this to a large-scale study at the contaminated site to find out more avenues of research in the removal of refractory pollutants (Figure 16.3).

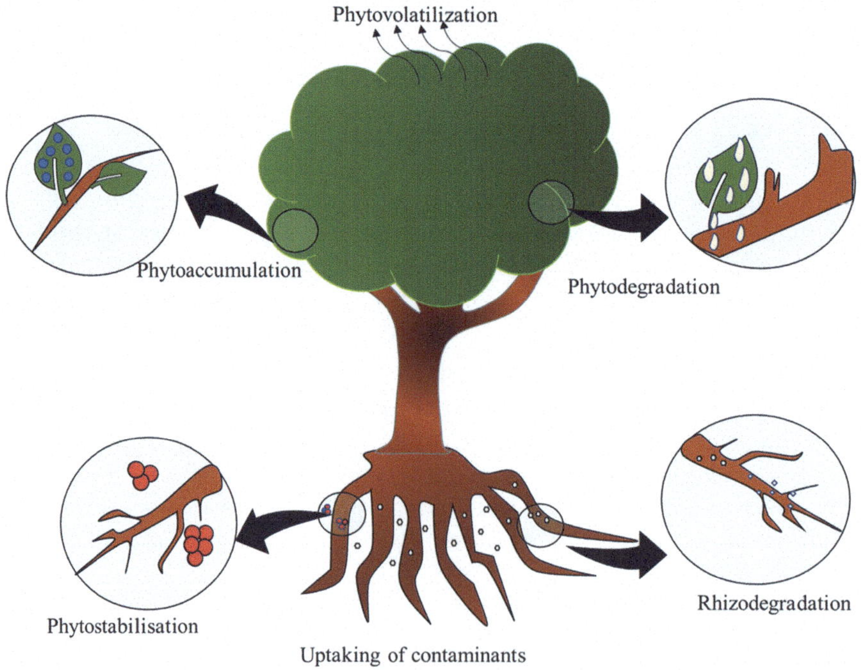

Figure 16.3 Contaminant uptake and degradation through different biological processes of phytoremediation.

16.4 OTHER TECHNOLOGICAL ADVANCEMENTS IN BIOREMEDIATION

Bioremediation is considered a feasible technology for the degradation of pollutants in soil and water. However, the rate and degree of deterioration of contaminants may often be constrained due to their limited solubility, and high hydrophobicity. In recent years, to overcome such problems many new promising hybrid technologies such as electro-bioremediation, nano remediation, and constructed wetlands (CWs) have been implemented in the cleanup of water/soil polluted with refractory contaminants. In the below section, the new techniques have been discussed.

16.4.1 Electro-bioremediation

A key microbial electrochemical technology (MET) that is emerging as a viable alternative for treating groundwater is electro-bioremediation (Ceballos-Escalera *et al.*, 2021). MET is based on bio-electrochemical systems that use solid electron conductors (e.g., electrodes) to execute oxidation and reduction processes by electroactive microorganisms. Thus, the anode and cathode can be used by these electroactive microorganisms as an endless source of electron acceptor or donor, respectively. Direct current (DC) is supplied across electrodes introduced into the polluted soil during the electro-bioremediation procedure for the generation of an electric field for boosting bioremediation efficiency (Li *et al.*, 2020). The structure of microbial population and activity is disturbed by electrochemical reactions and electrokinetic processes. Li *et al.* (2020) showed the outcome of an electric field on the microbes of soil through the examination of changes in the extent of PAH degradation and microorganisms. The outcomes indicated that the microbial activity and population structure was influenced by the intensity of voltage. In this study, the optimal electric strength for enhancing the activity of microbes and further degradation of PAHs was found to be 2 V/cm (voltage gradient). The microbial population structure was influenced by the voltage gradient as well as the electrode reaction. For efficient use of electro-bioremediation, choosing a suitable electric field that has no adverse effects on microbes and will encourage microbes to use their maximum degrading activity is crucial. Hence, to choose a suitable electric field, it is necessary to understand the changes in the microbe community and activity under different electric fields. Understanding the changes in microorganism communities and activity under different electric fields is a requirement for choosing a suitable electric field. Recently, the effects of electric fields on microorganisms have been studied thoroughly, such as polarity reversal, electrode configuration and the electric field strength (or current) (Barba *et al.*, 2017; Li *et al.*, 2015). Soil or water contaminated with refractory pollutants electro-bioremediation is a promising technology for remediation (Figure 16.4).

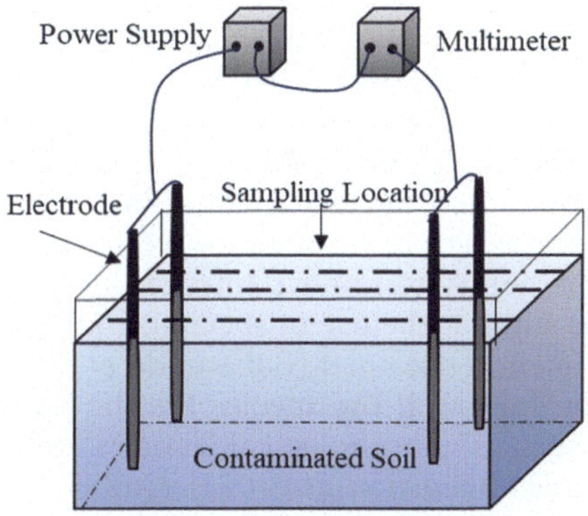

Figure 16.4 Schematic representation of experimental set-up of electro-bioremediation.

16.4.2 Nano bioremediation

The term 'nano remediation' refers to the use of nanoscale materials for the cleanup of the environment, and it provides quick, low-cost, and effective solutions for the remediation of affected sites (Kahraman *et al.*, 2022). Nanoparticles such as zero-valent iron (Karthick *et al.*, 2019) iron oxide (Karthick *et al.*, 2019) and silica (Chattopadhyay & Karthick, 2017) have been reported to be efficient for soil remediation. The advancement of nanotechnology and the incorporation of nanomaterials – defined as particles with sizes of 100 nm or less in at least one dimension and particular nanoparticles – signify a ground-breaking strategy to advance bioremediation beyond its limitations. This integrated method comprises a broad range of potential applications with less costs and the least negative impacts on the environment for treating pollutants in groundwater and wastewater (Cecchin *et al.*, 2017), residues contaminated with heavy metals and hydrocarbons (Darwesh *et al.*, 2021) and either organic or inorganic amalgams in soil (Cecchin *et al.*, 2017). Several studies showed that nanoparticles possess distinct properties including enhanced catalysis and adsorption as well as increased reactivity. Previously, only physical and chemical processes were used to create nanoparticles. Howbeit, in search of sustainable methods for the production microorganisms and plant extracts are being utilized. In a study, iron nanoparticles were mycosynthesized using *Fusarium oxysporum*, size ranging from 1 to 3 nm. This nanoparticle was significantly successful as an antimicrobial and adsorbing agent and was used for the redressal of municipal wastewater (Darwesh *et al.*, 2021). It is evident that this technique is more intelligent, safe, ecologically friendly, inexpensive, and green than either technology alone, the field of environmental remediation may be greatly altered in the long run by this amalgamation of the two technologies, known as nano-bioremediation.

16.4.3 Constructed wetlands

CWs are artificial, human-made wetlands that are created and designed to imitate natural wetland systems for the treatment of water. Intricate physical, chemical, and biological processes are used by these systems, which are mostly composed of vegetation, substrates, soils, microorganisms, and water, to remove various contaminants or improve the quality of the water. CW can be classified into two types according to wetland hydrology: free water surface (FWS) and subsurface flow (SSF). FWA is similar to a natural wetland, while in the SSF system, water flows vertically or horizontally through the substrates. SSF can further be categorized into vertical and horizontal flow. Early in the 1950s, Käthe Seidel made the initial attempt to explore the potential of CWs for wastewater treatment in Germany. Subsequent CW studies were conducted and used for wastewater treatments in the 1960s and 1970s. The application of CWs was initially applied to remediate conventional domestic and municipal wastewater. Currently, the use of CWs has been significantly increased to clear out industrial and agricultural effluents, mine drainage, landfill leachates, contaminated river and lake waters, and urban and highway runoff. It has also been developed in a variety of climatic conditions, including tropical, arid, and warm climates (Wu *et al.*, 2014). FWS CWs are more effective at removing organics and suspended particles than SSF CWs are at removing nitrogen and phosphorus. However, their therapeutic effectiveness and sustainable use are typically constrained in cold weather or after the degradation of the plant (Vymazal, 2011). In comparison to FWS CWs, SSF CWs are very effective in the removal of organics, microbial pollution, suspended solids, and heavy metals, and they are less cold-sensitive and easier to insulate for winter operation. However, these CWs are permanently underwater, and the ability to remove nitrogen depends on the availability of oxygen and carbon sources. In addition, without the use of specialized media with high sorption capacities, poor phosphorus removal is frequently obtained (Babatunde *et al.*, 2010). The remediating effect of macrophytes in CWs depends on how well the species fits the particular climatic and operational circumstances as well as on the species' morphological features and has been the subject of intensive research over the last three decades (Qadiri *et al.*, 2021). Qadiri *et al.* (2021) experimentally tested and compared the involvement of three macrophytes in eliminating toxic compounds from combined wastewater (graywater and septic effluent) using horizontal sub-surface flow type constructed wetlands

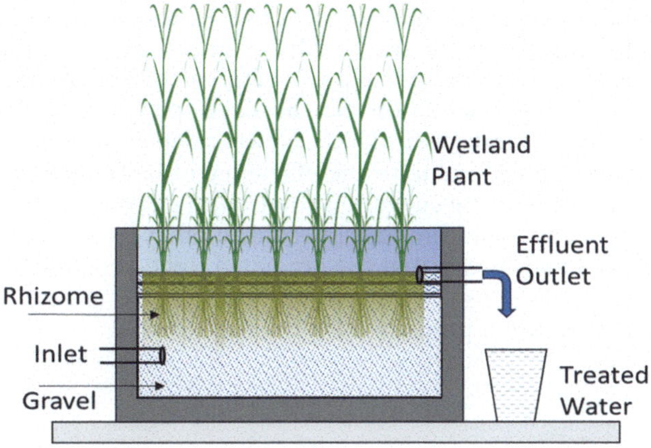

Figure 16.5 Schematic representation of constructed wetland system for wastewater treatment.

(HSSF-CW) in sub-tropical conditions. They monitored the standard wastewater parameters (COD, nitrogen, phosphorus, and solids) in the inlet and outlet of the CW microcosms, the fate of nutrients through the planted microcosms and the growth rate of the selected species. *Phragmites karka* exhibited better removal efficiencies with organic loading removal rates of $10 \pm 1.3/g^2/d$, nitrogen removal rates of $2.6 \pm 0.4/g^2/d$, phosphorus removal rates of $1.3 \pm 0.2/g^2/d$, and solids removal rates of $3.6 \pm 1.2/g^2/d$. This showed that in CWs, macrophytes with well-developed root systems have a greater ability to remove pollutants. Considering the successful and sustainable implementation of CWs, future research must concentrate on inclusive assessment of plants and substrates in field trials under real-world situations, optimization of environmental and operational parameters (Figure 16.5).

16.5 CHALLENGES AND PROSPECT

Bioremediation has emerged as a cost-effective alternative to remediation technologies. With advantages, they have several disadvantages such as harming the environment by producing secondary pollutants and having a detrimental impact on the ecosystem. From the aforementioned, it is obvious that bioremediation strategies are varied and have successfully restored sites that have been contaminated with various pollutants. Microorganisms are essential to the process of bioremediation. The variety, abundance, and population structure of microbes in contaminated environments provide a better understanding of the outcome of these strategies, as long as the other environmental factors that could hinder microbial activities that are kept at optimum levels (Azubuike *et al.*, 2016). Because some substances cannot be broken down by biological processes, bioremediation is severely constrained. If a material is biodegradable at all, it may in some situations undergo further processing and degradation that results in the production of harmful material. Due to the occurrence of some site-specific limiting variables, a specific bacterial strain that is effective at one site may occasionally fail to function well at other sites. The nature of toxic compounds, the presence of an optimum amount of nutrients, and the metabolic activities of microorganisms are thought to contribute to the complication of biological processes. The method is more time-taking and labor-intensive because it calls for soil excavation, construction, and modification of a specialized site plan (Vishwakarma *et al.*, 2020). Most of the time, the process takes place in remote, underground locations far from populated areas. In the case of soil bioremediation, aging is one of the greatest obstacles faced during remediation. Hence, a proper solution to exploit the bioavailability of pollutants in soil arises. Likewise in groundwater remediation, the geological settings need to be understood (Vishwakarma *et al.*, 2020).

The scientific community's persistent hard work in this field has been successful in pushing the envelope. It is anticipated that the rapidly evolving bioaugmentation technologies will improve our ability to control a cell's genetic makeup and help us overcome limits brought on by microbiological variables. New techniques for bioremediation can be developed with a deeper understanding of biotransformation at the ecological and genetic levels. Newer strategies are being developed to enhance the bioavailability of microbes to add value to the efficiency of bioremediation. The main goal is to assess the applicability of a theoretically suggested solution to guarantee that the cleanup objectives are achieved. Based on their understanding of molecular biology, researchers are also exhibiting a propensity to create improved techniques that are predicted to be accurate, efficient, and affordable. The *in-situ* characterization of the physiochemical parameters also appears promising and has the potential to transform field evaluation procedures in the future

16.6 CONCLUSION

The main problem of the era is pollution, and scientific societies are dedicating a major deal of attention to this problem. A microbial remediation is a powerful tool for enhancing the natural biodegradation processes, as microorganisms acclimatize speedily with lethal environments. For the formation of an ecologically stable, new, and feasible bioremediation approach, a thorough understanding of the microbial populations and their response to the environment and to contaminants is crucial. It is now generally known that many microorganisms are capable of degrading toxins found in the environment. The use of bioremediation to clean up contaminated soils and aquifers areas has become increasingly common. Furthermore, the applications of new emerging techniques could be a game changer along with the study of genetically modified microbes. A commonly recognized substitute for environmental cleanup is bioremediation, particularly because it promises to do so economically and with fewer risks of transmitting pollutants to other media.

REFERENCES

Abedin M. J., Feldmann J. and Meharg A. A. (2002). Uptake kinetics of arsenic species in rice plants. *Plant Physiology*, **128**(3), 1120–1128, https://doi.org/10.1104/pp.010733

Afrin S., Shuvo H. R., Sultana B., Islam F., Rus'd A. A., Begum S. and Hossain M. N. (2021). The degradation of textile industry dyes using the effective bacterial consortium. *Heliyon*, **7**(10), e08102, https://doi.org/10.1016/j.heliyon.2021.e08102

Agrawal N., Verma P. and Shahi S. K. (2018). Degradation of polycyclic aromatic hydrocarbons (phenanthrene and pyrene) by the ligninolytic fungi *Ganoderma lucidum* isolated from the hardwood stump. *Bioresources and Bioprocessing*, **5**(1), 1–9, https://doi.org/10.1186/s40643-018-0197-5

Ahmad K. S. (2020). Remedial potential of bacterial and fungal strains (*Bacillus subtilis, Aspergillus niger, Aspergillus flavus and Penicillium chrysogenum*) against organochlorine insecticide Endosulfan. *Folia Microbiologica*, **65**(5), 801–810, https://doi.org/10.1007/s12223-020-00792-7

Akhtar N. and Mannan M. A. (2020a). Mycoremediation: expunging environmental pollutants. *Biotechnology Reports*, **26**, e00452, https://doi.org/10.1016/j.btre.2020.e00452

Akhtar N. and Mannan M. A. U. (2020b). Mycoremediation: an unexplored gold mine. In: New and Future Developments in Microbial Biotechnology and Bioengineering: Recent Advances in Application of Fungi and Fungal Metabolites: Environmental and Industrial Aspects, J. Singh and P. Gehlot (eds). Elsevier BV, Amsterdam, Netherlands, https://doi.org/10.1016/B978-0-12-821007-9.00002-4

Al-Tohamy R., Ali S. S., Li F., Okasha K. M., Mahmoud Y. A. G., Elsamahy T., Jiao H., Fu Y. and Sun J. (2022). A critical review on the treatment of dye-containing wastewater: ecotoxicological and health concerns of textile dyes and possible remediation approaches for environmental safety. *Ecotoxicology and Environmental Safety*, **231**, 113160, https://doi.org/10.1016/j.ecoenv.2021.113160

Ali H., Khan E. and Ilahi I. (2019). Environmental chemistry and ecotoxicology of hazardous heavy metals: environmental persistence, toxicity, and bioaccumulation. *Journal of Chemistry*, **2019**, 1–14, https://doi.org/10.1155/2019/6730305

Amulya K., Reddy M. V., Rohit M. V. and Mohan S. V. (2016). Wastewater as renewable feedstock for bioplastics production: understanding the role of reactor microenvironment and system pH. *Journal of Cleaner Production*, **112**, 4618–4627, https://doi.org/10.1016/j.jclepro.2015.08.009

Azubuike C. C., Chikere C. B. and Okpokwasili G. C. (2016). Bioremediation techniques – classification based on site of application: principles, advantages, limitations and prospects. *World Journal of Microbiology and Biotechnology*, **32**(11), 1–18, https://doi.org/10.1007/s11274-016-2137-x

Babatunde A. O., Zhao Y. Q. and Zhao X. H. (2010). Alum sludge-based constructed wetland system for enhanced removal of P and OM from wastewater: concept, design and performance analysis. *Bioresource Technology*, **101**(16), 6576–6579, https://doi.org/10.1016/j.biortech.2010.03.066

Bankole P. O., Adekunle A. A., Jeon B. H. and Govindwar S. P. (2020). Novel cobiomass degradation of NSAIDs by two wood rot fungi, *Ganoderma applanatum* and *Laetiporus sulphureus*: ligninolytic enzymes induction, isotherm and kinetic studies. *Ecotoxicology and Environmental Safety*, **203**, 110997, https://doi.org/10.1016/j.ecoenv.2020.110997

Barba S., Villaseñor J., Rodrigo M. A. and Cañizares P. (2017). Effect of the polarity reversal frequency in the electrokinetic-biological remediation of oxyfluorfen polluted soil. *Chemosphere*, **177**, 120–127, https://doi.org/10.1016/j.chemosphere.2017.03.002

Bhadouria R., Das S., Kumar A., Singh R. and Singh V. K. (2020). Mycoremediation of agrochemicals. In: Agrochemicals Detection, Treatment and Remediation: Pesticides and Chemical Fertilizers, Majeti N. and V. Prasad (eds). Butterworth-Heinemann, UK, LTD, https://doi.org/10.1016/B978-0-08-103017-2.00022-2

Bilen Ozyurek S., Avcioglu N. H. and Seyis Bilkay I. (2021). Mycoremediation potential of *Aspergillus ochraceus* NRRL 3174. *Archives of Microbiology*, **203**(10), 5937–5950, https://doi.org/10.1007/s00203-021-02490-5

Ceballos-Escalera A., Pous N., Chiluiza-Ramos P., Korth B., Harnisch F., Bañeras L., Balaguer M. D. and Puig S. (2021). Electro-bioremediation of nitrate and arsenite polluted groundwater. *Water Research*, **190**, 116748, https://doi.org/10.1016/j.watres.2020.116748

Cecchin I., Reddy K. R., Thomé A., Tessaro E. F. and Schnaid F. (2017). Nanobioremediation: integration of nanoparticles and bioremediation for sustainable remediation of chlorinated organic contaminants in soils. *International Biodeterioration and Biodegradation*, **119**, 419–428, https://doi.org/10.1016/j.ibiod.2016.09.027

Cen Q., Wu X., Cao L., Lu Y., Lu X., Chen J., Fu G., Liu Y. and Ruan R. (2022). Green production of a yellow laccase by *Coriolopsis gallica* for phenolic pollutants removal. AMB Express, Vol. **12**, pp. 96–111, https://doi.org/10.1186/s13568-022-01434-6

Chandanshive V., Kadam S., Rane N., Jeon B. H., Jadhav J. and Govindwar S. (2020). In situ textile wastewater treatment in high rate transpiration system furrows planted with aquatic macrophytes and floating phytobeds. *Chemosphere*, **252**, 126513, https://doi.org/10.1016/j.chemosphere.2020.126513

Chattopadhyay P. and Karthick R. A. (2017). Characterization and application of surfactant foams produced from ethanol-sodium lauryl sulfate-silica nanoparticle mixture for soil remediation. *Macromolecular Symposia*, **376**(1), 1–4, https://doi.org/10.1002/masy.201600182

Darwesh O. M., Shalaby M. G., Abo-Zeid A. M. and Mahmoud Y. A. G. (2021). Nano-bioremediation of municipal wastewater using myco-synthesized iron nanoparticles. *Egyptian Journal of Chemistry*, **64**(5), 2499–2507, https://doi.org/10.21608/EJCHEM.2021.60921.3307

De Donno Novelli L., Moreno Sayavedra S. and Rene E. R. (2021). Polyhydroxyalkanoate (PHA) production via resource recovery from industrial waste streams: a review of techniques and perspectives. *Bioresource Technology*, **331**, 1–13, https://doi.org/10.1016/j.biortech.2021.124985

Dey P., Malik A., Mishra A., Kumar D., Von Bergen M. and Jehmlich N. (2020). Mechanistic insight to mycoremediation potential of a metal resistant fungal strain for removal of hazardous metals from multimetal pesticide matrix. *Environmental Pollution*, **262**, 114255, https://doi.org/10.1016/j.envpol.2020.114255

Dhananjayan V., Jayanthi P., Jayakumar S. and Ravichandran B. (2020). Agrochemicals impact on ecosystem and bio-monitoring. In: Resources use Efficiency in Agriculture, S. Kumar, R. S. Meena, and M. K. Jhariya (eds), Springer, Singapore, https://doi.org/10.1007/978-981-15-6953-1_11

Dias C., Reis A., Santos J. A. L. and Lopes da Silva T. (2020). Concomitant wastewater treatment with lipid and carotenoid production by the oleaginous yeast *Rhodosporidium toruloides* grown on brewery effluent enriched with sugarcane molasses and urea. *Process Biochemistry*, **94**, 1–14, https://doi.org/10.1016/j.procbio.2020.03.015

Erkan H. S. and Engin G. O. (2017). The investigation of paper mill industry wastewater treatment and activated sludge properties in a submerged membrane bioreactor. *Water Science and Technology*, **76**(7), 1715–1725, https://doi.org/10.2166/wst.2017.351

Gangola S., Joshi S., Kumar S. and Pandey S. C. (2019). Comparative analysis of fungal and bacterial enzymes in biodegradation of xenobiotic compounds. Smart Bioremediation Technologies: Microbial Enzymes, 169–189, https://doi.org/10.1016/B978-0-12-818307-6.00010-X

Gatidou G., Oursouzidou M., Stefanatou A. and Stasinakis A. S. (2017). Removal mechanisms of benzotriazoles in duckweed *Lemna minor* wastewater treatment systems. *Science of the Total Environment*, **596–597**, 12–17, https://doi.org/10.1016/j.scitotenv.2017.04.051

Godlewska-Żyłkiewicz B., Sawicka S. and Karpińska J. (2019). Removal of platinum and palladium from wastewater by means of biosorption on fungi *Aspergillus* sp. and yeast *Saccharomyces* sp. *Water (Switzerland)*, **11**(7), 1–17, https://doi.org/10.3390/w11071522

Gupta P. K. (2019). Concepts and Applications in Veterinary Toxicology – an Interactive Guide | Pawan Kumar Gupta. Springer, Switzerland, https://doi.org/10.1007/978-3-030-22250-5%0A

Haripriyan U., Gopinath K. P., Arun J. and Govarthanan M. (2022). Bioremediation of organic pollutants: a mini review on current and critical strategies for wastewater treatment. *Archives of Microbiology*, **204**(5), 1–9, https://doi.org/10.1007/s00203-022-02907-9

Jobby R., Jha P., Yadav A. K. and Desai N. (2018). Biosorption and biotransformation of hexavalent chromium [Cr(VI)]: a comprehensive review. *Chemosphere*, **207**, 255–266, https://doi.org/10.1016/j.chemosphere.2018.05.050

Justin M. Z., Pajk N., Zupanc V. and Zupančič M. (2010). Phytoremediation of landfill leachate and compost wastewater by irrigation of *Populus* and *Salix*: biomass and growth response. *Waste Management*, **30**(6), 1032–1042, https://doi.org/10.1016/j.wasman.2010.02.013

Kahraman B. F., Altin A. and Ozdogan N. (2022). Remediation of Pb-diesel fuel co-contaminated soil using nano/bio process: subsequent use of nanoscale zero-valent iron and bioremediation approaches. *Environmental Science and Pollution Research*, **29**(27), 41110–41124, https://doi.org/10.1007/s11356-022-18857-0

Kara M., Oztas E., Ramazanoğulları R., Kouretas D., Nepka C., Tsatsakis A. M. and Veskoukis A. S. (2020). Benomyl, a benzimidazole fungicide, induces oxidative stress and apoptosis in neural cells. *Toxicology Reports*, **7**, 501–509, https://doi.org/10.1016/j.toxrep.2020.04.001

Karthick A., Roy B. and Chattopadhyay P. (2019). Comparison of zero-valent iron and iron oxide nanoparticle stabilized alkyl polyglucoside phosphate foams for remediation of diesel-contaminated soils. *Journal of Environmental Management*, **240**, 93–107, https://doi.org/10.1016/j.jenvman.2019.03.088

Khan S. and Malik A. (2018). Toxicity evaluation of textile effluents and role of native soil bacterium in biodegradation of a textile dye. *Environmental Science and Pollution Research*, **25**(5), 4446–4458, https://doi.org/10.1007/s11356-017-0783-7

Khan H. K., Rehman M. Y. A. and Malik R. N. (2020). Fate and toxicity of pharmaceuticals in water environment: an insight on their occurrence in South Asia. *Journal of Environmental Management*, **271**, 111030, https://doi.org/10.1016/j.jenvman.2020.111030

Khandare R. V. and Govindwar S. P. (2015). Phytoremediation of textile dyes and effluents: current scenario and future prospects. *Biotechnology Advances*, **33**(8), 1697–1714, https://doi.org/10.1016/j.biotechadv.2015.09.003

Kishor R., Purchase D., Saratale G. D., Ferreira L. F. R., Bilal M., Iqbal H. M. N. and Bharagava R. N. (2021a). Environment friendly degradation and detoxification of Congo red dye and textile industry wastewater by a newly isolated *Bacillus cohnni* (RKS9). *Environmental Technology and Innovation*, **22**, 101425, https://doi.org/10.1016/j.eti.2021.101425

Kishor R., Purchase D., Saratale G. D., Saratale R. G., Ferreira L. F. R., Bilal M., Chandra R. and Bharagava R. N. (2021b). Ecotoxicological and health concerns of persistent coloring pollutants of textile industry wastewater and treatment approaches for environmental safety. *Journal of Environmental Chemical Engineering*, **9**(2), 105012, https://doi.org/10.1016/j.jece.2020.105012

Kishor R., Saratale G. D., Saratale R. G., Romanholo Ferreira L. F., Bilal M., Iqbal H. M. N. and Bharagava R. N. (2021c). Efficient degradation and detoxification of methylene blue dye by a newly isolated ligninolytic enzyme producing bacterium *Bacillus albus* MW407057. *Colloids and Surfaces B: Biointerfaces*, **206**, 111947, https://doi.org/10.1016/j.colsurfb.2021.111947

Kour D., Kaur T., Devi R., Yadav A., Singh M., Joshi D., Singh J., Suyal D. C., Kumar A., Rajput V. D., Yadav A. N., Singh K., Singh J., Sayyed R. Z., Arora N. K. and Saxena A. K. (2021). Beneficial microbiomes for bioremediation of diverse contaminated environments for environmental sustainability: present status and future challenges. *Environmental Science and Pollution Research*, **28**(20), 24917–24939, https://doi.org/10.1007/s11356-021-13252-7

Kulshreshtha S., Mathur N. and Bhatnagar P. (2014). Mushroom as a product and their role in mycoremediation. *AMB Express*, **4**(1), 1–7, https://doi.org/10.1186/s13568-014-0029-8

Kumar L. and Bharadvaja N. (2020). Microbial remediation of heavy metals. In: Microbial Bioremediation & Biodegradation, M. P. Shah (ed.), Springer, Singapore, https://doi.org/10.1007/978-981-15-1812-6_2

Kumar V. and Dwivedi S. K. (2021). Mycoremediation of heavy metals: processes, mechanisms, and affecting factors. *Environmental Science and Pollution Research*, **28**(9), 10375–10412, https://doi.org/10.1007/s11356-020-11491-8

Kumar V., Thakur I. S. and Shah M. P. (2020). Bioremediation approaches for treatment of pulp and paper industry wastewater: recent advances and challenges. In: Microbial Bioremediation & Biodegradation, M. P. Shah (ed.), pp. 1–48, https://doi.org/10.1007/978-981-15-1812-6_1

Kwon H. C., Sohn H., Kim D. H., Jeong C. H., Kim D. W. and Han S. G. (2021). Effects of flutriafol fungicide on the lipid accumulation in human liver cells and rat liver. *Foods*, **10**(6), 1–19, https://doi.org/10.3390/foods10061346

Li T., Guo S., Wu B., Zhang L. and Gao Y. (2015). Effect of polarity-reversal and electrical intensity on the oil removal from soil. *Journal of Chemical Technology and Biotechnology*, **90**(3), 441–448, https://doi.org/10.1002/jctb.4312

Li W., Jia M. X., Deng J., Wang J. H., Lin Q. L., Liu C., Wang S. S., Tang J. X., Zeng X. X., Ma L., Su W., Liu X. Y., Cai F. and Zhou L. Y. (2018). Isolation, genetic identification and degradation characteristics of COD-degrading bacterial strain in slaughter wastewater. *Saudi Journal of Biological Sciences*, **25**(8), 1800–1805, https://doi.org/10.1016/j.sjbs.2018.08.022

Li F., Guo S., Wang S. and Zhao M. (2020). Changes of microbial community and activity under different electric fields during electro-bioremediation of PAH-contaminated soil. *Chemosphere*, **254**, 126880, https://doi.org/10.1016/j.chemosphere.2020.126880

Lindholm-Lehto P. C., Knuutinen J. S., Ahkola H. S. J. and Herve S. H. (2015). Refractory organic pollutants and toxicity in pulp and paper mill wastewaters. *Environmental Science and Pollution Research*, **22**(9), 6473–6499, https://doi.org/10.1007/s11356-015-4163-x

Luo Q., Wang J., Wang J. H., Shen Y., Yan P., Chen Y. P. and Zhang C. C. (2019). Fate and occurrence of pharmaceutically active organic compounds during typical pharmaceutical wastewater treatment. *Journal of Chemistry*, **2019**, pp. 1–12, https://doi.org/10.1155/2019/2674852

Magnoli K., Carranza C. S., Aluffi M. E., Magnoli C. E. and Barberis C. L. (2020). Herbicides based on 2,4-D: its behavior in agricultural environments and microbial biodegradation aspects. A review. *Environmental Science and Pollution Research*, **27**(31), 38501–38512, https://doi.org/10.1007/s11356-020-10370-6

Mahajan P., Kaushal J., Upmanyu A. and Bhatti J. (2019). Assessment of phytoremediation potential of *Chara vulgaris* to treat toxic pollutants of textile effluent. *Journal of Toxicology*, **2019**, pp. 1–11, https://doi.org/10.1155/2019/8351272

Mandeep Gupta G. K., Liu H. and Shukla P. (2019). Pulp and paper industry-based pollutants, their health hazards and environmental risks. *Current Opinion in Environmental Science and Health*, **12**, 48–56, https://doi.org/10.1016/j.coesh.2019.09.010

Medfu Tarekegn M., Zewdu Salilih F. and Ishetu A. I. (2020). Microbes used as a tool for bioremediation of heavy metal from the environment. *Cogent Food and Agriculture*, **6**(1), pp. 1–19, https://doi.org/10.1080/23311932.2020.1783174

Nayak S. K., Dash B. and Baliyarsingh B. (2018). Microbial remediation of persistent agro-chemicals by soil bacteria: an overview. *Microbial Biotechnology*, **2**, 275–301, https://doi.org/10.1007/978-981-10-7140-9_13

Obire O., Aleruchi O. and Wemedo S.A. (2020). Fungi in biodegradation of polycyclic aromatic hydrocarbons in oilfield wastewater. *Acta Scientific Microbiology*, **3**(4), 220–224, https://doi.org/10.31080/asmi.2020.03.0572

Pandey C., Prabha D. and Negi Y. K. (2018). Mycoremediation of common agricultural pesticides. 155–179, https://doi.org/10.1007/978-3-319-77386-5_6

Piątkowska M., Jedziniak P., Olejnik M., Żmudzki J. and Posyniak A. (2018). Absence of evidence or evidence of absence? A transfer and depletion study of Sudan I in eggs. *Food Chemistry*, **239**, 598–602, https://doi.org/10.1016/j.foodchem.2017.06.156

Pironti C., Ricciardi M., Proto A., Bianco P. M., Montano L. and Motta O. (2021). Endocrine-disrupting compounds: an overview on their occurrence in the aquatic environment and human exposure. *Water (Switzerland)*, **13**(10), 1–32, https://doi.org/10.3390/w13101347

Polińska W., Kotowska U., Kiejza D. and Karpińska J. (2021). Insights into the use of phytoremediation processes for the removal of organic micropollutants from water and wastewater; a review. *Water (Switzerland)*, **13**(15), pp. 1–19, https://doi.org/10.3390/w13152065

Qadir I. and Chhipa R. (2015). Critical evaluation of some available treatment techniques for textile & paper industry effluents: a review. *American Chemical Science Journal*, **6**(2), 77–90, https://doi.org/10.9734/acsj/2015/15484

Qadiri R. Z. Z., Gani K. M., Zaid A., Aalam T., Kazmi A. A. and Khalil N. (2021). Comparative evaluation of the macrophytes in the constructed wetlands for the treatment of combined wastewater (greywater and septic tank effluent) in a sub-tropical region. *Environmental Challenges*, **5**, 100265, https://doi.org/10.1016/j.envc.2021.100265

Rai P. K., Kim K. H., Lee S. S. and Lee J. H. (2020). Molecular mechanisms in phytoremediation of environmental contaminants and prospects of engineered transgenic plants/microbes. *Science of the Total Environment*, **705**, 135858, https://doi.org/10.1016/j.scitotenv.2019.135858

Rajhans G., Barik A., Sen S. K., Masanta A., Sahoo N. K. and Raut S. (2021). Mycoremediation and toxicity assessment of textile effluent pertaining to its possible correlation with COD. *Scientific Reports*, **11**(1), 1–11, https://doi.org/10.1038/s41598-021-94666-8

Sadañoski M. A., Benítez S. F., Fonseca M. I., Velázquez J. E., Zapata P. D., Levin L. N. and Villalba L. L. (2019). Mycoremediation of high concentrations of polychlorinated biphenyls with *Pleurotus sajor-caju* LBM 105 as an effective and cheap treatment. *Journal of Environmental Chemical Engineering*, **7**(6), 1–28, https://doi.org/10.1016/j.jece.2019.103453

Sharma J., Kumar V., Kumar S. S., Malyan S. K., Mathimani T., Bishnoi N. R. and Pugazhendhi A. (2020). Microalgal consortia for municipal wastewater treatment – lipid augmentation and fatty acid profiling for biodiesel production. *Journal of Photochemistry and Photobiology B: Biology*, **202**, 111638, https://doi.org/10.1016/j.jphotobiol.2019.111638

Sharma P., Sirohi R., Tong Y. W., Kim S. H. and Pandey A. (2021). Metal and metal(loids) removal efficiency using genetically engineered microbes: applications and challenges. *Journal of Hazardous Materials*, **416**, 125855, https://doi.org/10.1016/j.jhazmat.2021.125855

Singh A. K. and Chandra R. (2019). Pollutants released from the pulp paper industry: aquatic toxicity and their health hazards. *Aquatic Toxicology*, **211**, 202–216, https://doi.org/10.1016/j.aquatox.2019.04.007

Singh J., Sharma D., Kumar G. and Sharma N. R. (2018). Microbial bioprospecting for sustainable development. Microbial bioprospecting for sustainable development. 1–397, https://doi.org/10.1007/978-981-13-0053-0

Singh J., Kumar V., Kumar P., Kumar P., Yadav K. K., Cabral-Pinto M. M. S., Kamyab H. and Chelliapan S. (2021). An experimental investigation on phytoremediation performance of water lettuce (*Pistia stratiotes* L.) for pollutants removal from paper mill effluent. *Water Environment Research*, **93**(9), 1543–1553, https://doi.org/10.1002/wer.1536

Vašinková M., Dlabaja M. and Kučová K. (2021). Bioaccumulation of toxic metals by fungi of the genus *Aspergillus* isolated from the contaminated area of Ostramo Lagoons. *IOP Conference Series: Earth and Environmental Science*, **900**(1), 1–9, https://doi.org/10.1088/1755-1315/900/1/012048

Vishwakarma G. S., Bhattacharjee G., Gohil N. and Singh V. (2020). Current status, challenges and future of bioremediation. In: Bioremediation of Pollutants: from Genetic Engineering to Genome Engineering, V. C. Pandey and V. Singh (eds), Elsevier, UK. INC, https://doi.org/10.1016/B978-0-12-819025-8.00020-X

Vymazal J. (2011). Plants used in constructed wetlands with horizontal subsurface flow: a review. *Hydrobiologia*, **674**(1), 133–156, https://doi.org/10.1007/s10750-011-0738-9

Wang R., Li F., Ruan W., Tai Y., Cai H. and Yang Y. (2020). Removal and degradation pathway analysis of 17-estradiol from raw domestic wastewater using immobilised functional microalgae under repeated loading. *Biochemical Engineering Journal*, **107700**, 1–31, https://doi.org/10.1016/j.bej.2020.107700

Wen X. J., Niu C. G., Zhang L., Liang C. and Zeng G. M. (2017). An in depth mechanism insight of the degradation of multiple refractory pollutants via a novel SrTiO3/BiOI heterojunction photocatalysts. *Journal of Catalysis*, **356**, 283–299, https://doi.org/10.1016/j.jcat.2017.10.022

Wu S., Kuschk P., Brix H., Vymazal J. and Dong R. (2014). Development of constructed wetlands in performance intensifications for wastewater treatment: a nitrogen and organic matter targeted review. *Water Research*, **57**, 40–55, https://doi.org/10.1016/j.watres.2014.03.020

Xie B., Tang X., Yong H., Deng S. and Shi X. (2020). Biological sulfamethoxazole degradation along with anaerobically digested centrate treatment by immobilized microalgal-bacterial consortium: performance, mechanism and shifts in bacterial and microalgal communities. *Chemical Engineering Journal*, **388**, 124217, https://doi.org/10.1016/j.cej.2020.124217

Xu M., Wu M., Zhang Y., Zhang H., Liu W., Chen G., Xiong G. and Guo L. (2022). Biodegradation of polycyclic aromatic hydrocarbons (PAHs) by bacterial mixture. *International Journal of Environmental Science and Technology*, **19**(5), 3833–3844, https://doi.org/10.1007/s13762-021-03284-4

Yu J., Shu S., Wang Q., Gao N. and Zhu Y. (2022). Evaluation of Fe^{2+}/peracetic acid to degrade three typical refractory pollutants of textile wastewater. *Catalysts*, **12**(7), 684, https://doi.org/10.3390/catal12070684

Zainith S., Purchase D., Saratale G. D., Ferreira L. F. R., Bilal M. and Bharagava R. N. (2019). Isolation and characterization of lignin-degrading bacterium *Bacillus aryabhattai* from pulp and paper mill wastewater and evaluation of its lignin-degrading potential. *3 Biotech*, **9**(3), 1–11, https://doi.org/10.1007/s13205-019-1631-x

Zhang H., Yuan X., Xiong T., Wang H. and Jiang L. (2020). Bioremediation of co-contaminated soil with heavy metals and pesticides: influence factors, mechanisms and evaluation methods. *Chemical Engineering Journal*, **398**, 125657, https://doi.org/10.1016/j.cej.2020.125657

Zhang J., Cao X., Yao Z., Lin Q., Yan B., Cui X., He Z., Yang X., Wang C. H. and Chen G. (2021). Phytoremediation of Cd-contaminated farmland soil via various *Sedum alfredii*-oilseed rape cropping systems: efficiency comparison and cost-benefit analysis. *Journal of Hazardous Materials*, **419**, 126489, https://doi.org/10.1016/j.jhazmat.2021.126489

IWA Publishing's authorised EU representative for General Product Safety Regulations is Diane D'Arras, 15 rue Duret, 75116 Paris, France, e-mail: safety@iwap.co.uk.

Printed and bound by CPI Group (UK) Ltd, Croydon, CR0 4YY
12/05/2026
02108877-0007